Plant Diseases Management in Horticultural Crops

Plant Diseases Management in Horticultural Crops

— *Editors* —

Dr. Shahid Ahamad

Programme Coordinator

Krishi Vigyan Kendra, Rajouri – 185 131

Sher-e-Kashmir University of Agricultural Sciences and Technology, Jammu, J&K

Dr. Ali Anwar

Associate Professor

Division of Plant Pathology

Sher-e-Kashmir University of Agricultural Sciences and Technology, Jammu, J&K

Dr. P.K. Sharma

Sr. Scientist (Plant Pathology)

ICAR Research Complex for North Eastern Hill Region, Manipur, Assam

2026

Daya Publishing House®

A Division of

Astral International Pvt. Ltd.

New Delhi – 110 002

First Published, 2011

Reprinted 2026

ISBN: 978-81-7035-702-5 (HB)

Publisher's note:

Every possible effort has been made to ensure that the information contained in this book is accurate at the time of going to press, and the publisher and author cannot accept responsibility for any errors or omissions, however caused. No responsibility for loss or damage occasioned to any person acting, or refraining from action, as a result of the material in this publication can be accepted by the editor, the publisher or the author. The Publisher is not associated with any product or vendor mentioned in the book. The contents of this work are intended to further general scientific research, understanding and discussion only. Readers should consult with a specialist where appropriate.

Every effort has been made to trace the owners of copyright material used in this book, if any. The author and the publisher will be grateful for any omission brought to their notice for acknowledgement in the future editions of the book.

Published by : **Daya Publishing House®**
A Division of
Astral International Pvt. Ltd.
– ISO 9001:2015 Certified Company –
4736/23, Ansari Road, Darya Ganj
New Delhi - 110 002
Phone: 011-4354 9197, 2327 8134
E-mail: info@astralint.com
Website: www.astralint.com

Printed by : **Replika Press Pvt. Ltd.**

Preface

A recent survey by the Economist revealed that the world population has increased by 90 per cent in the past 40 years while food production has increased only by 25 per cent per head. With an additional 1.5 billion mouth to feed by 2020, farmers worldwide have to produce 39 per cent more. The new millennium promises excitement and hope for the future by new advancement in ecofriendly technologies in integrated disease management of the crops. During past twentieth century, Plant Pathology has witnessed a dramatic advancement in management of plant diseases through in depth investigations of host–pathogen interactions, development of molecular diagnostic tools, integration of new concepts, principles and approaches.

The diseases of economic importance caused by fungi, bacteria, viruses and virus like organisms of each crop are covered, describing their history, distribution, losses incurred, symptoms, latest diagnostic tools, epidemiology and integrated management approaches including cultural, chemical, genetic resources, and bio control agents being adopted world-wide.

There has been a long felt need for a book exclusively on the "Recent Advances in Major Plant Diseases Management in India". An effort has been made to collect information on various aspects of diseases and their management. Chapters in this book have been contributed by noted Plant Pathologists of India. The Scientists, teachers, students, scholars, administrators and policy makers dealing with disease management in particular and plant protection in general expected this book will be very useful and informative. The book "Recent Advances in Major Plant Diseases Management in India", not only provides references but also serves as guide and inspiration for future research into the realm of biological, chemical, physical and quarantine aspects and onslaughts of modern agriculture on it. This book also serves as a very good references book on the IDM of Plant diseases. I am grateful and indebted to all the learned galaxy of contributors who have spent their considerable time in contributing the chapters on various important crops.

I am also thankful to *Shri Anil Mittal, Daya Publishing House,* New Delhi for his technical guidance and help in bringing out this book.

Finally, I wish to extend regards to Dr. A.K. Sarbhoy who was the source inspiration, Dr. D.K. Agarwal, Ex. Professor, Division of Plant Pathology, IARI, New Delhi and Dr. Nazir Ahmad Sudan, Director Research, SKUAST-J., Deepak Kher, V.K. Razdan, V.S. Verma, M.H. Chisti, and finally my beloved wife Bano Siddiqui for her continuous support and encouragement in bringing out this book.

Shahid Ahamad

Contents

List of Contributors

Ahamad, K.
Division of Plant Pathology, Sher-e-Kashmir University of Agricultural Sciences and Technology, Srinagar, J&K

Ahamad, S.
Department of Plant Protection, Allahabad Agricultural Institute, Deemed University, Allahabad, U.P.

Ahamad, Shahid
Programme Coordinator, K.V.K., Rajouri, Sher-e-Kashmir University of Agricultural Sciences and Technology, Jammu – 185 131, India

Ahmad, Shahzad
Associate Professor, Division of Plant Pathology, Sher-e-Kashmir University of Agricultural Sciences and Technology of Kashmir, Shalimar, Srinagar – 191 121, India

Akhtar, Jameel
Department of Plant Pathology, Birsa Agricultural University, Ranchi – 834 006, Jharkhand

Akram, M.
Department of Plant Pathology, C.S. Azad University of Agriculture and Technology, Kanpur, U.P.

Anandaraj, M.
Project Coordinator, AICRP on Spice Crops, IISR, Calicut

Anwar, Ali
Associate Professor, Division of Plant Pathology, Sher-e-Kashmir University of Agricultural Sciences and Technology of Kashmir, Shalimar, Srinagar – 191 121, India

Banday, Saba
Division of Plant Pathology, Sher-e-Kashmir University of Agricultural Sciences and Technology of Kashmir, Shalimar, Srinagar – 191 121, India

Benagi, V.I.
Department of Plant Pathology, UAS, Dharwad, Karnataka, India

Bhandarker, D.R.
IGAU, Raipur, C.G.

Bharat, Narender K.
Department of Mycology and Plant Pathology, Dr. Y.S. Parmar University of Horticulture and Forestry, Nauni, Solan – 173 230, H.P.

Bhat, Khurshid A.
Division of Plant Pathology, Sher-e-Kashmir University of Agricultural Sciences and Technology of Kashmir, Shalimar, Srinagar – 191 121, India

Bukhari, S. Asifa
Division of Plant Breeding and Genetics, Sher-e-Kashmir University of Agricultural Sciences and Technology of Kashmir, Shalimar, Srinagar – 191 121, India

Byadagi, A.S.
University of Agricultural Sciences, Dharwad – 580 005, Karnataka, India

Chesti, M.H.
RARS, Rajouri, Sher-e-Kashmir University of Agricultural Sciences and Technology of Jammu – 185 131, India

Devi, S.
Department of Plant Protection, Allahabad Agricultural Institute, Deemed University, Allahabad, U.P.

Farahnaz
Assistant Professor, Plant Protection, Directorate of Extension Education, Sher-e-Kashmir University of Agricultural Sciences and Technology of Kashmir, Shalimar, Srinagar – 191 121, India

Fatima, Nasreen
Division of Plant Pathology, Sher-e-Kashmir University of Agricultural Sciences and Technology of Kashmir, Shalimar, Srinagar – 191 121, India

Gupta, Dharmesh
Apple Scab Monitoring and Research Laboratory, Kotkhai, Shimla – 171 202, H.P.

Gupta, Sachin
Assistant Professor, Division of Plant Pathology, Faculty of Agriculture, Main Campus Chatha, Sher-e-Kashmir University of Agricultural Sciences and Technology of Jammu

Gupta, V.
Division of Plant Pathology, Faculty of Agriculture, Main Campus Chatha, Sher-e-Kashmir University of Agricultural Sciences and Technology of Jammu

Hussain, K.
Division of Olericulture, Sher-e-Kashmir University of Agricultural Sciences and Technology of Kashmir, Shalimar, Srinagar – 191 121, India

Iqbal, Shaheeda
Division of Plant Breeding and Genetics, Sher-e-Kashmir University of Agricultural Sciences and Technology of Kashmir, Shalimar, Srinagar – 191 121, India

Jamwal, Sonika
Dryland Research Sub Station, SKUAST-J, Dhiansar, Jammu, J&K

Kalha, C.S.
Associate Dean, Faculty of Agriculture, Main Campus Chatha, Sher-e-Kashmir University of Agricultural Sciences and Technology of Jammu

Kengnal, M.
University of Agricultural Sciences, Dharwad – 580 005, Karnataka, India

Khan, F.A.
Associate Professor, Division of Post Harvest Technology, Sher-e-Kashmir University of Agricultural Sciences and Technology of Kashmir, Shalimar, Srinagar – 191 121, India

Khar, M.S.
Sher-e-Kashmir University of Agricultural Sciences and Technology of Kashmir, Shalimar, Srinagar – 191 121, India

Khare, C.P.
Department of Plant Pathology, Indira Gandhi Krishi Vishwavidyalaya, Raipur, C.G.

Khare, N.
Indira Gandhi Krishi Vishwa Vidyalaya, Krishinagar, College of Agriculture, Raipur – 492 006 C.G., India

Kohli, Anshuman
RARS, Rajouri, Sher-e-Kashmir University of Agricultural Sciences and Technology of Jammu – 185 131, India

Kulkarni, Shripad
Department of Plant Pathology, UAS, Dharwad, Karnataka, India

Kumar, Atul
Department of Plant Pathology, B.A.U., Ranchi – 834 006, Jharkhand

Kumar, Bijendra
College of Forestry and Hill Agriculture, Ranichauri – 249 199, G.B.P.U.A.&T., Uttarakhand

Kumar, R.
M.L.K. (P.G.) College, Balrampur, U.P.

Kumar, V. Manoj
Department of Plant Protection, Allahabad Agricultural Institute, Deemed University, Allahabad, U.P.

Kunklikar, S.
University of Agricultural Sciences, Dharwad – 580 005, Karnataka, India

Lal, H.C.
Department of Plant Pathology, B.A.U., Ranchi – 834 006, Jharkhand

Masoodi, L.
Division of Plant Pathology, Sher-e-Kashmir University of Agricultural Sciences and Technology of Kashmir, Shalimar, Srinagar – 191 121, India

Mir, G.H.
Sher-e-Kashmir University of Agricultural Sciences and Technology of Kashmir, Shalimar, Srinagar – 191 121, India

Mir, Mohd. Nawaz
Division of Plant Pathology, Sher-e-Kashmir University of Agricultural Sciences and Technology of Kashmir, Shalimar, Srinagar – 191 121, India

Mishra, Anil
Division of Plant Pathology, IARI, New Delhi – 110 012

Najar, A.G.
Associate Professor, Division of Plant Pathology, Sher-e-Kashmir University of Agricultural Sciences and Technology of Kashmir, Shalimar, Srinagar – 191 121, India

Nargund, V.B
University of Agricultural Sciences, Dharwad – 580 005, Karnataka, India

Pandey, M.K.
Division of Plant Pathology, Sher-e-Kashmir University of Agricultural Sciences and Technology of Kashmir, Shalimar, Srinagar – 191 121, India

Persaud, R.
Guyana Rice Development Board, Rice Research Station, Burma, Mahaicony, East Coast Demerara, Guyana South America

Prasad, M. Santha Lakshmi
Crop Protection, Directorate of Oilseeds Research, Rajendranagar, Hyderabad

Rao, S. Chander
Crop Protection, Directorate of Oilseeds Research, Rajendranagar, Hyderabad

Rasool, F.
Division of Plant Pathology, Sher-e-Kashmir University of Agricultural Sciences and Technology of Kashmir, Shalimar, Srinagar – 191 121, India

Razdan, V.K.
Professor and Head, Division of Plant Pathology, Faculty of Agriculture, Main Campus Chatha, Sher-e-Kashmir University of Agricultural Sciences and Technology of Jammu

Roy, S.S.
Sr. Scientist, ICAR Research Complex for NEH Region, Imphal – 795 004, Manipur

Saxena, R.P.
N.D.U.A.T., Kumarganj, Faizabad, U.P.

Shailbala
Department of Plant Pathology, G.B.P.U.A.&T., Pantnagar – 263 145, Uttarakhand

Sharma, P.K.
Sr. Scientist, ICAR Research Complex for NEH Region, Imphal – 795 004, Manipur

Shivakumar, A.P.
University of Agricultural Sciences, Dharwad – 580 005, Karnataka, India

Singh, A.
Department of Genetics and Plant Breeding, T.D. College, Jaunpur, U.P.

Singh, A.K.
Directorate of Extension, C.S. Azad University of Agriculture & Technology, Kanpur, U.P.

Singh, M.
Department of Plant Pathology, C.S. Azad University of Agriculture & Technology, Kanpur, U.P.

Singh, Ranbir
Assistant Professor, Division of Plant Pathology, Faculty of Agriculture, Main Campus Chatha, Sher-Kashmir University of Agricultural Sciences and Technology of Jammu

Singh, V.K.
Dryland Research Sub Station, SKUAST-J, Dhiansar, Jammu, J&K

Singh, Vaibhav K.
Department of Plant Pathology, G.B.P.U.A.&T., Pantnagar – 263 145, Uttarakhand

Srivastava, J.N.
Jr. Scientist, Regional Horticultural Research Station, Bhaderwah, SKUAST, Jammu

Teli, M.A.
Professor Division of Plant Pathology, Sher-e-Kashmir University of Agricultural Sciences and Technology of Kashmir, Shalimar, Srinagar – 191 121, India

Tewari, J.P.
M.L.K. (P.G.) College, Balrampur, U.P.

Thakur, M.P.
Department of Plant Pathology, Indira Gandhi Krishi Vishwavidyalaya, Raipur, C.G.

Thrimurty, V.S.
Department of Plant Pathology, Indira Gandhi Krishi Vishwavidyalaya, Raipur, C.G.

Tiwari, P.K.
Department of Plant Pathology, Indira Gandhi Krishi Vishwavidyalaya, Raipur, C.G.

Upadhyay, J.P.
Department of Plant Pathology, T.C.A. Dholi – 843 121, Muzaffarpur, R.A.U., Bihar

Verma, V.S.
Dryland Research Sub Station, SKUAST-J, Dhiansar, Jammu, J&K

Vikram, P.
IRRI, Manila, Philippines

Wani, M. Azam
Associate Professor, Division of Plant Pathology, Sher-e-Kashmir University of Agricultural Sciences and Technology of Kashmir, Shalimar, Srinagar – 191 121, India

William, P.
Department of Plant Protection, Allahabad Agricultural Institute, Deemed University, Allahabad, U.P.

Plant Diseases Management in Horticultural Crops (2011) *Pages* **1–19**
Editors: **Shahid Ahamad, Ali Anwar and P.K. Sharma**
Published by: **DAYA PUBLISHING HOUSE, NEW DELHI**

Chapter 1

Fungal Diseases of Apple and Their Management

Mukesh Kumar Pandey, C.S. Kalha and V. Gupta

Division of Plant Pathology, Faculty of Agriculture,
Sher-e-Kashmir University of Agriculture Sciences and Technology,
Chatta campus, Jammu – 180 009

Marssonina Blotch (Premature Leaf Fall)

Introduction

Marssonina blotch is first reported from Japanese orchards as early as 1907 and was rated second in importance and severity as an apple disease in 1910's in northern part of the country by causing premature leaf fall (Sharma and Gautam, 1997; Sharma, 2001). Thereafter, this disease was also observed on apple in Canada, Korea, Rumania, China and Tiwan (Parmelee, 1971; Leite *et al.*, 1986; Jia, 1994). *Marssonina* blotch is responsible for causing widespread defoliation of the apple trees in midseason thereby affecting the tree vigour and fruit bearing capacity (Sharma and Gautam, 1997). The spots on the fruits also reduced their quality and marketability. Its incidence has also been reported from Uttarakhand, Himachal Pradesh, Jammu and Kashmir in India.

Symptom

The disease start with the development of dark green circular patches on the upper leaf surface during June end giving rise to 5 to 10 mm diameter brown leaf spots, which become dark brown in the due course. The leaf spot occurs initially on older leaves in the lower portion of the tree canopy. As the disease progresses, black pinhead-like fruiting bodies (acervuli) develop on the upper leaf surface and when the lesion are numerous, they coalesce to from large dark brown patches (blotches) and the surrounding areas turn yellow. Numerous leaf infections also results in the loss of chlorophyll giving rise to mosaic type appearance on the leaves. The infected leaves drop off the tree and produces the characteristic symptoms of premature leaf fall, the name by which the disease is commonly known (Figure 1.1). On the fruits, lesions are appears initially circular (4 to 5mm), which later become oval, depress and dark brown with age and almost black at harvest time. Small black acervuli are also visible on the fruit lesions (Sharma and Verma, 1999; Sharma and Kaul, 2000; Sharma, 2001; Sharma *et al.*, 2005).

Causal Organism

Marssonina blotch, is caused by a mitosporic fungus *Marssonina coronaria* (Ell. and J.J. Davis) J.J. Davis (Syn. *Marssonina mali* (P. Henn.) Ito). The fungus produces asexual fruiting bodies called acervuli which are 100 to 200 µm in diameter, subcuticular and irregularly rupturing the cuticle. Conidia are borne on short clavate conidiophores in acervuli which are typical hyaline, bi-celled guttulate with average size of 20-24 x 6.5-8.5 µm (Sharma *et al.*, 2005; Sharma and Sharma, 2005). There is a wide variability in its morphology with respect to the size and shape of the conidia (Sharma and Sharma, 2005). Spermatia are produced in acervuli, solely or mixed with conidia. *Diplocarpon mali* Harada and Sawamura is reported to be the perfect stage of the fungus (Harada *et al.*, 1974). Ascospore produces in the over-wintered leaves have been reported as the major sourse of primary infection in the orchards

Figure 1.1: Premature Leaf Fall

(Harada *et al.*, 1974; Takahashi and Sawamura, 1990). Cook (1979) reported the development of new acervuli in the original lesions, which released the spore to cause primary infection. The fungus produced small brown to dark brown colonies, which are round, sub-globoid and somewhat wrinkled. This pathogen showed maximum growth on minimal salts agar medium. Methionine and maltose support its best growth as nitrogen and carbon sources, respectively (Sharma and Sharma, 2005).

Epidemiology

Typical symptom of Marssonina blotch were observed 10-14 days after pathogen inoculation on the older leaves of the plants and younger are the last to get infected. Defoliation of leaves started after 20 days of disease appearence; however, complete defoliation are occurs after 45 days of inoculation (Bala, 1999; Sharma and Kaul, 2000; Sharma and Sharma, 2005). The fungus was capable of growing on a wide range of temperature from 5 to 30°C. Maximum mean radial growth and dry mycelial weight was recorded at 20°C followed by 15 to 25°C. The least growth and dry mycelial weight was observed at 30°C (Sharma and Sharma, 2005). Harada *et al.* (1974) reported that 20°C as the optimum temperature for the growth of *M. coronaria*. Bala (1999) reported maximum conidial germination and germtube length of *M. coronaria* at 20°C followed by 25°C. Dolan (1974) reported increased growth rate of two *Marssonina species* isolates with rising temperature from 9 to 24°C, whereas, Simpson and Hayes (1978) found 18 to 20°C temperature as the optimum for the growth of *M. brunnea*. Later Galea *et al.* (1986) reported 15 to 20°C as the cardinal temperature for the growth of *M. panattoniana*. This pathogen grew best 5.0 pH (Sharma and Sharma, 2005). Galea *et al.* (1986) reported optimum germination of conidia of *M. panattoniana* between pH 4.25-5.2 whereas, germination decreased rapidly above pH 6.0 and ceased at pH 8.0. While, Simpson and Hayes (1978) reported pH range of 7.0 to 7.5 as optimum for the growth of *M. brunnea* on solid medium.

Bala (1999) reported that asparagines as the best nitrogen source followed by potassium nitrate for the germination anf germ tube formation of conidia of *M. coronaria*. Simpson and Hayes (1978) reported best growth of *M. brunnea* on medium containing potassium nitrate and 4-amino *n*-butyric acid. Sharma and Sharma (2005) observed that among the different nitrogen sources, methionine (11.57mm) was the best. It was closely followed by potassium nitrate, ammonium nitrate and yeast extract. Sharma and Sharma (2005) also observed that among the different carbon sources, maximum radial growth of the fungus *M. coronaria* and closely followed by maltose, lactose and fructose. Bala (1999) reported maximum germination and germ tube length of conidia of *M. coronaria* on sucrose. Hong *et al.* (1998) have reported glucose as the best carbon source for the growth of *D. mali*.

The prevalence of blotch disease was influenced by leaf ageing followed by average temperature above 20°C and 70 per cent relative humidity at least continuously for 5-6 days which induce yellowing and defoliation process. Prolong hot and dry weather affect the primary infection and disease conditions adversely. Under favourable weather conditions, the defoliation begin within 3-5 days whereas under interrupted/unfavourable conditions the infected leaves are retained up to 3 months (Thakur *et al.*, 2005).

Management

Disease control can be achieved through orchard sanitation, adequate pruning and proper use of fungicides. As the disease is known to perennate in the infected leaves on the floor, removal of overwintering leaf litter may reduce primary inoculum level. Urea (5 per cent) spray can also be of great help in destroying the leaves. Proper pruning allows adequate air circulation in the tree canopy, thereby modifying the favourable microclimate and reduce the disease level. The fungicides *viz.* Mancozeb (0.3 per cent), copper oxychloride (0.3 per cent), zineb (0.3 per cent), dodine (0.075 per cent), mancozeb flowable (0.35 per cent), cCompanion (0.25 per cent) and ziram (0.3 per cent) provide 100 per cent disease control in a protective spray programme followed by dithianon (0.05 per cent) (Sharma and Sharma, 2005). Sharma (1997) recommended carbendazim (0.05 per cent) spraying in orchards showing disease symptoms followed by a mancozeb (0.3 per cent), propineb (0.3 per cent), dodine (0.075 per cent), carbendazim (0.05 per cent), thiophanate methyl (0.05 per cent), benomyl (0.05 per cent) and fluquinconazole (0.02 per cent) provided complete protection.

Alternaria Leaf Spot [*Alternaria alternata* (Fr.) Keissler, Hitherto]

Incidence

This disease frequently occurs in Himachal Pradesh during April and May (Gupta and Agarwala, 1968).

Symptom

This disease are characterize by the appearance of 1-3 mm diameter reddish brown necrotic spots on young apple leaves, later coalescing to form bigger blotch and giving blighted appearance to the foliage followed by defoliation. Dark brown to black elongated lesions also appears on leaf petioles and mid ribs in early part of the season causing severe yellowing and dropping of the infected leaves (Sharma *et al.*, 2005).

Causal Organism

Alternaria species had been reported from several countries attacking on apple leaves and fruit. Roberts (1924) identified the fungus as *Alternaria mali*, while Tailor (1996) considered it to be *Alternaria tenuis*. *A. mali* was regarded as a weak parasite on apple leaves and did not originate the leaf spot on apple leaves but got entry by areas killed by *Sphaeropsis malorum* (Roberts, 1924). Although *A. mali* and *A. tenuis* have been reported to be associated with the leaf spot, blight and defoliation symptoms on apple, but recent investigation indicated all *Alternaria* pathogens known to produce host specific toxin should be classified as *Alternaria alternate* because of the morphological similarity of their asexual state. Thus *A. mali* has been referred as apple pathotype of *A. alternate* (Fr.) Keissler, hitherto (Sawamura, 1990).

Disease Management

Sharma *et al.* (2005) reported that protective sprays of dodine (0.075 per cent) controlled the disease completely while zineb (0.3 per cent) and mancozeb (0.3 per cent) which provided 96.4 and 95.8 per cent respectively. Sawamura *et al.* (1993) reported Polyoxin AC (an antibiotic) and iprodione to be superior to copper and quinolinol, captan and dithiocarbamates in controlling Alternaria bloch of apple in Japan. Iprodione provided 69.1-79.8 and 71.6-75.0 per cent disease reduction in Alternaria blotch for higher (0.3 per cent) and lower application rate (0.15 per cent), respectively under natural infection conditions on different stages (Filajdic and Sutton, 1992). In India also, Sud and Agrawal (1971) found zirum (0.3 per cent) the most effective against *A. mali*, causing Alternaria blight of apple which was closely followed by ferbam, zineb and captan (0.15 per cent) in controlling the disease.White root rot (*Dematophora necatrix* Hartig) is an important soil born disease of apple nurseries and grown up trees. The fungus is widely distributed in temperate and sub-temperate regions of the world attacking more than 160 plant species belonging to 45 families and serious on apricot, pear, strawberry, plum, peach, grapes etc. (Behadad, 1975; Ito and Nakamura, 1984). Almost all the rootstocks are susceptible to the pathogen; however, the degree of susceptibility varies (Gupta and Verma, 1978). The root rot of fruit trees has been reported since 1883 from Germany but in apple, the disease was observed for the first time in 1900 from Norwich and later in 1913 from Canterbury (Salman and Wormald, 1913). In India, the disease was first observed in Uttar Pradesh hills in 1939 (Singh, 1943). The disease is considered a limiting factor in the production of apple in countries like India, Israel, California, Brazil, U.S.A., and Britain whereas in other countries it is serious on other host plants. The annual losses estimated in Himachal Pradesh due to this disease are about Rs 1.3 million (Agrawala and Sharma, 1966).

Symptoms

The above ground symptoms include premature yellowing of foliage, couples with bronzing and defoliation. The initial infection starts from the root hair and progresses through tertiary roots, secondary roots and then extends slowly to the primary root. The final diagnostic features of the disease are evident by the examination of root system. The latest roots turn dark brown and are covered with a greenish grey or white mat having flocculent web of whitish strands or ribbons during monsoon season. Appearance of pear shaped swellings near septa in the hyphae, the cortical cells are ruptured resulting in disruption of the plant system leading to the death of the tree and hence expressing disease on above ground plant parts.

Causal Organism

In 1925, *Rosellinia necatrix* Berl. Ex Prill, the perfect stage of *Dematophora necatrix* Hartig was identified as the cause of the death of apple trees at Winscombie-Somerset (Nattrass, 1927).The diagnostic feature of *D. necatrix* Hartig is the pear shaped swellings at each septum of the hyphae. The fungus produces scattered black, microsclerotia measuring 130 x 98 µm and does not infected xylem tissues of apple roots but bark is completely damaged. Two type of mycelium characteristic of the fungus. The greenish grey to white rhizomorph like, having fan shaped structures probably marking the site of destroyed roots and diffused fine white mycelium occupying the soil cavities. The later is responsible for the infection of new plants. The perfect stage, *R. necatrix* Berl. Ex Prill has not been observed under Indian conditions (Gupta and Sharma, 2000).

Disease Cycle and Epidemiology

The fungus survives either in the form of mycelium or sclerotia in the old infected roots or in soil for several years. Primary infection takes place from diffused mycelium present in root debris or soil or by contact of new plant roots with old dead roots. The fungus produces lot of pectolytic and cellulolytic enzymes which help in the degradation of host tissues in advance of fungus growth (Gupta and Gohain, 1982). Pathogen also produces toxins which help in killing the bark. The disease is most serious in water-logged acidic soils at pH 6.1 to 6.5. The disease severity is positively correlated with moderate temperature (15-25°C) and high moisture (Gupta and Sharma, 1999). Almost all the rootstocks have susceptibility to the pathogen, however, in the field the degree of susceptibility varies (Gupta and Verma, 1978).

Management

The management of white root rot consists of preventing the spread of the fungus in soil, removing primary inoculum sources such as root pieces, digging isolation trenches etc. The drained soils should be improved by following central drainage system. The acidic soils should be amended by applying lime for some years (Jain, 1961). Soil amendment with neem cake and deodar needles has been reported to reduce the incidence of the disease (Gupta and Jindal, 1989).

The fungicides should be applied through deep holes (15-20 cm) made in the basin of tree at 30 cm distance from each other. Prior to the application of chemicals the tree basin is brought to a soil moisture level of 30 to 40 per cent. The treatment is done 3-4 times every year during rainy season. Sousa (1985) observed preventive as well as curative effects of benomyl and carbendazim and reported that treatment of diseased plants in winter further improved the control.

Sharma *et al.* (2005) reported that *Pseudomonas fluorescens* IISR-8, T1R2K4 and *Bacillus subtilis* were more inhibitory giving 61.29, 59.62 and 56.11 per cent inhibition, respectively. Earlier several workers have also reported inhibitory effect of fluorescent pseudomonads and *Bacillus subtilis* against *Rosellinia necatrix* (Utkhede and Li, 1989; Gupta and Jindal, 1990; Cattelan, 1994; Lopez *et al.*, 1998). Gupta and Jindal (1990) recorded incidence of white root rot of apple in pots when *Bacillus subtilis* and *Enterobacter aerogenes* were applied as soil drench prior to the pathogen inoculation. Sharma (1993) have also reported that pre-inoculation of *E. aerogenes* gave good control of white root rot both in pot culture and field condition. Sharma *et al.* (20056) have reported that both bacterial antagonists (*Bacillus subtilis* and *Pseudomonas fluorescens*) significantly increase the growth of apple seedlings in comparison to control (Caesar and Burr, 1987; Utkhede and Li, 1989; Utkhede, 1999). *Trichoderma harzianum* isolate T-8 was found to significantly reduce *R. necatrix* colonization of avocado leaves in artificially as well as naturally infested soils (Freeman *et al.*, 1986). Soil solarization coupled with application of *T. harzanum* also improved the control over that achieved by either treatment alone (Sztejnbert *et al.*, 1987).

Scab (*Venturia inaequalis*)

Introduction

The disease was first reported from Sweden in 1819 and subsequently it spread throughout the world. In India, it was first detected in 1935 on the native cultivar Ambri in Kashmir valley (Nath, 1935) and from Himachal Pradesh in 1977 (Anonymous, 177). The disease is now present in almost all apple growing areas of India including Jammu and Kashmir, Himachal Pradesh (Gupta, 1978), Uttar Pradesh hills, Sikkim (Mallick *et al.*, 1984a), Arunachal Pradesh (Mallick *et al.*, 1984b) and Nilgiris hill of Tamil Nadu (Gupta 1987a). In 1973, this disease appeared in epidemic from in the valley whereby about 70,000 acres orchard area got infected. This epidemic resulted in a loss of rupees 54 lakhs in a season (Anon, 1975; Joshi *et al.*, 1975). The Indian Council of Agricultural Research and the Directorate of Plant Protection, Quarantine and Storage, Government of India therein declared it as one of the five main problems of national importance. The unsprayed trees may suffer losses upto 29.16 per cent due to scabby fruits (Din and Saksena, 1984).

The losses in terms of rupees were Rs. 54 lakh in a single season in Kashmir valley in 1973 the first epidemic. In Himachal Pradesh during 1983 epidemic, 10 per cent of apple crop (30,000 out of 3,00,000 mt) were made unfit for market and had to be destroyed on the spot resulting in a loss of Rs. 15 millions to state Government which paid compensation to farmers (Gupta *et al.*, 1984). Recently, Himachal Pradesh in the year 1996 alone suffered a loss of about Rs. 18.2 to 21.0 millions, inspite of the considerable expenditure (Rs. 30.6 millions) incurred for the control of this disease. Thus, the accumulated losses by the disease is around 50 millions *i.e.* 10 per cent of the total income.

Symptoms

The scab symptoms are normally observed on leaves, petioles, blossoms, sepals, fruits, pedicels and less frequently on bud scales. Symptoms first appear in the spring as small lesions on the lower surface, the side first exposed to fungal spores as buds open. At first, the lesions are usually small, velvety, olive green in color and have unclear margins but in some cases infection may be reddish in

Figure 1.2a: Scab on Leaf

Figure 1.2b: Scab on Fruits

color. As they age, the infections become darker and more distinct in outline. Lesions may appear more numerous closer to the mid-vein of the leaf (Figure 1.2a). If infection becomes heavily, the leaf distorted and drops early in the summer. Trees of highly susceptible varieties may be severely defoliated by mid to late summer. On the fruits, small superficial lesions or large patches developed. The margins of the spots are more distinct and legions becomes darker (black or scabby) with age (Figure 1.2b). Badly scabbed fruit becomes deformed and may fall before reaching good size.

Causal Organism

The apple scab caused by *Venturia inaequalis* (Cooke) Winter (imperfect stage : *Spilocaea pomi* Fr.) are encountered in nature in its parasitic and saprophytic forms. The pathogen is a polycyclic in nature and has two distinct phase in life cycle. The perfect (sexual stage) stage perpetuated on the fallen leaves in winter and spring and the imperfect (asexual) stage parasitize on leaves and fruits during the growing season. The mycelium in living tissues is located only between the cuticle and epidermal cells. It can be cultured in artificial medium. Modified glucose asparagines medium at pH 6.0 at 20°C is best for cultural growth and sporulation (Gupta and Lele, 1980b).

Epidemiology

The fungus perennates on the fallen leaves in the saprophytic phase. This phase starts with the entry of the fungus hyphae into the leaf tissues. The mycelium from the sub-cuticular stroma proliferate into the palisade and mesophyll cells of the leaf and form a network of dark brown hyphae from which pseudothecia develop (MacHardy and Gadoury, 1985). Most of the pseudothecia in leaves develop within 30 days of leaf fall in October and November (Gadoury and MacHardy, 1985a; Gupta, 1988). Moisture, either as rainfall or dew (Louw, 1951; Wilson, 1928) is essential and early development of pseudothecia which are initiated better in leaves kept alternately wet and dry than those kept continuously wet or dry (James and Sutton, 1982). Besides moisture, temperature has major influence on the growth of pseudothecium, pseudothecial number and rate of ascospore maturation (Gadoury and MacHardy, 1982b; James and Sutton, 1982). In the presence of adequate moisture, pseudothecia may be initiated at temperature below 0°C but their density is maximum at temperature between 4 to 8°C (Gupta, 1979a; Gadoury and MacHardy, 1982a) but optimum temperature for maturity was 15°C. Ascus development at all temperatures was continuous and was completed in approximately 9 weeks (Gadoury and MacHardy, 1982a). Prolonged dry weather conditions in the winter and spring did not allow normal development of pseudothecia and even the ascospore maturity get delayed by a month or so.

Adequate moisture is essential for the discharge of ripe ascospores from the pseudothecia. Hirst and Stedman (1962) demonstrated that as little as 0.13 mm of rain was sufficient to release a few spores but for large scale release atleast 0.20 mm was necessary. Gupta and Lele (1980b) observed that discharge of ascospores only takes place in presence under wet conditions. A few Ascospores are discharged within the first hour of wetting, but concentrations rapidly rise to a maximum during the second and third hour (Gilpatrick and Szkolnik, 1978). Low night temperature and darkness impose a black on ascospore releases mechanisms, which cannot triggered off by exposure of the leaves to 1 or 2 hours of early morning light and prevailing temperature. When the leaves get wet in the afternoon, after remaining exposed to several hours of day light and day time temperatures, all or most of the mature ascospores are released. The best conditions for ascospore discharge is a thin film of water over pseudothecia allowing asci to extend through ostiole and releasing the ascospores. Maturation and discharge of ascospores usually last for 3 to 11 weeks (Gupta, 1975; Szkolnik, 1981).

Ascospores after falling on susceptible host tissue, germinate under suitable temperature and moisture (leaf wetness) condition. Ascospores on germination produce a germtube which produces an appressorium at the tip. A peg like structure emerging out of appressorium punctured the cuticle of plant tissue to start primary infection. The fungus then grows beneath the cuticle and produces olive brown lesions within 2 to 4 weeks. Conidia produced on these lesions are mostly washed by rain or sometimes even get detached by wind (Sutton, 1981).

The most critical period for the development of epidemics is from the time the buds start swelling till 2 to 4 weeks after petal fall, provided the weather remains wet (Gupta, 1975). Another critical period for epidemic build up is prolonged late summer during monsoons.The outbreak of 1973 and 1983 apple scab epidemics in Kashmir and Himachal Pradesh, could be attributed to unusually wet conditions prevailing in early autumn (October-November) of the previous year, followed by similar conditions during spring (Gupta, 1985).

Management

The approach towards the management of apple scab delay these approaches : (1) eradication of primary inoculum by interrupting the life cycle of the pathogen (2) aquired protection against primary and secondary infection.

Eradication of Primary Inoculum by Interrupting the Life Cycle of the Pathogen

In order to interrupt the life cycle of the pathogen to reduce the extent of primary inoculum in spring season, orchard sanitation and/or chemical spray can be employed for quick decomposition of leaves harbouring the scab fungus (Gupta, 1985). Use of urea may be recommended to minimize the severity of scab. A spray of 5 per cent urea in autumn shortly before leaf fall in addition to orchard ploughing, raking and destroying the fallen leaves (Gupta, 1985). These pre-leaf fall sprays besides accelerating decomposition of fallen leaves, help in prevention of pseudothecial formation and increasing microbial activity on the fallen leaves (Gupta and Lele, 1980c; Gupta, 1987b). These microorganisms compete with the scab including gungus for nutrients and sometimes act antagonistically. Moreover, the application of urea at leaf fall stage further appreciate the plant health and productivity and provide early nutrition to flowering buds (Gupta, 1989). However, at low elevations the pre-dormant spray of urea, may make the plants prone to various fungal cankers.

Protection Against Primary and Secondary Infection

Protective Spray

A protective spray programme being adopted in apple orchard India includes 6 to 7 sprays of a non-systemic and systemic fungicides during the growing season and a single application of urea in autumn. The first spray of protective fungicide must start at silver tip to partially green tip stage. This spray is important to the low lying areas because there is early onset of season for the development of plant growth stages and release of primary inoculum. However, 3 or 4 sprays are required from bud break to petal fall to cover susceptible host tissues continuously. A single application technique had been suggested to give continuous protection to susceptible tissue for a longer duration. Earlier, captafol (Difolatan or Fltaf) at a massive dosage having double the recommended dose at silver tip stage was used in commercial orchards in USA and other countries. In India, since most of the tree bud development stages are over within 30 days a concentration of 0.3 to 0.5 per cent of formulated fungicide is commonly employed in single application technique. However, the optimum time of such massive dosage is at the appearance of green bud tissue but not later as such application may result in moderate damage to the cluster leaves and cause fruit russetting. This technique was introduced to save 1 or 2 subsequent sprays because of longer retention and redistribution properties.

Post-infection (Pre-symptom)

Most of the EBI fungicides have excellent after-infection curative activity against scab. These fungicides, particularly biteratanol, myclobutanil, prochloraz, etaconazole, penconazple and hexaconazole prevent the formation of visible lesions when applied within 3 days after inoculation (Brandes and Paul, 1981) and only chlorotic flecks, or spots developed if they were applied later than 3 days after the onset of infection (Thakur and Gupta, 1990a).

Where continous rainy weather, certain fungicides can be applied late in incubation period also but not beyond 2 to 3 days prior to symptom expression. In such situation, Szkolnik (1978) proposed the term pre-symptom to describe prevention, or inhibition, of spore production from chlorotic scab lesions that follow a spray of certain fungicides before the occurrence of such lesions. An equally high level of pre-symptom activity for EBI fungicides, *viz.*, fenarimol, bitertanol, triforine, fenapanil, hexaconazole, etaconazole and prochloraz has been reported (Brandes and Paul, 1981; Kelley and Jones, 1981; Thakur and Gupta, 1991, 1992a, 1992b). The post –infection spray programme with well-timed application of bitertanol and myclobutanil after 72 and 96 hours of prediction infection periods is not only more effective but is also more efficient requiring fewer sprays in a season (Thakur, 1987).

Eradication (Post-symptom)

Interruptio in carrying on sprays at regular intervals in inclement weather or delay in application by few days may result in the establishment of scab fungus on foliage and fruits. If this is accompanied by continuous wet weather, inoculum quickly builds up and increase the disease severity. To avoid this, fungicides having high level of eradicant or anti-sporulant activity or ability to completely inhibit the conidia production on the established lesions of 5 to 7 days are applied. Benzimidazoles have this post-symptom activity by almost completely inhibiting the production of germination of conidia in sporulating lesions. The EBI fungicides have distinct inhibitory effect on the growth of subcuticular hyphae and the conidia formation of apple scab fungus, thereby preventing development of scab symptoms (Thakur and Gupta, 1991). Dodine has advantage of being safe even when used in 3 to 4 or more sprays without the risk of resistance build up by scab fungus. Thus, these fungicides may be used once or twice alone and only in a mixture with protectant later on.

Since the onslaught of epidemics of scab, continuous efforts to find out resistant sources were made and seven scab resistant cultivars *viz.* Co-op-12, Co-op-13 (Red Free), Prima, Priscilla, Macfree, Liberty and Sir Prize with parentage developed by cooperative venture of three US Universities. Terton have been complete field resistance to apple scab in Kullu Valley (Sharma *et al.*, 1988). In addition, *Malus baccata* (Kashmir) and *M. baccata* (Shillong) were observed to remain completely free from scab and powdery mildew indicating presence of multiple disease resistance genes (Sharma *et al.*, 1996).

Resistant varieties: Since the first occurrence of apple scab, breedind programmes are in progress to find out or develop resistant sources or varieties with desirable horticultural traits. Cultivars such as Priscilla, Sir Prize, Jonafree, Redfree and Mcfree developed by cooperative venture of three US universities Purdu, Rutgers and Illinois have shown field resistance to scab in India (Sharma *et al.*, 1988). Several cultivars with resistance to *V. inaequalis* have been released through breeding programmes in Canada such as Macfree, Moira, Trent, Nova Easygro, Novamac, Richelieu and Rouville (Gupta and Sharma, 2000). Major breeding efforts for disease resistance were done in NewYork, where Liberty, Prima and Priscilla appeared to be resistant to scab (Anonymous, 2001).

Powdery Mildew

Powdery Mildew is a serious disease both in nurseries and grown up trees and attacks current year growth of the plant. The disease was first recorded on apple seedlings from Iowa, USA in 1871 (Bessey, 1877).

Symptoms

Symptoms first appear in the spring on the lower surface of leaves, usually at the ends of branches. Small, whitish felt-like patches of fungal growth appear and quickly cover the entire leaf. Diseased leaves become narrow, crinkled, stunted and brittle. By mid-summer, tiny, black round specks show up on the lower leaf surface, but more commonly on the twigs (Figure 1.3). The fungus spreads rapidly to twigs, which stop growing and become stunted. In some cases the twigs may be killed back. Leaves and blossoms from infected buds will be diseased when they open the next spring. Infected blossoms shrivel and produce no fruit. Fruit symptoms are not usually seen unless the disease has built up to high levels on susceptible cultivars. Diseased fruit has a fine network type surface blemish called russetting.

Economic damage occurs in the form of aborted blossoms, reduced fruit finish quality, reduced vigor, poor return bloom and yield of bearing trees, and stunted growth and poor form of nonbearing trees.

Causal Organism

The disease is caused by *Podosphaera leucotricha* (Ellis and Ever) Salm. The anamorph of the fungus is *Oidium farinosum* Cooke. Occurrence of *Erysiphe heraclei* and *Podosphaera oxycanthae* has also been reported.

Figure 1.3: Powdery Mildew

Disease Cycle

The mildew fungus overwinters mainly as mycelium in dormant blossom and shoot buds produced and infected the previous growing season. Conidia are produced and released from the unfolding leaves as they emerge from infected buds at about tight cluster stage. During spring or summer large number of conidia produced in infected tissues are dispersed by wind causing secondary infection. Conidial dispersal is low during night than day and is interrupted by heavy prolonged rains (Stephan, 1988). Conidia germinate in the high relative humidity usually available on the leaf surface at 10-25°C with an optimum of 19-22°C. Although high relative humidity is required for infection, the spores will not germinate if immersed in water (free moisture). Leaf wetting is, therefore, not conducive to powdery mildew development (Burr, 1980). Under optimum conditions, powdery mildew can be obvious to the naked eye 48 hr after infection. About 5 days after infection, a new crop of spores is produced. Non-germinated powdery mildew spores can tolerate hot dry conditions and may persist until favorable conditions for germination occur. Early-season mildew development is affected more by temperature than by relative humidity. Abundant sporulation from overwintering shoots and secondary lesions on young foliage leads to a rapid buildup of inoculum. Secondary infection cycles may continue until susceptible tissue is no longer available. Since leaves are most susceptible soon after emergence, infection of new leaves may occur as long as shoot growth continues. Fruit infection occurs from pink to bloom. Overwintering buds are infected soon after bud initiation. Heavily infected shoots and buds are low in vigor and lack winter hardiness, resulting in a reduction of primary inoculum at temperatures below -24 C. Increasing light intensity raised photosynthetic activity and inhibited disease development (Pathak, 1986).

Management

As the pathogen survives in vegetative buds, so removal of affected twigs, mildewed shoots and leaves helps in reducing severity of the disease in the current season growth as well as reduces the load of primary inoculum for the next season (Bose and Sindhan, 1977). Alekseeva *et al.* (1992) reported that summer pruning in late May is effective in controlling powdery mildew besides increasing fruit yield. Powdery mildew are best controlled with the application of sulphur fungicides, Wettable sulphur fungicides like Cosan, Solbar, Sulforix and Thiovit were found effective in reducing the disease (Kumar *et al.*, 1987). Recently, the use of ergosterol biosynthesis inhibitor fungicides such as triadimenol (0.18 l/ha), hexaconazole (0.28 and 0.34 l/ha), tetraconazole (1.0 l/ha) and bitertanol (0.4l/ha) were found effective (Mendoza *et al.*, 1991). Apple scab spray schedule also reduce intensity of powdery mildew. The demethylation inhiting fungicides such as fusilazole and myclobutanil effectively control powdery mildew under heavy inoculum pressure (Sholberg and Haag, 1994). These fungicides have inhibitory as well as antisporulent activities against *P. leucotricha* (Climanowski and Bielenine, 1996).

Delicious group of apple varieties is highly susceptible to this disease, however, crab apple cultivars *viz.* Renetka, Purpurnaya, Yantarka, Altarkya, Dolgoe and Saynets Karchenio and two apple cultivars Maharaja Chunth and Golden Chinese are resistant ro powdery mildew (Verma and Gupta, 1988). Some new resistant varieties namely Romus 1, Romus 2, Romus 3, Pionier, Vionea and Generos have been developed through crossing among different *Malus* spp. (Cociu *et al.*, 1989).

Sooty Blotch and Fly Speck

Sooty blotch and fly speck are two different diseases. Both diseases are commonly occurs together on the same fruit superficial blemishes or discoloration. The presence of disease reduces the grade and market value of the fruit.

Symptoms

Sooty blotch appears as sooty or cloudy blotches on the surface of the fruit. The blotches are olive green with an indefinite outline. The blotches are frequently a fourth of an inch in diameter or larger, and may coalesce to cover much of the fruit. The "smudge" appearance results from the presence of hundreds of minute, dark pycnidia that are interconnected by a mass of loose, interwoven dark hyphae. The sooty blotch fungus is generally restricted to the outer surface of the cuticle. In rare cases, the hyphae penetrate between the epidermal cell walls and the cuticle.

Fly speck Groups of a few to 50 or more slightly raised, black and shiny round dots that resemble fly excreta, appear on the apple fruit. They consists of definite circular black dots which are the sclerotia of the causal fungus. The individual "fly specks" are more widely scattered and much larger than the pycnidia of the sooty blotch fungus (Figure 1.4). The flyspecks are sexual fruiting bodies (pseudothecia) of the fungus, and are interconnected by very fine hyphae.

Causal Organism

The fungus *Gloeodes pomigena* (Schw) Colby causes sooty blotch, and *Schizothyrium pomi* (mont. And Fr.) Arx (*Microthyriella rubi*) causes fly speck. Both fungi overwinter on twigs of various wild woody plants, especially wild blackberry and raspberry canes. Both fungi require free water on the fruit surface to infect.

Figure 1.4: Sooty Blotch and Fly Speck

Disease Cycle

The "sooty blotch" or "smudge" appearance on affected fruit results from the presence of hundreds of minute, dark fungal fruiting bodies (pycnidia) that are interconnected by a mass of loose, interwoven dark hyphae (fungal filaments). In spring, pycnidia on wild plants produce large numbers of spores (conidia) that ooze out and collect in a gelatinous mass. The conidia are then spread by water splash or wind blown mists into orchards from late May or early June until fall. The fungus first affects apple twigs, then secondary colonies are initiated on the fruit. Cool, humid weather (optimum 18°C) is essential for disease development. The disease does not develop when temperatures reach 30°C. When May and June are cool and moist and are followed by a hot July and August, sooty blotch symptoms often do not appear for two to three months. The disease is absent or rare when hot, dry weather prevails until close to harvest time. The disease is most severe when cool, rainy weather in the spring is coupled with late summer rains and low temperatures in early fall. Under ideal conditions, the incubation period from the time the fungus reaches the fruit to the appearance of symptoms may be as short as five days. In the orchard, the incubation period usually lasts three to four weeks on fruits that are 42 to 45 days old.

The individual "fly specks" are sexual fruiting bodies (ascocarps) of the fungus. Starting in late spring, the fungus produces spores on wild hosts. These spores are carried by wind into the orchard. When spores come into contact with the fruit under the proper environmental conditions.

Mycelial growth that forms the sooty blotches can occur in the absence of free water at relative humidity greater than 90 per cent. Symptom development of both diseases is relatively slow, typically requiring 20 to 25 days in the orchard, but may occur in 8 to 12 days under optimum conditions. Optimum conditions for conidial production for the flyspeck pathogen are 16-21°C and relative humidity greater than 96 per cent.

Management

Select an orchard site that always has full sunlight, good air circulation, and good soil (water) drainage. Prune trees annually to an open center for maximum air circulation. Both diseases are most prevalent in the damp, low, shaded areas of the orchard. Any practice that opens up the trees to greater air movement and promotes faster drying greatly aids in control. Pruning to open up the tree canopy and thinning to separate fruit clusters may provide control of sooty blotch and fly speck by improving air flow and increasing drying during periods of heavy dew or rain. These practices will also allow for better spray coverage and improved fruit quality. Remove or destroy nearby wild or neglected apple trees. Post-harvest dips in commercial disinfectants were used to remove signs of the flyspeck pathogen (FS) and the sooty blotch complex (SB) : 200, 500, or 800 ppm buffered sodium hypochlorite or a mixture of hydrogen peroxide and peroxyacetic acid at 60 ppm/80 ppm, 120 pprn/160 ppm, or 360 pprn/480 ppm, then re-rated after brushing and rinsing for 30 s on a commercial grading line. Apple scab spray schedule also reduces the spots quite effectively. Especially in commercial plantings, fungicide sprays are important for controlling these diseases. Spray of zineb, captan, captafol and Carbendazim has been found highly effective in controlling these diseases (Agarwala, 1967; Gupta and Sharma, 1981).

References

Agarwala, R.K. (1967). Relative efficacy of fungicides for the control of *Gloeodes pomigena* causing sooty blotch disease of apple. *Proc. Acad. Sci., India*, 37: 171–178.

Agrawala, R.K. and Sharma, V. C. (1966). White root rot disease of apple in Himachal Pradesh. *Indian Phytopathol.*, 19: 82.–86.

Alekseeva, S.A., Berbekor, V.N. and Martynov, V.I. (1992). Summer pruning: A means of controlling powdery mildew in apple. *Sadovodstvo–i–Vinogradarstvo*, 8: 8–9.

Anonymous (1975). *Hot Spots of Diseases and Pests of Major Field and Horticultural Crops*. ICAR, New Delhi, pp. 63.

Anonymous (1977). New records. *Quarterly Newsletter, Plant Protection Committee for South East Asia and Pasific Region*, 20: 5–7.

Anonymous (2001). *Apple Scab: Integrated Pest Management for Home Gardeners*, Pest Notes, University of California, Agriculture and Natural Resources, pp. 1–3.

Bala, R. (1999). Studies on *Marssonina* leaf Blotch of Apple. *M. Sc. Thesis*, Dr. Y.S. Parmar Uni. of Horti. and Forestry, Nauni, Solan, 74 p.

Behadad, E. (1975). Morphology, distribution, importance and hosts of *Rosellinia nectrix* (Hart.) Berl. The cause of white root rot in Iran. *Iran. J. Plant Pathol.*, 11(1/2): 13–17, 30–71.

Bessey, C.W. (1877). *On injurious Fungi: The Blight (Erysiphe)*. Iowa State College of Agric. Bienn. Rep., pp. 185–204.

Bose, S.K. and Sindhan, G.S. (1977). Control of stem black disease of apple in Kumaon. *Punjab Hortic. J.*, 16: 68–70.

Brandes, W. and Paul, V. (1981). Studies on the effect of Baycor on apple scab pathogenesis. *Pflanzenschutz Nachr.*, 34: 48–59.

Burr, T.J. (1980). *Powdery Mildew of Apple*. Tree Fruit IPM. Sheet No. 4, Published by New York State Agric. Exp. Stn. Geneva.

Caesar, A.J. and Burr, T.J. (1987). Growth promotion of apple seedlings and rootstocks by specific strains of bacteria. *Phytopathol.*, 77: 1583–1588.

Cattelan, A.J. (1994). Antagonism of fluorescent pseudomonads to phytopathogenic fungi from soil and soybean seeds. *Rev. Brasi. Ciencia do Solo*, 18: 37–42.

Climanowski, I.J. and Bielenine, A. (1996). Antisporulent activity of new apple powdery mildew fungicides. *J. Fruit Ornamental Plant Res.*, 4: 21–25.

Cociu, V., Serboiu, L., Braniste, N. and Serboiu, A. (1989). Studies on genetic resistance to disease in apple. *Cercetari de Genetica Vegetala is Animala*, 1: 277–287.

Cook, R.T.A. (1979). Production of *Diplocarpon rosae* conidia on over-wintering rose leaves. *Trans. Br. Mycol. Soc.*, 73: 354–357.

Din, S.C. and Saksena, H.K. (1984). Control and assessment of losses of apple scab caused by *Venturia inaequalis* in Kashmir. *Indian Phytopathol.*, 37: 384.

Dolan, D. (1974). New anthracnose of melons. *Phytopathol.*, 37: 583–596.

Filajdic, N. and Sutton, T.B. (1992). Chemical control of *Alternaria* blotch of apples caused by *Alternaria mali*. *Plant Dis.*, 76: 126–130.

Freeman, S., Szrejnberg, A., Chet, I. and Katan, J. (1986). Solar and biological control of white root rot disease in apple orchards caused by soil born fungus *Rosellinia nectrix*. *Harradeh*, 66: 1608–1613.

Gadoury, D.M. and MacHardy, W.E. (1982b). A model to estimate the maturity of ascospores of *Venturia inaequalis*. *Phytopathol.*, 72: 901–904.

Gadoury, D.M. and MacHardy, W.E. (1985). Effect of temperature on the development of pseudothecia of *Venturia inaequalis*. *Plant Dis.*, 66: 468.

Galea, V.J., Price, T.V. and Sutton, B.C. (1986). Taxonomy and biology of the lettuce anthracnose fungus. *Trans. Br. Mycol. Soc.*, 84: 619–628.

Gilpatrick, J.D. and Szkolnik, M. (1978). Maturation and discharge of ascospores of apple scab fungus. *Proc. Apple and Pear Scab Workshop*, N.Y. Agric. Exp. Stn. Spec. Rep. 28: 1–6.

Gupta V.K.and Verma, K.D. (1978). Comparative susceptibility of apple rootstocks to *Dematophora nectrix*. *Indian Phytopathol.*, 31: 377–378.

Gupta, G.K. (1975). Canker of apple trees due to *Sphaeropsis malorum* Berk. and its control. *Himachal J. Agric Res.*, 3: 44–52.

Gupta, G.K. (1978). Present status of apple scab (*Venturia inaequalis*) in Himachal Pradesh and strategy for its control. *Pestisides*, 9: 141–149.

Gupta, G.K. (1979). Some observations on apple scab in Himachal Pradesh. *Indian Phytopathol.*, 32: 172.

Gupta, G.K. (1988). Effect of weather on the perennation and prediction of apple scab (*Venturia inaequalis*). In: *Perspectives in Mychology and Plant Pathology*, (Eds.) V.P. Agnihotri, A.K. Sarbhoy, and D. Kumar). Malhotra Publishing House, New Delhi, pp. 353–369.

Gupta, G.K. and Agarwala, R.K. (1968). Alternaria blight of apple. *Pl. Prot. Bull. F.A.O.*, 16: 32.

Gupta, G.K.; Verma, K.D. and Pal, J. (1984). Some observations on the prevalence and severity of apple scab in Himachal Pradesh during the year 1978 to 1983. *Indian J. Mycol. Plant Pathol.* 14: 12–14 (Abst.).

Gupta, V.K. (1987). Apple scab and its management. *Indian Hortic.*, 32: 48–52.

Gupta, V.K. and Gohain, R.N. (1982). *In vitro* and *in vivo* production of toxic compounds by *Dematophora necatrix*. In: *Contemporary Trends in Plant Science*, (Ed.) S.C. Verma. Kalyani Publishers, Ludhiana, pp. 103.

Gupta, V.K. and Jindal, K.K. (1989). Management of white root rot of apple by biological and cultural methods. *Proc. Int. Hortic. Cong.* Italy, 2: 3229.

Gupta, V.K. and Jindal, K.K. (1990). Management of white root rot of apple (*Rosellinia necatrix*) by biological and cultural methods. *Proc. 23rd Int. Hortic. Congr.*, Italy, August 2–11, Vil.2. 3229 (Abstr.).

Gupta, V.K. and Sharma, S.K. (1981). Studies on the effects of fungicides on apple powdery mildew. *Anz. Schandingskde Pflanzenschutz*, 54: 8–10.

Gupta, V.K. and Sharma, S.K. (1999). Fungal diseases of apple. In: *Diseases of Fruit Crops* (Ed.). Kalyani Publishers, Ludhiana, pp. 3–27.

Gupta, V.K. and Sharma, S.K. (2000). *Diseases of Fruit Crops*. Kalyani Publ., New Delhi, pp. 3–27

Gupta, V.K. and Lele, V.C. (1980). Morphology, physiology and epidemiology of the apple scab fungus, *Venturia inaequalis* (Cke.) Wint. in Kashmir valley. *Indian J. Agric. Sci.*, 50: 167–173.

Harada, Y., Sawamura, K. and Konno, K. (1974). *Diplocarpon mali* sp. Nov., the perfect state of apple blotch fungus *Marssonina coronaria*. *Ann. Phytopathol. Soc. Jpn*. 40: 412–418.

Hirst, J.M. and Stedman, O.J. (1962). The epidemiology of apple scab *Venturia inaequalis* (Cke.) Wint. III. The supply of ascospores. *Ann. Appl. Biol.*, 50: 551–567.

Hong, K.K.; Moon, C.Y. and Soon, Y.M. (1998). Cultural characteristic of *Diplocarpon mali*. *RDAJ. Crop Prot.*, 40: 120–123.

Ito, S.I. and Nakamura, N. (1984). An outbreak of white root rot and its environmental conditions in the experimental arboretum. *J. Jpn. For. Soc.*, 66: 262–267.

James, J.R. and Sutton, T.B. (1982). A model for predicting ascospores maturation of *Venturia inaequalis*. *Phytopathol.*, 72: 1081–1085.

Jia, X.Y. (1961). The identification and diagnosis methods for several easily confusable disease. *Bull. Agric. Sci. Technol.*, 2: 23.

Joshi, N.C., Mallik, A.G., Kaul, M.L. and Anand, S.K. (1975). Some observation on the epidemic of scab disease of apple in Jammu and Kashmir during 1973. *Indian Phytopathol.*, 28: 288–289.

Kelley, R.D. and Jones, A.L. (1981). Evaluation of two triazole fungicides for post-infection control of apple scab. *Phytopathol.*, 71: 737–742.

Kumar, S., Srivastava, K.K. and Pandey, J.C. (1987). Economic spray schedule for apple powdery mildew control. *Int. J. Trop. Plant Dis.*, 5: 59–65.

Leite, R.P. Jr., Tsuneta, M. and Kishino, A.Y. (1986). Apple leaf spot caused by *Marssonina coronaria*. *Fitopatol. Bras.*, 11: 725–729.

Lopez, J.C., Ropas, E.B., Rivas, P.G.G., Rivillas, O.C.A. and Perez, L.C.M. (1998). Isolation of fluorescent Pseudomonas potential antagonist of *Rosellinia necatrix* in solarization soil. *Eur. J. Plant Pathol.*, 105: 571–576.

Louw A.J. (1951). Studies on the influence of environmental factors on the overwintering and epiphytology of apple scab in the winter rain fall area of the Cape Province. *S. Afr. Dep. Agric. Sci. Bull.*, 310: 48.

MacHardy, W.E.and Gadoury, D.M. (1985). Forcasting the seasonal maturation of ascospores of *Venturi inaequalis*. *Phytopathol.*, 75: 381–385.

Mallick, F., Shukla, N.B. and Bhutia U. (1984a). Occurrence of apple scab in Sikkim. *Plant Prot. Bull.*, 36: 121.

Mallick, F., Singh, A.K. and Shukla, C.P. (1984b). Occurrence of apple scab in Arunachal Pradesh. *Plant Prot. Bull.*, 36: 123.

Nath, P. (1935). Studies on the diseases of apple in northern India. II. A Short note on apple scab due to *Fusicladium dendriticum* Puck. *J. Indian Bot.*, 14: 121–124.

Nattrass, R.M. (1927).The white root rot of fruit trees caused by *Rosellinia necatrix* (Hart.) Berl. Annu. *Rep. Agric. Hortic. Res. Sta. Long Ashton*, p. 66–72.

Parmelee, J.A. (1971). *Marssonina* leaf spot of apple. Can. *Plant Dis. Surv.*, 51: 91–92.

Pathak, V.N. (1986). *Diseases of Fruit Crops*. Oxford and IBH Publishing Co., New Delhi, pp. 309.

Roberts, J.W. (1924). Morphological characteristics of *Alternaria mali* Roberts. *J. Agric. Res.*, 27: 699–708.

Salman, E.S. and Wormald, H. (1913). Report on economic mycology. *J. South East Agric. Coll.*, 22: 453.

Sawamura, K. (1990). *Alternaria* blotch: A new disease causing premature leaf fall in apple. *Indian Phytopathol.*, 52: 101–102.

Sawamura, K., Harada, Y. and kikuchi, T. (1993). The apple Industry in Japan: A historic sketch and disease specific to the region. *Plant Dis.*, 77: 546–552.

Sharma, J.N. and Kaul, J.L. (2000). *Marssonina* blotch: A new disease hits apple crop in Himachal Pradesh. In: *Proc. International Conf. Integrated Plant Disease Management for Sustainable Agriculture*. New Delhi, pp. 775–776.

Sharma, J.N. and Verma, L.R. (1999). Premature leaf fall in apple: Diagnosis and control. In: *Diseases of Horticultural Crops: Fruits,* (Eds.) L.R. Verma and R.C. Sharma. Indus Publishing Co., New Delhi, pp. 80–88.

Sharma, J.N. (2001). Diagnosis and control of premature leaf fall in apple. *J. Mycol. Plant Pathol.*, 31: 89–95.

Sharma, J.N. and Gautam, D.R. (1997). Studies on premature leaf fall in apple: A new problem. *Indian J. Plant Prot.*, 25: 8–12.

Sharma, J.N., Gupta, D., Bhardwaj, L.N. and Kumar, R. (2005). Occurrence of *Alternaria* leaf spot (*Alternaria alternata*) on apple and its management, (Eds.) R.C. Sharma and J.N. Sharma. Scientific Publication (India), Jodhpur (Raj.), pp. 25–31.

Sharma, J.N.; Sharma, P.K. and Sharma, A. (2005). Studies on epidemiology and management of *Marssonina* blotch, the cause of premature leaf fall in apple, (Eds.) R.C. Sharma and J.N. Sharma. Scientific Publication (India), Jodhpur (Raj.), pp. 1–8.

Sharma, J.N., Sharma, Y.P. and Kishore, D. (1996). Reaction of apple germplasm to scab and powdery mildew diseases. *Nat. Sym. Mol. App. Plant Dis. Management*, Shimla, pp 27.

Sharma, R.C. and Sharma, J.N. (2005). *Integrated Plant Disease Management*. Scientific Publishers (India) Jodhpur, Raj., pp. 362.

Sharma, R.L., Kumar, J. and Ram, V. (1988). Performance of scab resistant cultivars of apple in Killu Valley, Himachal Pradesh. *J. Tree. Sci.*, 7: 45–49.

Sharma, S.K. (1993). Studies on management of white root rot of apple. *Ph.D. Thesis*, UHF, Nauni, Solan, 146 p.

Simpson, B. and Hayes, A.J. (1978). Growth of *Marssonina brunnea*. *Trans. Br. Mycol. Soc.*, 70: 249–255.

Singh, U.B. (1943). Pink disease of apple in Kumaon. *Indian Farm.*, 5: 566–567.

Sousa, A.J.T.D. (1985). Control of *Rosellinia necatrix* causal agent of white root rot, susceptibility of several plant species and chemical control. *Eur. J. Plant Pathol.* 15: 323–332.

Stephan, S. (1988). Studies on primary and secondary shoot infection with apple powdery mildew [*Podosphaera leucotricha* (EII. and Ev.) Salm.]. *Arch. Phytopathol. Pflanzenschutz*. 27: 491–501.

Sud, V.K. and Agrawal, R.K. (1971). Laboratory and field evaluation of fungicides for the control of *Alternaria* blight of apple. *Indian Phytopathol.*, 24: 201–204.

Sutton, T.B. (1981). Production and dispersal of ascospores and conidia by *Physalospora obtuse* and *Botryosphaeria dothidea* in apple orchards. *Phytopathol.*, 71: 584–589.

Szkolnik, M. (1978). Techniques involved ib greenhouse evaluation of deciduous trees fruit fungicides. *Annu. Rev. Phytopathol.*, 16: 103–129.

Szkolnik, M. (1981). Physical modes of action of sterol-inhibiting fungicides against apple diseases. *Plant Dis.*, 65: 981–985.

Sztejnbert, A., Freeman, S., Chet, I. And Katan, J. (1987). Control of *Rosellinia necatrix* in soil and in apple orchard by solarization and Trichoderma harzianum. *Plant Dis.*, 71: 365–369.

Tailor, J. (1996). Ghost spot of apple leaves by *Alternaria tenuis. Phytopathol.*, 56: 553–555.

Takahashi, S. and Sawamura, K. (1990). Marssonina blotch. In: *Compendium of Apple and Pear Diseases*, (Eds.) A.L. Jones and H.S. Aldwinckle. The American Phytopathol. Society, St. Paul (USA), pp. 33.

Thakur, V.S. (1987). Studies on the effect of fungicides on apple scab *Venturia inaequalis* (Cke.) Wint. Pathogenesis. *Ph.D. Thesis*, Univ. Hortic. For., Solan, 136 pp.

Thakur, V.S. and Gupta, G.K. (1990a). Evaluation of pre-symptom activity of fungicides on symptom expression, conidia production and viability of *Venturia inaequalis. Indian Phytopathol.*, 43: 520–526.

Thakur, V.S. and Gupta, G.K. (1991). Studies on the effect on sterol-inhibitors on apple scab. *Venturia inaequalis* pathogenesis. *Indian J. Plant Prot.*, 19: 185–190.

Thakur, V.S. and Gupta, G.K. (1992a). Curative and protective action of ergosterol biosynthesis inhibiting fungicides in relation to infection periods in the control of apple scab. *Pestology*, 16: 32–37.

Thakur, V.S. and Gupta, G.K. (1992b). Post-infection fungicidal inhibition of apple scab (*Venturia inaequalis*) sporulation. *Indian J. Agric. Sci.*, 62: 629–636.

Thakur, V.S.; Sharma, N. and Vaidya, S. (2005). Effect of weather parameters on the epidemic of apple blotch in Himachal Pradesh, (Eds.) R.C. Sharma and J.N. Sharma. Scientific Publication (India), Jodhpur (Raj.), pp. 19–24.

Utkhede, R.S. (1999). Biological treatments to increase apple tree growth in replant problem soil. *Alleopathy J.*, 6: 63–68.

Utkhede, R.S. and Li, T.S.C. (1989). Evaluation of *Bacillus subtilis* for potential control of apple replant disease. *J. Phytopathol.*, 126: 305–312.

Verma, K.D. and Gupta, G.K. (1988). Field reaction of apple germplasm to powdery mildew. *Indian J. Agric. Sci.*, 58: 233–234.

Wilson, E.E. (1928). Studies of the ascigerous stage of *Venturia inaequalis* in relation to certain factors of the environment. *Phytopathol.*, 18: 375–418.

Plant Diseases Management in Horticultural Crops (2011) *Pages* **20–28**
Editors: **Shahid Ahamad, Ali Anwar and P.K. Sharma**
Published by: **DAYA PUBLISHING HOUSE, NEW DELHI**

Chapter 2

Anthracnose and Fruit Rot of Chilli and their Management

P.K. Tiwari[1]*, *J.N. Srivastava*[2], *C.P. Khare*[1] *and V.S. Thrimurty*[1]

[1]*Department of Plant Pathology, Indira Gandhi Krishi Vishwavidyalaya, Raipur (CG)*
[2]*Regional Horticultural Research Station, Bhaderwah, SKUAST, Jammu*

Introduction

Chilli (*Capsicum annum* L.) is one of the most important vegetables cum spice crops. It is used as a principal ingredient of various dishes. Green chillies are rich in vitamin A and C and minerals. Chilli occupies 1,687,732.70 ha. area with the production of 24,947,587 tonnes in the world (Anonymous, 2005a). It is cultivated in China, USA, Brazil, Indonesia, Malaysia, Vietnam, Sri lanka, Thailand, Africa, Japan, Pakistan and Nigeria. In india the major chilli growing states are Andhra Pradesh, Karnataka, Maharashtra, Orissa, Tamil Nadu, Madhya Pradesh, West Bengal and Rajasthan in mentioned order and account for more than 80 per cent of the total area and production.

Chilli crop suffers every year due to anthracnose and fruit rot, leaf curl and bacterial wilt diseases. Among them anthracnose and fruit rot of chilli caused by *Colletotrichum capsici* has become a major problem in chilli cultivation. The disease seriously appears in every part of the country in the month of November and December, due to availability of favourable climatic conditions. *C. capsici* can attack the every part of chilli plant but the fruit rot is more conspicuous as it causes severe damage to mature fruit in the field as well as during harvesting, transit and storage. In India the losses up to 84 per cent were reported in the field as well as in storage due to this disease under favourable environmental conditions (Thind and Jhooty, 1985). In Assam 12-32 per cent of the fruits were found affected by Chowdhury (1957).

* Corresponding Author; E-mail: prdptiwari@yahoo.com.

Symptoms

On Seedlings

Symptoms appeared as necrosis of the tender twigs from the tip backwards. The entire top of the seedling may wither away. The twigs showed water soaked to brown discolouration becoming grayish white or straw coloured in advanced stage of the disease (Thind, 2001).

On Plants

Infection usually begins when the crop is in flowering stage. In diseased plants, flowers dry up; drying up spreads from the flower stalk to the stem and subsequently causes die-back of the branches and stem. Partially affected plants bear fruits, which are few and are of low quality. The dead twigs are water soaked to brown, becoming grayish white or straw coloured in advanced stage of the disease. Sometimes the necrotic areas are found separated from the healthy area by a dark brown to black band (Rajeshwari *et al.*, 2004).

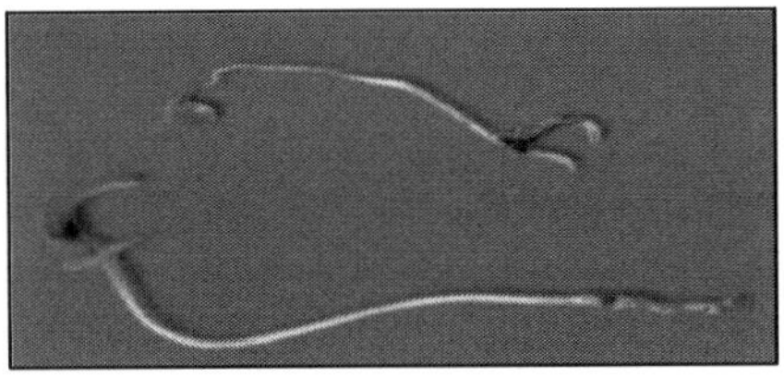

Figure 2.1: Die Back Symptoms on Seedlings H : Healthy; I : Infected

On Leaves

Symptoms on initial stage are characterized by production of small, circular brown spots on leaves. Later the spots develop into dark brown sickle shaped or circular to irregular spots in a concentric fashion with gray coloured center. After that spotted portion becomes papery and falls off from the leaves producing shot hole (Beniwal *et al.*, 1981).

On Fruits

The disease usually occurs on the ripe or mature fruits as circular to elliptical sunken spots with black margin. Severely diseased fruits turn straw colored from the normal red and on this discolored area, numerous black dots are present (Thind, 2001). Fruit symptoms initially begin as water-soaked that become soft, slightly sunken and become tan. The lesions can cover most of the fruit surface and multiple lesions may occur. The surface of the lesion becomes covered with the wet, gelatinous spores, from salmon coloured fungal fruiting bodies (acervuli) with numerous black setae. The lesions are initially brown and then black due to the formation of setae (Anonymous, 2001).

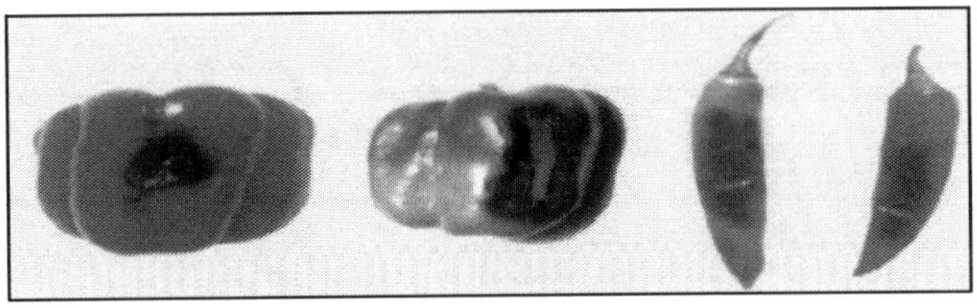

Figure 2.2: Fruit Rot Symptoms on Ripend and Green Fruits

Fruits affected with *C. capsici* become deformed, sometimes white in colour and loose their pungency. The attacked parts in fruit, turn black and become depressed or wrinkled. Ultimately the diseased fruits shrivel and dry up. When a diseased fruit is cut open the lower surface of the skin is found covered with minute, elevated, spherical, black, stromatic mass of the fungus. In advanced stage, a mat of fungal hyphae covers the seeds and seeds turn rusty in colour (Rajeshwari *et al.*, 2004).

Cherian and Verghese (2005) reported that the disease was characterized by dark brown or dirty black colour and oval to circular, depressed spots over the fruits. The mycelium penetrated the inner tissue and even the seeds and produced a peculiar pungent odour.

Etiology

Anthracnose and fruit rot of chilli is caused by *Colletotrichum capsici* (syd.) Butler and Bisby the aerial mycelium forms light to dark gray felt, consists of septate, inter and intracellular hyphae. acervuli formed on fruits, leaves and stems are round, elongate, intra and sub-epidermal. setae are abundant, brown, 1-5 septate, rigid, sometimes swollen at the base, slightly tapered towards the paler acute apex. conidia are falcate, fusiform, single celled, hyaline, uninucleate and conidiophores are hyaline to faintly brown, cylindrical, septate or aseptate and phialidic (Mah, 1985). Kar and Mahapatra (1981) reported that c. capsici produces acervuli which are amphigenous, black, punctiform, circular to irregular in out line, generally smaller in size, a few are larger up to 136.5 µm in diameter, setae many per acervulus (up to 40), scattered, stiff, bristle like, with pointed, pale hyaline apex and dark brown base up to 119

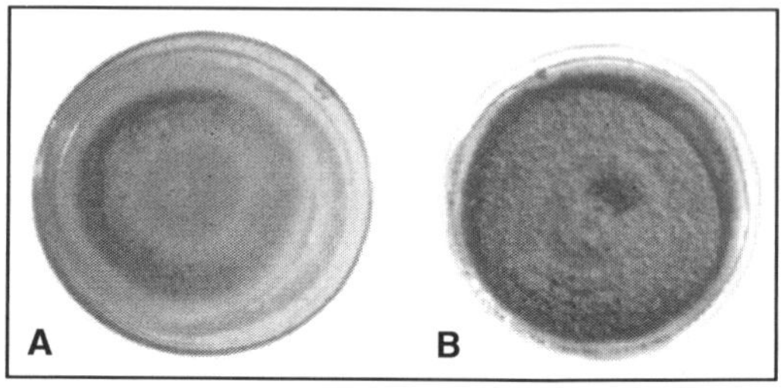

Figure 2.3: Pure Cultures of *C. capsici* Showing Pinkish Masses of Conidia

A: Pure Culture; B: Pure Culture Showing Masses of Conidia

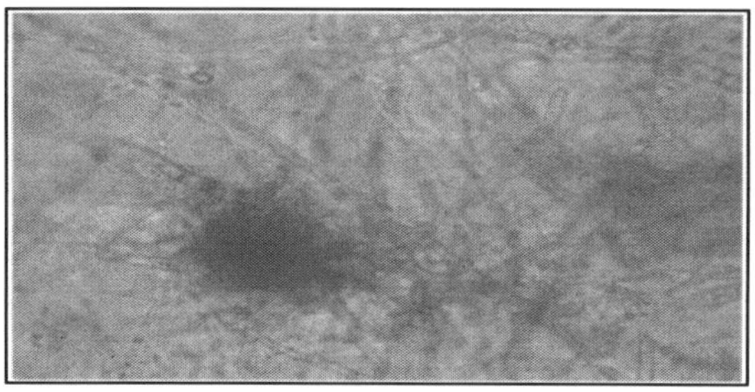

Figure 2.4: Mycelium Aggregation to Form Cushion

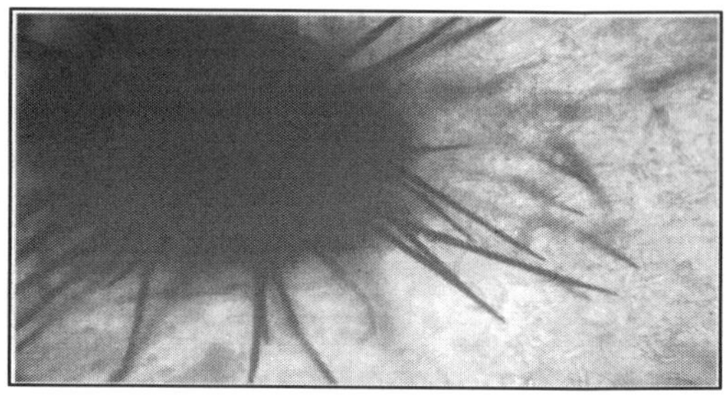

Figure 2.5: Acervulus Showing Setae and Conidia

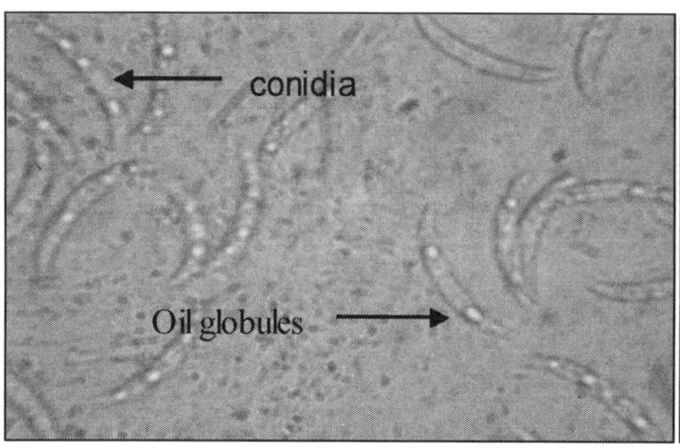

Figure 2.6: Sickle Shaped Conidia with Oil Globules

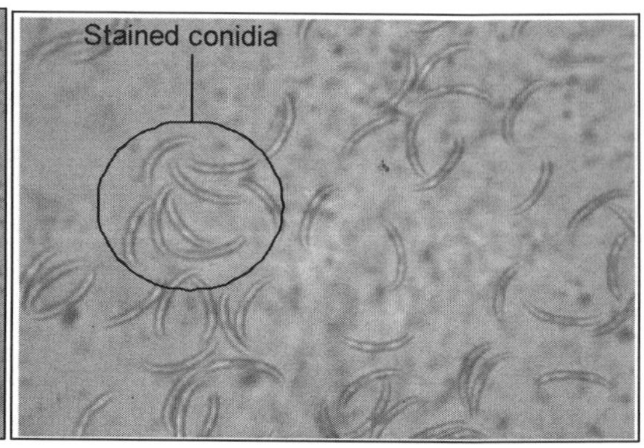

Figure 2.7: Stained Conidia with Oil Globules

x 6.5 µm. conidiophores are one celled, hyaline, almost cylindrical. Conidia are one-celled, hyaline, curved with two or more oil globules and measuring 14.64 -18.0 x 3.4-5.1 µm.

Disease Cycle and Influence of Environmental Factors on Disease Development

The disease is externally and internally seed borne. Seeds from severely diseased fruits also carry the primary inoculum. However, fungus can survive in plant debris and secondary infections take place through wind borne conidia. Seeds from badly diseased fruits also carry the primary inoculam *Colletotrichum capsici* has an extensive host range including vegetables, fruits and ornamental plants. The most common hosts are solanaceous, cucurbits, legumes and malvaceous plants (Dubey and Ekka, 2004).

Effect of Temperatures and pH Levels on *C. capsici*

Maximum growth and sporulation of *C. capsici* were recorded at 30°C and 5 pH under *in vitro* condition (Mazlan and Sariah, 1980). Astuti and Suharidi (1986) found that the temperature at 30°C favoured development of disease in storage condition. Whereas Alabi and Emechebe (1992) reported that the temperature at 25°C was optimum for growth and sporulation of *C. capsici*. They also observed 21.5°C as minimum and 31.3°C maximum temperature to be suitable for spread of the disease in field during July-October month. Similarly Ouyang *et al.* (1993) found that the culture filtrates of *C. capsici* were highly toxic at 25-28°C and pH 6-7. Datar (1995) found that temperature range between 20 to 30°C to be most conducive for fruit rot development in chillies caused by *C. capsici*. Beura and Acharya (2002) reported that 28°C is the optimum temperature for germination of spores of *C. capsici*, while no germination of spore was observed at 40°C. Optimum range of relative humidity for spore germination was between 95-100 per cent. The pH from 6-7 were most favourable, whereas the optimum pH for maximum degree of spore germination was 6.5. The temperature 20.5-29.1°C, relative humidity 62.5-91.3 per cent, rainfall 0.3-9.5 mm and 2-4 number of rainy days favoured the maximum anthracnose disease development during October month (Dubey and Ekka, 2004). The optimum temperature for conidia germination is 30°C and maximum disease development takes place at 28°C and 95.7 per cent relative humidity. The disease usually develops under high humid conditions, after fruits have started to ripen (Rajeshwari *et al.*, 2004).

Management

Antagonistic Effect of Bio-agents Against *C. capsici*

Studies on the effect of different formulation of *T. harzianum* Rifai isolates, Neemarin and Neem oil on *C. capsici* were conducted on two chilli cultivars *i.e.* Pusa sadabahar and Navjyoti. *T. harzianum* suppressed the symptoms expressions, conidial germination and mycelial growth of *C. capsici* up to 100 per cent and found that one pre harvest spray of *T. harzianum* two days before the harvesting should reduce the post harvest and storage losses caused by *C. capsici* (Sharma *et al.*, 2005). Pandey and Pandey (2005) reported that *T. harzianum* showed mycoparasitism and mycoparasitised the fungal colony, fully within 5-7 days of incubation by coiling and penetration over *C. gloeosporioides*. Jeyalakshmi *et al.* (1998) screened seven *Trichoderma* spp., seven Pseudomonas isolates, two isolates of *Bacillus subtilis* and one yeast *Saccharomyces cerevisiae* against *C. capsici* both *in vitro* and on the plant. Among the fungal antagonists *S. cerevisiae* exhibited the maximum growth reduction followed by *T. viride*, whereas among the bacterial antagonists, *B. subtilis* showed maximum growth reduction followed by *P. fluorescens* No. 27. The efficacy of various *P. fluorescens* isolates were tested for the management of fruit rot of chilli. Seed treatment plus soil application of talc-based formulations of the isolate Pf1 effectively reduced the disease incidence (Ramamoorthy and Samiyappan, 2001). Ekbote (2005) investigated the efficacy of *P. fluorescens* against anthracnose of chilli. The *P. fluorescens* treatments reduced the incidence of die-back and fruit rot and increased the yield of chilli compared to control. The seedling dip treatment was the most effective against die-back and fruit rot incidence and yield. Srinivas *et al.* (2006) found that pure culture of *P. fluorescens* and *T. harzianum* at the rate of 1 X 108 cfu/g and their talcum based formulations (28 X 107 cfu/g and 19 X 107 cfu/g, respectively) were effective in reducing *C. capsici* infection and increase the seed germination, vigour index and field emergence.

Effect of Medicinal Plant Products/Extracts Against *C. capsici*

The extracts of onion bulbs, neem leaves, *Pongamia glabra* leaves and *Tagetes erecta* leaves at 4 per cent concentration completely inhibited the mycelial growth and spore germination of *C. capsici* (Singh *et al.*, 1997a) Singh and Korpraditskul (1999) studied the use of crude plant extract of neem, garlic and tagak-tagak (*Rhinocanthus nasuta*) at 5000 ppm against Capsicum anthracnose. The garlic extract performed well under room humidity, while tagak-tagak extract showed good control of chilli anthracnose under high moisture condition, while neem extract minimizing the ripe fruit rot. Charigkapakorn (2000) found that sweet flag (*Acorus calamus* L.) crude extract at 1000 ppm was the most effective against anthracnose disease in chilli fruits, while betel pepper (*Piper betel* L.) crude extract and a mixture of these two herbal plants were ineffective. Total yield of sweet flag crude extract treatment was significantly higher than that of other treatments. Asha and Kannabiran (2001) reported that *Datura metel* leaf extract (10 per cent) was very effective, when sprayed at 8 days after sowing could protect chilli seedlings against *C. capsici* up to 35 days after sowing. Among the 15 plant extracts evaluated in vitro, Parthenium was found superior in inhibiting the germination of spore of *C. truncatum* (Madhusudan, 2002).The antimicrobial properties of Bitter Temru (*Diospyros cordifolia*), Datura (*Datura stramonium*), Amaltas (*Cassia fistula*), Brhati (*Salomum indicum*), Sandal (*Santalum album*), Mehandi (*Lawsonia inermis*) and Babool (*Acacia nilotica*) were tested for chilli fruit rot management at the rate of three concentration (1/500, 1/1000, 1/2000) against fungal growth *in vitro*. Maximum growth inhibition of the fungus was observed after inoculation with bitter Temru fruit extract and Datura leaves, respectively. Growth inhibition was minimum in Babool seed extract at 1/2000 concentration. Similar results were also obtained in event of spore germination and disease severity at the concentration level of 1/1000 in potted plants (Bagri *et al.*, 2004). *In vitro* evaluation of different plant extracts showed that duranta leaf extract and asafoetida extract as best botanicals for the management of *C. capsici* (Gorawar, 2004). Meera *et al.* (2004) reported that extracts of bulbs of *Allium sativum* (20 per cent) and *Allium cepa* (60 per cent) and seed extracts of *Azadirachta indica* (60 per cent) gave complete inhibition of mycelial growth of *C. capsici*. The leaf extracts of *Datura metel*, *E. globulus* and *P. julifora* at 60 per cent and *A. indica*, *V. negundo* and *Tactona grandis* at 80 per cent concentration totally inhibited the mycelial growth of *C. capsici*. Conidial germination of *C. capsici* was completely inhibited by A. sativum (20 per cent), D. metel (60) per cent and P. julifora (60 per cent). Under *in vivo* condition *A. sativum* (20 per cent) recorded maximum seed germination (80 per cent), higher yield and fruit character such as increased fruit length and fruit weight.

Screening of Different Genotypes of Chilli Against *C. capsici*

Perane and Joi (1986) screened 105 cultivars under natural infection in the field. Pant C-1 and B7-9 were resistant and 7262, Deglur, B79A, LIC 24, 764 Guntur and 574 Thirumalapurum were moderately resistant to the disease. The resistance to dieback stage has also been recorded in BG-1, Shahkoti and Laichi-11 and a F1 hybrid (CH-1) was also recorded and was released in 1991 for cultivation in the

Punjab state (Singh and Kaur, 1990). In a nursery experiment Kaur and Singh (1990) found that among 37 varieties at the leaf blight stage of anthracnose showed that only 5 varieties, that is, Lorai, JED 4-2-2, H6, S40-1 and C39 were resistant and 8 varieties like Feroj, California wonder, CW30, My 10-2, My12-6-3-3, My8-3-4, L16-1-1 and Perennial were moderately resistant to *Colletotrichum* spp. Fifty five cultivars/lines were tested under artificial inoculation, out of them five (Pusa sadabahar, 91-2, DC-9, DC-27 and Achar) were free from infection where as six (Aparna, Kalyanpur red, Sabour Anil, BG-1, Lorai and Perennial) were resistant and four (Pant C-1, DC-18, Suryamani and PS-1) were moderately resistant (Singh *et al.*, 1997b). Fifty two chilli genotypes were screened against *C. capsici* and LCA-301, LCA-324, K-1 and Byadgi Kaddi were found resistant, whereas KDSC-210-10 and S-32 were found highly susceptible (Hegde and Anahosur, 2001). Patil *et al.* (2002) also screened twenty cultivars out of them two varieties cv. Jayanti and Phule Jyoti appeared to be resistant and eleven cultivars were found to be moderate resistant.

Effect of Fungicides Against *C. capsici*

Seed should be obtained from spot less fruits.Seed treatment with thiram,bracicol and bisdithane has been found effective in elimination of seed borne inoculam (Growar and Bansal, 1970) The efficacy of Copper hydroxide at different concentration, Chlorothalonil at 0.2 per cent and Carbendazim at 0.1 per cent were experimented in controlling fruit rot of chilli. Among them the lowest disease index resulted from the application of 0.1 per cent carbendazim followed by 0.25 per cent copper hydroxide and crop yield was highest with application of 0.25 per cent Copper hydroxide (Ekbote, 2002). in vitro experiment Dubey and Ekka (2004) evaluated 13 fungicides *viz.* Carbendazim, Benomyl, Copper oxychloride, Captan, Hexaconazole, Mancozeb, Cholorothalonil, Iprobenphos, Ridomil MZ, Thiram, Tilt, Topsin M and Vitavax against *C. capsici* by poisoned food technique. They found that Tilt, Vitavax and Thiram completely inhibited the growth of pathogen. Tilt, Vitavax, Topsin M, Kitazin, Carbendazim and Captan completely inhibited the sporulation of the pathogen. Deshmukh *et al.* (2004) reported that Mancozeb (0.25 per cent) was highly effective in reducing disease intensity and fruit rot and produced maximum yield. Seven fungicides *viz.* Mancozeb (0.3 per cent), Carbendazim (0.1 per cent), Captacil (0.2 per cent), Calixin (0.2 per cent), Kitazin (0.1 per cent) and Captan (0.2 per cent) were evaluated *in vitro* against *C. capsici*. Carbendazim inhibited the radial growth up to 88 per cent followed by Mancozeb (77 per cent) (Kumar, 2004).The fungicides named as Penconazole 10 EC (0.05 per cent), Copper oxychloride 50 WP (0.05 per cent), Mancozeb + Metalaxyl 72 WP (0.25 per cent), Carbendazim 50 WP (0.10 per cent), Aureofungin 46.15 sol. (0.02 per cent), Hexaconazole 5 EC (0.05 per cent), Propiconazole 25 EC (0.10 per cent) and Captan 50 WP (0.25 per cent) were evaluated against anthracnose of chilli. Mancozeb + Metalaxyl was the best in controlling the disease. It was followed by Propiconazole and Carbendazim (Hingole and Kurundkar, 2004).

References

Alabi, O. and Emechebe, A.M. (1992). Effect of temperature on growth and sporulation of cowpea brown blotch pathogen *Colletotrichum capsici* (Syd.) Butler and Bisby. *Samara Journal of Agricultural Research*, 9 : 99–102.

Anonymous (2001). *Anthracnose Caused by Colletotrichum spp. on Pepper*. Department of Plant Pathology, Florida, 178 pp.

Anonymous (2005a). *FAO Yearbook*, 2005. 56 pp.

Asha, A.N. and Kannabiran, B. (2001). Effect of *Datura metel* leaf extract on the enzymatic and nucleic acid changes in the chilli seedling infected with *Colletotrichum capsici*. *Indian Phytopathology*, 54(3): 373–375.

Astuti, E.B. and Suharidi (1986). The effect of storage temperature and fruit maturity on the incidence of anthracnose on chilli pepper. *Bulletin Penelitian Hortikultura*, 13(1): 40–50.

Bagri, R.K., Choudhary, S.L. and Rai, P.K. (2004). Management of fruit rot of chilli with different plant products. *Indian Phytopathology*, 57(1) : 107–109

Beniwal, S.P.S., Saxena, G.C. and Tripathi, H.S. (1981). Natural occurrence of anthracnose of mungbean caused by *Colletotrichum capsici*. *Indian Journal of Mycology and Plant Pathology*. 13 (3) : 356–357.

Beura, S.K. and Acharya, B. (2002). Epidemiological studies of *Colletotrichum capsici* (Sydo.) Butler and Bisby causing anthracnose disease in cotton. *Journal of Research of Orissa University of Agriculture and Technology*, 20(2): 30–34.

Charigkapakorn, N. (2000). Control of chilli anthracnose by different biofungicides. *Report 2000* (www.google.com).

Cherian, T.T. and Verghese, K.I.M. (2005). Post harvest fungal rot of chilli (*Capsicum annuum* L.) fruits. *Journal of Mycology and Plant Pathology*, 35(1): 97–99.

Chowdhury, S. (1957). Studies on the development and control of fruit rot of chillies. *Indian Phytopathology*, 10: 55–61.

Datar, V.V. (1995). Pathogenicity and effect of temperature on six fungi causing fruit rot of chilli. *Indian Journal of Mycology and Plant Pathology*, 25(3) :196–197.

Dubey, S.C. and Ekka, S. (2004). Influence of weather on development of Colletotrichum blight of bitter guard. *Journal of Mycology and Plant Pathology*, 34(2): 404–406.

Deshmukh, G.P., Kurundkar, B.P. and Mehetre, N.M. (2004). Efficacy of Zetron, a biopesticides in the management of chilli anthracnose. *Journal of Mycology and Plant Pathology*, 34(2) : 301–303.

Ekbote, S.D. (2002). Bioefficacy of copper hydroxide against anthracnose of chilli. *Karnataka Journal of Agricultural Sciences*, 15(4): 729–730.

Ekbote, S.D. (2005). Effect of *Pseudomonas fluorescens* on anthracnose of chilli caused by *Colletotrichum capsici*. *Karnataka Journal of Agricultural Sciences*, 19(8): 162–165.

Growar, R.K. and Bansal, R.D. (1970). Seed born nature of *Colletotrichum capsici* in chlli seeds and its control by seed dressing fungicides. *Indian Phytopathology*, 23: 664–668.

Hegde, G.M. and Anahosur, K.H. (2001). Biochemical basis of resistance to fruit rot (*Colletotrichum capsici*) in chilli genotypes. *Karnataka Journal of Agricultural Sciences*, 14(3): 686–690.

Hingole, D.G. and Kurundkar, B.P. (2004). Fungicides evaluation and economics in controlling anthracnose of chilli. *Journal of Soils and Crops*, 14(1): 175–180.

Jeyalakshmi, C., Durairaj, P., Seetharaman, K. and Sivaprakasham, K. (1998). Biocontrol of fruit rot and die back of chilli using antagonistic microorganisms. *Indian Phytopathology*, 51(2): 180–183.

Kar, A.K. and Mahapatra, H.S. (1981). New host records of Colletotrichum species from India. *Indian Phytopathology*, 34(2): 214–221.

Kaur, S. and Singh, J. (1990). *Colletotrichum acutatum*: A new threat to chilli crop in Punjab. *Indian Phytopathology*, 43(2): 108–110.

Kumar, M.R. (2004). Compatibility of fungicides with insecticides and their effect on growth in *Colletotrichum capsici* and *Pythium aphanidermatum*. *Journal of Mycology and Plant Pathology*, 34 (2): 496–497.

Madhusudan, B.N. (2002). Studies on soybean anthracnose caused by *Colletotrichum truncatum* (Schw.) Andrus and Moore. *M.Sc. (Ag) Thesis*, University of Agricultural Sciences, Dharwad (Karnataka).

Mah, S.Y. (1985). Anthracnose fruit rot (*Colletotrichum capsici*) of chilli (*Capsicum annuum*): Causal pathogen, symptom expression and infection studies. *Technologi Sayur Sayuran (Malaysia)*, 1: 35–40.

Mazlan, S. and Sariah, M. (1980). Anthracnose of chilli in Malaysia: biology of the pathogen and varietal susceptibility. *Pertanika*, 3(1): 47–52.

Meera, T., Ancy, P.G. and Udhayakumar, R. (2004). Antifungal activity of plant products against Colletotrichum capsici, the incitant of fruit rot of chilli. In: *Paper Present in the 26th Annual Conference and Symposium*, Held at Goa, from Oct. 7–14.

Ouyang, F., Xie, B.Y. and Liu, F.C. (1993). *Colletotrichum capsici* toxin. *Acta Mycologica Sinica*, 12 (4): 289–296.

Pandey, B.K. and Pandey, M.K. (2005). Evaluation of biocontrol potential of *Trichoderma harzianum* against post harvest fungi of mango *in vitro*. In: *Paper Presented in the 58th Annual Meeting on Crop Disease Management in Dryland Agriculture*, Held at Parbhani (Maharashtra) from Jan. 12–14.

Patil, M.J., Ukey, S.P. and Ghoderao, B.N. (2002). Performance of chilli against fruit rot in natural condition. *Annals of Plant Physiology*, 16(2): 202–203.

Perane, R.R. and Joi, M.B. (1986). Reaction of chilli cultivars to fruit rot and die-back of chilli incited by *Colletotrichum capsici*. *Current Research Reporter*, Mahatma Phule Agricultural University, 2(1): 52–54.

Rajeshwari, E., Chezhiyan, N., Saraswathy, S. and Subbiah, A. (2004). Die-back and fruit rot of chilli. *Spice India*, 17(4): 33.

Ramamoorthy, V. and Samiyappan, R. (2001). Induction of Defense related genes in *Pseudomonas fluorescens*-treated chilli plant in response to infection by *Colletotrichum capsici*. *Journal of Mycology and Plant Pathology*, 31(2):146–155.

Sharma, P., Kadu, L.N. and Sain, S.K. (2005). Biological management of die-back and fruit rot of chilli caused by *Colletotrichum capsici* (Syd) Butler and Bisby. *Indian Journal of Plant Protection*, 33(20): 226–230.

Singh, H. and Korpraditskul, V. (1999). Evaluation of some plant extracts for the control of *Colletotrichum capsici* (Syd) Butler and Bisby, the causal agent of chilli anthracnose. *Azadirecta Indica* A Juss (Ed.) R.C. Saxena. Science Publishers, Inc. Enfield, USA, 131–138 pp.

Singh, J. and Kaur, S. (1990). Development of multiple resistance in chilli pepper. In: *Paper Presented in the 3rd International Conference on Plant Protection in the Tropics*, Held in Malaysia, pp. 51–54.

Singh, R.P., Singh, P.N. and Singh, P.R. (1977). Note on fruit rot disease of chillies. *Indian Journal of Agricultural Research*, 11(3): 188–190.

Singh, S.N., Yadav, B.P., Sinha, S.K. and Ojha, K.L. (1997b). Reaction of chilli genotypes against die-back disease caused by *Colletotrichum capsici*. *Journal of Applied Biology*, 7(1/2) : 62–64.

Srinivas, C., Niranjana, S.R., Kumar, L.P., Nayaka, S.C. and Shetty, H.S. (2006). Effect of fungicides and bioagents against *Colletotrichum capsici* on seed quality of chilli. *Indian Phytopathology*, 59(1): 62–67.

Thind, T.S. (2001). *Diseases of Fruits and Vegetables and their Management*. Kalyani Publishers, pp. 374–375.

Thind, T.S. and Jhooty, J.S. (1985). Relative performance of fungal diseases of chilli fruits in Punjab. *Indian Journal of Mycology and Plant Pathology*, 40: 543–545.

Plant Diseases Management in Horticultural Crops (2011) *Pages* **29–32**
Editors: **Shahid Ahamad, Ali Anwar and P.K. Sharma**
Published by: **DAYA PUBLISHING HOUSE, NEW DELHI**

Chapter 3

Root Rot of Apple and their Manangment

***Dharmesh Gupta*[1], *P.K. Sharma*[2], *S.S. Roy*[2] *and Anil Mishra*[3]**

[1]*Apple Scab Monitoring and Research Laboratory, Kotkhai, Shimla – 171 202, H.P.*
[2]*ICAR Research Complex for NEH Region, Imphal –795 004, Manipur*
[3]*Division of Plant Pathology, IARI, New Delhi – 110 012*

Apple is one of the most important temperate fruits of the world. With the increase in area under apple cultivation, year after year in the same region, the chances of development of several soil borne diseases have also increased. Amongst these soil borne diseases, white root rot (*Dematophora necatrix*) is the most prevalent and causes heavy losses in both nurseries and orchards. This disease is most important from loss point of view as the fungus can kill grown up trees. It infects a large number of plants consisting of annual and perennials, forest and fruit trees and around 158 host plants are susceptible to the root rot (Ito and Nakamura, 1984). *Dematophora necatrix* attacks number of temperate fruits, apple, apricot, cherry, kiwifruit, pear, plum, peach, pecan nut, persimmon, strawberry and walnut, causing white root rot throughout the world. The perfect stage of this pathogen has not been reported from India. In apple, the white root rot disease is most destructive during the rainy season. The disease was first noticed on apple in 1924 (Nattrass, 1927). In India, the disease was first reported from U.P. hills by Singh in 1939 (Singh, 1943). The annual losses estimated in Himachal Pradesh were found to be about Rupees 1.3 millions in 1966 (Agarwala and Sharma, 1966).

Symptoms

As the name indicates white root rot is a disease of root which remain covered with white fungal mycelial growth particularly in rainy season resulting in death of plants. The disease symptoms occur on the underground parts of the tree and the effects are noticed on the aboveground parts. The earliest aboveground symptoms of the disease is bronzing of leaves, reduction in their size and a stunted tree growth. Infected trees often persist for 2-3 years depending upon the severity of infection in the roots. The underground roots are completely rotten. The fibrous roots are completely devoured and infection spreads to main root through secondary root system. The lateral roots turn dark brown and the surface of roots may be covered with white cottony growth of the fungus. During rainy season white mycelial mat can be seen in the soil adhering to the roots.

Causal Organism

White root rot is caused by *Dematophora necatrix* Hartig (*Rosellinia necatrix* Prill.). The perfect stage of this pathogen has not been reported from India. Most distinguishing feature of the pathogen is the pear shaped swellings in its mycelium at each septum of the fungal hyphae.The fungus produces scattered black, microsclerotia measuring 130x98 µm and does not infect xylem tissues of apple roots but bark is completely damaged. The pathogen produces pectolytic and celllulolytic enzymes which degrade the host tissues. The pathogen also produces toxin which help in killing the cells. The ascomata of *Rosellinia necatrix* are superficial,ostilolate,often as dense swarms on a common mycelial mat,subglobose,smooth and dark.Ascospores are pigmented,aseptate,often with a minute hyaline appendage.

Disease Cycle and Epidemiology

The perithecia of this fungus are rarely produced in nature and probably play little role in the spread/survival of the fungus but microsclerotia evidently represent a resting stage and under suitable conditions can give rise to fine, white exploration hyphae. Thus the fungus survives either in the form of mycelium or scerotia in the infected roots. Infection of new roots is brought about by the mycelial growth in the soil or contact of plants with old dead roots left in the soil from previously infected plants. The fine roots are first to be infected by direct penetration. Waterlogged soil conditions favours this pathogen and the disease spread. The disease is most serious in acidic soil at pH 6.1 to 6.5.The disease severity is positively correlated with moderate temperature (15-25°C) and high moisture. The fungus produces pectolytic and cellulolytic enzymes which helps in the degradation of host tissues in advance (Gupta and Gohain, 1982).

Management

The control of this disease is difficult as the pathogen is deep in the soil and the fungicides when applied as a soil drench fail to reach the target pathogen due to adsorption on soil particles.however, increasing use of chemicals in agriculture is still a subject of concern to both the environmental and public health authorities. As a consequence, the interest in alternatives that reduce our dependence on agrochemicals, produce fewer residues, preserve the available host resistance and safeguard the environmental and human health has increased during the recent years.Therefore,integration of cultural,chemical,biological and host resistance methods of control is the best way to manage soil borne diseases of apple.

Cultural Practices

The cultural practices recommended for the management of white root rot of apple are hot water treatment of infected seedlings at 45°C for one hour before plantation, digging isolation trenches around infected trees and removal of rotten roots followed by application of disinfectant paste. Avoiding replanting in the same place for 5-6 years is also recommended. Replanting should be done in clean soil and infested soil should be followed with frequent cultivation to starve out the infection hyphae.Burning and heat drying of infected roots have also been reported to increase the life of the trees (Agarwala and Sharma, 1975; Jain, 1961). Soil solarization has proven to be most effective means for the management of soil borne pathogens (Freeman *et al.*, 1986; Katan, 1987; Sharma and Sharma, 2002).The transparent polyethylene plastic placed on the moist soil during the hot summer months increase soil temperatures to levels lethal to many soil borne plant pathogens. The acidic soils should be amended by applying lime for some years (Jain, 1961). Soil amendments with neem cake and deodar needles has been reported to reduce the incidence of the disease (Gupta and Jindal, 1989; Sharma and Gupta, 1996; Sharma and Sharma, 2005).

Chemical Control

Currently, the only approach to control root rot disease is the use of chemicals. The success in controlling root rot lies in the fact that the required and level of the fungicide should reach the point of infection in soil. Soil drenching with carbendazim (0.1 per cent) and Aureofungin (0.02 per cent) + Copper sulphate (0.02 per cent) was found effective against this disease (Gupta *et al.*, 1995).The fungicide should be applied through deep holes (15-20cm) made in the basin of tree at 30cm distance from each other. Prior to the application of chemical the tree basin is brought to a soil moisture level of 30 to 40 per cent and the treatment is repeated thrice a year during rainy season. Preventive as well as curative effects of benomyl and carbendazim were reported by Sousa (1985).

Biological Control

The antagonistic activity of various antagonists in the management of root rot has been explored (Freeman *et al.*, 1986; Sztejnberg *et al.*, 1987). *Trichoderma viride* and *T.harzianum* were the most effective in inhibiting mycelial growth of *D. necatrix* (Sharma and Sharma, 2001). *Enterobacter aerogenes* (Gupta and Jindal, 1989) have been found to protect the plant from root rot. *Bacillus* sp. was found highly effective against the root rot pathogen.

Resistant Rootstocks

Utilization of resistant host is another method to reduce disease but not much success has been achieved in root rot of apple. Almost all the rootstocks have shown susceptibility to the pathogen, however, in the field the degree of susceptibility varies (Gupta and Verma, 1978). Of the fifteen different rootstocks tested by Sharma *et al.* (2005), *Malus baccata* and *M. orientale* were found least susceptible.

References

Agarwala, R.K. and Sharma, V.C. (1966). *Indian Phytopath.*, 19: 82–86.

Agarwala, R.K. and Sharma, V.C. (1975). White root rot of apple. In: *Advances in Mycology and Plant Pathology*, (Eds.) S.P. Raychaudhary *et al.* Prof. R.N. Tandon's Birthday Celebration Committee, New Delhi, pp. 205–209.

Freeman, S., Sztejnberg, A., Chet, I. and Katan, J. (1986). Solar and biological control of white root rot disease in apple orchards caused by soil borne fungus *Rosellinia necatrix. Harradeh*, 66: 1608–1613.

Gupta, V.K. and Gohain, B.N. (1982). *Contemporary Trends in Plant Science*, (Ed.) S.C.Verma. Kalyani Publications, India, pp. 103–110.

Gupta, V.K. and Verma, K.D. (1978). Comparative susceptibility of apple rootstocks to *Dematophora necatrix. Indian Phytopath.*, 31: 377–378.

Gupta, V.K. and Jindal, K.K. (1989). Management of white root rot of apple (*Rosellinia necatrix*) by biological and cultural methods. *Proc. Int. Hortic. Cong., Italy*, 2: 3229.

Gupta, V.K., Gupta, P.K. and Gupta, S.K. (1995). Relative efficacy of some chemicals in the management of white root rot of apple. *Plant Dis. Res.*, 10(1): 83–85.

Ito, S.I. and Nakamura, N. (1984). An outbreak of white root rot and its environmental conditions in the experimental arboretum. *J. Japan For. Sci.*, 66: 262–267.

Jain, S.S. (1961). Root and collar rot diseases in Himachal Pradesh. *Himachal Hortic.*, 2: 19–23.

Katan, J. (1987). Soil solarization. In: *Approaches to Plant Disease Control*, (Ed.) I. Chet. Wiley Interscience Publication, New York, p. 77–105.

Nattrass, R.M. (1927). *Ann. Rept. Agric. and Hort. Res. Sta. Long Ashton*. Bristol For., 1926: 66–72.

Sharma, Anita and Sharma, S.K.(2005). Effect of oil cakes on soil borne disease management and soil micro flora in apple. *Acta Horticulturae*, 696: 341–347.

Sharma, Mamta and Sharma, S.K. (2001). Biocontrol of *Dematophora necatrix* causing white root rot of apple. *Plant Dis. Res.*, 16(1): 40–45.

Sharma, Mamta and Sharma, S.K. (2002). Effect of soil solarization on soil soil microflora with special reference to *Dematophora necatrix* in apple nurseries. *Indian Phytopathol.*, 55: 158–162.

Sharma, S.K. and Gupta, V.K. (1996). Management of root rot of apple through soil amendments with plant material. *Plant Dis. Res.*, 10: 164–167.

Sharma, S.K., Sharma, Y.P., Kishore, D.K. and Pramanick, K.K. (2005). A rapid method for evaluation of apple germplasm for resistance to *Dematophora necatrix* causing white root rot. In: *Integrated Plant Disease Management–Challenging Problems in Horticultural and Forest Pathology*, Solan, India 14–15 Nov. 2003, pp. 33–38.

Singh, U.B. (1943). Some diseases of apple fruits and fruit trees in Kumaon. *ICAR Misc. Bull.*, pp. 55–56.

Sousa, AJTD (1985). Control of *Rosellinia necatrix* causal agent of white root rot, susceptibility of several plant species and chemical control. *Eur. J. Plant Pathol.*, 15: 323–332.

Sztejnberg, A., Freeman, S., Chet, I. and Katan, J., 1987. Control of *Rosellinia necatrix* in soil and in apple orchard by solarization and *Trichoderma harzianum*. *Plant Dis.*, 71: 365–369.

Plant Diseases Management in Horticultural Crops (2011) *Pages 33–42*
Editors: **Shahid Ahamad, Ali Anwar and P.K. Sharma**
Published by: **DAYA PUBLISHING HOUSE, NEW DELHI**

Chapter 4

Powdery Mildew of Apple and their Management

Narender K. Bharat[1], P.K. Sharma[2], S.S. Roy[2] and Anil Mishra[3]

[1]Department of Mycology and Plant Pathology
Dr. Y.S. Parmar University of Horticulture and Forestry, Nauni, Solan – 173 230, H.P.
[2]ICAR Research Complex for NEH Region, Imphal – 795 004, Manipur
[3]Division of Plant Pathology, IARI, New Delhi – 110 012

Introduction

Apple powdery mildew caused by the fungus *Podosphaera leucotricha* (Ell. And Ev.) Salm. is an important foliar disease of apple present in all apple growing region of the world. The disease was first reported by Bessey in Iowa (USA) in 1871. The disease was once considered primarily a disease of nursery stock but is also of importance in commercial apple production system and requires many applications of fungicides for control in many countries. The relative prominence of the disease varies with annual precipitation and temperature patterns through out the region. The severity of powdery mildew and the need for control measures are related to susceptibility of the cultivars to powdery mildew.

Distribution and Economic Importance

Powdery mildew present in all the apple growing regions of the world. The disease occurs on all the continents where apple are grown: Africa (Angola, Ethiopia, Libya, Morocco, South Africa, Tanzania, imbabwe); Asia (Afghanistan, Bhutan, China, India, Iran, Iraq, Israel, Japan, Korea, Lebnon, Nepal, Pakistan, Saudi Arabia, Russia, Syria, Taiwan,Turkey,); Australia and Oceania (Australia, New ealand); Europe (Austria, Belgium, Britain, Ireland, Bulgaria, Cypruss, Czech, Denmark, Finland, France, Germany, Greece, Hungary, Iris Republic, Italy, Netherlands, Norway, Poland, Portugal, Romania, Russia, Slovakia, Sweden, Switzerland, Yugoslavia); North and South America (Argentina,

Belize, Brazil, Canada, Chile, Columbia, Costa Rica, Ecuador, El Salvador, Guatemala, Honduras, Mexico, Nicargua, Panama, Peru, USA) (Anon., 1987; Kapoor, 1967). In India the disease is present in all apple growing states *viz.*, Jammu and Kashmir, Himachal Pradesh, Uttarakhand, North-eastern states etc.

Losses from the disease vary depending upon the inherent level of susceptibility of cultivar, environmental conditions, and management practices. Economic damage from powdery mildew in bearing orchards results from reduction in tree vigour and blossom bud production, aborted blossoms, and fruit russetting (which result in loss of grade and value). Cumulative effects of powdery mildew infection on yield have been observed by van der Scheer (1980). Infection can reduce trunk growth, fruit size, crop weight, and value (Butt *et al.*, 1983). In nurseries and young plantings, powdery mildew stunts tree growth and causes poorly formed misshapen trees. Mildew infection reduces photosynthesis, transpiration, and carbohydrate content of the host (Ellis *et al.*, 1981) which affects tree vigour and blossom production.

Symptoms

The disease infects leaves, blossoms, green shoots, and the fruits. On the leaves, the fungus appears as whitish lesions, typically on the lower surface. Lesions on the lower surface may be accompanied by mild chlorosis in the corresponding position on the upper surface. Lesions expand into whitish or grayish felt-like patches often covering much of the leaf. Infection of young leaves causes curling, crinkling, reduction in leaf width, and longitudinal folding. Severely infected leaves become brittle and sometimes fall from the trees in mid-season.

Petals of infected flowers are pale yellow or green and covered with mycelium. The flowers are shriveled, may fail to set fruit, and are more susceptible to spring frosts. Severe blossom infection often involves the flower cluster leaves.

The symptoms on fruits are related to the severity of the infection and developmental fruit stage at which infection occurs. Severely infected blossoms emerging from infected buds may give rise to small, severely russetted fruit. Less severe blossom infection occurring later results in milder, net-like russetting.

Shoot infection occurs on succulent current season growth. Infected shoots are stunted, have shortened internodes, and are covered with grayish patches of mycelium, the colour of which may persist until the following season. Terminal and lateral buds are infected during the season they are produced. Infected buds and twigs are more susceptible to winter injury, resulting in poor tree form, particularly of young, non-bearing trees.

Causal Organism

The apple powdery mildew is caused by the ascomycetous fungus, *Podosphaera leucotricha* (Ell. and Ev.) Salm. The fungus was first named *Sphaerotheca leucotricha* by Ellis and Everhart in 1888. In 1892, Burrill changed the name to *S. mali*, but in 1900 E.S. Salmon established the name definitively as *P. leucotricha*.

The fungus produces conidia in long chains which measure 22-30 X 15-20 µm, and contain distinct fibrosin bodies. Cleistothecia (perithecia) are brown to dark brown in colour bear both apically and basally inserted appendages and measure 89-98 µm in diameter. The apical appendages are three to seven times as long as the diameter of the ascocarp, measure 300 – 650 X 4.5-8 µm and are usually straight and undivided. The basal appendages are rudimentary, rarely well developed, short, and more or less tortuous. The single ascus in the cleistothecium is hyaline, ovate to elliptical in shape and measures 48-62 X 46-55 μm. In each ascus eight ovate to elliptic ascospores measuring 20-24 X 10-16 μm are there (Kapoor, 1967; Bharat and Bhardwaj, 2000). The fungus is heterothallic (Coyier, 1973). After molecular characterization of different isolates of apple powdery mildew, Urbaniez and Dunemann (2005) have found chances of occurrence of physiological races in this fungus.

Disease Cycle

The apple powdery mildew fungus overwinters as mycelium in the dormant terminal and lateral shoot buds and in blossom buds produced and infected in the previous growing season (Woodward, 1927). The cleistothecia, which form the sexual stage of the fungus and produce ascospores, apparently do not play an important role in the disease cycle. Mycelia in the infected buds produce conidia to initiate primary infections and these infections can be found as early as the tight cluster stage. Conidial germination takes place at temperature range of 10-25°C (optimum 19 to 22°C) at 96 to 100 per cent relative humidity (Sharma and Gupta, 1991). Little germination occurs in free moisture (Berwith, 1936). In laboratory conditions Xu and Butt (1998) studied the growth of colonies on young apple leaves in relation to temperature and water vapour pressure deficit. They found that the colony growth responded non-linearly with in the range of 13-28°C and optimum temperature for growth occurred at 22°C and the vapour pressure deficit had little effect on colony growth.

Upon germination the conidium produces a germ tube which penetrates the cuticle (Woodward, 1927). Following penetration of the cuticle the hypha becomes thin and peg like, penetrates the epidermal cell and forms a haustorium. Fungal hyphae conformed closely to the surface of the leaf but were not firmly attached until haustoria had been formed. The infection process does not occur when leaf surface is covered with a water film.

Symptoms may develop as early as 5 days after infection. Numerous secondary cycles can occur under favourable environmental conditions. Because leaves are most susceptible soon after emergence, infection of new leaves may occur as long as shoot growth continues. Infections that result in fruit russet occur primarily during pink stage of bud development (Daines *et al.*, 1984).

Apparently, infection of lateral and buds occur within one month after they are formed. The infections remain latent until budbreak the following spring where they will serve as the initial source of inoculum. The lateral buds are susceptible to infection longer than the terminal buds, however, it is the terminal bud that are the likely source of overwintering of the fungus as infection can be greater than 50 per cent by terminal bud set.

Cleistothecia are produced on heavily infected shoots and leaves in mid-summer, but they are not regarded as an important inoculum source because the ascospores they contain do not germinate readily (Molnar, 1971). Numerous attempts to germinate ascospores or to inoculate foliage with them have been unsuccessful, leading to the conclusion that ascospores are not significant in spread of the disease. Tsuyama *et al.* (1967) observed some germination of ascospore in water and they concluded that the ascospores need atleast four to seven months to gain germinability from the initial time of cleistothecial formation. The maximum germination (12.5 per cent) was achieved at a temperature of 25°C.

Epidemiology

Apple powdery mildew fungus overwinters in infected buds and its survival is greatly affected by low temperature. Survival of healthy buds at -26°C was similar to infected bud survival at -22°C and terminal mildew infection averaged 27 per cent following winters warmer than -22°C but only 4 per cent following winters with a minimum of -24°C or colder (Spotts *et al.*, 1981). Lowest survival temperatures ranged from 2.5 to 9.8°C colder for healthy buds than diseased buds during individual winter months (Spotts and Chen, 1984). Although the deleterious effect of extremely cold temperatures on overwintering of apple powdery mildew has long been recognized, the basis for the reduction in disease incidence has remained somewhat elusive. Jeger and Butt (1983) provide tentative evidence that temperatures of near -12°C can kill mycelium in buds and allow them to produce healthy leaves.

Like other powdery mildews, *P. leucotricha* exhibits a diurnal periodicity in that the highest concentration of airborne conidia is found from midday to early afternoon (Xu *et al.*, 1995). The concentrations of conidia trapped were positively correlated with wind velocity, temperature and solar radiation and negatively correlated with relative humidity and leaf wetness (Sutton and Jones, 1979). Green house studies by Butt and Jeger (1986) found maximum number of attached conidia per conidiophore 7 to 12 days after the first sign of a colony. Numbers of conidia/conidiophore and conidiophores/mm^2 of colony on younger leaf surfaces were higher, however, total number of conidia/mm^2 colony, which also included detached conidia, was greater on lower leaf surface colonies.

Woodward (1927) found only 2 to 3 per cent germination of conidia in water, but recognized that conidia would germinate and infect a dry healthy leaf in a moist Petri dish. Berwith (1936) reported high relative humidity (90 per cent) was required for germination of conidia on dry glass slides but germination in hanging water drops never exceeded 1 per cent. The optimum temperature range for germination of conidia is 20-22°C. The minimum and maximum temperatures for germination are 4 and 30°C.

Several mathematical models have been presented on various aspects of the epidemiology of apple powdery mildew. These include incidence and severity relationships of secondary infection (Seem and Gilpatrick, 1980(a, b), the relationship between disease progress and cumulative numbers of trapped conidia (Jeger, 1984). Xu (1999) has developed a model to simulate powdery mildew epidemic. The model named Podem™ (short for Podosphaera, East Malling), has been incorporated into Adem™ (Apple Diseases, East Malling), more comprehensive forecaster for assisting growers in managing apple diseases. Podem simulates powdery mildew epidemic on vegetative shoots on a daily time step from vegetative bud break through the end of shoot expansion. The model consists of a series of submodels to generate daily forecasts of the severity of new infections and the total amount of infectious disease. To do this, the model calculates the per cent age of susceptible host tissue, the per cent age of infectious disease, the latent period, the rate of infection, and uses number of weather variables (temperature, relative humidity and rainfall). Podem will also generate risks of infection based only on the weather factors vapour pressure deficit and temperature (current and past) and tree phenology.

Management

Powdery mildew management strategies are related to cultivar susceptibility and presence and absence of other diseases to be controlled. For susceptible cultivars, the general management strategy is based on reduction of primary inoculum and protection from secondary inoculum. A major component of the management programme involves the timely protection of foliage, fruits and buds by fungicide spray applied throughout the period of susceptibility.

Because powdery mildew is already present in most of the major apple growing regions of the world, there is little effort to exclude by regulatory means.

The powdery mildew is managed by two basic strategies: resistance and fungicides. Biological and cultural control and control involving the disease forecasting are also being practiced in some areas.

Disease Resistance

Powdery mildew resistant varieties exist, however, this resistance has balanced with other factors such as markrt demand for that particular apple (Aldwinckle, 1974; Jeger and Butt, 1986). Cultivars which were ranked moderately to highly susceptible, for which regular spray schedule was not always adequate, include Jonathan, Cox Orange Pippin, Granny Smith, Gravenstein, Cortland, Idred and Romebeauty. The cultivars Golden Delicious, Red Delicious and Winesap were ranked as slightly susceptible, meaning that the regular spray schedule can sometimes be reduced for powdery mildew control on these cultivars.

Several cultivars have been shown to have heritable powdery mildew resistance potentially suitable for incorporation in breeding programme. These include Delicious progeny (Korban and Dayton, 1983), Winesap (Mowry, 1965), Lord Lambourne and Egri Red (Kovacs, 1989). Sources of high level resistance or immunity have been located in *Malus robusta* and *M. zumi* (Knight and Alston, 1968), crab apple cultivars David, White Angel, *M. x robusta* (Robusta 5), *M. x robusta* Korea, *M. zumi calocarpa*, *M. sargenti* x self and *M. baccata jackii*, *M. hupehensis*, *M. mandshurica*(Dayton, 1977; Gallott *et al.*, 1985; Caffier and Parisi, 2007). Crab cultivars namely, Renetka, Purrpurnaya, Yantaka, Altoskya, Dalgoe, Saynets, Karchemo and apple cultivars Maharaja, Chunth and Golden Chinese have been found resistant to the disease in India (Verma and Gupta, 1988). A number of varieties having resistance to this disease have been developed around the world include Coop 32 (Janick *et al.*, 1995), Nobella and Julia (Blazek *et al.*, 1995), Rosana (Varcammen, 1994) and Sunrise (Taylor, 1994).

Initial disease resistance emphasis was placed on apple scab in apple breeding programs, but resistance breeding for apple powdery mildew is now an objective of breeding program in many countries.

Cultural Practices

Pruning of dormant shoots infected with powdery mildew in the previous season has been recommended as a means of reducing primary inoculum. Removal of these infected shoots reduced secondary infection by half on mildewcide treated trees. Pruning of dormant tips of all shoots longer than 15 cm reduced the early summer mildew level to only 10-20 per cent of that in trees not pruned in this manner (Ketskhoveli, 1976). Winter and summer pruning have been found effective in checking apple powdery mildew (Holb and Abonyi, 2007). Summer pruning of the infected shoot terminal has also been recommended as a mean of controlling apple powdery mildew (Alekseeva *et al.*, 1992). The orchardists can reduce number of fungicide applications by adopting pruning of infected terminals. Although pruning of infected dormant tissues is an effective means of reducing inoculum, it should be considered as supplement to, but not the replacement for, routine fungicide application.

Biological Control

For biological control of apple powdery mildew all agents that have been reported to provide biocontrol have been fungal in nature. As powdery mildews are biotrophs and do not require exogenous nutrients for germination and initial penetration, control through competition for nutrients is not a viable strategy. Likewise as exposure of the pathogen on leaf surface after spore germination is limited, control through antibiosis is not likely to be a suitable mechanism for disease control. As such the greatest attention has been focused on the use of mycoparasites for the suppression of sporulation and dissemination of powdery mildews. These include *Ampelomyces quisqualis* (Novitskaya and Puzanova, 1992). The mechanism of biocontrol by this fungus has been established as hyperparasitism as the fungus possesses the ability to colonize the mycelium of powdery mildew and produce reproductive structures. This fungus is naturally occurring parasited of both sexual and asexual structures of powdery mildew pathogen. The parasitized colonies of powdery mildew are dull in appearance, flattened and off white to gray in colour with reduced spore production (Falk *et al.*, 1995). Natural occurrence of *A. quisqualis* on apple and other related plant species have also been observed in western Himalayan region by Vaidya and Thakur (2005).

Some commercial products based on mycoparasite *A. quisqualis* have also been released *viz.*, AQ 10 TM and Ampelomitsin for biological control of powdery mildew (Grove and Boal, 1997; Smolyakova *et al.*, 2004). With the use of Ampelomitsin 75-80 per cent control of apple powdery mildew has been achieved.

Chemical Control

A major effort in apple powdery mildew management on susceptible cultivars is in the timely application of fungicides to protect fruit and foliage from secondary infection. In region where apple scab control programme is adopted, it can be relatively easy to integrate a powdery mildew programme with that of scab. The fungicides currently used against powdery mildew include sulphur, dinocap, benomyl, thiophanate-methyl, carbendazim, fenarimol, triadimefon, triforine, bitertanol (Kumar *et al.*, 1987; Gupta and Sharma, 1981; Sharma *et al.*, 1992; Yoder, 2000). The sterol-biosynthesis inhibiting (SBI) fungicides are generally highly effective for control of apple powdery mildew (Hickey and Yoder, 1981). Use of ergosterol biosynthesis inhibitor (EBI) fungicides such as triademfon, penconazole, hexaconazole, fenarimol, triforine, flusilazole, difenoconazole and propiconazole was found very effective in the control of apple powdery mildew (Sharma and Sharma, 1996; Cimanowski and Bielenin, 1996; Sholberg and Haag, 1994). These fungicides are very effective in controlling powdery mildew and have suitable kickback activity and, when applied on a regular schedule, can effectively control the disease. Applications should begin at tight cluster and continue until terminal growth stops in mid-summer. The spray interval is generally 10 to 14 days. Recently, Gupta and Sharma (2005) and Xu *et al.* (2006) observed excellent control of powdery mildew with fungicides like hexaconazole, myclobutanil, fluquinconazole, kresoxim-methyl, bupirimate, fenarimol, triademafon, penconazole and pyrifenox in apple nurseries and orchards.

The aforementioned fungicides can provide effective mildew protection when applied at frequent intervals. Development of fungicide resistance by powdery mildew and other apple pathogens, as incase of benzimidazole and SBI fungicides, affects the desirability of the fungicides for the broad-spectrum disease control programme (Koller and Scheinpflug, 1987). The development of resistance to these fungicides are ongoing concerns for fungicide control of apple powdery mildew.Successful management of apple powdery mildew involves integration of many factors. These include disease forecasting based on inoculum assessment and weather conditions, potential crop loss, varietal composition and susceptibility and age of trees in the orchards, control of other diseases by mildew fungicides, and timing of broad-spectrum fungicide and insecticide applications to optimize the effectiveness of both. It is important to prevent mildew from becoming established in young trees one to three years after planting. Early exclusion of the disease is particularly critical for highly susceptible cultivars.

Conclusions

Apple powdery mildew is unique among apple diseases because infection occurs during dry weather, necessitating different management strategies than for wet weather diseases such as scab. Powdery mildew control strategy is related to varietal susceptibility and is based on reduction of primary inoculum and protection of fruits, foliage, and buds from secondary inoculum. General recommendations call for protection of susceptible tissue with application of fungicides from tight cluster stage until shoot growth ceases. Many of the sterol biosynthesis inhibiting fungicides are effective against powdery mildew.

Production of cleistothecia and rare germination of ascospores remains intriguing biological curiosity. The cleistothecium could serve as a means of long-term survival of extremely adverse climatic condition. The possibility of genetic recombination within the cleistothecium also holds implications for production of new races, development of resistance to fungicides and the durability of host resistance.

Development of resistance to fungicides especially benomyl and SBI would greatly affect apple powdery mildew management decisions and deserves vigilance.

The development and adoption of resistant cultivars may provide long term solution to management of apple powdery mildew. Hence identification of resistant sources and the selection of mildew-resistant progeny has become a part of apple-breeding programmes.

Potential for improved control remains in areas of orchard planting design, grower education, supervised management, continued fungicide development, development of materials capable of dormant season eradication of *P. leucotricha*, survey for possible effects of new fungicides on hyperparasites like *A. quisqualis*. Tactical application of fungicides at crucial times must be based on research involving individual cultivar susceptibility, growth stage and environmental factors.

References

Aldwinckle, H. (1974). Field susceptibility of 51 apple cultivars to apple scab and apple powdery mildew. *Plant Dis. Rep.*, 58: 625.

Alekseeva, S.A., Berbekor, V.N. and Martynov, V.I. (1992). Summer pruning: A means of controlling powdery mildew in apple. *Sadovodstvo-i-Vinogradarstvo*, 8: 8–9.

Anonymous (1987). *Distribution Maps of Plant Diseases*, Map No. 118. CMI, Kew, England.

Baker, J.V. (1961). Winter and spring pruning against apple mildew. *Proc. British Insecticide Fungicide Conf.*, 1: 179.

Berwith, C.E. (1936). Apple powdery mildew. *Phytopathology*, 26: 1071.

Bharat, N.K. and Bhardwaj, L.N. (2000). Occurrence of perfect stage of powdery mildew of apple in northwestern Himalaya. *Plant Dis. Res.*, 15: 81–82.

Blazek, J., Paprstein, F. and Janeckova, M. (1995). New apple variety, Julia. *Ved. Pr. Ovocnarske*, 14: 109–117.

Burchill, R.T. (1965). Seasonal fluctuations in spore concentrations of *Podosphaera leucotricha* (Ell. and Ev.) Salm. in relation to the incidence of leaf infections. *Ann. Biol.*, 55: 409.

Butt, D.J. and Jeger, M.J. (1986). Components of spore production in apple powdery mildew (*Podosphaera leucotricha*). *Plant Pathol.*, 35: 491.

Butt, D.J., Robinson, J.D., Souter, R.D. and Swait, A.A.J. (1983). Apple mildew crop loss study. *Rep. for 1982, East Malling Research Station*, Maidstone, England, p. 81.

Caffier, V. and Parisi, L. (2007). Development of apple powdery mildew on sources of resistance to *Podosphaera leucotricha* exposed to an inoculum virulent against the major resistant gene Pl–2. *Plant Breeding*, 126: 319–322.

Cimanowski, J. and Bielenin, A. (1996). Antisporulant activity of new apple powdery mildew fungicides. *J. Fruit Ornamental Plant Dis.*, 4: 593–594.

Coyier, D.L. (1968). Effects of temperature on germination of *Podosphaera leucotricha* conidia. *Phytopathology*, 58: 1047–1048.

Coyier, D.L. (1973). Heterothallism in apple powdery mildew fungus *Podosphaera leucotricha*. *Phytopathology*, 64: 246.

Daines, R., Weber, D.J., Bunderson, E.D. and Roper, T. (1984). Effect of early sprays on control of powdery mildew fruit russet on apples. *Plant Dis.*, 68: 326–328.

Dayton, D.F. (1977). Genetic immunity to apple mildew incited by *Podosphaera leucotricha*. *Hort Science*, 12: 225.

Ellis, M.A., Ferree, D.C. and Spring, D.E. (1981). Photosynthesis, transpiration and carbohydrate content of apple leaves infected by *Podosphaera leucotricha*. *Phytopathology*, 71: 392.

Falk, S.P., Gadoury, D.M. and Pearson, R.C. (1995). Partial control of grape powdery mildew by mycoparasite *Ampelomyces quisqualis*. *Plant Dis.*, 79: 483–490.

Gallot, J.C., Lamb, R.C. and Aldwinckle, H.S. (1985). Resistance to powdery mildew from some small fruited *Malus* cultivars. *Hort Science*, 20: 1085.

Grove, G.G. and Boal, R.J. (1997). Apple powdery mildew control trials using the mycoparasite *Ampelomyces quisqualis* (AQ 10). *F and N Tests*, 52: 7.

Gupta, D. and Sharma, J.N. (2005). Chemical control of powdery mildew of apple in warmer climates of Himachal Pradesh. *Acta Hortic.*, 696: 355–357.

Gupta, V.K. and Sharma, S.K. (1981). Studies on the effect of fungicides on apple powdery mildew. *Anz. Schandlingskde Pflanzenschutz*, 54: 8–10.

Hickey, K.D. and Yoder, K.S. (1981). Field performance of sterol-inhibiting fungicides against apple powdery mildew in the mid-Atlantic apple growing region. *Plant Dis.*, 65: 1002.

Holb, I. and Abonyi, F. (2007). Control of apple powdery mildew in integrated and organic apple production. *Novenyvedelem*, 43: 247–252.

Janick, J., Crosby, J.A., Pecknold, P.C., Goffreda, J.C. and Korban, S.S. (1995). Coop 32 (Pristine TM) apple. *Hortic. Sci.*, 30: 1312–1313.

Jeger, M.J. and Butt, D.J. (1983). The effect of weather during perennation on epidemics of apple mildew and scab. *EPPO Bull.*, 13: 79.

Jeger, M.J. and Butt, D.J. (1986). Epidemics of apple powdery mildew (*Podosphaera leucotricha*) in a mixed cultivar orchard. *Plant Pathol.*, 35: 498–505.

Jeger, M.J. (1984). Relating disease progress to cumulative numbers of trapped spores: apple powdery mildew and scab epidemics in sprayed and unsprayed orchard plots. *Plant Pathol.*, 33: 517–530.

Kapoor, J.N. (1967). *Podosphaera leucotricha, CMI Descriptions of Plant Pathogenic Fungi and Bacteria*, p. 158.

Ketskhoveli, E.B. (1976). Effectiveness of cultivar pruning against powdery mildew of apple *Podosphaera leucotricha* (Ell. and Ev.) Salm. *Rev. Plant Pathol.*, 55: 525.

Knight, R.L. and Alston, F.H. (1968). Sources of field immunity to mildew (*Podosphaera leucotricha*) in apple. *Canadian J. Genet. Cytol.*, 10: 294.

Koller, W. and Scheinpflug, H. (1987). Fungal resistance to sterol-biosynthesis inhibitors: A new challenge. *Plant Dis.*, 71: 1066.

Korban, S.S. and Dayton, D.F. (1983). Evaluation of *Malus* germplasm for resistance to powdery mildew. *HortScience*, 18: 219.

Kovacs, S. (1989). Breeding of new disease resistant apple varieties. *Fruit Var. J.*, 39: 26.

Kumar, S., Srivastava, K.K. and Pandey, J.C. (1987). Economic spray schedule for apple powdery mildew control. *Int. J. Trop. Plant Dis.*, 5: 59–65.

Molnar, J. (1971). Cleistothecia of the fungus *Podosphaera leucotricha* (Ell. and Ev.) Salm. under the conditions of Czechoslovakia. *Ceska Mykologia*, 25: 211.

Mowry, J.B. (1965). Inheritance of susceptibility of apple to *Podosphaera leucotricha*. *Phytopathology*, 55: 76.

Novitskaya, L.N. and Puzanova, L.A. (1992). Biological protection of fruit nursery from powdery mildew. *Zashch. Rast.* (Mosc.), 6: 25.

Seem, R.C. and Gilpatrick, J.D. (1980). Incidence and severity relationships of secondary infections of powdery mildew of apple. *Phytopathology*, 70: 851–854.

Sharam, J.N. and Sharma, S.K. (1996). Efficacy of different fungicides in controlling powdery mildew of apple. In: *Nat. Symp. Plant Dis. Holistic Approaches for their management, organized by Indian Soc. Plant Pathologists*, 17–18 Dec., p. 35.

Sharma, K.K. and Gupta, V.K. (1991). Effect of temperature and relative humidity on conidial germination and germ tube length of *Podosphaera leucotricha* on glass slide. *Plant Dis. Res.*, 6: 91–95.

Sharma, S.K., Gupta, S.K. and Kaith, D.S. (1992). Efficacy of fungicides in relation to number of sprays against powdery mildew of apple. *Indian J. Mycol. Plant Pathol.*, 22: 267–269.

Sholberg, P.I. and Hagg, P. (1994). Control of apple powdery mildew (*Podosphaera leucotricha*) in British Columbia by DMI fungicides. *Can. Plant Dis. Surv.*, 74: 5–11.

Smol-yakova, V.M., Podogornaya, M.E., Puzanova, L.A., Cherkezova, S.R. and Yakuva, G.V. (2004). Ecologization of production of stone fruits from diseases. *Sadovodstvo-i-Vinograderstvo*, 5: 5–8.

Spotts, R.A. and Chen, P.M. (1984). Cold hardiness and temperature response of healthy and mildew-infected terminal buds of apple during dormancy. *Phytopathology*, 74: 542.

Spotts, R.A., Covey, R.P. and Chen, P.M. (1981). Effect of low temperature on survival of apple buds infected with the powdery mildew fungus. *Hort Science*, 16: 781.

Sutton, T.J. and Jones, A.L. (1979). Analysis of factors affecting dispersal of *Podosphaera leucotricha*. *Phytopathology*, 69: 380–383.

Taylor, P. (1994). Sunrise offers option as early apple. *Orchardist N.Z.*, 67: 1–26.

Tsuyama, H., Nagai, M. and Aizawa, T. (1967). Germination of ascospore of apple mildew. *J. Faculty Agric. Iwate Univ.*, 8: 235.

Turechek, W.W. (2004). Apple diseases and their management. In: *Diseases of Fruits and Vegetables Vol. 1*, (Ed.) S.A.M.H. Naqvi. Kluwer Acad. Pub., Netherlands, pp. 1–108.

Urbaniez, A. and Dunemann, F. (2005). Isolation, identification and molecular characterization of physiological races of apple powdery mildew (*Podosphaera leucotricha*). *Plant Pathol.*, 54: 125–133.

Vaidya, S. and Thakur, V.S. (2005). *Ampelomyces quisqualis* Ces.-mycoparasite of apple powdery mildew in western Himalayas. *Indian Phytopath.*, 58: 250–251.

Van der Scheer, H.A. (1980). Threshold of economic injury for apple powdery mildew and scab. In: *Integrated Contol of Insect Pests in the Netherlands*, p. 49.

Vercammen, J., Daele, G.V. and Van Daele, G., (1994). Apple varieties resistant to scab and powdery mildew. *Fruit Belge.*, 62: 132–135.

Verma, K.D. and Gupta, G.K. (1988). Field reaction of apple (*Malus pumila*) germplasm to powdery mildew (*Podosphaera leucotricha*). *Indian J. Agric. Sci.*, 58: 233–234.

Woodward, R.C. (1927). Studies on *Podosphaera leucotricha* (Ell. and Ev.) Salm. *Trans. Br. Mycol. Soc.*, 12: 173.

Xu, X.M. and Butt, D.J. (1998). Effects of temperature atmospheric moisture on the early growth of apple powdery mildew (*Podosphaera leucotricha*) colonies. *Europian J. Plant Pathol.*, 104: 133–140.

Xu, X.M., Butt, D.J. and Ridout, M.S. (1995). Temporal patterns of airborne conidia of *Podosphaera leucotricha*, causal agent of apple powdery mildew. *Plant Pathol.*, 44: 944–955.

Xu, X.M., Robinson, J. and Berrie, A. (2006). Chemical control of apple powdery mildew (*Podosphaera leucotricha*): mode of actions. *Bull. OILB/SROP*, 29: 95–106.

Xu, X.M. (1999). Modelling and forecasting epidemics of apple powdery mildew (*Podosphaera leucotricha*). *Plant Pathol.*, 48: 462–471.

Yoder, K.S. (2000). Effect of powdery mildew on apple yield and economic benefits of its management in Virginia. *Plant Dis.*, 84: 1171–1176.

Plant Diseases Management in Horticultural Crops (2011) *Pages* **43–62**
Editors: **Shahid Ahamad, Ali Anwar and P.K. Sharma**
Published by: **DAYA PUBLISHING HOUSE, NEW DELHI**

Chapter 5

Diseases of Coriander and their Management

R.P. Saxena[1] and M. Anandaraj[2]

[1]N.D. University of Agriculture and Technology, Kumarganj, Faizabad
[2]Project Coordinator, All India Coordinated Research Project on Spices Crops, Indian Institute of Spices Research, Calicut

Introduction

India possess many innate advantages over other spice producing countries because of its large size, varied climatic conditions, relatively low cost skilled human power and is recognized as one of the traditional producer, consumer, exporter of spices and its value added products. According to Bureau of India Standard 63 spices are grown in India out of 109 spices listed by ISO. Among 17 seed spices grown in India, coriander, cumin, fennel and fenugreek are the major seed spices being cultivated in an area about 800,000 hectares with an annual production of 500,000 tonnes (Anonymous, 2007a). Export of coriander has extended all the set records of 2007-2008 exporting 20,500 t of coriander seeds valuing 306.20 crores of Rupees (DASD, 2008). The medicinal value of seed spices is also gaining importance in view of their inbuilt extractable constituents and their convenient use (Rama Murthy and Sridhar, 2001; Thakral, 2002 and Jakhar and Rajput, 2005).

India is blessed with varied agro climatic condition such as tropics, subtropics, and nearly temperate zones, which is advantageous for cultivation and production of large variety of seed spices. Although productivity of seed spices have increased in last two decades as a result of development of new high yielding varieties but problems encountered, have also posed a threat to cultivation of these crops due to high incidence of disease. Wilt, powdery mildews, and blights are the major diseases in seed spices crops like coriander, cumin, fennel, and fenugreek. Other diseases of minor importance caused by virus, phytoplasma, bacteria, and nematodes either alone or in combination with diseases of major importance also play important role in cultivation. Moreover, incidence of disease(s) is localized to each zone.

The seed spices are grown in almost every state in the country. The principal growing states of these crops are Rajasthan, Maharashtra, Madhya Pradesh, Karnataka, Gujarat, Uttar Pradesh, Andhra Pradesh, Tamil Nadu, Haryana, Punjab, to some extent Bihar and Orissa as rain fed crops in about 3,85,300 hectares. On the other hand, these states also face problems in cultivation of seed spices either due to drought, salinity, or alkalinity in few pockets and arid or semi arid regions in Rajasthan.

Since, Indian conditions are very conducive for fungal growth and their proliferation, occurrences of high incidences of diseases in many seed spices may cause natural contamination of mycotoxins in seed spices, biochemical composition and quality of seed including essential oils (Christensen and Kaufmann, 1969; Hashmi and Thirana, 1990; Prasad and Narayan, 1988; Samajpati, 1983; Singh, 1983). The alteration in biochemical composition of seeds makes the produce either unfit for export, germination and reduction in prices due to quality deterioration.

Coriander is an annual herb belonging to family Apiaceae. Seeds possess medicinal value being carminative, diuretic, and tonic in properties. Volatile oil is used as flavoring agent. A number of diseases in coriander have been reported in different agro-climatic zones in India and abroad by the organizations located at major seed spices research centers either as part of All India Coordinated Research Project in Spices crops, State Agriculture Universities, Deemed Universities, Research Institutes. Efforts have been made to compile the information on various aspects of fungal diseases of coriander and disease management of these valuable crops to achieve high yield and quality produce. (Mukherjee and Bhasin, 1986). The views expressed in diseases of major importance are based on the information collected from the research project reports/publications etc.

Stem Gall (*Protomyces macrosporus* Unger)

Stem gall disease of coriander caused by *Protomyces macrosporus* was reported for the first time in Madhya Pradesh in 1952-53 near Gwalior (Gupta, 1954) and in Uttar Pradesh in 1984. According to Lakra (2000), Sydow and Butler reported stem gall disease in 1911 from Pusa, Bihar for the first time. The systemic stem gall disease appears in the form of small tumor like hypertrophied swelling on all the vegetative parts of affected plant staring from stem to leaves, veins of leaves, petiole, flower stalks and fruits (Gupta, 1962). The galls usually first appear on stem, as elongated swellings which vary in size according to severity of disease (Figure 5.1). The galls are soft and fleshy when young and later become hard. Often thin walled parenchymatous cortical cells may replace the collenchyma in hypertrophied tissue. Sections of infected galls show presence of numerous golden color chlamydospores. Micro and mega-sporogenesis are inhibited and generative cells are destroyed by pathogenic diffusates (Pavgi and Mukhophadyay, 1972). Fungus is restricted to tumors only. The infection becomes systemic in host stem before or during the pre flowering period, inducing organogenic changes in flower and fruits due to hypertrophy and hyperplasy (Figure 5.2). The contaminated hypertrophied seeds are larger in size, lightweight and show reduced or no germination of seeds, converting it unfit for use as seed material or as spice (Gupta, 1954; Prakash and Pratap, 1995). Seedlings shows stem infection via integument thus working as a source of primary inoculum. Infected seeds transmit disease from one season to another (Pavgi and Mukhopadhyay, 1969b). Light weight infected fruits causes severe seed yield losses and deterioration of quality of seeds for flavour and other properties (Lakra, 1993; Singh *et al.*, 1984; Tripathi, 2005). The disease is both seed and soil borne in nature (Tripathi, 2003).

Figure 5.1: Long Elongated Swelling at the Base of Stem

Figure 5.2: Hypertrophied Pedicel and Seeds of Coriander

Infection of stem gall disease takes place through chlamydospores in pericarp of seeds, in the plant debris in soil. Chlamydospores can survive in soil for long time after their release in the monsoon rains from infected crop debris/infected seeds or unclean seeds (Pavgi and Mukhopadyay, 1969a, 1972). Germination of chlamydospores is dependent on environmental conditions. Ten per cent of chlamydospores have been reported to germinate in soil even after 8 years when stored at room temperature and 18-30 per cent relative humidity even after 6 years (Gupta, 1973). Rise and fall in soil pH, addition of fertilizer and manures, soil moisture by irrigation are conducive for permeability of chlamydospores walls to break for germination at 15-24° C and maximum germination in a temperature range of 22-24°C. Germination is also favored by exposure to high and low temperatures (Hawker, 1950). The over summering resting spores and desiccated blastospores in soil are effective after the onset of favorable environmental conditions (Pavgi and Mukhopadhyay, 1972). Systemic invasion of pathogen in seeds considerably reduce the germination of seeds collected from stem gall infected plants. Biotypes from North India have also been reported (Gupta and Sinha, 1963).

The disease is prevalent in most of coriander growing areas of Delhi, Madhya Pradesh, Rajasthan, Uttar Pradesh, Tamil Nadu, West Bengal, Andhra Pradesh, Bihar, Himachal Pradesh, and Punjab (Gupta, 1954, 1962; Naqvi, 1985; Paul, 1992; Singh *et al.*, 1991). The pathogen is also reported to cause diseases in other seed spices (Awasthi *et al.*, 1978).

Stem gall disease was earlier regarded as a disease of minor importance (Butler, 1918). Later, Mundkar (1949) considered this disease causing more economic losses in India and since then disease has been recorded as the disease of major significance responsible for severe losses in seed yield to a tune of 15 to 71.66 per cent at different locations (Gupta, 1954, 1962; Jain *et al.*, 1995; Lakra, 1993;

Naqvi, 1985; Prasad, 1983; Saxena, 2002; Singh *et al.*, 1984, 1996; Srivastava *et a.l*, 1971; and Tripathi, 2001). In Gwalior, 23 per cent of mean disease intensity that was correlated with average yields loss of 15 per cent per plant. In Kangra district, 10.76 to 60.00 per cent of disease incidence was reported, 0.32-7.3 per cent in Mandi and 0.2-2.3 per cent in Kulu district. A loss of 50 per cent in seed yield was recorded in Kota, Bharatpur, Dholpur and Sawai Madhopur districts of Rajasthan (Awasthi *et al.*, 1978). 0.25 –12 per cent of hypertrophied seeds with chlamydospores were reported in 11 dry seed samples of Rajasthan (Singh *et al.*, 1991). The spore load in stem gall infected seeds varied from 1802-10681 spores/gm seed at different locations. Seed treatments and foliar spray of Carbendazim, Bayleton, Karathane, Wettable Sulphur and foliar spray of onion plant leaf extract showed one to seven per cent of hypertrophied seeds. Lowest hypertrophied seeds were observed in seeds treated with Carbendazim and foliar spray of Bayleton (Shukla, 2005).

Crop sown on different dates exhibited variable degree of disease incidence due to the change in varieties, date of sowing, variation in agro-climatic condition and other inputs at different locations. (Gupta, 1954, 1976; Gupta and Sinha, 1973; Lakra, 1999; Naqvi, 1985; Patel *et al.*, 1998; Saxena, 2002; Singh *et al.*, 1996, 2003; Tripathi, 2003). Disease incidence has been reported lower in early sown crop in October (Keshwal *et al.*, 1979). A positive correlation has been reported with date of sowing of crop, disease incidence, and losses in yield. Crop sown during 16 Oct. to 16 Nov. had the minimum disease incidence and yield losses and considered as the best time of sowing to minimize yield losses due to stem gall disease in Sehore in Madhya Pradesh and in most of coriander growing areas of U.P. The rainfall had positive but non-significant correlation with disease intensity in various dates of sowing. The disease intensity also increased with corresponding increase in relative humidity. The rainfall had positive but non-significant correlation with disease intensity in various dates of sowing in Gwalior (Tripathi, 2003). On the other hand, Lakra (1999) at Hisar, observed significantly high incidence of stem gall disease during 16th Oct and 1st Nov. sown crop to a tune of 71.66 and 66 per cent respectively and minimum disease incidence of 26 per cent in late sown crop of 16 December with yield loss of 16.6 per cent. Formation of gall on stem was less in early sown crop with lowest seed yields of 16 per cent. In late sown crop, plant height, number of branches per plant, number of umbels/umbelets, number of seeds per umbel and seed yield were most affected components. Toncer *et al.* (1998) reported highest seed yield in crop sown on Nov 15.

Studies have shown that time of infection vary with the predisposing factors like inoculum potential in soil, location, seasonal changes, its microclimate, air around the host plant, susceptibility of host, and tenderness or age of the plant. The stem gall disease was mainly influenced by inoculum potential in soil. The severity of infection and magnitude of damage to the crop proportionally increases with heavy rain fall, the period over which moisture is retained in the soil in low lying shady areas, early plant age (Mukhopadhyay and Pavgi, 1971; Pavgi and Mukhopadhyay, 1972; Tripathi, 2003). Rise in temperature also adversely affect incidence of disease by retarding growth of pathogen with in the host. Presence of low temperature with abundant moisture deposits during early morning and late night hours associated with high day temperature were conducive to disease development under Varanasi conditions during February and a change in incidence at other places (Sharma and Anandaraj, 2000). In Faizabad conditions, time of appearance of disease was recorded after 72 days of sowing *i.e.* on 17th January 2005 in 2nd Nov. sown crop when sunshine ranged between 6.6 – 6.9 hours, high soil moisture, temperature range of 2.7 to 23.8 °C and relative humidity between 45.5 to 89.7 per cent accompanied 2mm of rains (Shukla, 2005).

In late sown crop, severities of stem gall disease in heavy and richly irrigated soil was correlated with environment conditions and susceptible plant age. Escape of stem gall disease in early-planted crop was explained due to low threshold level of quantum of inoculums. The basis of genetic and physiological susceptibility of the host is still unknown. Most of the factors are variable (Mukhopadhyay and Pavgi, 1975).

Levels of fertilizers or ill balanced nutrient status of host are another pre disposing factors for disease development and loss in yield. Application of balanced dose of nitrogen and phosphatic fertilizers decreases the disease and more seed yield while application of potassium had no remarkable impact on disease intensity under Gwalior conditions (Gupta, 1976a,b). Application of nitrogen, phosphatic fertilizers, and manganese in naturally pathogen infested field soil bear positive correlation with nutrient status, disease intensity, and seed yield. The highest disease intensity and yield loss of 14.85 per cent was recorded after application of 22.42 kg P_2O_5 per hectare and $MnSO_4$ @17.94 kg hectare without application of nitrogen at Gwalior. Application of nitrogen decreased the disease intensity. Application of phosphatic fertilizers increases particularly in absence of nitrogen. Application of manganese reversed the reaction. Disease can be managed up to 25 per cent by applying balanced fertilizers (Lakra, 2000; Tripathi, 2001). Heaviest soil moisture (IW/CPE=1.33) resulted in 46 per cent decrease in disease intensity where as lowest soil moisture (IW/CPE=0.70) developed disease intensity by 57.2 per cent and yield losses by 10.4 per cent at Hisar (Lakra, 2000).

The stem gall disease is favoured by soil pH of 7.4- 8.4 accompanied by other predisposing factors like enough inoculum potential, susceptible growth stage of plant and favourable microclimate. Least infection of stem gall disease has been observed at soil pH of 4.8 (Mukhopadhyay and Pavgi, 1971). The severity disease was fully dependent upon environmental factors like prevailing temperature, relative humidity, photoperiod and frequency of rays (Mathur, 1962).

Infection of coriander with *Protomyces macrosporous* alters biochemical constitution of plants and/ or infected seeds. Fat content was reduced by 14 per cent (Gupta, 1956b) Decrease in free amino acids and sugar content have been recorded (Prasad, 1983). In all affected parts, oil canals was greatly affected and collapsed (Rao, 1972). Galls had high concentration of amino acids and reducing sugars, low non-reducing total sugars, starch and protein, high invertase, protease, acid phosphatase, peroxidase activity and decreased poly phenol activity (Goel *et al.*, 1983; Maheswari *et al.*, 1984; Tayal *et al.*, 1981). A two-fold increase in the protein and 34 per cent decrease in starch content were observed in diseased hypertrophied tissue of stem over normal tissues. The disease also causes considerable losses in minerals like calcium, manganese phosphorus, potassium, and iron (Goel *et al.*, 1983).

Management Strategies

Stem gall disease is both seed and soil borne in nature. The pathogen survives through chlamydospores, which perpetuates through infected plant debris in soil as systemic infection. The control measure involves adoption of recommended cultural practices, seeds treatment, soil and foliar application of fungicidal chemicals and use of bio-control agents to manage the disease.

Cultural Methods

Removal, destruction and burning of crop debris carrying all tumor-bearing plant parts may be an effective measure in reducing the incidence of this disease (Butler, 1918). It has been observed that early and late sown crop contracted less incidence of stem gall than normal crop (Lakra, 2000). The disease can be managed to some extent (25 per cent) by applying balanced dose of fertilizers and amount of moisture in the soil (Gupta, 1976 a, b; Lakra, 2000).

Chemical Methods

In early studies, organo-mercurials were used to control this disease but with the advent of new fungicidal chemicals and considering its toxicity, use of organo-mercurials has been discontinued or restricted. sulphur drugs *e.g.* sulphathiazole and sulphadiazine) have been observed to reduce the germination of chlamydospores at 0.025 and 0.05 per cent respectively and total inhibition at 0.1 per cent (Gupta 1956a). Treatment of formalin at 0.5 and 1.0 per cent stopped chlamydospore germination. Fungicides such as femosam, copper sandoz, dithane Z-78, spersul, blitox, coppesan, dithane Z-78, aretan, agaloll, achromycin, and chloromycetin also inhibited the chlamydospore germination to extent of 50 per cent or more (Narula, 1962; Mathur and Narula, 1963). Studies have also shown that Captafol at the rate of 2g/kg was effective in inhibiting chlamydospore germination and improving seedling vigour with partial control of disease in field (Nene *et al.*, 1966).

Spraying of penicillin and streptomycin at 400 µl/ml every three weeks in sand cultures decreased disease intensity from 87.0 to 85.4 per cent. Partial control of disease was observed with seed treatment of thiram (0.25kg/100kg seed), sulphathiazole,sulphadiazine and tetra chloro nitro anisole (TCNA) at the rate of 0.5 kg/100kg of seeds (Gupta and Sinha 1963a,b).

A combination of seed treatment and foliar spray of fungicides have shown enhanced disease control and yield economics over single treatment. Efficacy of seed treatment in control of stem gall disease has been reported using bavistin, thiram and vitavax (Lakra, 2000). Seed treatment with thiram at the rate of 2g/kg followed by three sprays of calixin (0.1 per cent) at different stages of crop gave best results for control of disease. Soil and seed treatment with thiram were also very effective treatments (Nene *et al.*, 1966). Seed treatment with agrosan GN at the rate of 2g/kg, thiram + bavistin(1:1 proportion) at the rate of 2g/kg of seeds and foliar spray of 0.1 per cent bavistin (500-700 lit of water) at the time of appearance of disease and repeat spray twice or thrice after 20 days interval have been reported as effective measure to control the disease (Singh *et al.*, 2001). Spray of carboxin, carbendazim, and tridemorph could not control the disease effectively when sprayed at 30 days interval. Delay in spray schedule resulted in more yield losses. Seed treatment with Carbendazim + foliar spray of bayleton have been found as the best treatment for the effective disease control and seed yield (Lakra, 2000). Seed treatment with thiram @ 3g/kg of seeds along with soil application of neem cake @1.5t/ha and two sprays with thiram 0.2 per cent beginning from 35 days of sowing (DAS) at a interval of 15 days was recorded as most effective treatment. (Tripathi, 2005). Four foliar sprays with thiram, captafol or carboxin when applied at fortnightly intervals from the time of disease manifestation were more effective in control of diseases and increase in yield than either seed treatment alone or combined seed or soil treatment at Rampur, Nepal (Bhardwaj and Shrestha, 1985).

Total seed yield/ha, per cent yield loss and yield of diseased seeds were positively correlated with decrease in intensity while yield of healthy seeds were negatively correlated (Tripathi, 2001).

Seed and soil treatment of bio inoculant Azospirillum in combination with 100 per cent nitrogen and FYM 5.0 t/ha reduced incidence of stem gall disease to 8 per cent under semi arid conditions of Rajasthan and seed yield of 10.8 q/ha (Malhotra *et al.*, 2006).

Resistant Cultivars

Varietal resistance for stem gall disease have also been searched by number of research workers (Gupta, 1973; Gupta and Sinha, 1973; Narula and Joshi, 1963; Tripathi, 2002; Tripathi *et al.*, 1998, 2001; Pandey and Dangey, 1998., Patel *et al.*, 1998; Saxena, 2002; Shukla, 2005; Singh *et al.*, 1996, 2003). Naqvi (1985) found UD-48, UD-41, UD-46, UD-03 as resistant to stem gall disease. GC-88-8, G5356-91, Pant-1 and UD-20 have been reported as resistant to stem gall disease (Tripathi *et al.*, 1998). Singh *et al.* (1998) found lowest stem gall incidence in Panipat local and CS-193. RCr-41 has been reported as highly resistant or tolerant to stem gall disease under Rajasthan conditions and UD-48, UD-46 were highly resistant and UD-1, UD-3, UD-12, UD-12, UD-20, UD-21, UD-373, UD-374 were found resistant to stem gall disease in sick plots under Jobner conditions (Edison, 1989; Sastry and Sharma, 2001; Singh *et al.*, 2001). Out of 70 accessions screened, PH-7, Pant Haritma, COR-17, COR-2 was recorded as highly resistant (Singh *et al.*, 2003). Pant –1, UD-20, RCr-41 had less than 10 per cent of stem gall disease under Gwalior conditions (Tripathi, 2001) (Table 5.1).

Table 5.1: Disease Tolerant and Resistant Varieties of Coriander

Sl.No.	*Variety (Year of Release)*	*Area of Adoption*	*Tolerant/Resistant to*	*Centre Responsible for Developing*
1.	RCr-41* (1988) (UD-41)	Rajasthan, Gujarat, Haryana	Resistant to stem gall, wilt and moderately tolerant to powdery mildew.	RAJAU, Jobner
2.	RCr-435* (1995)	Rajasthan	Moderately resistant powdery mildew	RAJAU, Bikaner
3.	RCr-446*	Rajasthan	Moderately resistant stem gall and wilt	RAJAU, Bikaner
4.	CO 3* (1991)	Tamil Nadu, Andhra Pradesh and Gujarat	Field tolerant to wilt, powdery mildew and grain mould	TNAU, Coimbatore
5.	CS-28	Tamil Nadu	Tolerant to wilt, powdery mildew and grain mould	TNAU, Coimbatore
6.	Gujarat Coriander-1* (Guj CO-1)	Gujarat	Tolerant to wilt and powdery mildew.	GAU, Jagudan
7.	Gujrat Coriander-2* (Guj CO-2) (1974)	Gujarat	Moderately tolerant to wilt and powdery mildew.	GAU,Jagudan
8.	Rajendra Swathi* (RD-44) (1988)	Plains of North Bihar	Moderately resistant to wilt, stem gall.	RAU,Dholi
9.	Sadhna (CS-4) (1989)*	Andhra Pradesh		APAU,Guntur
10.	Swathi (CS-6) (1989)*	Andhra Pradesh	Field tolerant to wilt and powdery mildew	APAU,Guntur
11.	Sindhu (CS-2) *	Andhra Pradesh	Tolerant to wilt and powdery mildew.	ANGRAU, Guntur
12.	RCr-20 (UD-20) 1966	Rajasthan	Moderately powdery mildew wilt and stem gall.	RAJAU, Jobner
13.	Karan (UD-41)	Punjab	Resistant to stem gall, moderately resistant to wilt.	
14.	Pant Haritma	Uttarakhand and Uttar Pradesh	Tolerant to stem gall	GBPUAT, Pant nagar
15.	CO 1 (1981) *	Tamil Nadu	Resistant to grain mould.	TNAU, Coimbatore
16	CS-287 *	Andhra Pradesh	Tolerant to wilt and grain mold.	TNAU, Coimbatore.
17.	RCr-684 *	Rajasthan	Resistant to stem gall	RAJAU, Jobner
18.	Hisar Sugandh (2002)*	Haryana and Rajasthan	Resistant to stem gall	CCS, Hisar
19.	NRCSS-A Cr-1	Rajasthan	Resistant to stem gall and tolerance to powdery mildew.	NRCSS, Ajmer
20.	Lam Selection (CS-2)*	Andhra Pradesh	Tolerant to harmful diseases.	APAU,Guntur
21.	CS-287 (CS-6)*	Tamil Nadu and Andhra Pradesh	Field tolerance to wilt and gain mold	TNAU, Coimbatore.

*: Improved varieties of coriander developed under aegis of AICRP.

The Stem gall disease cannot be managed effectively by any single mode. A collective approach for both seed treatment and foliar spray at scheduled interval is needed. Selection of disease free seeds, collection and destruction of affected plants, growing of coriander variety RCr-41, seed treatment with agrosan GN/thiram (2.5g/kg) or bavistin (2g/kg), spraying of baylenton/systhane (0.4 ml/l of water) after 20 days interval has been recommended for effective disease management (AICRPS recommendation, Parthasarathy, 2005). In addition, use of balanced fertilizers and avoidance of heavy and frequent irrigation may be additional effective measures in the management of disease (Lakra, 2000).

Wilt (*Fusarium oxysporum* Schlecht.f sp *corianderii*. Kulkarni.)

Wilt caused by *Fusarium oxysporum* Schlecht.f sp. *corianderii* was reported in central India for first time in 1936 (ICAR Report, 1953) and from Madhya Pradesh in 1953 (Joshi and Nikam 1953). The pathogen is soil as well as seed borne. The cultivation of coriander in Gwalior, Guna division of Madhya Pradesh, Kota division of Rajasthan and in Gujarat severely damaged by wilt disease. The yield losses has been estimated from 25 to as high as 60 per cent in Madhya Pradesh and 100 per cent in few pockets of Rajasthan (Mall, 1968; Mathur, 1963).

The plants of coriander are attacked by wilt at all stages of plant growth from seedlings to maturity. It can cause pre emergence mortality in germinating seeds and wilting of germinating seedlings. Old plants show drooping of terminal portion to base followed by withering and drying up of leaves eventually leading to death of plants. The disease is systemic and also soil borne in nature. Discoloration of vascular bundles of root system and vascular plugging with fungus can be seen (Chattopadhyay, 1967; Chhatpadhyay and Maity, 1990; Anonymous, 1971). Sterility is a major and common feature in wilted plants. Seeds if formed, are light, immature, shriveled with out any symptoms. Localization of pathogen is restricted to carpophores, pericarp, seed coat and endosperm in asymptomatic seeds infected with *Fusarium oxysporum* Schlecht. f sp *corianderii*. In moderately discolored seeds, it was observed in carpophores, pericarp seed coat and endosperm. In heavy infected discolored seeds, pathogen colonizes embryo also besides all these components (Agarwal *et al., 2005*). Hashmi and Gaffar (1991) isolated 56.8 per cent seed samples infected with *Fusarium moniliforme* from African and Asian countries. Species of *Fusarium* was observed as seed mycoflora of which *Fusarium solani* was found to be pathogenic. A species of Fusarium has also been reported from Beltsville (U.S.A) to cause localized lesions in plants at soil surface and stem causing wilting and drooping of foliage (Stevenson, 1943).

The colonization of pathogen was both as inter and intra cellular and cause depletion in cell contents (Agarwal *et al.*, 2005). The disease does not appear in patches and infection occurs at early stage of growth, entire field is affected. This results in total failure of crop. The wilt-affected field appears like a brownish burnt crop. Wilting may be observed sudden, instead of being gradual. The leaves of some affected plants wilts and turn brown while entire plants may remain alive. Infected seeds loose germination by 49-62 per cent and above 70 per cent of relative humidity. (Komaraiah and Reddy, 1986).

The fungus produces macro, micro conidia and chlamydospores. The micro conidia are scattered freely in mycelial mats and macro conidia in sporodochia. The fungus is capable of growing at 12-35°C and best growth is observed at 19.5°C. Factors like irrigation, soil pH between 5.8-6.9, temperature between 29-8 to 33°C and soil moisture content between 60-70 per cent favours the disease development. (Srivastava, 1969; 1972).

Cultural Practices

Fertilizer levels also play an important role in incidence of wilt disease. Application of fertilizers at low levels of nitrogen, and high levels of phosphorus and potash (40:80:80) gave low wilt per cent age (8.3) while combination of NPK (80:40:40) gave higher incidence of 59.7 per cent. Addition of phosphorus @ 111.21 kg/ha decreased the disease intensity while doubling the phosphorus level increased (Joshi *et al.,* 1945-1952)

Summer ploughing, soil solarization and proper three years crop rotation (e.g cluster bean, cumin–cluster bean, wheat-cluster bean-mustard) may be an effective measure to reduce the wilt incidence (Singh *et al.,* 2001). Sowing of seeds in first to second week of November and use of resistant variety CO-3, Sadhana and Surabhi have been recommended to minimize the disease incidence.

Control Measures

The pathogen is specific to coriander, therefore, crop rotation may result reduced wilt incidence. The incidence of wilt may be reduced by soil amendment with organic manure and oil cake and adjustment of soil pH to 8.2 (Srivastava, 1972). It has also been recorded that sowing of the crop between first to second weeks of November reduced the disease incidence. However, a variation in this respect has been experienced from year to year. Seed treatment with agrosan GN, benomyl (benlate) or carbendazim (bavistin @ 3-4 g/kg of seeds) proved better to reduce wilt incidence. Use of disease free seeds of a certified variety collected from wilt free areas may be preferred for sowing in wilt sensitive areas/variety. Seed treatment with bavistin @ 1.0g/kg of seeds to check spread of disease to new areas and rotation of crop has been recommended to reduce the wilt incidence. According to Singh *et al.* (2001) incidence of wilt disease can be reduced by use of healthy and disease free seeds, seed treatment with bavistin 1.0g/kg of seeds, or *Trichoderma* @ 4.0g/kg of seeds or seed treatment with 1:1 mixture of bavistin + Captan @ 4.0g/kg of seeds.

Biocontrol Agent

Seed treatment with *P. fluorescens* @ 10g/kg seed + soil application of *P. fluorescens* @ 5kg/ha was found effective as biocontrol agent against wilt at Coimbatore and use of *T. harzianum* as seed treatment was effective at Jobner (Anonymous, 2005). Treatment of *Pseudomonas fluorescens* strain –2 at the rate of 10g/kg of seed has been found effective in controlling wilt incidence (Potty and Krishnakumar, 2001).

Trichoderma have been used for management of wilt incidence. Seed treatment of *T. harzianum* and soil application recorded minimum wilt incidence (Anonymous, 2005). Use of *Trichoderma viridae* @ 4g/kg seeds alone as seed treatment was effective in reducing the wilt incidence from 28.3 to 5.3 per cent at Coimbatore with significantly high seed yield and cost benefit ratio (Muthulakshmi, *et al.,* 2002). *Trichoderma* @ 4g/kg of seeds or soil solarization of field during summer also reduce the disease incidence (Chaudhury, 2005). Fusarium wilt complex in coriander can better be managed by seed pelleting with *Trichoderma viridae* (CFU 10^6) @ 4g/kg of seeds, plus application of neem cake (150 kg/ha) followed by foliar spray of Hexaconazole @ 0.05per cent or thiophenate methyl on 25th,40th, and 55th DAS (AICRP recommendations; Parathasarathy, 2005).

Resistant Cultivars

Local varieties like Gwalior Seoni and MP-5365 has been found susceptible to attack of wilt disease (Chattopadhyay and Maiti, 1990). Other variety include UD-373 (high yielding) Cultivation of Sindhu, Swathi (CS-6), CS-287, Rajendra Swathi (RD-44), UD-20, Gujarat Coriander –1 and 2, CO-3, RCR-41, and Sadhna (Prakasam *et al.,* 1987; Potty and Krishnakumar, 2001; Rajpurohit, 2007).

Other disease management practices include adoption of crop rotation, deep summer ploughing, use of disease free seeds, seed dressing as well soil drenching with carbendazim (0.2 per cent), seed pelleting with *Trichoderma viridae* (CFU 10^6 @ 4g/kg of seeds) plus neem cake application (150kg/ha) and seed dressing with 1.5 g bavistin + 1.5g thiram/kg seeds and foliar spray twice with carbendazim @ 0.1 per cent have been recommended.

Powdery Mildew (*Erysiphe polygoni* DC)

Powdery mildew disease of coriander was considered earlier to be a disease of minor importance (Butler, 1918). But now it has been considered as disease of major importance. Losses in seed yield due to powdery mildew disease is now reported to 15-20 per cent in Rajasthan and 9.0 to 33.8 per cent yield loss in disease prevalent areas with intensity of 7.5 to 24.4 per cent in seven districts of U.P. (Singh *et al.*, 2003; Srivastava *et al.*, 1971). *Erisiphe polygoni* is a most common pathogen causing powdery mildew in India. However, occurrence of *E. chicaracearum*, *E. umbelliferum* has also been reported from Afghanistan, Burma, Iran, Russia and other countries. (Hirata, 1966).

The disease first appears as small white circular patches on young parts of stem to leaves usually in Feb-March. These patches increase in size and often coalesce to cover whole area of leaf surface extending even to seeds. Considerable damage is caused by distortion and reduction in leaf size in Rajasthan (Chattopadhyay and Maiti, 1990). Severely affected plants dry prematurely and show poor seed setting. Seeds produced are light in weight, shriveled and fetch less market value (Srivastava *et al.*, 1971). If the disease infection occurs at early stage, seeds are not formed. With late appearance of disease although seed formation take place but they remain small and discolored (Chaudhury, 2005). The disease is prevalent not only in Rajasthan but also in other coriander growing areas.

Incidence of powdery mildew disease is favoured by dry season, high day temperature, and humid conditions (Mathur, 1949). The incidence of disease is lower in early sown crop (October) than the late sown. Disease management varied with respect to cultivars differing in maturing period (Kalra *et al.*, 1995a,b). Increase in temperature and decrease in relative humidity have been correlated with high incidence of powdery mildew particularly in case of late sown crops flowering in last week of January. High humidity and low temperature, dews deposition during night may be the reason for low incidence of disease by pathogen

The disease can be managed by two dusting of elemental sulphur 300 mesh (Nikam *et al.*, 1959) or two sprays of wettable sulphur/sulfex (0.3 per cent) or ecosan, cosan, thiovit etc. Dusting of crop regularly with sulphur or with mixture of hydrated lime and finely divided sulphur has been recommended for its control (Chattopadhyay and Maiti, 1990). Two sprays of cosan, sultaf (0.25 per cent), karathane W.D (0.5 per cent), elosal, thiovit and moricide, carbendazim (0.5per cent) alone or in combination with mancozeb (0.2 per cent), thiophenate methyl and carbendazim gave effective control of disease (Ali *et al.*, 1998; Srivastava *et al.*, 1971). Other fungicide i.e dinocap (karathane) @ 0.5 per cent alone or in combination with mancozeb (0.2 per cent) were also effective for disease management (Adivar and Ranjana, 1991; Keshwal *et al.*, 1979; Raju *et al.*, 1982; Srivastava *et al.*, 1971; Vinyagamoorthy *et al.*, 1985). Three application of sultaf or moresan gave effective control with high incidence of disease in late sown crops in December at Jabalpur (Keshwal, *et al.*, 1979). Studies conducted by Raju and co workers (1982) application of sulphur dust, wettable sulphur suspension (0.6 per cent) and sulfex gave maximum control of disease with high seed yield. Spraying of karathane (0.1 per cent) or bavistin (0.1 per cent) or wettable sulphur (0.25 per cent) at the time of flower initiation followed by second foliar spray after ten days controlled the disease (Khader and Vedamuthu, 1989). Three to four application of dinocap provided satisfactory control of powdery mildew disease in late and early maturing cultivars (Kalra *et al.*, 2000). The fungicides namely hexaconazole (0.1 per cent) and propiconazole (1per cent) not only reduced the disease severity but also increased the yield 2.54 and 2.00 times and cost benefit ratio of 1:5.3 and 1: 4.0 over control respectively in Chattishgarh (Singh, 2006). Maximum incidence of powdery mildew incidence was recorded with hexaconazole (0.1 per cent) that was at par with propiconazole (0.1 per cent) and wettable sulphur (0.3 per cent). Powdery mildew could effectively managed by applying carbendazim as soil drench and foliar spray at the rate of 0.01 per cent (Anonymous, 2005). Spraying of 0.1 per cent solution of karathane I.C. or 0.05 per cent of calixin or 0.1 per cent of bavistin @ 400-500 l/ha controls the disease (Singh *et al.*, 2001). Seed treatment with carbendazim (1g/kg of seeds) along with foliar spray with wettable sulphur (0.25 per cent) was the most effective and profitable treatment in controlling the disease to minimum (10.72 per cent) among other treatments with of seed treatment with carbendazim and foliar sprays of onion leaf extract (5 per cent), and 0.4 per cent of bayleton (Shukla *et al.*, 2006)

Two sprays of (250 l each) with karathane 48 E C/dinocap/bavistin (0.01 per cent)at disease initiation and 10 days after with Carbendazim (0.1per cent).or Two sprays of wettable sulphur @2.5g/ l or dusting of sulphur thrice have been recommended. The ecofriendly technique of using 5 per cent onion leaf extract as foliar spray three times can also protect the plants from powdery mildew at Coimbatore have been recommended in All India Coordinated Research Project on Spices Crops (AICRP, Parthasarathy *et al.*, 2005). Spraying of neem kernel extract (NSKE, 5 per cent) thrice, first spray after appearance of powdery mildew and second and third spray at 15 days interval was effective in checking powdery mildew incidence at Coimbatore (AICRP recommendation- Anonymous, 2005).

Biocontrol Agent

T.harzianum when used both as seed treatment and soil application effectively reduce the wilt incidence at Jobner (Anonymous, 2003) Seed treatment with *Pseudomonas fluorescens* IISR- 6 @ 10g/kg of seeds followed by foliar spray @ 10^8 CFU on 60 days after sowing was found to be effective to control the powdery mildew disease (18PDI) and recorded cost benefit ration of 1:2.5 (AICRP recommendation, Anonymous, 2007).

Resistant Cultivars

Cultivation of resistant/tolerant varieties like UD-48, UD-41,RCr-41,UD-20,Guj Cor-1, 2,RCr-41, Sadhna, Sindhu, CO-3, have been recommended (Keshwal and Khatri, 1998; Potty and Krishnakumar, 2001; Rajpurohit, 2007). RCr-41was free from powdery mildew symptoms in sick plot at Kota (Sastry and Sharma, 2001).

Blight of Coriander (*Alternaria poonensis*)

Coriander crop is affected by blight disease. Continuous cloudy weather after flowering is conducive for incidence of disease. The disease appears in the form of small brown spots on leaves and stem with the symptoms of shoot tips bending downwards. The spread of disease is very fast and is difficult to control after its appearance. Prophylactic measures are recommended in disease prone areas for effective prevention of disease.

If cloudy weather prevails with increase in atmospheric humidity during flowering period, spray of 0.1 per cent solution of bavistin in 400-500 lit solutions per hectares or 0.2 per cent solution of either of indofil M-45, cupramar, fytolan and blue copper is recommended. The foliar sprays are repeated once or twice at 10-15 days interval or even earlier if there is rain shower or heavy dews deposition (Chaudhury, 2005).

Blight of Coriander [*Colletotrichum gleosprioides* (Penz. and Sact.)]

The diseases incited by *C. capsicea* (Syd.) Butler (Bibsy) was first reported from Rajasthan in Kota district and later from Anantpur in Andhra Pradesh. Later, *Colletotrichum gleospoioides* {*Gloremella cingulata* (Stonem.) Spauld and Schrenk} was isolated as the causal organism of the disease. The fungus is ploуphagous in nature and can infect other hosts due to production of non specific thermo stable toxin. Plants are infected during the flower initiation stage in inflorescence. Early seed setting stage is more susceptible to the disease. (Bhatta and Hiramath, 1989). Infected inflorescence turns pale yellow and droop down on account of infection at the juncture of young pedicel bearing flowers. In advance stage whole inflorescence and leaves gets damaged inhibiting the seed formation The crop infected at later stages produced shriveled and light weight seeds.

The disease can be managed by spraying triademefon (bayleton, 0.05per cent) or foltaf (0.03 per cent) and cuman-L (0.30 per cent) at flowering stage. Two sprays of RH-2161 (0.05 per cent) before flowering and another after 8 days were found effective in reducing blight incidence (Bhatta and Hiramath, 1989). Persistent fungicide diflotan can also be used but due its persistent nature and grown for vegetable purposes, non-persistent fungicides like captan is recommended for control of *Colletotrichum*. (Naik, *et al.*, 1998).

Diseases of Minor Importance

Stem Rot (*Rhizoctonia solani.* Kuhn)

The stem rot disease was first reportedly by Mall (1970). The disease mainly manifest seedlings near the radicle as brown necrotic girdles slowly extending to plumule leading to pre emergence mortality. In post emergence damage lower parts of the hypocotyls shows brown necrotic rings, which leads to collapse or mortality of plant. Stem rotting extends to whole part of upper part of stem. The root system usually starts rotting only after total damage of aerial parts of plants, the disease is found to be severe in rainy season only under the conditions of high soil moisture. The incidence of disease is less frequent during dry climate of winter. The disease cause complete loss in seed germination above 70 per cent moisture (Komaraiah and Reddy, 1986).

Mahor and associates in 1982 recorded similar disease of coriander caused by *Macrphomina phaseolina*.

Root Rot (*Curvularia pallescens.* Boedjin).

The root rot disease in coriander was reported at frequency of 20 per cent (Dwivedi *et al.*, 1982). Young plants were more susceptible than older ones. Rotting at seedling stage may extend to 50-75 per cent. Symptom of disease appears as brownish to blackish roots lacking secondary root system with yellowing of tips of young leaves. Yellowing of tips of young leaves spreads downward and to leaf blade. Ultimately, entire plant turns yellow and collapse due to rotting of plant roots (Chattopadhyay and Maity, 1990).

Grain Mold

It is minor diseases caused by *Helminthosporium* species, *Alternaria* species, *Fusarium* species and *Curvularia* species.

Management

The disease can be managed by spraying Carbendazim (0.1 per cent), 20 days after the grain set twice at 15 days interval (Mehra *et al.*, 2005; Parthasarathy, 2005).

A number of other diseases reported by include leaf spot caused by *Ramularia corianderii*, foot rot caused by aecial and pycnidial stages of *Uromycis graminis* and *Sclerotinia sclerotium*, stem rot caused by *Phoma applicata*, leaf spot caused by *Cercospora corianderii*, root rot caused by *Torula herbarum* f sp *quaiternella*, and *Rhizoctonia solani*, leaf blight caused by *Alternaria poonensis*, *Alternaria alternata,and Curvularia pallenscence* Boedijn and phyllody in coriander (Agnihotri, 1991, 2005; Kumar, 2002; Prakasham *et al.*, 1987; Agarwal *et al.*, 2001). Sharma and Chaudhury (1981) reported damping off and root rot caused *by Pythium irregulare* in some fields at Varanasi (U.P).

The management strategies developed by various centres of All India Coordinated Research Project on Spices are summarized below.

Coriander

Table 5.1: Management of Major Diseases and Insect Pests of Coriander

Disease/Pest	*Management Schedules*
Powdery mildew	(1) Two sprays of 250 litres with karathane 48EC/dinocap/carbendazim (0.01per cent) at disease initiation and 10 days after with carbendazim (0.1per cent) or (2) Two spray of wettable sulphur (@ 2.5 g/lits.), one at the time of flower initiation and seed 15 days later or (3) dusting of sulphur @ 25 kg/ha thrice. (4) Ecofriendly techniques of using 5per cent onion leaf extract as foliar spray (Coimbatore) three times can also protect the plants. (5) A cost effective ecofriendly biocontrol method by spraying of neem seed kernel extract (NSKE) @ 5per cent thrice, first spray immediately after the appearance of disease and the subsequent two sprays at 15 days interval.
Stem gall	Selection of disease free seeds, collection and destruction of affected plants, growing coriander variety RCr-41, seed treatment with agrosan GN/thiram (2.5 g/kg) or carbendazim (1 g/kg) of seed. Spraying of bayleton (0.4 ml/L of water) repeat after 20 days.
Grain mould	Spray carbendazim (0.2per cent) 20 days after flowering (grain formation stage and again at grain hardening stage or captan (0.2per cent) twice at grain set and at 15 days interval.
Wilt complex	Sow in the first and second week of November for reducing disease incidence. Seed treatment with *Trichoderma viride* (formulation with CFU 10^6 @ 4 g/kg seed and neem cake application @ 150 kg/ha following foliar spray of hexaconazole (0.05per cent). Biocontrol agent *T. viride* in combination with foliar spray of either hexaconazole or thiophenate methyl on 25th, 40th and 55th DAS and soil application of neem cake @ 150 kg/ha. Use resistant varieties–Co-3, Sadhana and Surabhi. Adopt crop rotation, give deep summer ploughing, and use disease free seeds. Seed dressing as well as soil drenching with carbendazim 0.2per cent. Seed dressing with 1.5 g carbendazim + 1.5 g thiram/kg seed and spraying twice with carbendazim 0.1per cent. Another ecofriendly biocontrol method by seed treatment with *Pseudomonas fluorescens* @ 10 kg/ha plus soil application of neem cake @ 150 kg/ha is also advocated.
Root rot	Drenching carbendazim 0.1per cent twice at 20 days intervals.
Aphids and white fly	Endosulphan (0.05per cent) twice just at the time of floral initiation and 15 days there after with monocrotophos or quinalphos (0.1per cent) or dimethoate (1.5 per cent) or methyl demethon (2.01 per cent) or phosphomidon (0.5per cent) thrice @ 400-500 lits./ha at 10 days interval.
Heliothis Caterpillar	Foliar application of 0.03 per cent phosphomidon or dimethoate or methyldematon (0.05 per cent)
Red mites	Foliar application of sulfex 3 g or dicofol 3 ml/l of water, 2-3 times at 10 days interval or spraying monocrotophos/quinalphos (0.1per cent) at 15 days interval.
Frost	One light irrigation can save the crop in irrigated areas. In rainfed areas, spray 0.1per cent H_2SO_4 can save the crop and smoking around the border of the coriander field.

Researchable Issues for Disease Management, Enhancing Productivity and Quality of Coriander/Future Prospects

- Collection evaluation and characterization of germplasm resistant to major diseases like stem gall, powdery mildew and wilt with improved quality attributes.
- Mutation and biotechnological approaches to enhance variability.
- Breeding for high yielding disease resistant varieties coupled with high quality parameters.
- Identification of specific sites and gene responsible for disease resistance and its incorporation in high yielding varieties using biotechnological tools.
- Identification of region wise strains of pathogen for disease management using biocontrol agent.
- Standardize integrated disease management practices.
- Work out toxic levels of fungicide residues in plant produce to minimize misuse of agrochemicals, their accumulation tendencies, and toxicity.
- Thrust on focusing attention on ecofriendly organic farming and integrated disease management using herbal extracts.
- Search for effective biocontrol agent region wise as ecofriendly measure, its promotion and popularization among farmers.
- Generate database computer aided programs for organic agriculture, disease forecasting models using new satellite agristat to delineate organic farming zone and disease forecasting.
- Formulate region wise package of practices for adoption of IDM.

References

Adivar, S.S and Rajanna, K.M. (1991). Control of powdery mildew in Coriander. *Univ Agri Sci. Bangalore*, 20(4): 59.

Agnihotri, J.P. (1991). Diseases of arid zone seed spices : Present status and future strategies for their management. *Dryland Resources and Technology*. 6: 1–40.

Agnihotri, J.P. (2005). Management of diseases of Apiaceae family seed spices. In: *Winter School Proceedings: Advances in Seed Spices Production*. National Research Centre on Seed Spices, Ajmer, p. 114–122.

Ali, S.A., Pathak, R.K. and Saraf, R.K. (1998). Fungicidal seed treatment in coriander. *Adv. Pl. Sci.*, 11: 303–304.

Agarwal, K., Dwivedi, M. and Sharma, J. (2005). Colonization of coriander (*Coriandrum sativum* L.) by *Fusarium Oxysporum* f sp *corianderi*. *J. Mycol. Pl. Path.*, 35(2): 253–256.

Agarwal, S., Sastry, E.V.D. and Sharma, R.K. (2001). *Seed Spices Production, Quality and Export*. Pointer Publication, Jaipur.

Anonymous (2003). Annual report. In: *All India Coordinated Research Project on Spices Crops*. Indian Institute of Spices Research, Calicut.

Anonymous (2005). Annual report. In: *All India Coordinated Research Project on Spices Crops*. Indian Institute of Spices Research, Calicut.

Anonymous (2007a). http://www.indianspices.com/html/s0420sts.htm.

Anonymous (2007b). Annual report. In: *All India Coordinated Research Project on Spices Crops*. Indian Institute of Spices Research, Calicut.

Awasthi, P.B., Mishra,U.S. and Pandey, B.N. (1978). A new record of *Protomyces macrosporus* Ung. on *Foeniculam vulgare* L. *Curr. Sci.*, 47: 823–824.

Bharadwaj, L.N. and Shrestha, S.M. (1985). Efficacy of fungicide application in control of stem gall of coriander. *Agric. Ecosyst. and Environment*, 13 (3/4). :319–323.

Bhatta, G.K.M.B. and Hiremath, P.C. (1989). Age of seedling in relation to coriander blight caused by *Colletotrichum gleosporiodes*. *Karnataka Journal of Agricultural Sciences*, 2(4): 336–338.

Bhatta *et al.* (1988). *Pl. Path. Newsletters*. Univ. of Agric. Sci., Dharwad, p. 47.

Butler, E.J. (1918). *Fungi and Diseases in Plants*. Thacker Spink and Co., Calcutta, p. 359–361.

Chattopadyay, S.B. (1967). *Diseases of Plants Yielding Drugs, Dyes and Spices*. Indian Council of Agriculture Research, New Delhi.

Chattopadhyay, S.B and Maity, S. (1990). *Diseases of Betelwine and Spices*. Indian Council of Agricultural Research, New Delhi.

Chaudhury, G.R. (2005). *Advances in Seed spices Production*. National Research Centre on Seed Spices, p. 43.

Christensen, C.M. and Kaufmann, H.H. (1969). Grain storage. In: *The Role of Fungi in Quality Loss*. Minnesota Press, Minneapolis, USA, p. 7–153.

Dwivedi, D.K., Shukla, D.N. and Bhargava, S.N. (1982). Two new root rot diseases of spices. *Curr. Sci.*, 51: 243–244.

Edison, S. (1989). *Proceeding First National Seminar on Seed Spices*, Jaipur, .p 195–196.

Evoor, S. (2008). Coriander: Cultivation practices and seed production. *Spice India*, 21(2): 15–17.

Goel, A.K., Kumar, S. and Tayal, M.S. (1983). Biochemical alteration induced in *Coriandrum sativum* L. by *Protomyces macrosporus* Unger. *Indian Bot. Reporter*, 2(1): 62–64.

Gupta, J.S. (1954). Disease appraisal of stem gall of *Coriandrum sativum* L. *Indian Phytopath.*, 7: 53–60.

Gupta, J.S. (1956a). Effect of certain antibiotics and fungicides on germination of chlamydospores of *Protomyces macrosporus* Ung. *Proc 43rd Indian Sci Cong.*, 3: 217–218.

Gupta, J.S. (1956b). Chemistry of diseased fruits of Coriandrum sativum affected by *Protomyces* causing stem gall disease of coriander. *Proc 43rd Indian Sci Cong.* 3: 217.

Gupta, J.S. (1962). Pathological anatomy of floral parts and fruits of coriander affected with *Protomyces macrosporus* Ung. *Agra Univ. J. Res. (Sci.)*, 11: 307–11.

Gupta, R.N. (1973). Longivity of chlamydospores of coriander gall fungus. *Indian Phytopath.*, 26: 581–582.

Gupta, R.N. (1976a). Impact of nitrogen, phosphorus and manganese on stem gall disease of coriander. *Acta Botanica Indica*, 4(1): 30–35.

Gupta, R.N. (1976b). Impact of nitrogen, phosphorus and manganese on stem gall disease of coriander *Acta Bot. India*, 16: 104 (Abs).

Gupta, J.S. and Sinha, S. (1963a). Biochemical grouping in Protomyces macrosporus Ung. causing stem gall disease of coriander. *Proc. Nat. Acad. Sci.*, 33(13): 507–510.

Gupta, R.N. and Sinha, S. (1963b). Some therapeutic control trials on stem gall disease of coriander. *Indian Phytopathology*, 16: 106.

Gupta, R.N. and Sinha, S. (1973). Varietal field trials in the control of stem gall disease of coriander. *Indian Phytopath.*, 21: 337–40.

Hashmi, M.H. and Thirana, U. (1990). Mycotoxins and other secondary metabolites in spices of *Fusarium* isolated from seeds of Capsicum, Coriander, and Fenugreek. *Pak. J. Bot.*, 22: 106–116.

Hashmi, M.H. and Gaffar, A. (1991). Seed mycoflora of *Corianderum sativum* L. *Pakistan J.*, 23(2): 115–172.

Hawker, L.E. (1950). *Physiology of Fungi.* University of London Press. London.

Hirata (1966). *Host Range and Geographical Distribution of Powdery Mildew.* Faculty of Agriculture, Nigata Univ., Nigata, Japan, p. 472.

ICAR (1953). *Scheme on Studies on Wilt Diseases of Coriander in Madhya Pradesh.* New Delhi.

Jain, S., Srivastava, A., Arya, H.C. and Kant, U. (1995). *In vitro* studies on dual cultures of *Protomyces macropsorus* on *Coriandrum sativum*. *Ind. J. Bot. Sci.*, 74: 269–271.

Jakhar, M.L. and Rajput, S.S. (2005). Medicinal properties of seed spices. In: *Advances in Seed Spices Production*. National Research Centre on Seed Spices, Ajmer, p. 229–230.

Joshi, L.K. and Nikam, B.G. (1953). *Studies on Wilt Diseases of Coriander in Madhya Pradesh.* Govt Central Press, Gwalior, p. 20.

Joshi, L.K., Sankaran, R.S. and Nikem, B.G. (1945–52). *Studies on Wilt Diseases of Coriander in Madhya Pradesh.* Govt Press Gwalior.

Kalra, A., Parameshwaran, T.N., Ravindra, N.S. and Dimri, B.P. (1995). Effect of powdery mildew (*Erisiphae polygoni*). on yield and yield components of early and late maturing coriander (*Coriandrum sativum* L.). *J. Agric. Sci.*, 125: 395–398.

Kalra, A., Parameshwaran, T.N., Ravindra, N.S., Dimri, B.P and Rao, M.G., Sushil Kumar and Kumar, S. (2000). Effect of planting date and dinocap application on the control of powdery mildew and yield of seed and seed oil in coriander. *J. Agric. Sci.*, 135: 193–197.

Kamal, M. and Khan, S.A. (1962). Identification of *Cercospora foeniculi. P. Magnus* causing infection of *Foeniculam vulgare* in Indo-Pakistan sub continent. *Bilogia Lahore*, 8:61–62.

Keshwal, R.L., Choubey, P.C. and Singh, K. (1979a). Effect of different sowing dates and fungicidal spray on incidence of Powdery mildew of coriander. *Pesticides*, (Oct): 15.

Keshwal, R.L., Choudhury, P.C. and Singh, K. (1979b). Effect of different dates of sowing and fungicidal spray on Powdery mildew of Coriander. *Veg. Sci.*6 (2):135–136.

Keshwal, R.L. and Khatri, R.K. (1998). Reaction of some high yielding varieties of coriander to powdery mildew. *Journal of Mycology and Plant Pathology*, 28: 58–59.

Khader, Mohd Abdul and Vedamuthu, G.B. Peter (1989). *Proceeding First National Seminar on Seed Spices*, Oct 24–25, Jaipur, p. 288–291.

Komaraiah, M. and Reddy, S.M. (1986). Influence of humidity on seed deterioration of Methi (*Trigonella foenicum-graceum*). seeds by seed borne fungi. *Indian J. Myco. Plant Pathol.*, 16: 77–79.

Kumar, S. (2002). Management of pests and diseases in cumin, coriander and fenugreek crops. *Ind. J. of Arecanut Spices and Medicinal Plants*, 4(2): 92–95.

Lakra, B.S. (1977). Stem gall of coriander – an intro spection and strategies for control. *J. Mycol. Mycol. Plant Pathol.*, 27(1): 98.

Lakra, B.S. (1993). Quantitative and germination losses in coriander seeds due to *Protomyces macrosporus* infection. *Indian J. Mycol. and Pl. Pathol.*, 23: 93 (Abs).

Lakra, B.S. (1995). Evaluation of dormancy and viability of the chlamydospores of *Protomyces macrosporus* and its effect on seed health of coriander. *Haryana J. Hortic. Sci.*, 24(3–4): 325–327.

Lakra, B.S. (1997). Stem gall of coriander: Introspection and strategies for control. *J. Mycol. Pl. Pathol.*, 27(1): 98.

Lakra, B.S. (1999). Assessment of losses due to stem gall in coriander. *Pl. Dis. Res.*, 14(1): 85–87.

Lakra, B.S. (2000). Management of stem gall of coriander (*Coriandrum sativum* L.) incited by *Protomyces macrosporus*. *Indian J. Agric. Sci.*, 70(5): 338–340.

Maheswari, D.K., Chaturvedi, S.N. and Yadav, B. (1984). Qualitative and quantitative changes in starch and protein content induced by *Protomyces macrosporus* in coriander stem. *Indian Phytopath.*, 37(1): 1070–1073.

Mahor, R., Naqui, S.M.H. and Gupta, R.N. (1982). A new fungal disease in coriander in India. *Acta Botanica Indica*, 10: 323.

Malhotra, S.K., Vashishtha, B.B. and Apparao, V.V. (2006). Influence of nitrogen, *Azospirillum* and farmyard manure on growth, yield and incidence of stem gall disease of coriander. *J. Spices and Aromatic Crops*, 15(2): 115–117.

Mall, O.P. (1968). Studies on wilt disease of coriander with special reference to variability of the pathogen. *Indian Phytopath.*, 21: 379–388.

Mall, O.P. (1970). Rhizoctonia stem rot of coriander. *Indian Phytopath.*, 23: 699–700.

Mathur, R.L. (1949). Plant protection: Notes and news. *Pl. Prot. Bull. Govt. of India*, 1: 30.

Mathur, S.B (1962). Soil pH in relation to stem gall disease of coriander caused by *Protomyces macropsporus* Unger. *Indian Phytopath.*, 15: 75–76.

Mathur, B.L. (1963). Studies on Fusarium wilt of coriander in Rajasthan. *Ph.D. Thesis*, University of Rajasthan, Jaipur.

Mathur, S.B. and Narula, P.N. (1963). Effect of certain fungicides, antibiotics and synthetic phytohormones on the germination of *Protomyces macrosporus* Unger causing stem gall disease of coriander. *Proc. Indian Acad. Sci. Sec.* B33: 615–617.

Mehra, R. (2005). Seed spices : Diseases and their management. *Indian J. Arecanut, Spices and Medicinal Plants*, 7(4): 134–143.

Mukherji, K.G. and Bhasin, J. (1986). *Plant Diseases of India*. Tata McGraw Hill Pub. Co. Ltd., New Delhi p. 88.

Mukhopadyay, A.N. and Pavgi, M.S. (1971). Environment in relation to infection of coriander by *Protomyces macrosporus* Unger. *Ann. Phytopath Soc. Japan*, 37: 215– 216.

Mukhopadyay, A.N. and Pavgi, M.S. (1975). Environmental control of chlamydospores germination in *Protomyces macrosporus*. *Friesia*, 11: 14–23.

Mundkar, B.B. (1949). *Fungi and Plant Diseases*. Macmillan and Co. Ltd., London, p. 107–109.

Muthulakshmi, Cheziyan, N., Muthukrishnan, K. and Doraiswamy, S. (2002). Management of coriander wilt using biocontrol agents. *J. Spices and Aromatic Crops*, 11(2): 138–140.

Naik, K.S., Hiramath, P.C. and Hedge, R.K. (1998). Persistence of some fungicide on coriander foliage against *Collectotrichum gleosporoides. Karnataka J. Agric. Sci.*, 11: 1288–1289.

Narula, P.N. (1962). Efficacy of different fungicides against stem gall disease of *Coriandrum sativum* L. *Sci. and Cult.*, 28: 481–482.

Narula, P.N. and Joshi, M.G. (1963). Differential strains of coriander strains of stem gall disease. *Sci. and Cult.*, 29: 206.

Nikem, B.G., Kulkarni, S.N. and Pawar, P.K. (1959). Coriander and its diseases. *Indian Farming*, 9: 12–13.

Naqvi, S.A.M.H. (1985). Varietal screening of coriander against stem gall in relation to disease intensity and crop losses. *Indian J. Mycol. and Pl. Pathol.*, 16: 270–276.

Nene, Y.L., Siddiqui, L.A. and Kharbanda, P.D. (1966). Control of stem gall of coriander by fungicides. *Mycopathologia et Mycolgia et. Applicata*, 29(1–2):142–144.

Pandey, R.N. and Dangey, S.R.S. (1998). Diseases of coriander and Fennel. *Agric. Rev.*, 19(2): 120–125.

Parthasarathy, V.A., Shiva, K.N. and Johny, A.K. (2005). Research and developments in major seed spices under AICRPS. In: *Winter School on Advances in Seed Spices Production*. National Research Centre on Seed Spices, Ajmer, Oct., p. 255– 271.

Patel, K.C., Pandya, R.K. and Sharma,B.L. (1998). Resistance against diseases in coriander varieties. *Proc. 5th National Scientist Conference.* Indian Agriculture Research Society, p. 45.

Paul, Y.S. (1992). Studies on seed borne mycoflora of coriander with special reference to stem gall in Himachal Pradesh. *Pl. Dis. Res.*, 7(1): 19–23.

Pavgi, M.S. and Mukhopadhyay, A.N. (1969a). Transmission of the incitant of stem gall of coriander *Protomyces macrosporus* Unger. *Ann. Phytopath. Soc., Japan*, 35: 265– 270.

Pavgi, M.S. and Mukhopadhyay, A.N. (1969b). Viability of seeds of coriander infected by *Protomyces macrosporus* Unger. *Ann. Phytopath. Soc., Japan*, 35: 271–34.

Pavgi, M.S. and Mukhophadyay, A.N. (1972). Development of coriander fruit infected with *Protomyces macrosporus* Ung. *Cytologia*, 37: 619–27.

Pawar, A.D. and Singh, Jasvir (2002). Integrated pest management (IPM) package for coriander (*Coriandrum sativum*). L. *Indian Journal of Arecanut, Spices and Medicinal Plants*, 4(4): 150–154.

Potty, S.N. and Krishna, V. (2001). Organic farming in seed spices in relation ot quality and export. In: *Seed Spices Production, Quality and Export*, (Eds.) S. Agarwal, E.V.D. Sastry and R.K. Sharma. Pointer Publications, Jaipur.

Prakash, S. and Pratap, (1995). Effect of inoculation of plants and contamination of seeds with *Protomyces macropsorus* Unger on stem gall disease severity and seed germination. *Haryana J. Hortic. Sci.*, 24: 70–73.

Prakasham, V. *et al.* (1987). A preliminary report occurrence of phyllody in coriander. *Veg. Sci.*, 15(2): 198–199.

Prasad, B.K. (1983). Free sugar and amino acids contents of coriander infected with stem gall disease. *Indian Phytopath.*, 36: 150–51.

Prasad, B.K. and Narayan, N. (1988). Influence of storage fungi on physio-characteristics and germination of coriander seeds. *Geobios* (Jodhpur), p. 141–142.

Raju, K.S., Rao, G.S., Rao, T.S. and Babu, M.K. (1982). Note on fungicidal control of coriander powdery mildew. *Ind J. Agri. Sci.,* 52(4): 262–263.

Rama Murthy and Sridhar, V. (2001). Seed spices and ayurveda. In: *Seed spices Production, Quality and Export*, Pointer Publication, Jaipur, p. 290–302.

Rao, G.G.P. (1972). Anatomical studies on abnormal growth caused by *Protomyces macrosporus* on *Coriandrum sativum* L. *Indian Phytopath.,* 25: 483–486.

Rajpurohit, T.S. (2007). Integrated disease management in seed spices. *Seed Spice,* 20(2): 17–22.

Samajpati, N. (1983). Incidence of aflatoxin in somecommon spices of Calcutta market. In: *Proc. of Symp. on Mycotoxins in Food and Feed,* (Eds.) K.S. Bilgrami, T. Prasad, K.K. Sinha. Allied Press, Bhagalpur, p. 8–9.

Sastry, E.V.D. and Sharma, R.K. (2001). Status of seed spices research in Rajasthan. In: *Seed Spices Production, Quality and Export,* (Eds.) S. Agarwal, E.V.D. Sastry and R.K. Sharma. Pointer Publication, Jaipur.

Saxena, R.P., Dixit, J., Pandey, V.P. and Singh, V.K. (2002). Germplasm screening for stem gall disease in coriander. In: *National Symposium on Integrated Management of Plant Diseases of Mid Eastern India with Cropping System Perspective,* Held at N.D. University of Agriculture and Technology, Faizabad.

Sharma, Y.R. and Anandaraj, M. (2000). Diseases of spices crops and their management. *Indian J. Arecanut Spices and Medicinal Plants,* 2(1): 6–19.

Sharma, Y.R. and Chaudhary, K.C.B. (1981). Damping off and root rot of Coriander. *Natn, Acad. Sci. Lett.,* 4: 155– 156.

Shukla, S. (2005). Studies on stem gall and powdery mildew diseases of coriander. *M.Sc. Thesis,* N.D. University of Agriculture and Technology, Faizabad.

Shukla, S., Saxena, R.P., Awasthi, L.P. and Kumar, P. (2006). Effect of fungicide and onion extract on powdery mildew disease and yield of coriander. In: *Proceedings National Symposium and IPS Meeting (MEZ)*, Held at D.D.U. Gorakhpur University, Gorakhpur during 18–19 Dec. 2006, Abstract, pp. 98–99.

Singh, A. (1983). Mycotoxin contamination in dry fruits and spices. In: *Seed Spices Production, Quality and Export,* (Eds.) S. Agarwal, E.V.D. Sastry and R.K. Sharma. Pointer Publication, Jaipur.

Singh, A.K. (2006). Evaluation of fungicides for the control of powdery mildew disease in coriander (*Coriandrum sativum* L.). *J. of Spices and Aromatic Crops,* 15(2): 123– 124.

Singh, S.P., Gupta, J.S. and Sharma, A.K. (1984). Disease appraisal and crop loss estimate in coriander attacked by *Protomyces macrosporus* Unger. in Uttar Pradesh. *Geobios,* 11: 276–278.

Singh, B.K., Singh, T. and Singh, D. (1991). Contamination of coriander seeds by *Protomyces macrosporus* Unger in Rajasthan. *J. Ind. Bot. Soc.,* 70: 413–414.

Singh, D., Chaudhary, G.R., Singh, Shyam and Singhania, D.L. (2001). Production technology for seed spices. In: *Seed Spices Production, Quality and Export,* (Eds.) S. Agarwal, E.V.D. Sastry and R.K. Sharma. Pointer Publication, Jaipur, pp. 89–114.

Singh, H.B., Singh, A., Tripathi, A., Rai, S.K., Katiyar, R.S., Johri, J.K. and Singh, S.P. (2003). Evaluation of Indian coriander accessions for resistance against stem gall disease. *Genetic Resource and Crop Evaluation*, 50(4): 339–343.

Singh, U.C., Singh, R. and Jadav, M.R. (1998). Incidence of *Protomyces* gall and aphid on coriander. *Flora and Fauna*, Jhansi, 2: 123–130.

Singh, U.C., Singh, R. and Jadhav, M.R. (1996). Incidence of Protomyces gall and aphid on coriander. *Flora and Fauna*, Jhansi, 2: 129–130.

Spice India (2005). All time record in export of spices. *Spice India*, 18(5): 11–14.

Srivastava, U.S. (1969). Effect of inoculam potential on wilt development of coriander caused by *Fusarium oxysporum* f.sp *corianderii*. *Indian Phytopath.*, 22: 406– 408.

Srivastava, U.S. (1971a). Edaphic factors and wilt of coriander. *Indian Phytopath.*, 24: 679–683.

Srivastava, U.S. (1972). Effect of interaction of factors on wilt of coriander caused by *Fusarium oxysporum* f.sp corianderi. Kulkarni, Nikem and Joshi. *Indian J. Agric. Sci.*, 42: 618–621.

Srivastava, U.S., Rai, R.A. and Agarwal, J.M. (1971). Powdery mildew of coriander and its control. *Indian Phytopathology*, 24: 437–440.

Srivastava, U.S. and Sinha, S. (1977). *Indian J. Agric. Sci.*, 41: 779–782.

Stevenson, E.C. (1943). Diseases of drugs and related plants at the plant industry station, Beltsville, Maryland in 1943. *Pl. Dis. Rep.*, 27: 700–703.

Sydow, H. and Butler, E.J. (1911). Fungi Indiae Orientales. *Annals Mycologici.*, 9: 372– 421.

Tayal, M.S., Kumar, S., Goel, A.K. and Maheswari, D.K. (1981). Role of IAA, Oxidase *o*-dihydroxy-phenyl polyphenol oxidase and peroxidase in stem gall disease of *Coriandrum sativum* L. *Curr. Sci.*, 50: 785–786.

Thakral, K.K., Mangal, J.L. and Thakral, R. (2002). Medicinal Importance of Spices grown in Northern India. *Indian J. Arecanut and Medicinal Plants*, 4(1): 53–55.

Tripathi, A.K. (2001). Evaluation of coriander cultivars for resistance against stem gall diseases and yield potential. *Crop Res. Hisar*, 22(3): 485–488.

Tripathi, A.K. (2002). Coriander varietal reaction of stem gall disease caused by *Protomyces macrosporus* Unger. *Ind. J. Pl. Protection*, 30(1): 97–98.

Tripathi, A.K. (2003). Effect of sowing dates and weather parameters on stem gall disease of coriander *Indian Phytopath*, 56(2): 191–193.

Tripathi, A.K. (2005). Efficacy of fungicide and plant products against stem gall disease of coriander. *J. Mycol. Pl. Pathol.*, 35(2): 388–389.

Tripathi, A.K., Bartaria, A.M., Pandya, R.K., Chauhan, Shasi and Chauhan, S. (1998). Evaluation of coriander cultivars against stem gall diseases. *Flora and Fauna Jhansi*, 4: 98.

Tripathi, A.K., Pandya, R.K. and Tripathi, M.L. (2001). Effect of nitrogen, phosphorus and potassium on stem gall disease and yield of coriander. *Pl. Protec. Sci.*, 9(2): 304–366.

Toncer, O., Tansi, S. and Kizil, S. (1998). The effect of different sowing time on essential oil of coriander in GAP region. *Anadolu*, 8(2): 101–105.

Vinayagamoorthy, A., Mihodeen, M.K., Jeyarajan, R. and Prakasham, V. (1985). Fungicidal control of powdery mildew of coriander. *South Indian Hort.*, 133: 347–348.

Plant Diseases Management in Horticultural Crops (2011) *Pages* ***63–80***
Editors: **Shahid Ahamad, Ali Anwar and P.K. Sharma**
Published by: **DAYA PUBLISHING HOUSE, NEW DELHI**

Chapter 6

Mango Malformation: Epidemiology and Management

M.K. Pandey and V.K. Razdan

Division of Plant Pathology, Sher-e-Kashmir University of Agricultural Sciences and Technology, Jammu – 180 002

Malformation is a serious disease of mango (*Mangifera indica* L.) having a history of more than hundred years and has assumed a serious proportion both in intensity and space adversely affecting the mango industry all over the world. The disease was first reported in 1891 by Maries from Darbhanga (Bihar) (Watt, 1891) and since then it has been reported from the entire mango growing areas in India (Burns, 1910; Singh and Chakaraverty, 1935; Singh *et al.*, 1940) and world over: Egypt (Hassan, 1944; Attiah, 1955; Hifny *et al.*, 1978), South Africa (Schwartz, 1968), Pakistan (Khan and Khan, 1960), Middle East (Hassan, 1944), Sudan (Minessy *et al.*, 1971), Cuba (Padron and Soroa, 1983), Australia (Peterson, 1986), Bangladesh (Meah and Khan, 1992), United Arab Emirates (Burham, 1991), Israel, Central America, Mexico and USA (Malo and McMillan, 1972).

In India the disease is particularly threatening the mango industry in Northern India. According to estimation by the Uttar Pradesh Council of Agricultural Research, Lucknow, in the state of Uttar Pradesh, where 50 per cent of the mango orchards of the country occur, 50 to 80 per cent of the trees are infected with the disease (Schlosser, 1971; Anon, 1983; Anonymous, 1994). The incidence and loss vary from place to place and cultivar to cultivar. Verma *et al.* (1969) recorded highest incidence from Punjab, Delhi and western UP, where almost 50 per cent plants were affected, whereas, in eastern UP, Maharashtra, Andhra Pradesh and Tamil Nadu the incidence was hardly 10 per cent, while in Mysore it was only 5 per cent and in Kerala and around Kanyakumari typical malformation was not observed at all. In Jammu, 21-45 per cent floral malformation was recorded by Chib *et al.* (1984). According to Majumdar and Sharma (1990) the region beyond Hyderabad was free from this malady.

Diagnostic Symptoms

The disease is characterized by two distinct types of malformations, vegetative and floral. Vegetative malformation is more pronounced on young seedlings in comparison to mature trees (Nirvan, 1953). According to Singh *et al.* (1986), trees between 4 to 8 years age suffer the more (90.9 per cent) from vegetative malformation. Vegetative malformation occurs both on grafted as well as seedling types (Garg, 1951). Typical symptom of vegetative malformation is loss of apical dominance resulting in numerous small shootlets arising from apex of shoot or leaf axil bearing small scaly leaves, thicken and clustering together giving the twig a broom like appearance, hence commonly known as 'Bunchy top' (Nirvan, 1953). Young plants frequently sprout vegetative buds all over the internodes making witches broom like appearance. Vegetative buds or shoots also sprout in grown up trees where small bunches are produced at the terminals. The terminals which change into bunch ultimately dry leading to death of growing shoots. Kumar *et al.* (1993) stated that axillary buds get swollen and persist on mother tree for a long time as dry masses.

Symptom of floral or blossom malformation appear with the emergence of inflorescence. In the early stages of panicle formation there is no differentiation between the healthy and diseased panicles. In some cases, the flower buds never open and remain dull green in colour. Branches with the diseased inflorescence can produce both malformed as well as healthy panicles in the following bearing season. The affected inflorescences are of three types, heavy, medium and light.

Heavy

Panicles are very compact due to crowding of flowers and keep growing forming a heavy panicle. The head sometimes looks like cauliflowers and usually persists on the tree without fruit setting. This usually persists on the tree, with low per cent age of perfect flowers (<1 per cent).

Medium

Panicles are slightly less compact, the growth is not continuous. They persist on tree longer than the normal panicles. The per cent age of perfect flowers varies between one and five.

Light

Panicles are difficult to distinguish from normal ones at a later stage though during development they can be recognized by their compactness. The per cent age of perfect flowers is almost half that of the normal panicles. Fruit setting is appreciable.

Due to large bracts, malformed panicles give leafy appearance unlike normal panicles. Usually, the sex ratio of flowers in a malformed inflorescence decreases. Khan and Khan (1962) remarked that the per cent age of hermaphrodite flowers in malformed panicles is very low, therefore such panicles rarely bear the fruits and if fruit setting takes place, at all, the fruit drops drown at the pea stage size. Mallik (1963) reported that the pistils in malformed hermaphrodite flowers are usually non-functional and pollens exhibit poor viability. According to Hifny *et al.* (1978), the thickened rachis of panicle produces large number of flowers having 1 to 4 ovaries per flowers as against 1 to 2 in normal of flowers. The affected flowers show hypertrophied stigma, style and ovary. These flowers show high degree of embryo degeneration (Singh and Dhillon, 1990). All the branches produced on a malformed branch may not bear malformed panicles and even in malformed panicles some secondary rachis may be normal. The abnormal inflorescence may persist long after the normal one has fallen off (Singh *et al.*, 1961) and may finally become vegetative. The malformed heads dry up in black masses and persist on the tree for along time even till the next flowering season (Pandey and Chakraborty, 2004).

Causal Organism

There has been a lot of controversy about the cause of the disease due to the complex nature of the malady, which has been attributed to many factors like cultural practices, nutrition, mites, fungal, hormonal and physiological imbalance (Kumar and Beniwal, 1992). Sttar (1946) had suggested that the disease was of viral origin, however, results from electron microscopy, transmission, cultural and serological studies in India and Egypt have convincingly disproved the involvement of a virus or MLO with the disease (Kishtah *et al.*, 1985a and b; Kumar and Beniwal, 1992). The disease was also attributed to the feeding injury caused by eriophyid mite (*Aceria mangiferae*) but later it has been concluded that the mite does not cause the disease (Wahba *et al.*, 1986).

In 1966, Summanwar *et al.* reported the association of *Fusarium moniliforme* Sheld. with malformed mango tissues which was subsequently reported from other countries also. Even then for a long time people were hesitant to accept it as the cause of the disease, but after the physiology of pathogenesis was worked out with improved techniques, *F. moniliforme* has been gradually accepted as the cause of the disease. Varma *et al.* (1974) and Kumar and Beniwal (1987) reported *F. moniliforme* var. *subglutinans* Wollenweb. and Reinking to cause both vegetative and floral malformations. Ploetz and Prakash (1997) reported that *Fusarium subglutinans* (Wollenweb. and Reinking) Nelson, Toussoun and Marass comb. nov was the basionym of *F. moniliforme* Sheldon var. *subglutinans* Wollenweb and Reinking: (synonym: *F. moniliforme* Sheldon emend. Snyd and Hans. 'Subglutinans' sensu Snyd., Hans. and Oswald). Chakrabarti and Kumar (1998) proposed to considered *F. moniliforme* Sheldon var. *subglutinans* as a special form and be identified as *F. moniliforme* f. sp. *mangifera*. Recent Zheng and Ploetz (2002) described the isolate of *Fusarium* for mango as member of a new species, *F. mangiferae* Britz, Wingfield and Marasas sp. nov.

Epidemiology

Pathogen Factors

Sporulation

The pathogen *Fusarium moniliforme* var. *subglutinans* was found to colonize the cortex (Chakrabarti and Ghosal 1989, Varma *et al.*, 1974), phloem (Varma *et al.*, 1974a) and parenchymatous pith cells (Bhatnagar and Beniwal 1977, Varma *et al.*, 1974) of the host but the pathogen did not sporulate *in situ*, it did so on necrotic malformed panicle (Varma *et al.*, 1974; Chakrabarti and Ghosal 1989). Pandey (2003) observed conidial germination on emerging buds and colonization in epidermal and cortical cells. Mitra and Lele (1981) recorded maximum sporulation of *F. moniliforme* var. *subglutinans* at 30°C. Rotem (1988) reported that survival of spores was strongly affected by sun light. For conidia production, 35°C temperature and 90 per cent relative humidity (RH) were highly favourable. A hyphae took about 24 h to start conidia production but once production started, subsequent conidia from the same hyphal tip were produced at about 1 h interval and conidia germinated within 5-6 h. Temperature regime of 30°C and high RH (100 per cent) favored germination. Conidia showed maximum viability in the early morning hours; later it was reduced. Noriega-Cantu *et al.* (1999) reported July, October-November and February as the peak periods for conidia production. According to them for highest spore production, high relative humidity and moderate temperature were required. They trapped greater number of air borne micro-conidia by morning hours. Pandey *et al.* (2005) observed that in nature, the peak period of conidia production was July-August.

Growth

Chattopadhya and Nandi (1981) found 27-28°C as suitable temperature for growth of *F. moniliforme* var. *subglutinans in vitro,* which was inhibited at approximately 35°C (Varma *et al.*, 1974). Mitra and

Lele (1981) cultured the fungus at temperature range of 10-40°C, of which 30°C was found best for the growth. Kumar (1992) observed that the fungus isolated from malformed mango panicle grew well at 25-30°C, similar observations were made by Friday *et al.* (1989) and Lee *et al.* (1988). Singh *et al.* (1999) while studying the effect of different temperatures on growth of the fungus over period observed good growth at 25°C and the growth decreased with the decrease in incubation temperature from 10-25°C.

Host Factors

Host Age

It appears from various reports that host age is linked with the type of malformation and the disease severity. Singh and Chakravarty (1935) found the disease to be confined to the younger mango trees. Nirvan (1953) observed that the initial symptoms, mainly of bunchy top type, appeared when the plants were hardly 3-4 years old. Singh *et al.* (1961) reported that plants between the age of 4 to 8 years suffered from vegetative malformation varying from 2.7 to 80.4 per cent, being more pronounced in young plants than the older plants. The intensity of floral malformation on plants of 15 years age varied from 32.12 to 70.01 per cent. Pandey *et al.* (2003) has observed that incidence of vegetative malformation was maximum on 5-year old plants, disease decreased from 5 to 20 year old plants with an average rate of 10.82 per cent per annum and floral malformation was maximum on 10 year old plants. Using polynomial model they concluded that the optimum host age 11.29 years for the floral malformation to reach the peak (26.47 per cent). Similar observations were recorded by Puttarudriah and Channa-Basvanna (1961), Mallik (1963) and Singh *et al.* (1992). Chadha *et al.* (1979) surveyed the incidence of floral malformation on Dashehari plants of different age groups and observed highest per cent age in plants which were 6-10 years of age, with increase in age the disease incidence lowered down. Sharma and Badiyola (1990) observed that disease incidence was highest in 10-year-old trees and decreased with increasing age irrespective of cultivar. Puttoo *et al.* (1975) surveyed the disease incidence among 13 commercially important mango cvs. and recorded the disease incidence ranging from 2.8 per cent in young trees to 100 per cent in trees over 10 years of age. Ram *et al.* (1990) found trees 6-15 years of age were more susceptible than older trees of 16-26 years.

Not only the host age even the age of flowering shoots was reported to influence the incidence of floral malformation, as Shawky *et al.* (1980) observed that in variety Taimur more malformed panicles developed on flowering shoots of spring flush than of summer and autumn flushes of the previous year.

Bearing Habit (Alternate/Regular Bearing Cultivar)

Most of the north Indian popular mango varieties are of irregular bearing habit. They either flower erratically or at one year interval which is known as alternate bearing phenomena. Mallika and Neelum were found highly susceptible to malformation (Kumar and Beniwal, 1992). Neelum which is regular bearer in southern India was severely affected by both off year phenomenon and malformation in the north (Mallik 1963), but the alternate bearing cultivars like Dashehari (Singh and Jawanda, 1961; Sharma, 1953) and Langra (Jagirdar and Shaik 1968) less susceptible. Badliya and Lakhanpal (1990) found that incidence of malformation was highest in Mallika (55.0 per cent) followed by Amrapali (50.0 per cent) and lowest in Langra (4.37 per cent) (Ram *et al.*, 1990). Singh *et al.* (1961) and Kumar and Chakrabarti (1997a) observed that the intensity of malformed panicles was greater in years when the trees bloomed profusely but in 'off' years' when there were fewer inflorescences, the incidence of floral malformation was also less. Thus, during 'on' year there might be lack of initial inoculum level at threshold level, consequently, there was lesser infection. Whereas, in regular bearing cultivars there was steady build up of the inoculum and in every year sufficient fusarial population was available to initiate the disease in the next season.

Time of Flowering (Early/Late Flowering Cultivar)

Among various factors that influence the occurrence and severity of the disease, time of flower bud initiation plays an important role. Nirvan (1953) pointed out that mango varieties that flowered earlier were more susceptible to malformation in comparison to late flowering varieties. Singh *et al.* (1961,) Jawanda (1963) and Jagirdar and Shaik (1968) confirmed that most of the mid and late season varieties such as Chausa and Langra were less susceptible. Majumdar and Sinha (1972) recorded 60 per cent malformed panicles in cultivar Neelum during the flowering of February-March, whereas, the tree had only 4.5 per cent malformation during off season flowering in June. Singh *et al.* (1979) has also reported that early emerging inflorescences were severely affected, whereas, the buds that emerged later escaped the disease. It is presumed that increase in ambient temperature during late emergence of flower buds affected the pathogen as well as the physiology of the host that favoured the disease development. Chadha *et al.* (1979) recorded 84 per cent incidence of floral malformation in early flowering Dashehari while it was only 60 per cent in Chausa which flowered late. Majumdar and Sinha (1972) concluded that during flowering in June the average minimum and maximum temperatures were considerably higher than those of February-March, hence, the disease incidence was less. The varieties from Southern India, where the disease incidence was sporadic, were found to be more severely affected when introduced in north (Mallik 1963). The constant higher temperature particularly absence of severe cold in Southern India kept the malformation in check. The disease is serious in the north-west region where temperature lies between 10 to 15°C during December-January (winter) before flowering. The disease is mild in the areas where temperature lies from 15 to 20°C, sporadic from 20 to 25°C and nil beyond 25°C. However, Prasad *et al.* (1965) had not observed any correlation between time of flowering and disease incidence.

Mangiferin: A Natural Source of Resistance

Mango cultivar Elaichi has been found tolerant towards the malformation syndrome (Misra *et al.*, 2000). The *mangiferin* (a phenolic compound) content in Elaichi was found higher than the susceptible cultivars Beauty Mc-lin, Amrapali and Dsahehari. Its content reaching the peak first in September prior to autumn flushing (new growth) and flower bud differentiation and then in February, before spring flush and complete flowering. Higher level of mangiferin content in Elaichi might be responsible for checking the fusarial attack and correct the hormonal imbalances, as a result of which the symptom of malformation does not express (Singh and Prasad, 2003). The vegetative growth-promotive activity of mangiferin in mango and other angiospermic plants is well documented (Ghosal *et al.*, 1997), the compound has been found very effective in controlling the vegetative malformation in mango saplings (Chakraborty *et al.*, 1994) and also increased IAA content in mango (Chand, 2000). *In vitro* experiment showed that the mangiferin affects the growth of the *Fusarium*, which is a biotic factor for occurrence of malformation (Chakraborty *et al.*, 1994). The antifungal property and curing effect of mangiferin is also reported by Ghosal *et al.* (1979). Singh and Prasad (2003) also observed that a strong inverse relationship between mangiferin content and malformation incidence.

Environmental Factors: Effect on Host Invasion, Disease Incidence and Severity and Host Physiology

Invasion

Koti-Babu and Rao (1998) studied histopathology of malformed shoots and floral apices and observed that the pathogen invaded the buds in October-November. However, in the subsequent summer months (May and June) the apices were without any mycelium of the fungus, but the apical meristems of March-April were greatly invaded with hyphal filaments of the pathogen. Noriega-Cantu *et al.* (1999) also made similar observations reporting that vegetative shoots emerging from June to November were extensively colonized by the pathogen and first symptoms appeared during October-November. In this vegetative state disease incidence had the highest rate of incremental increase and usually these vegetative shoots developed into diseased panicles. Panicles that were infected after this period did not turn into heavily malformed panicles.

Incidence

The incidence of malformation varies considerably from season to season, the highest being in spring flush (Majumdar and Sinha, 1972; Singh *et al.*, 1979; Shawky *et al.*; 1980). The summer growth, even of diseased plants, escapes the infection (Varma *et al.*, 1971; Chand and Chakrabarti, 2000). Majumdar and Sinha (1972) observed lesser malformed inflorescences in summer and autumn flushes in comparison to spring flushes. Shawky *et al.* (1980) noted that panicles on spring shoots were most susceptible, as summer and autumn flushes carried almost the same per cent age of malformed inflorescences (cv. Hindy Be-Sinara) during 'on' year, whereas, in 'off' year, the spring flushes carried more malformed inflorescences.

Severity

The development of maximum number of malformed shootlets was recorded from July to October followed by the period of January to April (Chakrabarti and Kumar, 2000). The disease in the first place appeared as vegetative malformation while in the latter period as floral one. Nirvan (1953) observed that the disease incidence increased rapidly from the month of November to March. Malo and McMillan (1972) observed development of mango malformation in South Florida under very humid climatic condition.

The interaction among *F. moniliforme* population, mangiferin content, RH and temperature with the resultant sequence of development of floral malformation were recorded by Chand (2000). The results showed that from December to January when temperature was higher no symptom appeared on the inoculated buds, whereas, during January-February. Subsequently, the fungal population increased logarithmically and within the next fortnight manifestation of the disease symptoms was complete. During this period temperature was mild (8-19°C) and RH was high (>87 per cent). Since end of February, the average temperature became hotter (16.35°C) and RH went down (<64 per cent), consequently the progress of malformation was affected. Further developed parts of the malformed panicles (during mid March – mid April) escaped the disease and looked normal. The agroclimatic conditions also considerably affected the virulence of the pathogen as well as the expression of the disease symptoms (Chakrabarti and Kumar1999). For example, in Uttar Pradesh, where both summer and winter temperatures were extreme, the pathogen produced malformed shoots as well as panicles. The malformed panicles were mainly of 'heavy type' and survived for a prolonged period (up to June-July). But in eastern state of India, West Bengal, where the temperature fluctuation was very less, the fungus produced predominantly malformed shoots. The malformed panicles which were occasionally developed were of 'light type' and survived only up to March. The fungus under north Indian conditions also showed more virulence.

Host Physiology

Seasonal variation in the incidence of the malformation has been attributed to various factors, such as, amount of photosynthetase in the flowering shoots (Shawky *et al.*, 1980) and effects of high temperature that were reported to increase the femaleness of the flowers or inhibition of the growth of the pathogen (Varma *et al.*, 1971). Singh *et al.* (1998) reported that with increasing atmospheric temperature the internal temperature of leaf also increased which concomitantly altered the physiological attributes resulting in an early emergence of panicle with faster growth and inhibition of floral malformation in them.

The population of the pathogen, *F. moniliforme,* on developing mango buds was reported to be at its peak during spring flush (Chakrabarti *et al.*, 1997; Chakrabarti and Kumar, 1998). This was because content of the anti-fusarial compound, mangiferin, in the host cells remained low during this period, thus causing less resistance to the fungal invasion (Chakrabarti *et al.*, 1997). The carbohydrate content *in situ* during February-March was reported to be maximum (Khan and Khan, 1963) which facilitated the colonization by the fungus. *F. moniliforme* grew optimally at 27-28°C (Varma *et al.*, 1971; Chattopadhyay and Nandi, 1981; Kumar, 1992; Chakrabarti and Kumar, 1999) and at high humidity (Campbell and Marlatt, 1986). In February the fungus got this optimal environmental condition which pushed the fungal density to its highest peak (Kumar *et al.*, 1993; Chakrabarti, 1996; Chakrabarti *et al.*, 1997; Chakrabarti and Kumar, 1998).

Symptoms of this malady seem to be linked with imbalances between growth promoters and inhibitors. Pandey *et al.* (1974) observed higher level of inhibitors and lower levels of all the four fractions of auxins, *viz.*, free neutral free acidic, bound neutral and bound acidic, in malformation panicles of Dusheri mango. They were also found to have lower level of abscisic acid-like inhibitory substances and higher promoter activity than the healthy panicle (Mishra and Dhillon 1978). There were lower level of indole-3-acetic and (IAA), 3-indole acetonitrile (IAN) and gibberellin-like substances and higher level of inhibitors in 'bunchy top' affected than in the healthy tissue (Beniwal *et al.*, 1979). Kumar *et al.* (1980) also found decreased amount of IAA and IAN in malformed than in the healthy tissues. The incidence of malformation also resulted in increased level of gibberellin-like substances, which may account for the production of solely male flowers and continuous growth of the malformed panicles (Mishra and Dhillon, 1980). The malformation stimulated ethylene production in the malformed tissues of mango (Singh and Dhillon, 1989a; 1990a). Malformation-like symptoms are also produced by the synthetic plant regulator, derived from α-hydroxy flurence-9-carboxylic acid (Chadha *et al.*, 1979c, Pal *et al.*, 1983) which cause hormonal imbalance in plants through enzyme systems (Pal *et al.*, 1983). Higher activity of peroxidase and polyphenol oxide in malformed tissues than healthy ones has been observed (Chattopadhyay and Nandi, 1976). The activity of IAA-oxidase peroxidase, polyphenol oxidase in malformed seedlings was 350, 118 and 32 per cent, respectively, which were higher than in healthy seedlings (Beniwal *et al.*, 1979).

The occurrence of malformin like substances in fully grown malformed panicles of mango has been reported (Ram and Bist, 1984). The activity of malformin like substances along with other malformins were detected in malformed panicles at the fully swollen bud, bud inception and fully grown stages prior to full bloom and at full bloom but not in the healthy panicles (Singh, 1986; Singh and Dhillon 1989a; 1993b). Leaves of the twigs with malformed flowers generally had a lower rate of photosynthesis, lower concentration of sugars, starch and chlorophyll and a higher rate of respiration (Yadava and Singh, 1995).

Environment

The disease is serious in the north-west region where temperature ranges between 10-15°C during December-January before flowering, mild in areas where temperature is between 15-20°C, sporadic between 20-25°C and nil beyond 25°C (Puttarudriah and Channabasavanna, 1961). It is widely spread in northern India where temperatures remain low during winter while its occurrence is rare in South

India (Varma *et al.*, 1969) where temperature remain relatively high even in winter. Maximum (20 per cent) floral malformation was recorded at an altitude of 400 m and the disease incidence was almost nil at an altitude of 1250 m and above. The night temperature below 10°C for long duration was found responsible for suppressing the incidence of floral malformation at higher altitudes (Singh *et al.*, 1999). The density of *Fusarium moniliforme* f. sp. *subglutinans* reaches maximum in mango shoot when temperature ranges between 12 to 27°C along with humidity over 80 per cent (Nath *et al.*, 1987; Misra and Singh, 1998). Kumar and Chakraborty (1997) have reported that temperature above 28°C and humidity above 80 per cent favour the growth of *F. moniliforme* f. sp. *subglutinans* but its growth in inhibited at 35°C.

Management Strategies

Sanitation

Narasimhan (1959) was first to attempt sanitation as a tool to control the malformation. Conducting a trial on 20 heavily infected mango trees the malformed panicles were excised 1 to 2 feet below the inflorescence and were burnt. The removal of malformed inflorescences completely freed 15 trees from malformation, while five other developed only one or two malformed panicles. Desai *et al.* (1962) confirmed the results of Narsimhan (1959) when he successfully controlled the disease up to 90 per cent in the cultivar Rajapuri. They pruned the malformed shoots and panicles in July and August following the procedure of Narshimhan. The success of controlling the disease by pruning was reported by others also (Mallik, 1963; Doval *et al.* (1976; Singh *et al.*, 1983; Campbell and Marlatt, 1986; Darvas, 1987; Manicom, 1989). Bindra and Bakhetia (1971) reported that pruning of malformed panicles below 30 cm during July may help in reducing the incidence of malformation. Pandey and Chakraborti (2004) reported that removal of malformed panicles successfully reduced the disease level in the next crop season but in the second year due to an increased rate of disease progress in the pruned plants effects of sanitation was not sustained. Kumar (1992) reported pruning of shoots bearing malformed panicles (two-year-growth) was effective treatment in suppressing the disease satisfactorily. Prasad (1965) failed to confirm the findings of Desai *et al.* (1962), similarly, Varma *et al.* (1971) headed back a severely malformed plant, but in the next flush he observed the malformed shoots to reappear. Deblossoming at bud stage was reported to check the disorder (Singh *et al.*, 1979). Chib *et al.* (1984a) observed that pruning and deblossoming, with or without fungicidal and acaricidal applications were successful in reducing floral malformation in cv. Dashehari. Since 1966, when *Fusarium moniliforme* was reported as causal organism of the disease, inflicting the host injury by colonizing the xylem vessels of the host, it was presumed that systemic fungicides may be the right remedy. Benlate (Varma *et al.*, 1971) and carbendazim (Kumar, 1992) were highly effective against the fungus. found was highly effective under *in vitro* conditions. Later, Mehta *et al.* (1986) and Siddiqui *et al.* (1987) tested the effects of carbendazim under field condition and reported 95 per cent reduction of the disease in cultivar Dashehari and in Chausa. Misra *et al.* (2000) recommended spraying of Bavistin (0.1 per cent) at 15 days interval to control the disease. However, there are many conflicting reports on the efficacy of the systemic fungicides. Sharma and Tiwari (1975) found that Bavistin did not work systemically against mango malformation, though, spraying of the fungicide could reduce the disease. Chadha *et al.* (1979) sprayed Bavistin on malformed Dashehari plants. They did not find any reduction in disease intensity rather the disease level in the treated plats was more than the untreated control. Later Kumar *et al.* (1987) reported the plausible cause of failure of bavistin and benlate in controlling the disease, *in vivo* although it was highly effective against pathogen *in vitro*. It was due to the poor translocation of the fungicide within the plant. The bark of mango contained high amount of tannin which rapidly absorbed about 80 per cent of the fungicide when sprayed over leaves or trunk, thus rendered the compound ineffective. However, recently Chakrabarti *et al.* (2001) got success in using Bavistin as a part of integrated disease management strategy. Pandey and Chakraborti (2004) have reported that the application of carbendazim reduced the rate of disease progress and its effect became pronounced only in the second year. Muhammad *et al.* (1999) observed that Benlate was more effective than Topsin-M to reduce the problem of mango malformation.

Micro-nutrients

Various workers have studied the effects of exogenous application of micro- and macro nutrients on the reduction of the malady. The similarities of symptoms of malformed shootlets and leaflets with those observed in Zn deficient mango saplings (Lynch and Ruckle, 1940) mooted the idea that the disease might be caused by micronutrient deficiency. Saeed and Schlosser (1972) suspected some role of micronutrients in disease development. Tripathi (1955) applied Zn and Cu in the form of sulphate by injecting the trunks and shoots of malformed plants and by foliar spraying but did not find any improvement in the treated plants. But, Minessy *et al.* (1971) could correct the deformity when Fe in chelated form (50-100 g/tree) at the rhizosphere of soil of malformed plant was applied. Abu-El-Daheb (1977) also could check malformed shoots by the application of 50 or 100 g Sequestrene 350 containing 6 per cent metallic Fe. Azzouz *et al.* (1981) in 2-season trials with 18-year-old trees of the cv. Taimour applied Zn, Mn, Cu, B, via foliage. The best reduction of flower malformation was obtained with Zn and Mn (0.3 per cent) each applied in May and September. According to Abdel-Mottaleb *et al.* (1983) the panicle malformation was attributed to the nonavailability of Cu and Zn in subsurface soil layers, especially in sandy soils, and or with poor soil and water management. Mitra and Lele (1981) reported that Cu^{++} did not support growth of the fungus *in vitro* and suppressed sporulation considerably. Pandey *et al.* (1973) reported that the transport system in malformed shootlets was not well developed. Chakrabarti and Ghosal (1982) reported that due to accumulation of mangiferin (natural chelating agent of mango) at the site of infection, the transport system of plant became disrupted. Therefore, better response of micronutrient treatment may be obtained when they were used in form of chelates. Chand *et al.* (2003) tested effectively of three chelating agents (*viz.* mangiferin, EDTA and amino acids) for carrying micronutrients in mango. The best result was obtained when mangiferin was used as chelate since it is the natural metabolite of mango. Bayfolan (containing N, P, K, Fe, Cu, Mn, B, Zn, Co and Mo) at 100 ml/tree applied to the soil in February in an 'on' year reduced the number of panicles, both total and malformed, but had no significant effect on the number of healthy panicles and did not affect the per cent age of malformed panicles (El-Beltagy *et al.*, 1980). Chakrabarti and Ghosal (1989), Chakrabarti *et al.* (2001) and Chand *et al.* (2002) successfully used micronutrient metal chelates as the component of the integrated disease management strategy against malformation. Under *in vitro* test, Pandey and Chakrabarty (2004) reported that in mangiferin Cu^{++} chelate treatment only 14 per cent of spores germinated while in carbendazim and control the germination was 76 and 83 per cent, respectively. The germ tubes of treated conidia were extremely small, shriveled and distorted. Chattopadhay and Nandi (1977) and Gafar *et al.* (1980) completely controlled the disease by spraying of copper oxychloride. However, Chand *et al.* (2003) suggested that micro-nutrient should be used after removal of malformed panicles and shoots vis-a-vis the pathogen from the host surface.

Bioagent

Aspergillus niger has been reported to be natural mycoflora of malformed mango panicles and shoots particularly when they dried up in summer and under go rotting after rains and with increase in population of *A. niger*, the population of *F. moniliforme* var. *subglutinans* declined (Dam, 1992). Ali

(1980) isolated *A. niger* along with *F. moniliforme* var. *subglutinans* and observed antagonistic activity against the *Fusarium*. reported The presence of *A. niger* on malformed shoots has been reported by Rath (1982) and Noriega-Cantu *et al.* (1999) also. Pandey and Chakrabarti (2004) reported that in field, application *A. niger* has been effective in bringing down the inoculum level on malformed dead panicles. Michel-Aceves *et al.* (2001) suggested that *Trichoderma* species isolated from soil in mango orchards represent a resource for evaluation as biocontrol agents against mango malformation.

Plant Growth Regulators and Phenolic Compound

Spray of NAA (100 to 200 ppm) during the first week of October has been worked out to reduce floral malformation (Majumdar *et al.*, 1970, Mishra, 1976, Chadha *et al.*, 1979). Later, Mishra and Dhilon (1979) suggested four sprays of NAA at weekly intervals from 20th October onwards to reduce the intensity of malformation to the maximum possible limits. Singh and Dhillon (1986) claimed that spraying of 100 ppm NAA in first week of October was more effective. Hence, spray of 100 ppm NAA was suggested instead of 200 ppm. They also observed that spraying 200 ppm NAA in the first week of October followed by spraying of 500 ppm of ethrel at bud inception stage (February) was highly effective in reducing floral malformation. Galla and EI-Masry (1991) observed increase in floral branches and yield by the spraying of ethrel (250 or 500 ppm), or uniconazole 500 ppm in the month of December. EI-Beltagy *et al.* (1980) and Das *et al.* (1989) observed that gibberellic acid at 50 ppm caused delayed panicle emergence, increased number of perfect flowers and increased pollen viability. According to Das *et al.* (1989) sprays in the months of October and November caused 50 per cent reduction in malformation. Dhillon and Singh (1989) observed that spraying of 1000 ppm Paclobutrazol (10-60 g/tree), prior to FBD, during the first week of October, reduced malformation, increased number of healthy flowers and increased the yield. Tahir *et al.* (2002) observe the effect of three growth retardants *i.e.* paclobutrazol, alar and cycocel, out of three paclobutrazal was found more effective in which least number of growth flushes were observed. As a result of decreased number of flushes, the production of malformed panicles was also minimized.

Deblossoming

Deblossoming at the bud burst stage alone and in combination with the spraying of 200 ppm NAA was very effective. Deblossoming alone at bud brust stage gives substantial reduction in malformation, and alleviates panicle malformation (Singh, 1986; Singh and Dhillon, 1988). However, Bains (1989) reported that there was no effect of delossoming in reducing the intensity of malformation. Chadha *et al.* (1979a) reported application of 200 and 500 ppm ethrel to get complete control. Pal and Chadha (1982) deblossomed the panicle by applying 250 ppm of cycloheximide.

Anti-malformins: Ram and Bist (1984) reported 2240 ppm glutathione and 2110 ppm ascorbic acid to be very effective, as malformed panicles turned healthy after 15 days of application. The chemicals may be applied three times after appearance of malformation. Bist and Ram (1986a) reported the application of above compounds, when the panicles are 4-6 cm long with the same results as reported by them in 1984.

References

Abdel-Mottaleb, M., Ibrahim, A.E.M. and Bayoumi, N.A. (1983). Preliminary study on the relation between soil properties and mango malformation. *Mihufiya J. Agric. Res.*, 7: 395–406.

Abou-EL-Daheb, M.K. (1977). Correcting malformation symptoms of mango trees in Egypt by soil application of iron chelates. *Egypt. J. Phytopath.*, 7: 97–99.

Ali, Z.A. (1980). Fungi and bacteria as antagonists against *Fusarium* species on tomato, pepper and cucumber. *Rastiteuna Zashchita.*, 28: 18–21.

Annonymous (1996). *Annual Research Report*, CIHNP, Lucknow.

Anonymous (1983). *Mango Cultivation*. CMRS, IIHR, Bangalore, India, Extension Bull., 9: 21.

Attiah, H.H. (1955). A new eriophyid mite on mango from Egypt. *Bull. Sci. Ent. Egypt*, 30 : 379–383.

Azzouz, S., Hamdy, Z.M. and Dahshan, I.M. (1981). Studies on malformation inflorescence of mango, the degree of susceptibility among different varieties. *Agric. Res. Review*, 56: 17–27.

Badliya, S.D. and Lakhanpal, S.C. (1990). Reaction of some mango cultivars to floral malformation under Poanta valley conditions of Himachal Pradesh. *South Indian Hortic.*, 38: 152.

Bains, H.S. (1989). Studies on control of floral malformation in mango CV. Dashehari. *M.Sc. Thesis*, PAU, Ludhiana.

Beniwal, S.P.S., Bhatnagar, S.S. and Shabey, S.N. (1979). Studies on causes of mango malformation. *Res. Report. Mango Workers Meeting*, Goa, p. 410–414.

Bhatnagar, S.S. and Beniwal, S.P.S. (1977). Involvement of *Fusarium oxysporum* in causation of mango malformation. *Plant Dis. Report.*, 61: 894–898.

Bindra, O.S. and Bakhetia, D.R.C. (1971). Investigations on the etiology and control of mango malformation. *Indian J. Hortic.*, 28: 80–85.

Bist, L.D. and Ram, S. (1986) Effect of some antimalformins and growth regulators on the control of floral malformation in mango. *Indian J. Hort.*, 42: 161–174.

Burham, M.J. (1991). Mango malformation disease recorded in United Arab Emirates. *FAO Plant Protection Bull.*, 39: 46–47.

Burns, W. (1910). A common malformation on mango inflorescence. *Poona Agric. College Magazine*, 2: 38.

Campbell, C.W. and Marlatt, R.B. (1986). Current status of mango malformation disease in Florida. *Pro. Interam. Soc. Trop. Hortic.*, 30: 223–226.

Chadha, K.L., Pal, R.N., Prakash, O., Tandon, P.L. and Singh, H. (1979a) Studies on mango malformation, its causes and control. *Indian J. Hort.*, 36: 359–368.

Chadha, K.L., Sahay, R.K. and Pal, R.N. (1979b). Induction of floral malformation like symptoms on mango by a morphactine. *Indian J. Hort.*, 36: 238–244.

Chakrabarti, D.K, Kumar, R., Kumud and Kumar S. (1997a). Interaction among *Fusarium moniliforme, Tyrolichus casei* and mangiferin as related to malformation of *Mangifera indica.Trop. Agric.*, 74: 317–320.

Chakrabarti, D.K, Prasad, J. and Singh K. (1994). Effect of mangiferin on vegetative growth of mango. *Indian J. Hortic.*, 6: 51–53.

Chakrabarti, D.K. (1996). Etiology and remedy of mango malformation. In: *Disease Scenario in Crop Plants: Fruits and Vegetables*, Vol. 1, (Eds.) V.P. Agnihotri, Om Prakash, Ram Kishun and A.K. Misra. International Books and Periodicals supply Service, Delhi, pp. 49–59.

Chakrabarti, D.K. and Ghosal, S. (1982). Control of malformation disease of mango incited by *Fusarium moniliforme* var. *subglutinans* with mangiferin metal chelates. *Proc. 69th Indian Sci. Cong Pt. III* (Ag. Sci.), (ISCA Young Sci. Award Lect.), Mysore, pp. 70–71.

Chakrabarti, D.K. and Ghosal, S. (1989). Disease cycle of mango malformation induced by *Fusarium maniliforme* var. *subglutinans* and the curative effects of mangiferin metal chelates. *J. Phytopathol.*, 125: 238–246.

Chakrabarti, D.K. and Kumar, R. (1998). Mango malformation: role of *Fusarium moniliforme* and mangiferin. *Agric. Rev.*, 19: 126–136.

Chakrabarti, D.K. and Kumar, R. (1999). Effects of agro-climatic condition on floral malformation of mango and its pathogen, *Fusarium moniliforme* Sheld. *Sci. and Cult.*, 65: 11–12.

Chakrabarti, D.K. and Kumar, R. (2000). Epidemiological principles of control on mango malformation: a review *Agric. Rev.*, 21: 129–132.

Chakrabarti, D.K. and Kumar, R. and Ali, S. (2001). An integrated management strategy for mango malformation *Proc. Int. Conf. on Integrated Plant Dis. Management for Sustainable Agri.*, IARI, New Delhi, Nov. 10–15, 1997, Vol. II, pp. 753–754.

Chakrabarti, D.K., Kumar, R. and Ali, S. (1997). Mango Malformation seasonal variation in *Fusarium moniliforme* population in relation to environmental factors, mangiferin content and flushings in *Mangifera indica* L. *Indian J. Plant. Prot.*, 25: 146–148.

Chand, G. and Chakrabarti, D.K. (2000). Techniques to reproduce floral malformation in mango (*Mangifera indica* L.). *Indian J. Mycol. Pl. Pathol.*, 30: 296.

Chand, G. (2000). Studies on mango malformation induction and its management through chelates. *M.Sc. (Ag.) Thesis*, N.D. University of Agriculture and Technology, Faizabad.

Chand, G., Kumar, R. and Chakrabarti, D.K. (2002). Mango malformation: Chelated micro-nutrients in the disease management. *Indian J. Mycol. Plant Pathol.*, 32: 144.

Chand, G., Pandey, M. K. and Chakrabarti, D.K. (2003). Role of metal chelates in integrated managemant of mango (*Mangifera indica*) malformation. *Indian J. Agric. Sci.*, 72: 612–614 (in press).

Chattopadhya, N.C. and Nandi, B. (1977) Chemical control of malformation of mango. *Indian Agriculturist*, 21: 217–221.

Chattopadhyay, N.C. and Nandi, B. (1976). Peroxidase and polyphenol oxidase activity in malformation mango inflorescence caused by *Fusarium moniliforme* var. *subglutinans*. *Biologia Plant*, 18: 321–336.

Chattopadhya, N.C. and Nandi, B. (1981). Nutrition in *Fusarium moniliforme* var. *subglutinans* causing mango malformation. *Mycologia*, 73(3): 414–407.

Chib, H.S., Andotra, P.S. and Gupta, B.R. (1984). Survey report on the incidence and extent of mango malformation in mango growing areas of Jammu and Kashmir state. *Indian J. Mycol. Plant Pathol.*, 14: 86–88.

Chib, H.S., Andotra, P.S. and Gupta, B.R. (1984a). Studies on chemical and cultural control of mango malformation. *Research and Development Reptr.*, 3: 121–124.

Dam, S. K. (1992). Studies on interaction between *Fusarium moniliforme* and *Aspergillus niger* and its effects on development of malformation disease of mango. *M.Sc. Ag. Thesis*, N.D. University of Agriculture and Technology, Faizabad.

Darvas, J.M. (1987) Control of mango blossom malformation with trunk injection. *S. Afr. Mango Growers' Assoc. Year Book*, 7: 21–24.

Das, G.C., Rao, D.P. and Lenko, P.C. (1989). Studies on mango malformation in mango clone–Chiratpudi. *Acta Hort.*, 231: 866–872.

Desai, M.V., Patel, V.K., Patel, M.K. and Thrumalachars, M.J. (1962). Control of mango malformation in Gujarat. *Curr. Sci.*, 31: 392–394.

Dhillon, B.S. and Singh, Z. (1989). Depletion of indole-2-acetic acid in malformed tissue of mango, *Mangifera indica* L. and its alleviation. *Acta Hort.*, 239: 371–374.

Doval, S.K., Kaul, C.K. and Mathur, B.P. (1976). Note on control of mango malformation. *Indian J. Agric. Sci.*, 46: 545–546.

EL-Beltagy, M.S., EL-Ghandour, M.A. and EL-Hanawi (1980). Effect of bayfolan and some growth regulators on modifying flowering and the incidence of flowering malformation of mango. *Mangifera indica* L. *Egypt. J. Hortic.*, 6: 125–130.

Friday, D., Tutie, J. and Stroshine, R. (1989). Effect of hybrid and physical damage on mold development and carbondioxide production during storage of high moisture shelled corn. *Cereal Chem.*, 66: 422–426.

Gafar, K., Abd-EL-Monem, S. and Abd-EL-Megid Mv. (1980). Pathological studies on the phenomenon of mango fruit dropping. *Agric. Res. Rev.*, 57: 11–17.

Galila, A.S. and El-Masry, H.M. (1991). Effect of ethrel and S–3307 (a new plant growth retardant) on flowering, malformation, fruit set and yield of Ewais mango cv. *Annals of Agric. Sci. (Cairo)*, 36(2): 645–650.

Garg, D.N. (1951). Some new disease of economic plants in U.P. *Agric. Animal Husbandry*, 1: 12–14.

Ghosal, S., Chakrabarti, D.K. and Kumar, Y. (1979). Toxic substances produced by *Fusarium* Concerning malformation disease of mango. *Experientia*, 35: 1633–1634.

Hassan, A.S. (1944). Note of Eriophyis mangiferae (Acaria). *Bull. Soc. Found. Ent.*, 28: 179–180.

Hifny, H.A.A., El-Barkouki, M. and El-Banna, G.S. (1978). Morphology and physical aspect of the floral malformation in mangoes. *Egypt. J. Hort.*, 5: 43–52.

Jagirdar, S.A.P. and Shaik, M.R. (1968). Control and malformation of mango inflorescences. *Souvenir Mango and Summer Fruit Show*, Mirpur Khas, West Pakistan.

Jawanda, J.S. (1963). Studies on mango malformation. *Punjab Hortic. J.*, 3: 281–285.

Khan, M.D. and Khan, A.H. 1962. Relation of growth to malformation of inflorescence in mangoes, West Pakistan. *Punjab Hort. J.*, 3: 229.

Khan, M.D. and Khan, A.H. 1960. Studies on malformation of mango inflorescence in West Pakistan. *Punjab Fruit J.*, 23: 247–258.

Khan, M.D. and Khan, A.H. 1963. Some chemical studies on malformation of mango influences in West Pakistan. *Punjab Hortic. J.*, 3: 229–234.

Kishtah, A.A., Nyland, G., Nasr El-Din, T.M., Tolba, M.A. and Lowe, S.K. (1985a). Mango malformation disease in Egypt. I. Electron microscopy, effect of antibiotics, cultural and serological studies. *Egypt J. Phytopathol.*, 17: 151–157.

Kishtah, A.A., Nyland, G., Nasr El-Din, T.M., Tolba, M.A. and Lowe, S.K. (1985b). Mango malformation disease in Egypt. II. Studies on transmission and thermotherapy. *Egypt. J. Phytopathol.*, 17: 159–164.

Koti Babu, A.C. M. and Rao, K.S. (1998). Mango malformation. histopathological and histochemical studies on shoot and floral apices. *Indian Phytopathol.*, 51: 349–352.

Kumar, J. (1992). Mango malformation: Current status and future prospects. In: *4th International Mango Symp.*, Miami, 5–10th July, pp. 133 (Abst.) p. 17.

Kumar, J., Beniwal, S.P.S. and Ram, S. 1980. Depletion of auxins in mango seedlings affected with bunchy top stage of mango malformation. *Indian J. Exp. Biol.*, 18: 286–289.

Kumar, J. and Beniwal, S.P.S. (1987). Vegetative and floral malformation: Two symptoms of the same disease of mango. *FAO Plant Prot. Bull.*, 35: 21–23.

Kumar, J. and Beniwal, S.P.S. (1992). Mango malformation. In: *Plant Disease International Importance. Disease of Fruit Crops*, Volume III, (Eds.) J. Kumar, H.S. Chaube, U.S. Singh and A.N. Mukhopadhyay. Prentice Hall, New Jersey, pp. 357–393.

Kumar, J., Singh, U.S. and Beniwal, S.P.S. (1987). Binding of carbendzim by lignin. *Indian J. Mycol. Plant Pathol.*, 17: 24.

Kumar, J., Singh, U.S. and Beniwal, S.P.S. (1993). Mango malformation: One hundred years of research. *Annu. Rev. Phytopathol.*, 31: 217–232.

Kumar, R. (1992). Studies on Host Specificity of *Fusarium moniliforme* Sheld. the incitant of malformation disease of mango (*Mangifera indica L.*). *M Sc (Ag.) Thesis*, N.D. University of Agriculture and Technology, Faizabad.

Kumar, R. and Chakrabarti, D.K. (1997). Assesment of loss of mango (*Mangifera indica*) caused by mango malformation. *Indian J. Agric. Sci.*,67: 130–131.

Kumar, R. and Chakrabarti, D.K. (1997a). Spatial patterns of spread of floral malformation of mango. *Acta Hortic.*, 455: 600–608.

Lee, Y.W., Kim, K.H. and Chung, H.S. (1988) Toxicity of *Fusarium* isolates obtained from corn producing area in Korea. *Korean J. Plant Pathol.*, 4: 40–48.

Lynch, J.L., Ruehle, G.D. (1940). Little leaf of mango: A zinc deficiency. *Proc. Florida State Hortic. Soc.*, 53: 167–169.

Majumdar, P.K. and Sharma, D.K. (1990). In: *Fruit: Tropical and Sub-tropical*, (Ed.) T.K. Bose. Naya Prakash, Calcutta, pp. 1–62.

Majumdar, P.K. and Sinha, G.C. (1972). Seasonal variation in the incidence of malformation in *Mangifera indica. Acta Hortic.*,. 24: 221–223.

Majumdar, P.K., Sinha, G.C. and Singh, R.N. (1970). Effect of exogenous application of α-naphthyl acetic acid on mango (*Mangifera indica* L.) malformation. *Indian J. Hortic.*, 27: 130–131.

Mallik, P.C. (1963). Mango malformation, symptoms, cause and cure. *Punjab Hortic. J.*, 3: 292–299.

Malo, S.E. and McMillan, R.T.J. (1972). A disease of *Mangifera indica* L. in Florida similar to mango malformation. *Proc. Florida State Hortic. Soc.*, 85: 264–268.

Manicom, B.Q. (1989). Blossom malformation of mango. S. Af. *Mango Grower's Assoc. Year Book.*, 10: 11–12.

Meah, B. and Khan, A.A. (1992). Mango diseases in Bagladesh. In: *International Mango Symp. Miami*, 4th–10th July, 133 pp. (Abst.) p. 20.

Mehta, N., Sandooja, J. K., Madaan, R.I. and Yamdagni, R. (1986). Role of different chemicals in mango malformation and related physiological factors. *Pesticides*, 20: 17–18.

Minessy, F.A., Biely, M.P. and El-Fahl, A. (1971). Effect of iron chelates in correcting malformation of terminal bud growth in mango. *Sudan Agric. J.*, 6: 71–74.

Mishra, K.A. and Dhillon, B.S. (1978). Levels of abscissic acid like substances in healthy and malformed panicles of mango *Mangifera indica* L. *Science and Culture*, 44: 419–420.

Mishra, K.A. and Dhillon, B.S. (1980). Levels of endogenous gibberellins in healthy and malformed panicles of mango (*Mangifera indica* L). *Science and Culture*, 44: 419–420.

Mishra, K.A. (1976). Studies on bearing behaviour of *Mangifera indica* L. and its malformation. *Ph.D. Thesis*, Punjab Agric. Uni., Ludhiana, India.

Misra, A.K. and Singh, V.K. (1998). *Fusarium moniliforme* var. *subgliutinans* in relation to mango malformation. *Proc. Natl. Symp. on Mango Production and Export*, June 25–27, pp. 71.

Misra, A.K. Pandey, P. and Singh, V.K. (2000). Mango malformation. In: *Advances in Plant Disease Management* (Eds.)U. Narain, K. Kumar and M. Srivastva. Advance Pub. Concept, Delhi, pp. 185–214.

Mitra, A. and Lele, V.C. (1981). Morphological and nutritional studies on mango malformation fungus *Fusarium moniliforme* var. *subgliutinans*. *Indian Phytopathol.*, 34: 475–483.

Narasimhan, M.J. (1959). Control of mango malformation disease. *Curr. Sci.*, 28: 254–255..

Nath, R., Kamalwanshi, R.S. and Sachan, I.P. (1987). Studies on mango malformation. *Indian J. Mycol. Plant Pathol.*, 17: 29–33.

Nirvan, R.S. (1953). Bunchy top of young mango seedlings. *Sci. and Cult.*, 18: 335–336.

Noriega-Cantu, D.H., Teliz, D., Mora-Auguilra, G., Rodriguez-Aleazar, J., Zavaleta-Mejia, E., Otero-colinas, G. and Campbell, C.L. (1999). Epidemiology of mango malformation in Guerrero, Mexico with traditional and integrated management. *Plant Dis.*, 83: 223–228.

Pal, R.N. and Chadha, K.L. (1982). Deblossoming mangoes with cycloheximide. *J. Hort. Sci.*, 57: 271–277.

Pal, R.N., Kalra, S.K., Tandon, D.K. and Chadha, K.L. (1983). Activity of IAA oxidase, catalase and amylase in morphactin-induced malformations of mango inflorescence. *Scientia Hort.*, 19: 271–277.

Pandey, M.K. (2003). Some aspects of epidemiology and principles of control of mango malformation. *Ph.D. Thesis*, N.D. Uni. Agric. Tech., Faizabad, UP.

Pandey, M.K. and Chakrabarti, D.K. (2004) Management of malformation of mango (*Mangifera indica* L.). *J. Mycol. Pl.Pathol.*, 34(3): 881–884.

Pandey, R.M., Singh, R.N. and Sinha, G.C. (1973). Usefulnes of ethrel in regulating flower bearing in mango. *Sci. and Cult.*, 39(3): 148–150.

Pandey, R.M., Rathore, D.S. and Singh, R.N. (1974). Hormonal regulation of mango malformation. *Curr. Sci.*, 43: 694–695.

Pandey, M.K., Chakrabarti, D.K. and Kumar, S. (2003) Analysing mango (*Mangifera indica*) malformation in relation to the host age. *Indian J. Agric. Sci.,*: 73 (7) 395–396.

Pandey, M.K., Chakrabarti, D.K. and Kumar, S. (2005). Production and germination of conidia of *Fusarium moniliforme* var. *subglutinans* incitant of malformation in mango (*Mangifera indica*). *J. Mycol. Pl. Pathol.*, 35 (1): 163–166.

Ploetz, R.C. and Prakash, O. (1997). Foliar, floral and soil borne disease. In: *The Mango: Botany, Production and Uses*, (Ed.) R.E. Litz. CAB International, U.K., pp. 297–325.

Prasad, A., Singh, H. and Shukla, T.N. (1965). Present status of mango malformation disease. *Indian J. Hortic.*, 22: 254–265.

Prasad, A., Singh, H. and Shukla, T.M. (1965). Present status of mango malformation disease. *Indian J. Hortic.*, 22: 254–265.

Puttarudriah, M. and Channa-Basavanna, G.P. (1961). Mango bunchy-top and eriophyid mite. *Curr. Sci.*, 30: 114–115.

Puttoo B.L., Gupta, B.K. and Singh, Harbajan (1975).Extent of mango maformation in Jammu. *Indian J. Mycol. Plant. Pathol.*, 5: 181–182.

Ram S. and Bist, L.D. (1984). Occurrence of malformin-like substances in malformed panicles and control of floral malformation in mango. *Scientia Hort.*, 23: 333–336.

Ram, S., Singh, B.P. and Singh, S.P. (1990). Studies on malformation of mango *Mangifera indica* L. inflorescence with reference to varieties and age of the trees. *Hortic. J.*, 31: 31–36.

Rath, G.C., Swain, N.C. and Mohanan, M.K. (1982). A note on die back of mango in Orissa. *Indian Phytopathol.*, 31: 384–386.

Rotem, J. (1988). Climatic and weather influences on epidemics. In: *Plant Disease: An Advanced Treatise*, Vol. II, (Ed.) J.K. Horsfall and E.B. Cowling. Academic Press, New York, pp.19–31.

Saeed, A. and Schlosser, E. (1972). Effect of some cultural practices on the incidence of mango malformation, *Z. Pflazenkrankh and Pflanzenzchuz*, 79: 349–351.

Scholosser, E. (1971). Mango malformation: Symptoms, occurrence and varietal susceptibility. *FAO Plant Prot. Bull.*, 19: 12–14.

Schwartz, A. 1968. A new mango pest. *Farming in S. Africa*, 9: 7.

Sharma, B.B. (1953). Studies in the disease of *Mangifera indica*. In: *Proc. 40th Indian Sci. Cong.*, Part IV, Abst. p. 70.

Sharma, I.M. and Badiyala, S.D. (1990). Incidence of mango malformation in different locations of Himachal Pradesh. *Indian J. Mycol. Plant. Pathol.*, 20: 179–181.

Sharma, O.P. and Tiwari, A. (1975). Studies on mango malformation. *Pesticides*, 9: 44–45.

Shawky, I., Zidan, Z., El-Tomi, A. and Dahshan, A.D.J. (1980). Flowering in relation to vegetative growth of Taimour mangoes. *Egypt. J. Hortic.*, 7: 1–8

Siddiqui, S., Sandooja, J.K., Mehta, N. and Yamadagni, R. (1987). Biochemical changes during malformation in mango cultivars as influenced by various chemicals. *Pesticides*, 21: 17–19.

Singh, Zora and Dhillon, B.S. (1986). Effect of plant regulators in floral malformation, flowering, productivity and fruit quality of mango (*Mangifera indica* L.). *Acta Hort.*, 174: 315–319.

Singh, Zora and Dhillon, B.S. (1989). Relationship of endogenous and exogenous ethylene with floral malformation of mango (*Mangifera indica* L.). *Acta Hort.*, 231: 367–370.

Singh, Zora and Dhillon, B.S. (1993). Mango malformation. In: *Advances in Hort*iculture*, Vol. 4: Fruit Crops, Part–4,* (Eds.) K.L. Chadha and O.P. Pareek. Malhotra Pub. House, New Delhi, India.

Singh, Zora and Dhillon, B.S. (1988). Influence of deblossoming on floral malformation of mango (*Mangifera indica* L.). Tropical Agric., 65: 295–296.

Singh, Zora and Dhillon, B.S. (1990). Floral malformation, yield and fruit quality of *Mangifera indica L.* in relation to ethylene. *J. Hort. Sci.,* 65: 215–220.

Singh,V.K. and Prasad, S. (2003). Mangiferin: A natural source of resistance to mango malformation. *Indian J. Plant Physiol. (Special Issue),* p. 508–511.

Singh, L.b., Singh, S.M. and Nirvan, R.S. (1961). Studies on mango malformation: Review, symptoms, extent, intensity and cause. *Hort. Adance,* 5: 197.

Singh, L., Bajwa, S.S.S., Bal, S. and Khan, A.A. (1940). Mangoes. *Punjab Fruit J.,* 4: 13, 678.

Singh, B.N. and Chakravarty, S.C. (1935). Observations on a disease of mango at Banaras. *Sci. and Cult.,* 1: 294–295.

Singh, D.S., Pathak, P.A. and Singh, R.D. (1983). Studies on control of malformation in mango cv. Bombay Green. *Punjab Hortic. J.,* 23: 220–221.

Singh, J. N., Rajput, C.B.S. and Maurya, A.N. (1992). Survey of the intensity on mango malformation in Varansi area. *Prog. Hortic.,* 24: 9–12.

Singh, K. and Jawanda, J.S. (1961). Malformation in mangoes. *Punjab Hortic. J.,* 1: 18–22.

Singh, U.R., Dhar L. and Gupta, J.H. (1979). Note on the effect of time of bud burst on the incidence of floral malformation in mango. *Prog. Hortic.,* 11: 41–43.

Singh, V.K., Saini, J.P. and Mishra, A.K. (1998). Mango malformation in relation to physiological parameters under elevated temperature. *Indian J. Plant Physiol,* 3: 231–233.

Singh, V.K., Saini, J.P., Misra, A.K. and Pandey, D. (1999). Role of temperature on the occurrence of mango floral malformation in Kumaon hills. *Biol. Memoirs,* 25: 47–49.

Sattar, A. (1946). Disease of mango in Punjab. *Punjab Farmer J.,* 10: 56–58.

Summanwar, A.S., Raychaudhuri, S.P. and Pathak, S.C. (1966). Association of fungus *Fusarium maniliforme* Sheld with the malformation in mango. *Indian Phytopathol.,* 19: 227–228.

Tripathi, R.D. (1955). Malformation disease of the mango as related to deficiency of mineral nutrients. *Indian J. Hortic.,* 12: 173–179.

Varma, A., Lele, V.C., Majumdar, P.K., Asha Ram, Sachidananda, J., Shukla, U.S., Sinha, G.C., Yadava, T.D. and Raychaudhuri, S.P., 1969. Mango malformation. *Proc. ICAR Workshop on Fruit Research,* Ludhiana, April, 28–30, 1969, pp. 1.

Varma, A., Lele, V.C. and Goswami, B.K. (1974). Mango malformation. In: *Current Trends in Plant Pathology,* (Eds.) S.P. Raychaudhuri and J.P. Varma. Lucknow University, Lucknow, pp. 196–206.

Varma, A., Lele, V.C., Raychaudhuri, S.P., Ram, A. and Sang, A. (1974a). Mango malformation: A fungal disease. *Phytopath. Z.,* 79: 254–257.

Varma, A., Raychaudhuri, S.P. Lele, V.C. and Ram, A. (1971). Preliminary investigation on epidemiology and control of mango malformation. *Proc. Indian Nat. Sci. Acad.,* 37B: 291–300.

Watt,G. (1891). Dictionary of the economic products of India. *The Mango Tree*, 5(1): 149.

Yadava, R.B.R. and Singh, V.K. (1995). Extent of floral malformation and itsrelationship with physico-chemical components in mango cvs. *Indian J. Plant Physiol.*, 38(4): 328–330.

Zheng, Q. and Ploetz, R. (2002). Genetic diversity in the mango malformation pathogen and development of a PCR assay. *Plant Pathol.*, 51: 208–216.

Uttar Pradesh Council of Agricultural Research (1993). A News item in "*Dainik Jagaran* (Vernacular), Lucknow, April 15, pp. 3.

Plant Diseases Management in Horticultural Crops (2011) *Pages* ***81–93***
Editors: **Shahid Ahamad, Ali Anwar and P.K. Sharma**
Published by: **DAYA PUBLISHING HOUSE, NEW DELHI**

Chapter 7

Management of Alternaria Leaf Spot Disease of Cabbage

Mohd. Nawaz Mir, M. Azam Wani, Nasreen Fatima, Shahzad Ahmad and Ali Anwar

Division of Plant Pathology,
S.K.University of Agricultural Sciences and Technology of Kashmir,
Shalimar, Srinagar – 191 121, J&K

Vegetables constitute a major portion of human diet. According to the international dietary standards, human diet must be included 250 to 300 g vegetables per day, out of which 150 to 200 g should be leafy materials. Cabbage (*Brassica oleracea* var. *capitata* L.) is one of the most important vegetable crops and belongs to family cruciferae.

Tremendously cabbage is grown throughout the world. India is one of the important cabbage growing countries in Asia covering an area of 245.4 thousand hectares with an annual production of 5617.1 thousand tonnes (Anonymous, 2002). Cabbage is grown in almost all vegetable growing areas of Kashmir valley covering an area of 0.5 thousand hectares with an annual production of 17 thousand metric tonnes (Anonymous, 2004).

It is mostly used as culinary and dietic article, salad, pickling, boiled vegetables, cooked in curries, dehydrated vegetables and can also be used for feeding live stock and chicken as well. The taste in cabbage is due to "Sinigrin glucoside" which is purgative. Nutritionally cabbage is a rich vegetable owing to the fact that each 100 g of its edible portion contains moisture 91.9 per cent, 4.6 per cent carbohydrates, 1.8 per cent protein, 0.1 per cent fibre and 0.6 per cent minerals. It also contains vitamins A, B, B_2 and C_1. Young and tender leaves are rich in vitamin A as compared with old leaves (Singh *et al.*, 2004). Besides being highly nutritious, cabbage has also an appreciable medicinal value, as it helps to prevent constipation, increases appetite, speed up digestion and is very useful for patients of diabetes.

Production Constraints

Being perishable, shortage of vegetables can not be met by imports. This emphasizes the need for increasing the production and availability from existing area. Inspite of favourable environmental conditions for cabbage cultivation in the valley, the crop is affected by a number of fungal, bacterial and viral diseases of which leaf spot (*Alternaria* spp.), downy mildew (*Perenospora parasitica*), club root (*Plasmodiophora brassica*), black leg (*Phoma lingum*), white rust (*Albugo candida*), black rot (*Xanthomonas* sp.) etc. drastically reduce the yield of the crop. Among these diseases Alternaria leaf spot caused by *Alternaria brassicae* and *Alternaria brassicicola* is of serious nature resulting 40-70 per cent yield losses from many parts of the world (Degenhardth *et al.*, 1974).

Occurrence and Status of the Disease

Alternaria leaf blight of cabbage is world wide in occurrence and has been reported from Australia, Bangladesh, China, England, Germany, India, Pakistan, Sri-lanka and Taiwan (Duhan and Suhag, 1989; Humpherson and Phelp's 1989; Fazal *et al.*, 1994). Humpherson (1983) reported that Alternaria leaf blight might be caused up to 100 per cent pod infection on *Brassica oleracaea*. The yield losses in India ranging from 10-70 per cent due to Alternaria leaf blight (Kolte, 1982). Maude and Humperson (1983) found that *Alternaria brassicicola* was the most prevalent pathogen attacking *Brassica oleracea* crops, whereas, Duhan and Suhag (1989) reported that the occurrence of *Alternaira brassicae* was more frequent than *Alternaria brassicicola*. Humpeson (1983) found that 88 per cent of *Brassica oleracea*, seed samples were infected with *Alternaria brassicicola* and 55 per cent with *Alternaria brassicae*. Alternaria leaf spot of cabbage has been observed frequently in field (Plate 7.1) at some important vegetable growing areas of Kashmir valley. Since the disease can become extensively destructive under favourable conditions and render large per cent age of cabbage plants incapable of producing marketable heads (Mir, 2007).

Causal Organisms and Morphology

Much of the past research has been devoted in distinguishing two species *i.e.*, *Alternaria. brassicae* and *Alternaria brassicicola* on the basis of their morphology (Plates 7.5 and 7.6), pathogencity and symptoms expression (Plate 7.3). Later the taxonomic confusion between *Alternaria brassicicola* and *Alternaria brassicae* was distinguished by Wiltshire (1947) and Changastri and Weber (1963) who concluded that absence of conidial beak in the former distinguishes it from the latter. Besides *Alternaria brassicicola* forms darker lesions in contrast to pale brown lesions by *Alternaria brassicae* and produce sooty black colony, while *Alternaria brassicae* forms white mycelial growth on potato-dextrose agar.

Comparative morphological characteristics of *Alternaria brassicae* and *Alternaria brassicicola* revealed that septate mycelium of *Alternaria brassicae* was brown in colour, whereas, in *Alternaria brassicicola*, it was greenish grey and becomes dark olive on ageing. The length of *Alternaria brassicicola* spores measured 45-55 µm which was almost half the spore length of *Alternaria brassicae i.e.*, 98-114 µm (Koul, 1996; Changstri and Weber, 1963;). Mir, 2007 has been reported that colonies of *Alternaria brassicicola* were usually dark blackish brown, conidiophores singly or in groups, straight or flexuous, pale to mid olivaceous brown and measured 34-76 x 6-10 µm with one or several conidial scars. Conidia were often in chains and measured 40-62 x 11-20 µm in size. Whereas, the hyphae of *Alternaria brassicae* were septate, hyaline to mid olive grey, conidiophores were mid pale to olive grey and measured 29-236 µm in length and 8-17 µm in width, conidia obclavate, pale to olive grey and measured 63-172 x 10-21 µm in size. *Alternaria brassicae* isolates from diverse agro-climatic zones of India has been studied.

Healthy Crop

Alternaria blighted

Plate 7.1: Cabbage (*Brassica oleracea* var. *capitata*) Plantation

Plate 7.2: Categorization of Alternaria Blighted Leaves of Cabbage Showing 6 Grades (0–5)

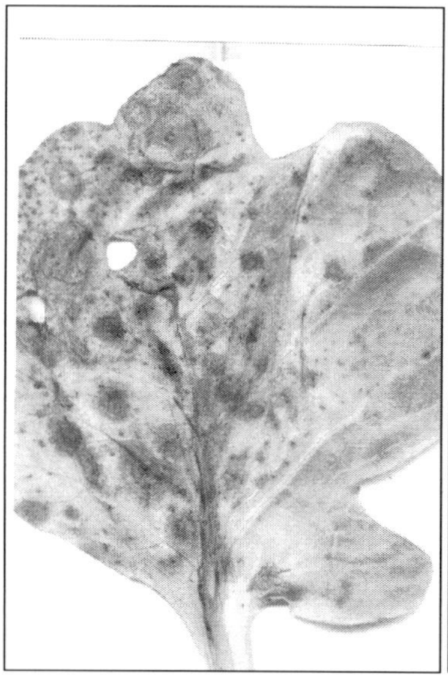

Alternaria Blighted Leaf of Cabbage

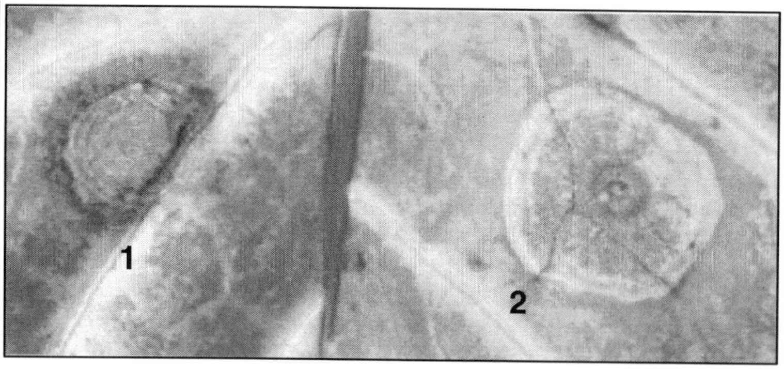

Plate 7.3: Typical Symptoms of *Alternaria brassicicola* (1) and *Alternaria brassicae* (2)

The morphological characteristics of each isolate such as length, breadth, number of septa, beak length and beak septa etc. were recorded from 15 days old culture (Mehta *et al.*, 2003).

Pathogenicity

Alternaria generally a weak pathogen require injury or weakened tissues to penetrate and establish (Mc Collach and Worthington, 1952). Increasing inoculum concentration, increases disease severity

Plate 7.4: Diseased Leaf Debris Buried at Three Depths

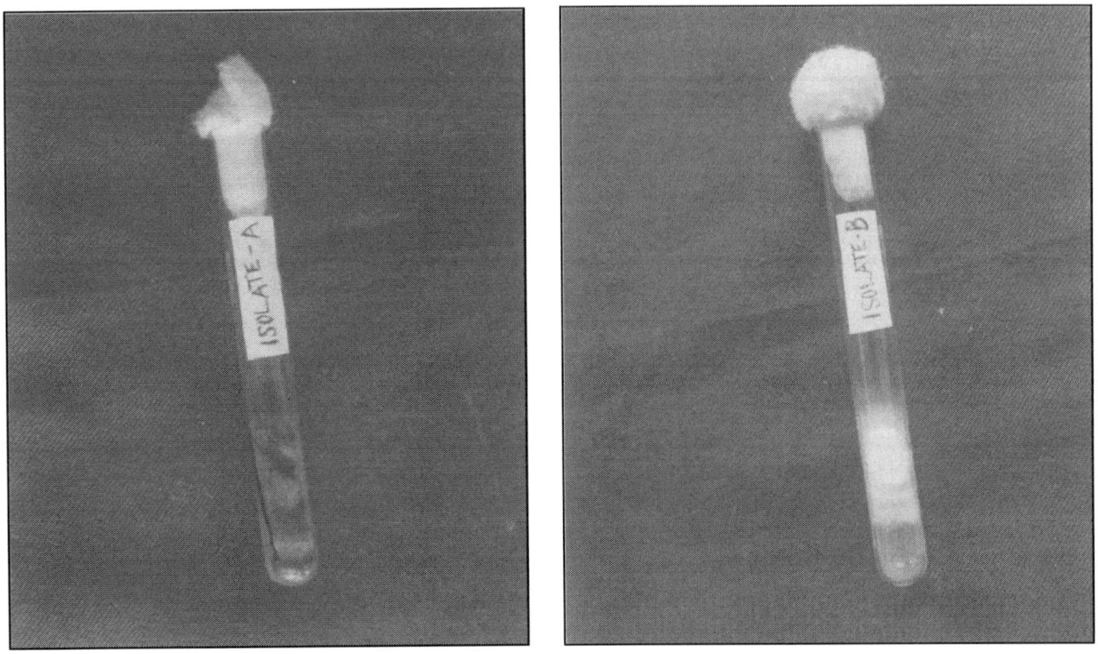

(*a*) *Alternaria brasssicicola* **(*b*) *Alternaria brassicae***

Plate 7.5: Culture's of *Alternaria* spp. Isolated from Infected Cabbage Leaves

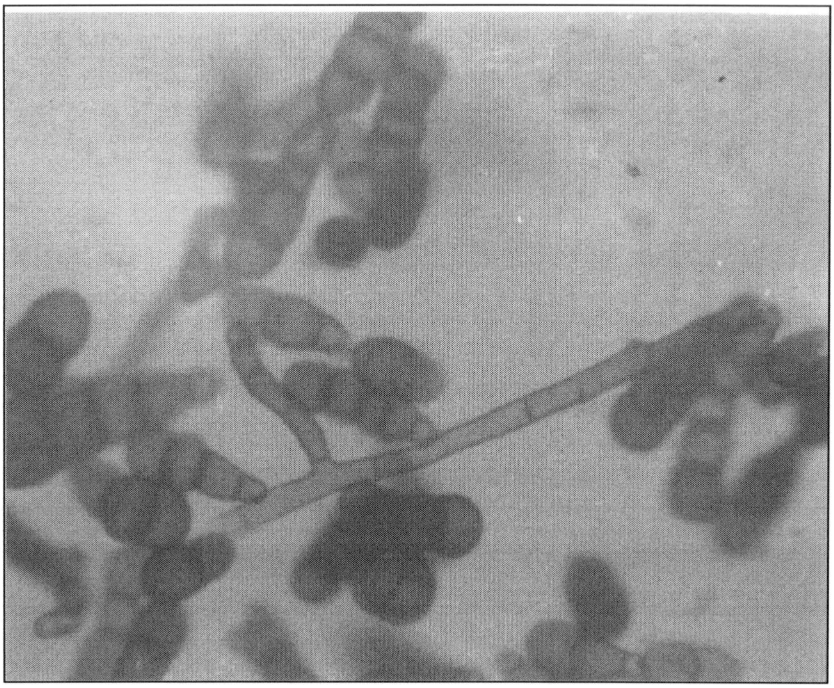

(*a*) ***Alternaria brassicicola*** **(Schw.) Wilts**

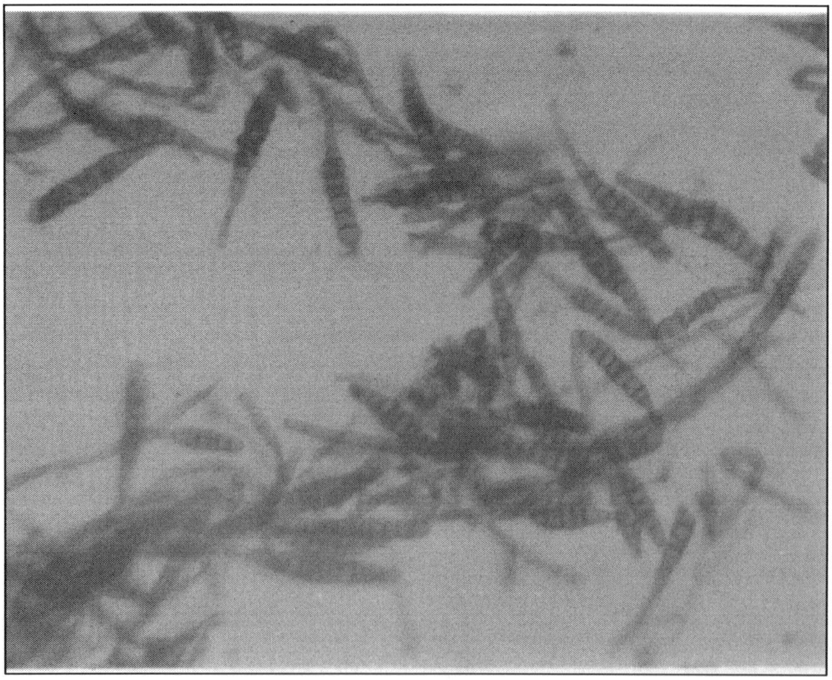

(*b*) ***Alternaria brassicae*** **(Berk.) Sacc.**

Plate 7.6: Morphological Appearance of Condia and Conidiophore (in culture)

(*a*) *Alternaria brassicicola*

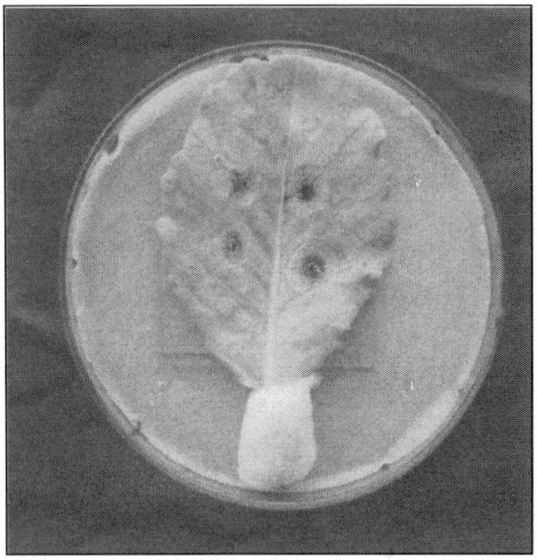

(*b*) *Alternaria brassicae*

Plate 7.7: Pathogenicity Test by Detached Leaf Technique

(Hong and Flitt, 1995) the pathogenicity of *Alternaria brassicicola* and *Alternaria brassicae* has been established on young healthy leaves of *Brassica juncea* separately, typical leaf spot symptoms developed in five and four days of inoculation, respectively (Koul, 1996). The spores of Alternaria isolated from necrotic lesions of cabbage (*Brassica oleracea* var. capitata), cauliflower (*Brassica oleracea* var. botrytis) and Kale (*Brasica oleracea* var. *acephela*) were proved by Koch's postulates, as the causal agents of dark leaf spot disease (Pattanmahakul and Strange, 1999). Mir (2007) has been ascertained and determined the severity level of the pathogen on the leaves of cabbage (Plates 7.2 and 7.7).

Vishwanath and Kolte (1999) reported a method of inoculation, to establish pathogenicity, by inoculating the detached leaves with needle soaked in spore suspension of *Alternaria brassicae*, inoculated leaves maintained in moist chambers at 20 ± 1 °C were scored after seven days of inoculation. *Alternaria brassicae* produced typical symptoms on detached leaves of oil seed rape (*Brassica napus* cv. Zephyr) within 24 hours of inoculation at controlled temperature (Degenhardth *et al.*, 1982).

Perpetuation

Weed hosts and crop residues are important sources for the primary inoculum in many Alternaria diseases through which the pathogen overwinter (Rotem, 1994). Both the pathogens (*Alternaria brassicae* and *Alternaria brassicicola*) survive longer and produced viable spores on the out door soil for as long as leaf tissues of the host crop remained intact (Murthy *et al.*, 2003). *Alternaria brassicae* remained viable in diseased rapeseed and mustard plant debris and seeds of infested plants, which serve as primary source of inoculum (Ansari *et al.*, 1989). Mehta *et al.* (2002) studied that survival of *Alternaria brassicae* was better in autoclaved soil as compared to unautoclaved soil. Both the pathogens (*Alternaria brassicae* and *Alternaria brassicicola*) survive as resting structures in the seed coat and embryos of rapeseed and mustard (Shrestha *et al.*, 2000; Kumar and Gupta, 1994). They are readily transmitted from seed to seedlings represents a source of disease in the crop (Shrestha *et al.*, 2000; Maude and Humpherson 1983). Mir (2007) has also investigated the perpetuation of the pathogen and found that it can be survived upto 15 cm depth in the debris of the crop (Plate 7.4).

Host Range

Black leaf spot of crucifers was found to be caused by *Alternaria brassicicola* and *Alternaria brassicae*. *Alternaria brassicicola* and *Alternaria brassicae* showed the same host range among eight cultivated species of crucifers and six species of cruciferous weeds tested (Huang and Chung, 1993). Ansari *et al.* (1990) have been reported that *Alternaria bassicae is* infected all the crucifers but only three species from other families *viz., Chenopodium album, Convolvulus arvensis* and *Anagallis arvensis*. Two collateral weed hosts *viz. Anagallis arvensis* and *Convolvulus arvensis* of Alternaria pathogens constitute new host records (Saharan *et al.*, 1982). Kadian and Saharan, (1983) observed that *Alternaria brassicae* isolates from cabbage, cauliflower and radish infected all eight test hosts, however, isolates from rai, brown sarson and toria infected seven test crops but not cabbage. Mir (2007) has artificially been inoculated the leaves of knol khol and cauliflower by both the pathogens and he found the symptoms of the disease (Plate 7.8).

Management

Botanical Extracts

Studies conducted on the use of plant extracts have opened a new avenue for the control of plant diseases. Several plant extract are known to possess antimicrobial properties (Nene and Thapliyal, 1993). Sheikh and Agnihotri (1972) have screened the extracts of 71 commonly occurring plants for antimycotic activity and observed that extracts of *Canna indica, Cenchrus cathoriticus, Allium cepa, Allium sativum, Lawsonia inermis, Argemone mexicana* and *Datura stramonium*, completely inhibited the spore germination of *Alternaria brassicae.*

Chand and Singh (2004) have been studied the effect of plant extract of *Calotropics procera, Eucalyptus globulens, Jatropha multifida, Azadiracta Indica* and *Allium sativum* under laboratory conditions, against *Alternaria brassicae*. Foliar spray with bulb extract of *Allium sativum* showed lowest disease intensity (2.87 per cent) followed by *Eucalyptus globulens* (5.3 per cent) and *Azadiracta indica* (7.4 per cent) as compared to control (20 per cent).

Chemical Control

Ziram and Cerosan completely inhibited the mycelial growth of *Alternaria brassicae* and *Alternaria. Brassicicolain the host.* The alterate sprays of Dithane Z-78 and Dithane M-45 at the rate 0.2 per cent for are to be most effective against *Alternaria brassicicola* (Saha, 1989). Many other fungicides *viz.*, Chlorothaloni, thiram and mancozeb were also inhibited spore germination completely at 500 ppm, whereas, thiram and copper oxychloride reduced the mycelial growth of test fungus significantly at 1000 ppm followed by captan and captafol (Pandey *et al.*, 2000).

Andres and Austin (1978) have been reported that chlorothalonil at the rate of 0.2 per cent is most effective in controlling Alternaria leaf spot of cauliflower when plants were sprayed. Captafol (0.2 per cent) provided maximum disease control and persisted for longer period on all the hosts, whereas, copper oxychloride enhanced the disease intensity to various levels. (Thind and Jhooty, 1988). Three sprays of iprodione applied to *Brassica* seed crops in field trials, significantly reduced severity of Alternaria diseases caused by *Alternaria brassicae* and *Alternaria brassicicola*, seed yields were increased and their germination was also improved (Humpherson and Maude, 1982; Babadoost, *et al.*, 1993). *Alternaria brassicae* and *Alternaria brassicicola* infection of *Brassica* seeds was most effectively controlled by seed treatment with iprodione (Maude and Humpherson 1980a; Shrestha *et al.*, 2000).

***Alternaria brassicae* Inoculated on Cabbage Leaf**

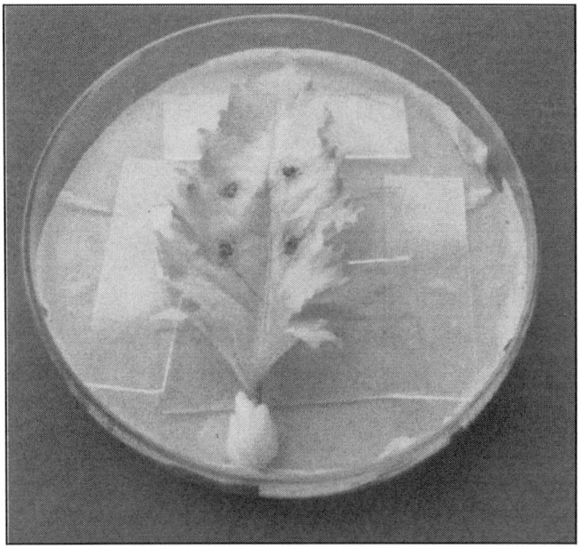

***Alternaria brassicicola* Inoculated on Knol Khol**

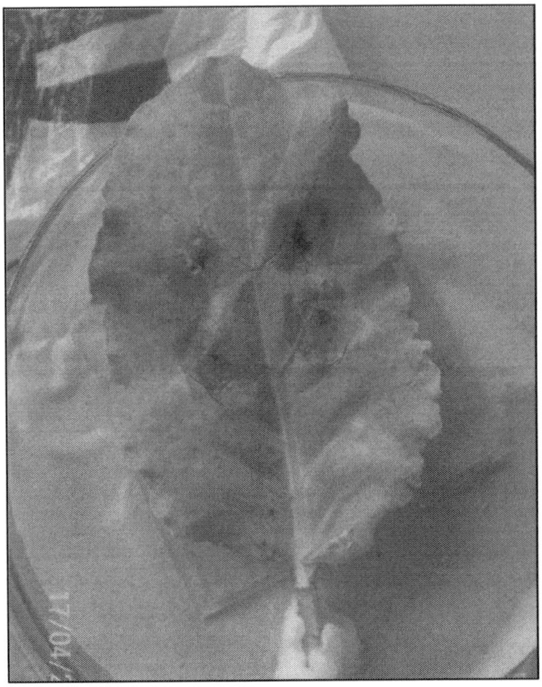

***Alternaria brassicae* Inoculated on Cauliflower Leaf**

***Alternaria brassicicola* Inoculated on Cauliflower Leaf**

Plate 7.8: Host Range by Detached Leaf Technique

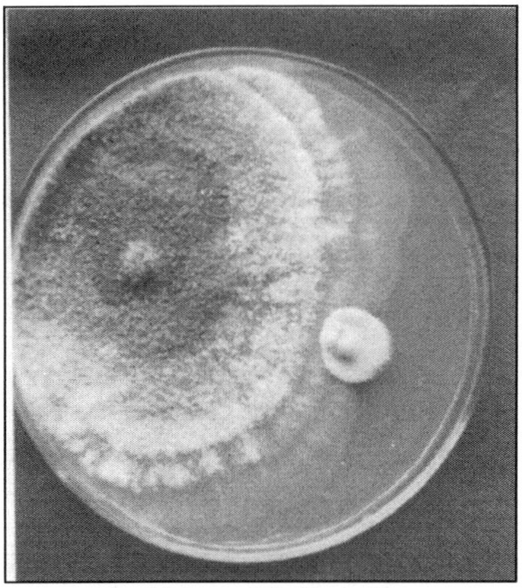

T. viride* and *A. brassicae

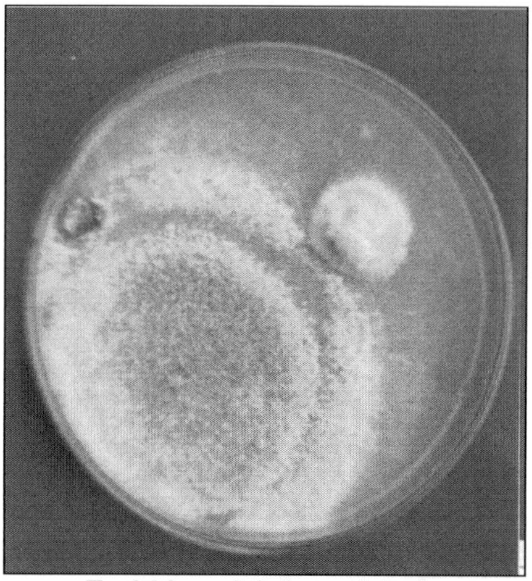

T. viride* and *A. brassicicola

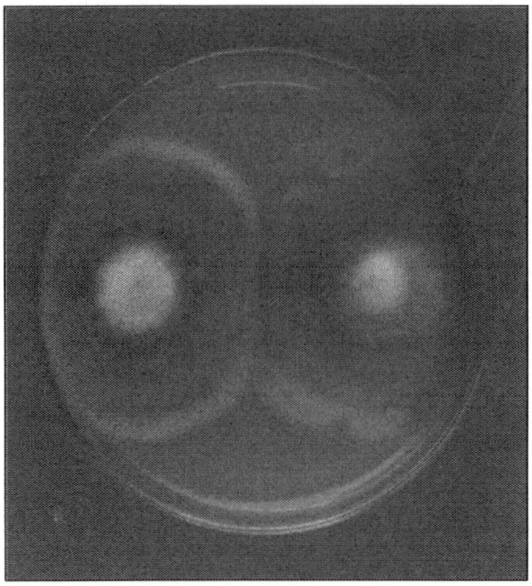

Aspergillus flavus* and *A. brassicicola

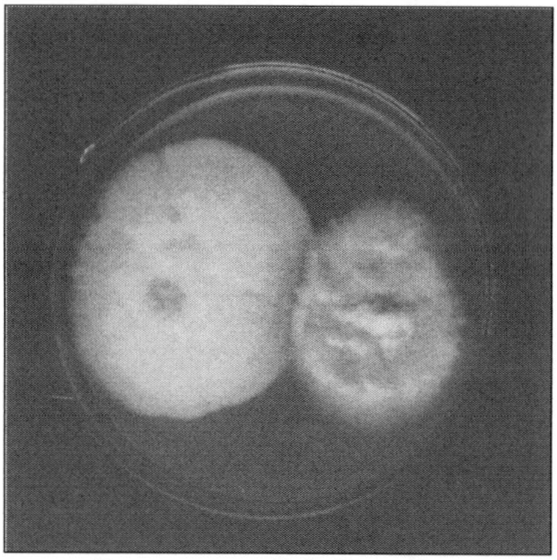

G. roseum* and *A. brassicae

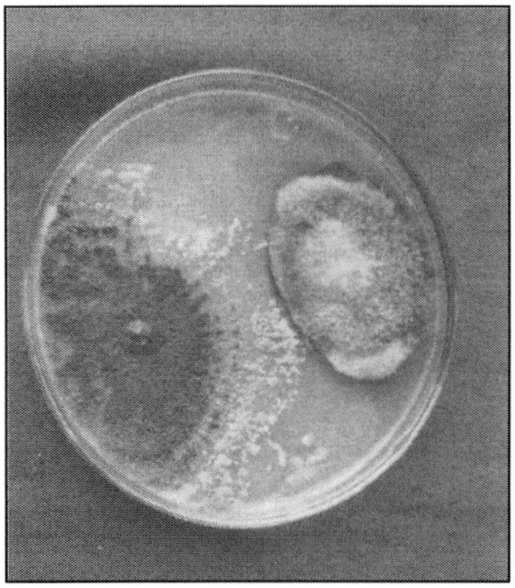

T. harzianum* and *A. brassicae

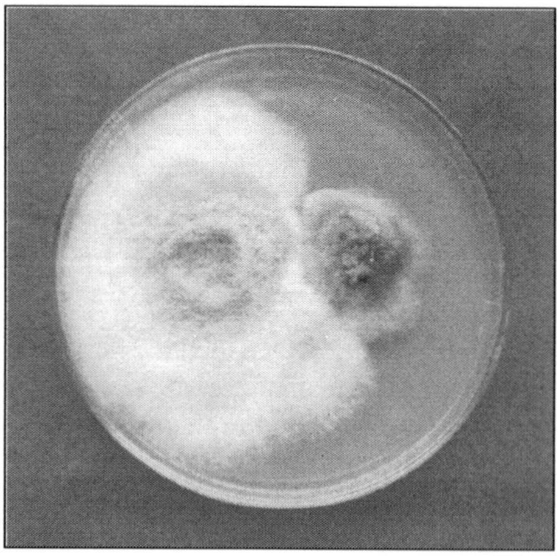

T. harzianum* and *A. brassicicola

Plate 7.9: Interaction of Different Antagonists with *Alternaria brassicae* and *Alternaria brassicicola*

Mir (2007) has been found antagonistic effect of the native bioagents *viz. Trichoderma harzianum, T. viride, Gliocladium roseum* and *Aspergillus niger* against both the pathogens causing Alternaria leaf spot disease in cabbage (Plate 7.9).

References

Andres, A.R. and Austin, E.M. (1978). The effect of chlorothalonil on Alternaria leaf spot of crucifers under laboratory conditions. *Canadian Plant Disease Survey*, 58: 75–76.

Anonymous, (2002). *Indian Horticulture Database–2002*. National Horticultural Board, Ministry of Agriculture, Govt. of India, pp. 126.

Anonymous, (2004). *Information on Status of Vegetable Species, Mushroom and Flowers of Kashmir*. Department of Agriculture, Govt. of Jammu and Kashmir, p. 2.

Ansari, N.A., Khan, M.W. and Muheet, A. (1989). Survival and perpetuation of *Alternaria brassicae* causing Alternaria blight of oil seed crucifers. *Mycopathlogia*, 105: 67–70.

Ansari, N.A., Khan, M.W. and Muheet, A. (1990). Host range of *Alternaria brassicae*. *Acta Botanica Indica*, 18: 104–105.

Babadoost, M., Gabrielson, R.L. Olson, S.A. and Mulanax, M.W. (1993). Control of *Alternaria* disease of brassica seed crops caused by *Alternaria brassicae* and *Alternaria brassicicola* with ground and aerial fungicide applications. *Seed Science and Technology*, 21: 1–7.

Chand, H. and Singh, S. (2004). Effect of plant extracts on Alternaria blight of mustard caused by *Alternaria brassicae*. *Indian J. Plant Prot.*, 32: 143–144.

Changstri, W. and Weber, G.F. (1963). Three *Alternaria* species pathogenic on certain cultivated crucifers. *Phytopathology*, 53: 643–648.

Degenhardth, K.J., Petrie, G.A. and Morall, R.A.A. (1982). Effect of temperature on spore germination and infection of rapeseed by *Alternaria brassicae, Alternaria brassicicola* and *Alternaria raphani*. *Canadian J. Plant Path.*, 4: 115–118.

Degenhardth, K.J., Skoropod, W.P. and Kondra, Z.P. (1974). Effect of Alternaria black spot on yield, oil content and protein content of rape seed. *Canadian J. Plant Sci.*, 54(1): 795.

Duhan, J.C. and Suhag, C.S. (1989). Studies on leaf and pod blight of cauliflower. Pathogenicity and distribution in Haryana. *Indian Phytopath.*, 42: 87–94.

Fazal, A. Khan, M.I. and Sexena, S.K. (1994). The incidence of *Alternaria* spp. in different cultivars of cabbage and cauliflower seeds. *Indian Phytopath.*, 47: 419–421.

Hong, C.X. and Flitt, B.D. L. (1995). Effect of inoculum concentration, leaf age and wetness period on the development of dark leaf and pod spot (*Alternaria brassicae*) on oilseed rape. *Ann. Applied Bio.*, 9: 283–295.

Huang, J.W. and Chung, W.C. (1993). Studies on survival of cruciferous black spot pathogens. *Alternaria brassicicola* and *A. brassicae*. *Plant Prot. Bulletin*, 35: 39–50. [cf. CAB Abstract, 1993–1994].

Humpherson, J.F.M. (1983). The occurrence of *Alternaria brassicicola, Alternaria brassicae* and *Leptosphaeria maculans* in brassica seed samples in South England between 1976–1980. *Plant Path.*, 32: 33–39.

Humpherson, J.F.M. and Maude, R.B. (1982). Control of dark leaf spot (*Alternaria brassiccicola*) of *Brassica oleracea* seed production crops with foliar sprays of Iprodione. *Ann. Applied Biol.*, 100: 99–104.

Humpherson, J.F.M. and Phelps, K. (1989). Climatic factors influencing sporulation of *Alternaria brassicae* and *Alternaria brassiciola*. *Ann. Applied Biol.*, 114: 449–458.

Kadian, A.K. and Saharan, G.S. (1983). Symptomatology, host range and assessment of yield losses due to *Alternaria brassicae* infection in rapeseed and mustard. *Indian J. Mycol. Plant Path.*, 13: 319–323.

Kolte, S.J. (1982). Disease management strategies for rapeseed mustard crop in India. *National Seminar on Production Problem of Rapeseed Mustard in India*, Held at Indian Agricultural Research Institute, New Delhi, 9–10 August.

Koul, T.K. (1996). Epidemology and cultural management of major foliar diseases of *Brassica juncea*. *M.Sc. Thesis*, Sher-e-Kashmir University of Agricultural Sciences and Technology of Kashmir, Shalimar Srinagar (SKUAST–K), pp. 76.

Kumar, R. and Gupta, P.P. (1994). Survival of *Alternaria brassicae, Alternaria brassicicola* and *Alternaria alternata* in the seeds of mustard (*Brassica juncea*) at different temperatures and relative humidities. *Ann. Bio.*, 10: 55–58.

Maude, R.B. and Humpherson, J.F.M. (1980). The effect of iprodione on the seed-borne phase of *Alternaria brassicicola*. *Ann. Applied Bio.*, 95: 321–327.

Maude, R.B. and Humpherson, J.F.M. (1980). Study on the seed borne phase of dark leaf spot (*Alternaria brassicicola*) and grey leaf spot (*Alternaria brassicae*)) of brassicas. *Ann. Applied Biol.*, 95: 311–319.

Mc Collach, C.P. and Worthington (1952). Salad crops. In: *Postharvest Pathology of Fruits and Vegetables*. Academic Press, London, pp. 162–163.

Mehta, N., Sangwan, M.S. and Srivastava, M.P. (2003). Morphological and pathological variations in rapeseed and mustard isolates of *Alternaria brassicae. Indian Phytopathology*, 56: 188–190.

Mehta, N., Sangwan, M.S., Srivastava, M.P. and Kumar, R. (2002). Survival of *Alternaria brassicae* causing Alternaria blight in rapeseed mustard. *J. Mycol. Plant Path.*, 32: 64–67.

Murthy, K.K., Shenvi, M.M. and Sreenivas, S.S. 2003. Perpetuation and host range of *Alternaria alternata* causing brown spot disease of tobacco. *Indian Phytopathology*, 56: 138–141.

Nene, Y.L. and Thapliyal, P.N. (1993). *Fungicides in Plant Disease Control*, 2nd Edn. Oxford and IBH Publication, New Delhi (India), pp. 413–414.

Pandey, K.K., Vishwakarama, S.N. and Chaube, H.S. (2000). Fungicidal control of brinjal leaf spot caused by *Alternaria alternata*. *Pestology*, 24: 42–46.

Pattanamahakul, P. and Strange, R.N. (1999). Identification and toxicity of *Alternaria brassicicola*, the causal agent of dark leaf spot disease of *Brassica* species grown in Thailand. *Plant Pathology*, 48: 749–755.

Rotem, J. (1994). *The Genus Alternaria: Biology Epidemiology and Pathogenicity*. APS Press. The American Phytopathological Society St. Paul, Minnesota, USA pp. 326.

Saha, L.R. (1989). Efficacy of some fungicides against *Alternaria brassicae* and *Alternaria brassicicola*, the cause of leaf blight of rapeseed and mustard. *Pestology*, 6: 19–21.

Saharan, G.S., Kaushik, J. and Kaushik, C.D. (1982). Two new host records of *Alternaria brassicae*. *Indian Phytopathology*, 35: 172.

Sheikh, R.A. and Agnihotri, J.P. (1972). Antifungal properties of some plant extracts. *Indian J. Mycol. Plant Path.,* 2: 143–146.

Shrestha, S.K., Mathur, S.B. and Munk, L. (2000). *Alternaria brassicae* in seeds of rape seed and mustard, its location in seeds, transmission from seeds to seedlings and control. *Seed Sci.Tech.,* 28: 75–84.

Singh, N.P., Bhardwaj, A.K., Kumar, A. and Singh, K.M. (2004). *Modern Technology on Vegetable Production.* International Book Distributing Co., Charbagh, Lucknow, U.P., India, p. 366.

Thind, T.S. and Jhooty, J.S. (1988). Differential behaviour of fungicides against Alternaria blight of some cruciferous hosts. *Indian J. Mycol. Plant Path.,* 18: 122–127.

Vishwanath, J. and Kolte, S.J. (1999). Methods of inoculation of resistance to Alternaria blight of rapeseed and mustard. *J. Myco. Plant Path.,* 29: 96–99.

Wiltshire, S.P. (1947). Species of Alternaria on brassica. *Mycological Papers,* 20: 1–15.

Plant Diseases Management in Horticultural Crops (2011) *Pages* **94–103**
Editors: **Shahid Ahamad, Ali Anwar and P.K. Sharma**
Published by: **DAYA PUBLISHING HOUSE, NEW DELHI**

Chapter 8

Management of *Fusarium solani* f.sp. *melongenae* Causing Wilt Disease in Egg Plants

A.G. Najar, Ali Anwar, M.A. Teli and M.S. Khar

Division of Plant Pathology,
Sher-e-Kashmir University of Agricultural Sciences and Technology of Kashmir,
Shalimar, Srinagar – 191 121

Egg plant (*Solanum melongena* L.) is one of the most popular and principal solanaceous vegetable crop in India as well as other parts of the world. It is a perennial crop but grown as an annual. In hilly and temperate regions, it is grown only in summer. Brinjal fruits are fairly good sources of Ca, P, Fe and vitamins particularly 'B' group and it is reported to stimulate the interapeptic metabolism of blood cholesterol (Singh and Kalda, 2001). Major countries growing brinjal include China, India, Turkey, Japan, Egypt and Italy (Salunkhe *et al.*, 1987). In India it is grown over an area of 472100 hectares with the annual fruit production of 7676900 metric tones (Anonymous, 2003). India ranks next to China in its cultivation and covers 8.14 per cent of total vegetable area consisting 9.0 per cent of total vegetable production. In Jammu and Kashmir, it is grown as summer crop over an area of 460 hectares producing 4600 metric tonnes of fruit (Singh and Kalda, 2001). Compared to per hectare yield of 17.4 tonnes in the country, Jammu and Kashmir produces only 10.0 tonnes of fruits.

A number of diseases such as damping-off (*Pythium* spp.; *Phytophthora* spp; and *Fusarium* spp.); root-rot (*Rhizoctonia* spp. and *Sclerotium* spp.); blight (*Phomopsis* spp.); fruit rot (*P. vexans*; and *Rhizopus stolonifer*) and wilt (*Verticillium* spp., *Fusarium* spp.) are also caused a considerable declined of the produce annually. Brinjal wilt-complex is known to be caused by a number of fungal genera such as *Fusarium*, *Verticillium*, *Rhizoctonia*, *Sclerotium* and *Phytophthora* in different parts of the world (Rangaswami, 1979).

Status of Disease

The wilt disease of brinjal was reported for the first time by Montemartini in the year 1907. Since then, it has been reported cause exorbitant losses round the globe-in Argentina (Pontis, 1940), Armania (Babayan and Shakhnubayan, 1969) Netherland (Steeketenbug, 1976) and Bulgaria (Koleva and Vitanov, 1990). In India, it has been reported by several workers from different parts of the country (Mitra, 1953; Mandhare *et al.*, 1989). *Fusarium semitectum* Berk and Rav. {Syn, *F. pallidoroseum*} (Cook.) Sacc. has been found responsible for seedling wilt and fruit rot of brinjal in Punjab (Kaur *et al.*, 1985). FAO of United Nations started giving major emphasis to this group of plant disease since mid of last century (Gangopadhyay and Gill, 1976). Infection by pathogen during various stages has been reported to prevent fruiting cause defoliation, chlorosis, stunting, reduced fruit size, deteriorated fruit and death of whole plant during fruit development stage (Mandhare *et al.*, 1989). Species of *Fusarium* are associated with severe wilts, fruit and root rots of important solanaceous crops in India and abroad. Severe incidence of wilt in some cultivars of chilli in Punjab has been caused 100 per cent crop loss (Thind and Jhooty, 1985) while in other states of India such as Delhi, Mahrashtra and Tamil Nadu has also been appeared in variable range of 3-100 per cent.

Etiology

Casual Organism

The etiology of the disease in India has already been established (Vimala *et al.*, 1994). First time in Jammu and Kashmir state of India, the studies conducted by Lolpuri in the year 2002 have confirmed *Fusarium* solani f.sp. m*elongenae* as a causal agent of brinjal wilt. The predominant pathogen associated with wilt syndrome had been reported to be different species of *Fusarium* (Kaur *et al.*, 1985). Different workers ascribed the wilt to different species of *Fusarium* such as *F. oxysporum* (Mandhare *et al.*, 1989), *F. equiseti* (Singh *et al.*, 1997), *F. semitectum* (Kaur *et al.*, 1985; Kapoor and Sharma, 1988) and *F. solani* (Fournet and Jacqua, 1978; Kumar *et al.*, 1983),. The pathogens have been found causing similar symptoms on other solanaceous crop like tomato, chilli, pepper and potato. *F. oxysporum*, *F. semitectum* and *F. equiseti* have been associated with tomato wilt syndrome (Kapoor, 1988), stem rot and wilt of chillies (Srivastava and Diwakar, 1987), root- rot and wilt of pepper (Koleva and Vitanov, 1990 Rivelli and Black, 1992), and potato root rot and wilt (Jones and Woltz, 1981, Saxena, 1989). Fusarium solani has been isolated and found poathogenic to wide host range in India and other parts of the world (Raj, *et al.*, 1974; Bhaskar and Ahmad, 1991).

Symptoms

Wilt (*F. solani*) of brinjal is characterized by yellowing of lower leaves gradually progressing upwards to upper leaves. The apical shoot was found to drop followed by withering of the shoot and fruit leading to ultimate drying and death of whole plant, owing to cortical decay, vascular browning and necrosis of underground stem (Gangopadhyay, 1984). The brinjal wilt caused by *F. oxysporum* f.sp. *melongenae* under field conditions showed vein clearing of young leaflets and epinasty of older leaves followed by general yellowing, sudden wilting and defoliation starting from lower plant parts progressive upwards often at flowering stage. The infected xylem vessels often turn brown (Mandhare *et al.*, 1989). The authors also observed partial wilting and stunting owing to infection by the pathogen, causing complete wilting within 10-12 days.

Predisposition

Wilts of solanaceous crops are favoured by some environmental conditions like poor drainage

(Matta and Garibaldi, 1972), acidic soils (Jones and Woltz, 1970) and higher levels of nitrogenous fertilizers (Weinke, 1962; Woltz and Jones, 1973).

Perpetuation of Pathogen

Seed borne nature and soil transmission of wilt-causing *Fusarium* spp. have been reported by many workers.

In/on Seeds

Many researchers established the seed-borne nature of *Fusarium* spp. causing wilt of brinjal and other crops. Babayan and Shakhnubaryan (1969) from Armania frequently isolated *F. oxysporum* from brinjal seeds, and the infection reduced seed germination, growth rate and fruit yield, and induced wilting in transplanted crop. The seed-borne nature of *Fusarium* spp. in India has also been established (Kumar *et al.*, 1983; Kaur *et al.*, 1985; Singh *et al.*, 1997). Khar(2004) has also been investigated and isolated the pathogen from seeds of brinjal under temperate conditions of Kashmir region of the Jammu and Kashmir state of India.

Soil

Fusarium spp. have long been reported to be facultative saprophytes inhabiting the soil (Beckman, 1987). The soil-borne nature has also been established by Nedumaran and Vidhyasekaran (1982).

Disease Management

Being monocyclic nature of disease, management of wilt has been attempted by reducing its inoculum levels in soil or in/on seed, thus making host to evade or defend the attack of the pathogen.

Cultural

Managing plant diseases of soil-borne nature by manipulating cultural practices, such as crop rotation and fallowing, field sanitation deep ploughing, time and method of planting, irrigation and soil pH, have been attempted successfully by many researchers (Fournet and Jacqua, 1978; Kannaiyan and Nene, 1979; Sen, 1986).

Fournet and Jacqua (1978) recommended drainage improvements for badly drained soils of egg plants attacked by *F. solani*. *Fusarium* brown rot of Chinese Yam has long been controlled by incorporation of 30 metric tones of compost per hectare by some vegetable growers of Japan, and the decrease in the disease correlated with increase in population of *Trichoderma* spp. and suppressed populations of *Fusarium* spp. (Hoitink *et al.*, 1976). However, high levels of potash fertilization was found to cause marked increase in resistance to *Fusarium* spp. and a significant decrease in the population of *F. oxysporum* (Walker and Foster, 1946; Schneider, 1985). It was also decreased by raising the soil pH from 6.0 to 7.5 (Jones and Woltz, 1972) these practices are sustainable through some are labour-intensive (Thurston, 1990).

Chemical

Since Fusarium wilt pathogens are mainly soil and seed-borne, disinfection of soil and seeds with chemicals/fungitoxicants has yielded encouraging results. Prior to 1970,some gaseous fumigants such as methyl bromide and chloropicrin were the only commercially available measures of vascular wilt control. Methyl bromide is being proved most effective compared to ethylene dibromide and propane-propene mixture in controlling *Fusarium* wilt disease (Wensley, 1953). Wilt complex (*F. solani*;

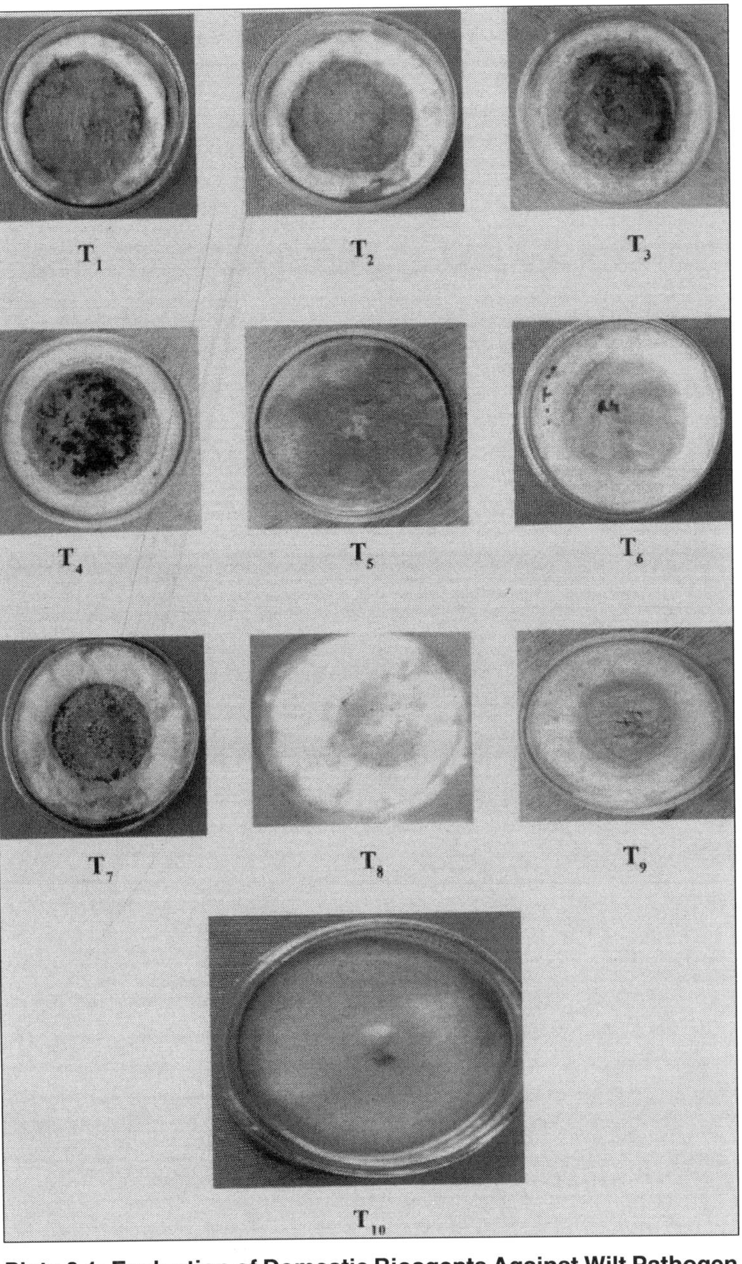

Plate 8.1: Evaluation of Domestic Bioagents Against Wilt Pathogen

T_1: *T.harzianum* strain-I showing complete destruction of pathogen; T_2: *T.harzianum* strain-II expressing growth pattern similar to strain-II; T_3: T. viride strain-I exhibiting clear zone of inhibition; T_4: *T. harzianum* strain-II expressing growth pattern similar to strain-II; T_5: *Paecilomyces varioti* depecting overgrowth on pathogen; T_6: *Gliocladium roseum* exhibiting no further advancement of growth beyond; the point of contact with wilt pathogen; T_7: *A.flavus* expressing growth pattern similar to *G. roseum*; T_8: *T. roseum* showing less growth than wilt pathogen; T_9: *P. fluorescens* showing no further advancement of growth.

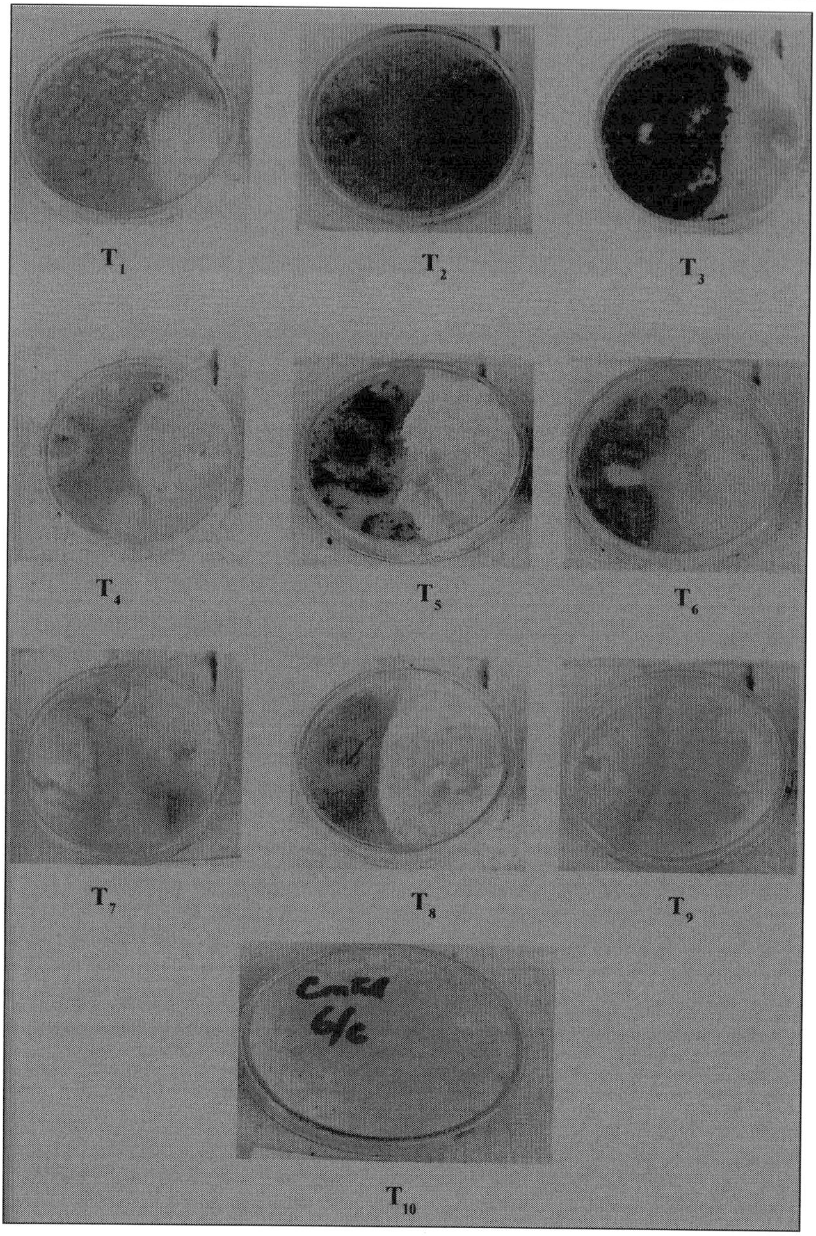

Plate 8.2: Sreening of Domestic Bioagents Against *F. solani* f.sp. *melongenae* by Dual Culture Method

T_1: *T.harzianum* strain-I showing complete destruction of pathogen; T_2: *T.harzianum* strain-II expressing growth pattern similar to strain-II; T_3: *T. viride* strain-I exhibiting clear zone of inhibition; T_4: *T. harzianum* strain-II expressing growth pattern similar to strain-II; $T_{5:}$ *Paecilomyces varioti* depecting overgrowth on pathogen; T_6: *Gliocladium roseum* exhibiting no further advancement of growth beyond; the point of contact with wilt pathogen; T_7: *A.flavus* expressing growth pattern similar to *G. roseum*; T_8: *T.roseum* showing less growth than wilt pathogen; T_9: *P.fluorescens* showing no further advancement of growth; T_{10}: Check only wilt pathogen.

R. solani and *P. capsici*) was controlled by application of diazomet (70 g m^{-2}) or drenching with metalaxyl @ 0.1 gm^{-2} (Ahmed *et al.*, 2000).

Sugha *et al.* (1995) were reported that carbendazim and benomyl are most effective *in vitro* against *Fusarium* spp. compared to captafol, thiram, thiophanate methyl and captan. Carbendazim and thiophanate methyl are inhibited mycelial growth and spore germination (Pandey and Upadhyay, 1999).

Seed treatment with fungicides is regarded as the most economical and convenient method of chemical control of any seed borne plant disease. Treatment of seeds with either carbendazim or captafol at 1 g kg^{-1} seed controlled *F. semitectum* infection, and effectivity of captan, thiram and benomyl used in soaking or slurry seed treatment was enhanced by storing the treated seeds for at least 24 h (Nedumaran and Vidhyasekaran, 1982).

Biological

Biological control of soil-borne plant pathogens is receiving tremendous attention throughout the world on account of the growing public concern about the health hazardous and pollutant effects of the pesticide usage. Plant ecosystem or antagonists to pathogenic forms and are, therefore, viable and potent alternatives to conventional chemicals for the management of plant diseases (Jeyarajan and Angappan, 1998). In sustainable agriculture biological control assumes major importance especially in cases where soil-borne plant disease problems are serous. Unlike foliar disease pathogens, biological control of soil-borne plant pathogens *viz.*, *Sclerotium rolfsii, Rhizoctonia solani, Sclerotinia sclerotiorum, Pythium aphanidermatum* and *Fusarium* spp. of high value plantation crops have been extensively studied (Prasad *et al.*, 2000). A tremendous breakthrough in the research efforts on biocontrol of important diseases in India has been witnessed by using spp. of *Trichoderma, Gliocladium, Aspergillus, Chaetomium, Paecilomyces, Pseudomonas, Bacillus* and some non-pathogenic strains of disease-causing fungi and bacteria (Singh and Hussaini, 1998). Seed treatment with *T. harzianum, T. virens* and *Streptoverticillum* spp. were effective in suppressing *F. solani* in brinjal and *R. solani* in paddy under greenhouse conditions (Mishra and Narain, 1994; Anwar *et al.*, 2008). Apart from spp. of *Trichoderma* and *Gliocladium*, some other fungal antagonists like *Aspergillus nidulans* (Upadhyay and Rai, 1983); *A. niger* (Chattopadhyay and Sen, 1996) and *Paecilomyces* sp. (Munshi, 1998; Masoodi, 2000) have also been successfully used for wilt disease control.

Anuratha and Gnanamanickam (1990) successfully controlled bacterial wilt (*Pseudomonas solanacearum*) of banana, egg plant and tomato by inoculation with *P. fluorescens*. Seed treatment with talc based and pea based formulations of *P. fluorescens* effectively controlled wilt disease and increased the yield under field conditions (Vidhyasekaran and Muthamilan, 1995).

For successful expression, the proper biocontrol agent needs to be delivered in its effective method. Bio-agent propagules in water suspensions or dry powder has been used to coat seed (Mukhopadhyay, 1994). Soil borne disease of eggplant, potato, tomato and pea were significantly reduced by amending the soil with *Trichoderma* spp. and natural substrates (Elad *et al.*, 1980; Sreenivasaprasad and Manibhushanroa, 1990). On the mode of action of different bio-control isolates phenomena of competition, mycoparasitism, antibiosis and predation have been considered responsible (Mukhopadhyay, 1994; Mukherjee *et al.*, 1995) Mukhopadhyay (1994) has suggested that production of antifungal metabolites by *G. virens* was responsible for bio-control of *P. aphanidermatum*. Mukherjee *et al.* (1995) suggested that the parasitization of sclerotia by *G. virens* was the principal mechanism of suppressing *S. rolfsii* and *R. solani*. Pandey and Upadhyay (2002). Khar (2004) has also ascribed his investigations and revealed that biocontrol agents *viz.*, *Trichoderma harzianum* (S.I and S.II), *T. viride*

(S.I and S.II), *Paecilonyces varioti*, *Gliocladium roseum*, *Aspertillus flavus*, *Trichothecium roseum* and *Pseudomonas fluorescens* were much effective against *Fusarium soalni* f.sp. *melongenae but Trichoderma* spp. were highly antagonistic to wilt pathogen exhibited zones of inhibition followed by *P. varioti* and *A. flavus*, *P. fluorescens*, *G. roseum* and *T. roseum* proved least effective against the wilt pathogen. Under field conditions *Trichoderma* spp. applied as seedling dip prior to transplanting exhibited least disease incidence was recorded in untreated control. In addition the bio-agent applications resulted improvement of plant height, weight and fruit yield.

References

Ahmed, S., Hamid, K., Tariq, A.H. and Jamil, F.F. (2000). Chemical control of root rot and collar rot of chillies. *Pakistan J. Phytopath.*, 12: 1–5.

Ali Anwar, Bhat, G.N. and Bhat, K.A. (2008). Mycoparasitic behavior of certain bio-agents against sheath blight pathogen (*Rhizoctonia solani*) or rice. *J. Mycol.Pl. Path.*, 38(1): 135–137.

Anonymous, (2003). *Agricultural Resrearch Databook*. Indian council of Agricultural Research (ICAR), New Delhi, pp. 245.

Babayan, D.N. and Shakhnubaryan, S.T. (1969). Presence of infection by fungal wilt in the seeds of tomato, pepper and egg plant under conditions of Arafat plain, the Armanian SSR. *Uchen. Zap. Erevan, Univ.*, 1(110): 136–147.

Beckman, C.H. (1987). *Nature of Wilt Disease.* American Phytopathological Society, St. Paul, pp. 174.

Bhaskar, R.B. and Ahmad, S.T. (1991). Root rot diseases of berseem and its control. *Indian Phytopath.*, 43: 589–590.

Chattopadhyay, C. and Sen, B. (1996). Integrated management of *Fusarium* wilt of muskmelon caused by *F. oxysporum*. *Indian J.Mycol. Pl. Path.*, 26: 162–170.

Chet, I. (19878). *Innovative Approaches to Plant Disease Control.* John Wiley and Sons, New York, pp. 372.

Cole, J.S. and Zvenyika, Z. (1988). Integrated control of *Rhizoctonia solani* and *Fusarium solani* in tobacco transplants with *Trichoderma harzianum* and triadimenol. *Plant Pathology*, 37: 271–277.

Dube, H.C. (2000). Rhizobacteria in biological control and plant growth promotion. *J. Mycol. Plant Path.*, 31: 9–21.

Dwivedi, R.S. and Pathak, S.P. (1981). Effect of certain chemicals on population dynamics *of Fusarium oxysporum* f.sp. *lycopersici* in tomato field soil. *Proceedings in Indian National Science Academy*, 47: 751–755.

Fournet, J. and Jacqua, G.. (1978). Note on attack of *Fusarium solani* on egg plant (*Solanum melongena* L.). *Farm Research*, 27: 13.

Gangopadhyay, S. (1984). Fusarial wilts of crucifers, leguminaceous solanaceous and umbelliferous vegetables. In: *Advances in Vegetable Diseases*. Associated Publishing Co., New Delhi, pp. 41–52.

Gangopadhyay, S. and Gill, H.S. (1976). Is fusarium still a problem for growers in the hills. *Delhi Garden Magazine*, 5: 14–16.

Hoitink, H.A., Heer, L.J. and Schmitthenner, A.F. (1976). Survival of some plant pathogens during composting of hard tree bark. *Phytopathology*, 66: 1369–1372.

Jacob, C.K. and Sivaprakasam, K. (1994). Evaluation of some plant extracts and antagonists for the control of pre-emergence damping-off of brinjal. In: *Crop Diseases: Innovative Techniques and*

Management, (Eds.) K. Sivaprakasam and K. Seetharaman. Kalyani Publishers, New Delhi, India, pp. 289–294.

Jeyarajan, R. and Angappan, K. (1998). Mass production technology for fungal antagonists and field evaluation. In: *Biological Control of Plant Diseases, Phytoparasitic Nematodes and Weeds*, (Eds.) S.P. Singh and S.S. Hussaini. Project Directorate of Biological Control, Bangalore, Karnataka, India, pp. 48–58

Jones, J.P. and Woltz, S.S. (1970). *Fusarium* wilt of tomato; interaction of soil liming and micro-nutrient amendments on disease development. *Phytopathology*, 60: 812–813.

Jones, J.P. and Woltz, S.S. (1972). Effect of soil pH and micronutrient amendments on Verticillium and Fusarium wilt of tomato. *Plant Disease Reporter*, 56: 151–153.

Jones, J.P. and Woltz, S.S. (1981). *Fusarium*-incited diseases of tomato and potato and their control. In: *Fusarium Diseases, Biology and Taxonomy*, (Eds.) P.E., Nelson, T.A. Tousson, and R.J. Cooke. Pennsylvania State University Press, Pennsylvania, pp. 157–158.

Kapoor, I.J. (1988). Fungi involved in tomato wilt syndrome in Delhi, Maharashtra and Tamil Nadu. *Indian Phytopathology*, 41: 208–213.

Kapoor, K.S. and Sharma, S.R. (1988). Soil application of fungicides against wilt of egg plant. *Capsicum Newsletter*, 7: 90–91.

Kaur, S., Kaur, R., Kaur, P. and Singh, D. (1985). Studies on wilt and fruit rot of brinjal caused by *Fusarium semitectum*. *Indian Phytopath.*, 38: 736–737.

Kannaiyan, J. and Nene, Y.L. (1979). Effect of crop rotation and planting time on the incidence of wilt in lentil. *Indian Journa of Plant Protection*, 7: 114–118.

Khar, M.S. (2004). Biological control of *Fusarium solani* f.sp. *melongene* (Mart.) Sacc. incitant of brinjal wilt in Kashmir. *M.Sc. Thesis*, SKUAST–K, Shalimar Srinagar, pp. 18–40.

Koleva, K. and Vitanov, M. (1990). *Fusarium species* related to root rot of pepper. *Rasteniev dni–Nauki*, 27: 61–63 [c.f. RPP (1990) 69: 819].

Kumar, R.U., Safeeulla, K.M. and Shetty, H.S. (1983). Seed borne nature and infectivity of *Fusarium solani* (Mart.) Sacc.–The causal agent of wilt and fruit rot of brinjal (*Solanum melongena* L.). *Proceedings of the Indian National Science Academy*, 49: 134–140.

Lolpuri, Z.A. (2002). Management of fungal wilt complex of brinjal (*Solanum melongena* L.). *M.Sc. (Agri.) Thesis*, Sher-e-Kashmir University of Agricultural Sciences and Technology of Kashmir, Srinagar, pp. 63.

Mandhare, V.K., Ruikar, S.K. and Donde, B.K. (1989). New report of wilt disease in India. *Current Sciences*, 58: 1036–1037.

Masoodi, M.A. (2000). Biological Management of Chilli Wilt [*Fusarium pallidoroseum* (Cooke) Sacc.] in Kashmir. *M.Sc. Thesis*, Sher-e-Kashmir University of Agricultural Sciences and Technology of Kashmir Shalimar, Srinagar, pp. 63.

Matta, A. and Garibaldi, A. (1972). Effect of different moisture conditions on certain plant disease caused by soil fungi. *Agriculture Italiana*, 72: 237–253. [c.f. RPP (1973) 52: 64].

Mishra, D.K. and Narain, A. (1994). *Gliocladium virens* and *Streptoverticillium* as sources of biocontrol of few phytopathogenic fungi. *Indian Phytopath.*, 47: 236–240.

Mitra, A. (1935). Investigation on the wound parasitism of certain fusaria. *Indian Journal of Agriculture Sciences*, 5: 632–637.

Mukherjee, P.K., Mukhopadhyay, A.N., Sarmah, D.K. and Shrestha, S.M. (1995). Comparative antagonistic properties of *Gliocladium virens* and *Trichoderma harzianum* on *Sclerotium rolfsii* and *Rhizoctonia solani*. *Journal of Phytopathology*, 143: 275–279.

Munshi, N.A. (1988). Studies on the fusarial blight of mulberry. *Ph.D. Thesis*, Sher-e-Kashmir University of Agricultural Sciences and Technology of Kashmir, Srinagar, pp. 45–50.

Nedumaran, S. and Vidhyasekaran, P. (1982). Damage caused by *Fusarium semitectum* in tomato. *Indian Phytopathology*, 35: 322–323.

Neelamegam, R. and Govindarajalu, T. (2002). Integrated application of *Trichoderma viride* and farm yard manure to control damping-off of tomato. *Journal of Biological Control*, 16: 65–69.

Pandey, K.K. and Upadhyay, J.P. (1999). Comparative study of chemical, biological and integrated approach for management of *Fusarium wilt* of pigeon pea. *J. Mycol. Plant Path.*, 29: 214–216.

Pandey, K.K. and Upadhyay, J.P. (2000). Microbial population from rhizosphere and non-rhizosphere soil of pigeon pea, screening for resident antagonist and mode of mycoparasitism. *J. Mycol. Plant Path.*, 30: 7–10.

Papavizas, G.C. and Lumsden, R.D. (1980). Biological control of soil borne fungal propagules. *Annual Review of Phytopathology*, 18: 389–413.

Pontis, R.E. (1940). El marchitonatto' del pimiento (*Capsicum annuum*) en la provincial de Mendoza. *Review of Argentina Agronomy* 7: 113–127. [c.f. RAM (1941) XIX: 676].

Prasad, R.D., Rangeshwaran, R. and Sankaranarayanan, C. (2000). *Biological Control of Plant Pathogens.* Project Directorate of Biological Control, Bangalore, India. pp 57.

Raj, J.N., Tewari, J.P., Singh, R.P. and Saxena, V.C. (1974). A report on the occurrence of wilt of cauliflower in the Nilgiris. *Botany and Plant Pathology of Vegetables*, 29: 425–426.

Rangaswami, G. (1979). *Diseases of Crop Plants in India*, 2nd edn. Prentice Hall of India Pvt. Ltd., New Delhi, pp. 298–302.

Rivelli, V. and Black, L.L. (1992). Pathogencity of *Fusarium oxysporum* f.sp. *capsici* to *Capsicum* spp. and the effect of temperate and seedlings age on disease severity. *Plant Disease*, 76: 340–346.

Saxena, S.K. (1989). Soil moisture regimes in relation of Fusarium wilt of potato. In: *Proceedings of National Symposium on Epidemiology and Forecasting of Plant Disease*, Held at IARI, New Delhi on 28th Feb. 2nd March, pp. 35–40

Schneider, R.W. (1985). Suppression of Fusarium yellows of celery with potassium chloride and nitrate. *Phytopathology*, 75: 40–48.

Singh, N. and Kalda, T.S. (2001). Brinjal (*Solanum melongena* L.). In: *Vegetable, Tuber crops and Spices*, (Eds.) S. Thamburaj and N. Singh. Indian Council of Agricultural Research, New Delhi, p. 29.

Singh, R.K., Bhandari, T.P. and Singh, B.K. 1997. *Fusarium equiseti:* A new record on brinjal from India. *Progressive Horticulture*, 29: 100–101.

Singh, S.P. and and Hussaini, S.S. (1998). *Biological Suppression of Plant Diseases: Phytoparasitic Nematodes and Weeds.* Project Directorate of Biological Control, Bangalore, Karnataka, India, pp. 285.

Srivastava, V.K. and Diwakar, M.C. (1987). Note on stem rot and Wilt of chillies in Haryana. *Plant Protection Bulletin*, 39: 35.

Stecketenbug, N. A.W.V. (1976). *Fusarium wilt* of egg plant in the Netherlands. *Netherlands Journals of Plant Pathology*, 82: 191–192.

Sugha, S.K., Kapoor, S.K. and Singh, B.M. (1995). Management of chick pea wilt with fungicides. *Indian Phytopathology*, 48: 27–31.

Thind, T.S. and Jhooty, J.S. (1985). Relative prevalence of fungal diseases of chilli fruits in Punjab. *Indian J. Mycol. Plant Path.*, 15: 96–102.

Thurston. H.D. (1990). Plant disease management practices of traditional farmers. *Plant Diseases*, 74: 96–102.

Upadhyay, R.S. and Rai, B. (1983). Mycoparasitism with reference to biological control of plant disease. In: *Recent Advances in Plant Pathology*, (Eds.) A. Hussain, K. Singh, B.P. Singh and V.P. Agnihotri. Print House, Lucknow India, pp. 112–115.

Upadhyay, R.S. and Rai, B. (1992). Wilt of pigeon pea. In: *Plant Diseases of International Importance, Vol. 1*, (Eds.) U.S. Singh, A.N. Mukhopadhyay, J. Kumar and H.S. Chaube. Prentice Hall, New Jersey, pp. 388–414.

Plant Diseases Management in Horticultural Crops (2011) *Pages* **104–119**
Editors: **Shahid Ahamad, Ali Anwar and P.K. Sharma**
Published by: **DAYA PUBLISHING HOUSE, NEW DELHI**

Chapter 9

Management of Chilli Wilt Disease in Kashmir

A.G. Najar, M.A. Teli and Ali Anwar

Division of Plant Pathologym
Sher-e-Kashmir University of Agricultural Sciences and Technology of Kashmir, Shalimar, Srinagar – 191 121

Status of Disease

Wilt disease of chilli was reported for the first time in New Mexico (Arizona) as early as 1908 (Chupp and Sherf, 1960). Since then, it has been reported causing exorbitant losses round the globe-in Peru and Arizona (Brown and Gibson, 1926), West Indies (Ciferri, 1926), India (McRae, 1932), Spain (Benlloch and Dpminguez, 1934), Argentina (Machionatto, 1935), New Mexico (Crawford, 1934), Uganda (Hansford, 1940), Brazil (Chaves, 1947), Chile (Fernandez, 1983), Iraq (Sarhan and Sharif, 1986), Albania (Ibrahimillari, 1987), Cuba (Camino *et al.*, 1987) and Tunisia (Moens and Benaicha, 1990). Chilli plants showing severe wilt in fruiting stage destroyed the crop before fruit ripening (McRae, 1932). Thomas (1938) has also recorded considerable damage to chilli pepper (*C. annuum* L.) after flowering stage due to the disease. Early infection of chilli by wilt pathogen prevented fruiting and caused complete destruction of the crop in Argentina (Pontis, 1940). Severe incidence of the disease in some genotypes in the Punjab state of India up-to 100 per cent crop loss (Thind and Jhooty, 1985). In Haryana too, chilli wilt and stem rot occurred in mild to severe form in most areas, and the disease incidence in general varied from 20 to 60 per cent. In heavily infected fields, however, 60-80 per cent wilt incidence was recorded (Anonymous, 1987). The disease assumed devastating proportions in Kashmir valley exhibiting wilt incidence of 58.88–81.00 per cent in district Srinagar alone and causing 26-54 per cent loss (Wani, 1994; Dar and Mir, 1995). Najar(2003) has also been reported that disease was found to affect upto 85.56 per cent areas showing maximum extent of incidence 62.67 per cent in Kashmir region of the Jammu and Kashmir state inflicting yield losses upto 52.92 ha^{-1} under epidemic form of the disease.

Causal Organism

A number of pathogens have been found associated with the wilt complex of chilli in different parts of the world. Various workers ascribed the wilt to different species of *Fusarium* such as *F. annuum* (Crawford, 1934), *F. oxysporum* Schl. (Garofalo, 1957; Vidhyasekaran and Thiagarajan, 1981). *F. solani* (Moens and Benaicha, 1990; Koleva and Vitanov, 1990), *F. moniliforme, F. oxysporum* and *F. solani* (Hashmi, 1989) and *F. pallidoroseum* (Thind and Jhooty, 1985). *F. pallidoroseum* (Cooke.) Sacc. (*Syn. F. semitectum* Berk. and Rav.) is known to cause wilt of many solanaceous crops especially in chilli crop. Nedumaran and Vidhyasekaran (1982) have been reported that the fungus produced wilt symptoms on healthy seedlings. Najar(2003) has also been reported that *Fusarium pallidoroseum* is the most predominant pathogen causing wilt disease in the chilli-cropped soil.

Symptoms

Slight variation in the development and expression of wilt in chilli by different pathogens has been observed in different parts of the world. Crawford (1934) reported that *F. annuum* infected the base of chilli stem and caused more or less sudden wilting of lower leaves followed by death of the whole plant within a short period of time, given favourable temperature and moisture conditions (Plate 9.1).

The older leaves of *Fusarium* infected chilli plants turn yellow first followed by younger leaves, petioles and stem; the leaves become chlorotic and dessiccated, and the whole plant withered and died slowly (Dimond, 1970). Sarhan(1982) has been observed wilting on capsicum plants caused by *F. oxysporum* and *F. solani* was characterized by above-ground stem browning peeling -off of epidermis and vascular browning (Plate 9.1). He also reported that *F. solani* occurred mostly in roots and lower stem where there was vascular browning and that it could also infect fruit and spread slowly than *F. oxysporum* which could be found in all parts of inoculate plants and cause outright killing of seedlings. Infection of chilli plants with *F. pallidoroseum* in the Punjab were symbolized by tan-brown to brownish-black stems, withering of leaves and ultimate death of the affected plants (Kaur, 1993). The fungus also attacked tip of mature chilli fruit which started drying and shriveling upwards (Thind and Jhooty, 1985). Wani (1994) reported sudden wilting of leaves with loss of green colour, vein clearing and epinasty of older leaves often at flowering stage due to root infection with *F. pallidoroseum* in Kashmir. He further reported that the infected roots were brown within and the wilting started in the lower-plant parts and spread upwards.

F. pallidoroseum is also known to cause wilting in other plant species. Kaur *et al.* (1985) observed that infection of brinjal seedlings with *F. pallidoroseum* caused withering and drooping of its leaves 5-6 days after inoculation followed by wilting of the whole plant 10 days later. This fungus was also responsible for black gram wilting where it caused browning of stem at collar region and filling of stem within by cottony mycelial growth (Swain *et al.*, 1989).

Predisposition

The disease is known to be favoured by many environmental conditions Crawford (1934) reported that chilli wilt (*F. annuum*) become severe in low-lying and un-drained fields. The infection of capsicum plants by *F. oxysporum* and *F. solani* occurred only when the plants were weakened by un-favourable weather conditions (Lukacs and Sjarka, 1988). Most of *Fusarium* spp. responsible for wilts in solanaceous crops remained confined to acidic soils (Albert, 1946). Application of higher levels of nitrogenous fertilizers have also been found to predispose the hosts to infection by *Fusarium* spp. (Chakarabarti and Basu-chaudhary, 1978; Huber, 1981; Alois and Mace, 1981).

Perpetuation of Pathogen in/or Seed

Many researchers have been established the seed-borne nature of *Fusarium* spp. causing wilt in chilli crops. Babayan and Shakhnubaryan (1969) frequently isolated *F. oxysporum* from chilli seeds, and the infection reduced seed germinating, growth rate and fruit-yield, and induced wilting in transplanted crop. Pandey (1976) isolated *F. moniliforme* from chilli seeds even after surface sterilization. Vidhyasekaran and Thiagarajan (1981) isolated *F. oxysporum* form wilted chilli plants raised from infected seeds. The *Fusarium* spp. isolated from *C. annuum* seeds colleted from 30 different countries include *F. semitectum, F. moniliforme, F. solani, F. equiseti, F. oxysporum* and *F. moniliforme* of which *F. moniliforme, F. solani, F equiseti, F. oxysporum* were found to cause root rot and wilt of capsicum seedlings (Hashmi, 1989). In China, samples of *C. annuum* seeds also yielded *F. equiseti, F. moniliforme, F. oxysporum, F. semitectum* and *F. solani* with an average infection of 1.0, 2.6, 0.1, 0.8 and 0.2 per cent respectively (Liang, 1990). Basak *et al.* (1996) have been detected highest per cent age of *Fusarium* spp from chilli seeds.

Perpetuation through Soil

Fusarium spp. have long been reported to be facultative saprophytes inhabiting the soil (Beckman, 1987). Units of singly or several chlamydospores of the fungus remain embedded in humus or plant debris (Nash *et al.*, 1961) and get transmitted to the transplanted crop (Vidyaskaran and Thiagarajan, 1981). The soil borne nature of the fungus has also been established by other researchers (Lukacs and Sjarka, 1988; Nasreen *et al.*, 1988; Kim *et al.*, 1990). Once in the soil, fusaria enter a new phase of existence in which they must complete for suitable growing niches and energy, competitive with many micro-organisms that are particularly adapted to saprophytic growth (Nelson, 1981).

The dormant immobile structures, the chlamydospores, produced in decaying hosts in soil by most *Formae speciales* of *F. oxysporum* get stimulated to germinate by host roots and the invasion, usually through the roots of a susceptible plant, is followed by the development of systemic vascular wilt symptoms (Griffin, 1969). In advanced stages of disease development, the fungus grows out of vascular system and produces conidia and chlamydospores in soil (Vishwakarma and Basu-chaudhary, 1982).

Disease Management

Being simple interest of single-cycle diseases, wilts and root rots have been managed by attempting at eliminating or at least reducing the primary inoculum in soil or in/on seed and propagating material besides making the host evade or defend the attack of the pathogen.

Host Resistance

Host resistance is by far the most economical and ecofriendly disease management venture. Being soil and seed-borne, Fusarial wilts can be management by cultivation of resistant genotypes to wilt pathogen. Camino *et al.* (1987) who have identified 'Keystone','Resistant Grand', Mild 'Californian Wonder' and 'Export' susceptible to *F. solani*, whereas genotypes 'Bell Boy' Citeina' and 'New Ace' were reported susceptible to *F. oxysporum* f.sp. *lycopersici*; the genotype 'Hungarian Sweetwax' was, however, susceptible to both the pathogens. The pepper accessions- Pimiermente, Longtam and Cayanne proved resistant reaction with wilt infection of less than 20 per cent, whereas other eight

Plate 9.1: Chilli Plants Showing Wilt Symptoms with Clinging Leaves and Fruits

1: Root and lower shoot cut longitudinally to exposing browning of vascular bundles in wilted plant; 0: Healthy vascular bundles.

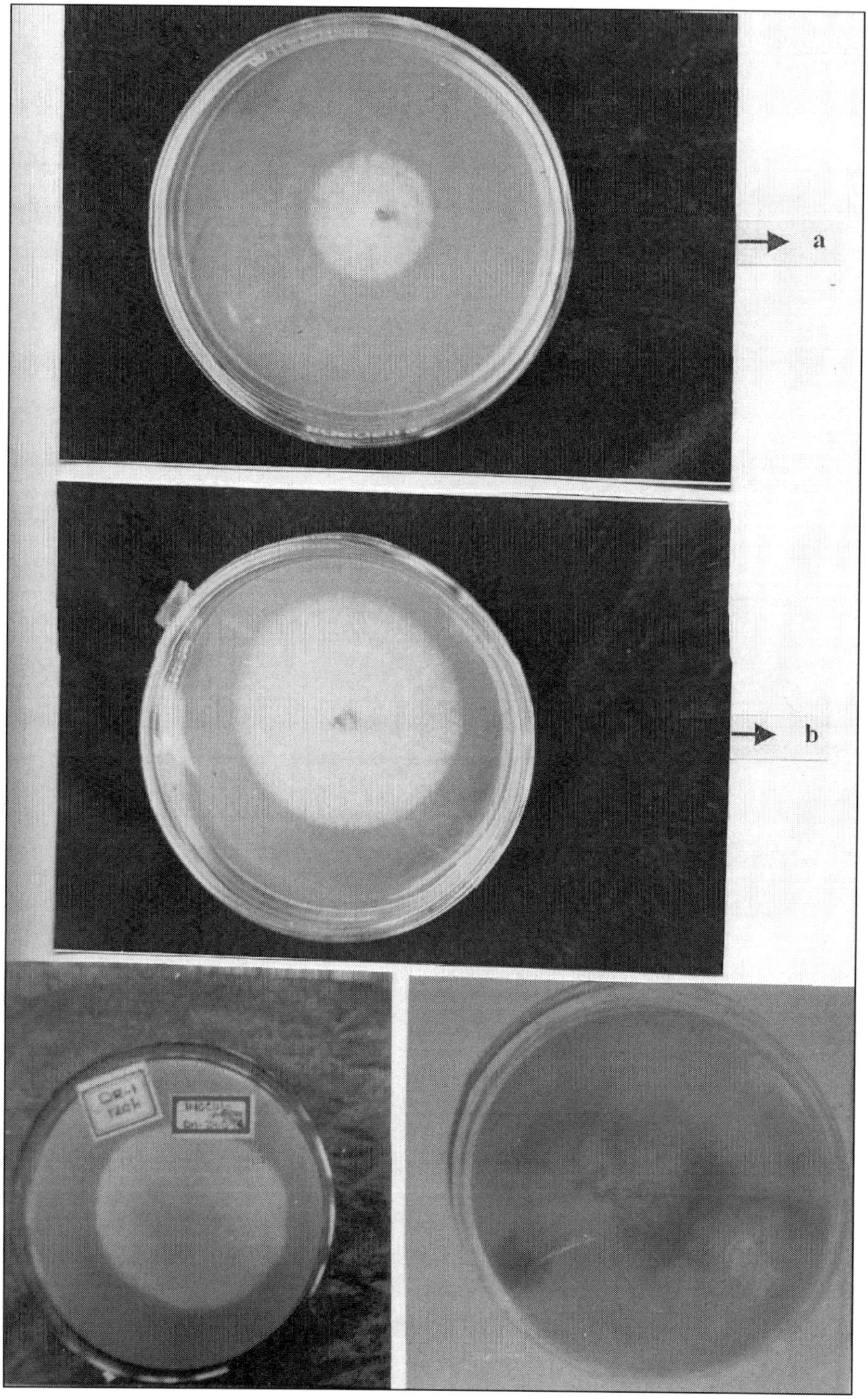

Plate 9.2: Mycelial Growth of *Fusarium pallidoroseum* on PDA (a,b and c)

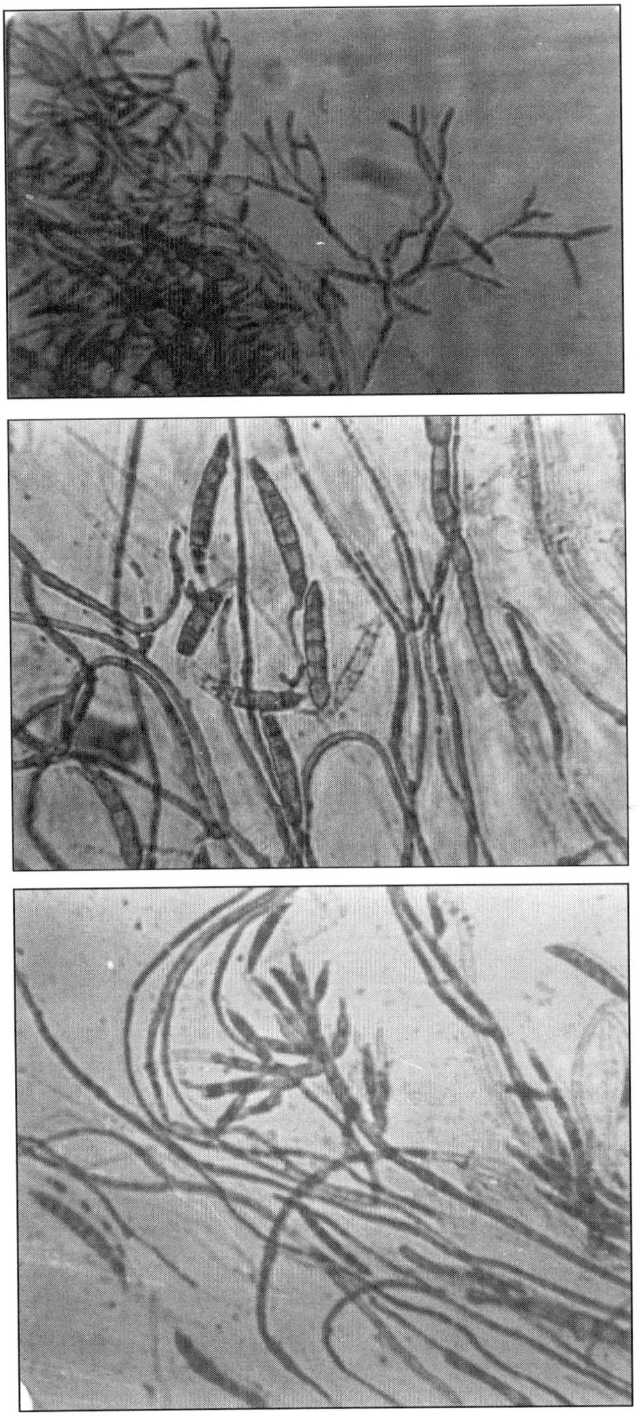

Plate 9.3: (a) Primary and Secondary Conidiophores; (b) Primary Macro Conidia with Wedge Shaped Foot Cell; (c) Simple Phialids with Secondary Condia

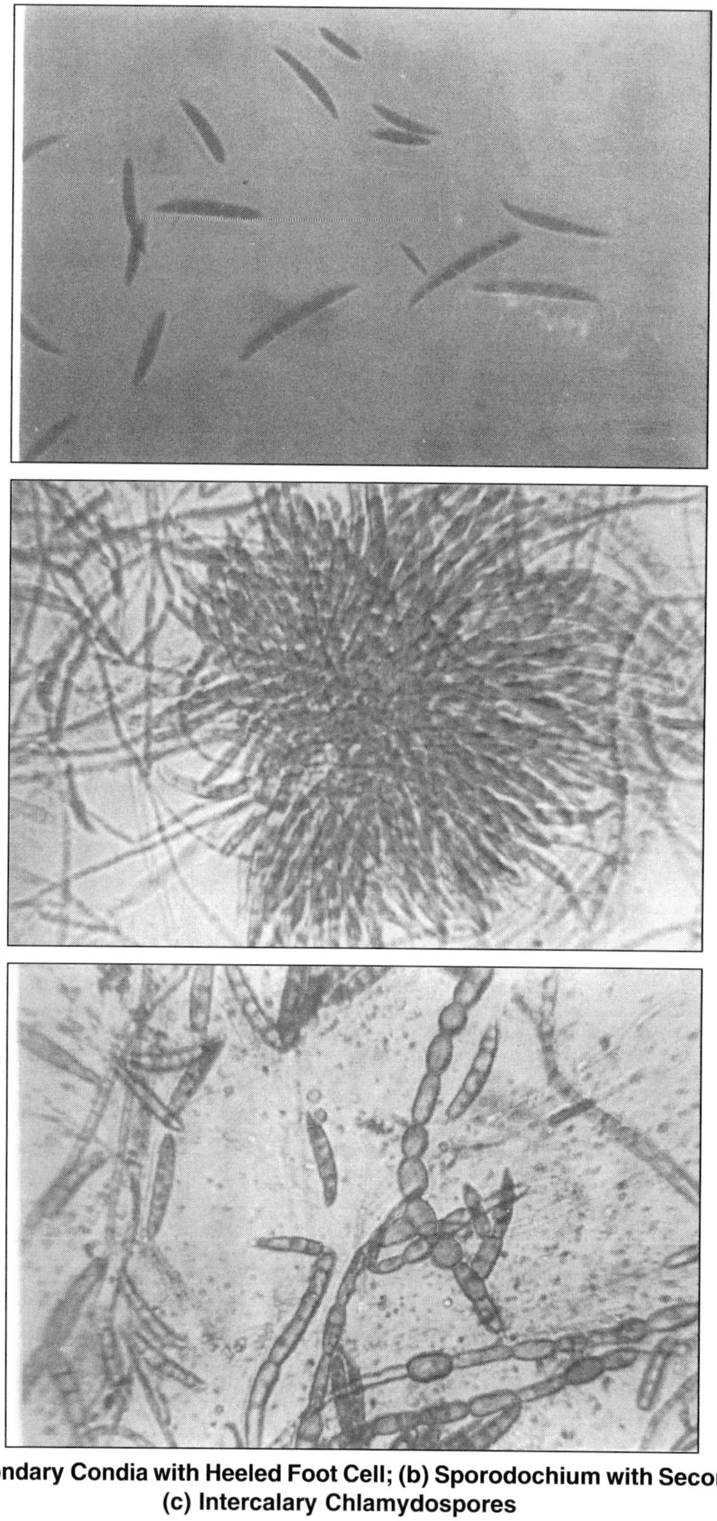

Plate 9.4: (a) Secondary Condia with Heeled Foot Cell; (b) Sporodochium with Secondary Conidia; (c) Intercalary Chlamydospores

genotypes *viz.*, Pimiermente 111-1-2, Pimiermente 36-2-1, Pimiermente 54-1-1, Canyanne 5-8, and Chile are to be moderately resistant reaction (Wang and Wang, 1989). Ahmad *et al.* (1994) also identified SC-101, SC-102, SC-108, SC-111, SC-114, SC-118, SC-122, SC-125, SC-126, SC-137, SC-152, SC-158, SC-161, SC-166, SC-185 and SC-191 chilli lines are resistant to *Fusarium* wilt under natural epiphytotic conditions. They also recorded 12 hot pepper lines X235, LCA 248, Pant C-2, Masalawadi, Phule C-5, C-586, Jawahar 218, Pant C-1, LCA 206, Sel-1, DPLC 1 and Solan Red and two sweet pepper lines Rokash and Veradale resistant to the pathogen Nayeema *et al.* (1995) identified Masalawadi cultivar of hot pepper immune to *F. pallidoroseum* under field and controlled conditions in Kashmir, whereas the lines SC-120, Phule C-5, SC-335, SC-502, SC-415, SC-107, SC-348, SC-108, LCA-304, Aekalohit, SC-101, SC-371, SC-137, SC419, SC-451, SC-31, LCA-248, Pusa Jawala, Pant C-8 and Jawahar-218, where moderately to highly resistant under these conditions. Out of 30 genotypes for chilli wilt resistance, Punjab Lal, Solan Red, Solan yellow, Pant C-1, IC-4, IC-21, IC-8, Sweet Banana and Pachhad yellow showed moderately resistant reaction, while the remaining 21 showed susceptible to highly susceptible reaction to *F. oxysporum* (Singh and Singh, 1995).

Different isolates of *Fusarium* spp. are also known to show different reactions on different capsicum species and varieties. Based on differential disease reaction of isolates *viz.* Ps-Ia and H-71 on Canyanne, Cajun-IA, PI-201232, PI-159241 and PI-439431 have been representing distinct physiological races (Rivelli and Black, 1992). Jones and Black (1992) rated one accession each belonging to *C. annuum* and *C. chacoense* as highly resistant and 16 others as resistant to the highly virulent isolate Ps-Ia of *F. oxysporum* f.sp. *capsici*. Najar(2003) has been evaluated 59 genotypes for resistance to pathogen revealed only two genotypes *viz.*, SC-335 and SC-415 were highly tolerant under both field and pot house conditions; nine other genotypes *viz.*, SC-107, HC-44, Phule C-5, Pant C-2, SC-108-1, SC-101, SC-263, SC-108 and SC-348, exhibited tolerant reaction to the pathogen, whereas all other genotypes were susceptible to *F. pallidoroseum* under epiphytotic conditions in the Kashmir.

The varieties identified as resistant to a particular pathogen may not necessary have other desirable traits and may not be directly introduced for wide-scale cultivation. The two chilli cultivars PBC 408 and PBC 199, even though identified as resistant with only 6-10 per cent wilt incidence, had undeniable horticultural traits and, thus, could be used only as donors of resistance under breeding programme, however, PBC-2529 and PBC-2530 were found to possess acceptable horticultural characters and yield in addition to having good degree of field tolerance to wilt pathogen (Kaur *et al.*, 1999).

Cultural Methods

Managing plant diseases of soil-borne nature by manipulating cultural practices such as crop rotation and fallowing, field sanitation, deep ploughing, time and method of planting, irrigation and soil pH are reported by many researchers (Kannaiyan and Nene, 1979; Nelson, 1981; Sen, 1986; Baris *et al.*, 1986). These practices are sustainable, through some are labour-intensive (Thurston, 1990).

Irrigation and Planting Method

Restricted irrigation and planting on raised beds or ridges have been found to help reduce the incidence of wilts and root/collar rots (Garcia, 1933, Baris *et al.*, 1986; Dar and Mir, 1995). Garcia (1993) has recorded seven per cent wilt of chilli pepper by planting on ridges compared to 42 per cent observed in native leveled plantation. Pontis (1940) and Beattie *et al*, (1944) also reported successful management of *Fusarium* wilt of chilli by setting plants on ridges and avoiding excessive irrigation. Significant reduction in *Fusarium* wilt of chilli by combination of drainage with planting on ridges has also been reported by Verma and Sharma (1995). They were reported that flood irrigation resulted in heavy losses and that the disease appeared in patches at fruit formation stage before the onset of rains and assumed serious proportion after heavy rainfall followed by hot dry days in poorly drained conditions. Fusarium wilts can be reduced by planting on ridges and raised beds with good drainage system and less frequent irrigation under temperate zones of Kashmir (Dar and Mir, 1999).

Najar (2003) suggested raised bed plantings, irrigated every 10-15 days resulting in minimum wilt incidence with corresponding increase in fruit yields compared to other planting methods and caused maximum increase in root length. A negatively highly significant correlation existed between wilt incidence and fresh fruit yield during both the years under study by separately trying different combinations of NPK, planting methods and irrigation frequencies and different organic amendments and fungicides treatments, thereby, suggesting that decrease in wilt incidence resulted in increase in fresh fruit yields.

Soil Amendment with Composts

Composts are being used in agriculture with beneficial effects for years (Kelman and Cook, 1977). During the past few decades, several reports have discussed suppressive effects of composts on variety of soil-borne plant pathogen (Baker and Cook, 1974; Singh, 1983). Composted high-wood-content larch bark incorporated into field soil at 30 metric tones per hectare is used by vegetable growers in the Nagano valley of Japan for the control of *Fusarium* brown rot of Chinese Yam, and the decrease in the disease correlated with increase in populations of *Trichoderma* spp. and suppressed population development of *Fusarium* spp. by the compost treatment (Hoitink *et al.*, 1976). Najar(2003) has also been advised for soil amendments with different composts, sand,lime and cow dung @ 2 kg m^{-2} for impeding he wilt incidence. The soil amendments with compost found to cause significant reduction in *F. pallidoroseum* population with simultaneous increase in total fungal and bacterial counts also in soil.

Considerable success in the management of wilt and root rots has been achieved with cow dung and chicken manure as soil amendments. The populations of *F. solani* and *F. oxysporum* f. sp. *conglutinans* in soil were considerable reduced by incorporating in soil sum-dried organic matter prepared by admixing cow dung manure and chopped *Sesbania aculeate* 'Dhaincha' (Chattopadyay and Mustafee, 1978). Soil amendment with 1 per cent animal residue reduced *F. oxysporum* and *F. solani* populations responsible for root rot diseases (Zakaria and Lockwood, 1980).

Soil pH

Higher pH levels are known to retract the growth and development of *Fusarium* spp. in soil, and hydrated lime has often been used for the purpose. Jones and Wolt (1970) showed that the fungus remained confined to acidic soils. In 1972, the authors observed reduction in wilt incidence by raising the pH from 6.0 to 7.0. Similar observations are also made by Jones and Overman (1971) and Sarhan (1982) who succeeded in reducing wilt by raising soil pH to 81.1 from 6.2 through hydrated time [$Ca(OH)_2$]. The number of the fungal propagules witnessed significant reduction, whereas its incubation period was prolonged on increasing the pH to 7.0 from 4.5 (Nirwanto and Djajati, 1994). Other plant disease which are reported to be reduced/controlled by increasing the soil pH are wilts solanaceous crops (Albert, 1946), chrysanthemum (Woltz and Engelhard. 1973), cucurbits (Jones and Woltz, 1975). Safflower (Chakrabarti and Basu-Chaudhry, 1978), chilli (Sarhan and Sharif, 1986) and carnation (Garles *et al.*, 1995).

Soil Fertilization

The role of nitrogen, phosphorus and potassium fertilizers in disease development depends upon the nature of the pathogen and host interaction on one hand and the quality and quantity of the fertilizers on the other. Nitrogen, an important constituent of most important bio-molecules such as amino acids, proteins, growth hormones, enzymes, phytoalexins and phenols and a determinant of cell wall characteristics, is known to exhibit significant influence on dieses development especially the ones caused by soil-borne fungi (Toussoun *et al.*, 1960; Kaufman and Williams, 1964). The disease severity has in general, been found to increase as the rate of nitrogen application is increased. The influence of nitrogen fertilizers on a particular disease has been found to vary depending upon whether the available form of nitrogen in soil is nitrate or ammonia. Jones and Woltz (1969 and 1970) recorded lesser decrease in resistance to wilts by nitrate nitrogen than by ammonical nitrogen or a mixture of the two in crops. The influence of nitrogen fertilization on soil-borne pathogens is modified by soil pH. Sarhan and Sharif (1986) noticed reduction of chilli wilt incidence (*F. oxysporum* f. sp. *redolens*) when soil was amended separately with nitrate nitrogen and lime.

Phosphorus, the second most important element applied to soil as fertilizer is essentially required for meristematic tissue and root proliferation. The vigorous root growth promoted by adequate phosphorus applications helps plants bear the damage caused by root pathogens (Huber, 1980). Woltz and Jones (1973 and 1981) reported that a high level of phosphorus increased the severity of Fusarium wilt disease with combination of high lime and low phosphorus greatly curtailed it.

Potassium is the third major nutrient supplement in soils for plant growth. Besides regulation enzyme activity in most cellular functions such as photosynthesis, phosphorylation, protein synthesis and reproduction, it induces formation of thicker cuticular and epidermal cell wall (Huber, 1980) and modifies disease reactions in plants both directly or indirectly. The direct effects involve formation of adequate structural barriers and to reduce penetration, multiplication, survivability and aggressively of invading pathogen; whereas indirectly, the delayed initiation of senescence by potassium avoids infection by facultative parasites (Kiraly, 1976). The host resistance to pathogen, however, remains unchanged if K_2O level is enhanced beyond certain optimum dose. The excess K_2O level is known to inhibit uptake of other cations like Ca^{++}, Mg^{++} and Mo^{++} so essential for plant growth (Dick and Tisdale, 1938; Balaji and Vaitheeswaran, 1988). It is believed that the synthesis of high molecular weight compounds, like proteins, starch and cellulose, in K-deficient plants get impaired and low molecular weight compounds accumulated which increase susceptibility to pathogen infection (Kiraly, 1976; Huber, 1980).

Applications of slaked lime reduced the total fungal population including that of *F. pallidoroseum* while simultaneous boosting bacterial counts in soil. Increase in the levels of nitrogen and phosphorus is also increased the wilt incidence, without exerting any direct influence on *F. pallidoroseum* population in soil Najar (2003).

References

Ahmad, N., Tanki, M.I. and Mir, N.M. (1994). Screening of advance breeding lines of chilli, sweet and hot pepper cultivars against Fusarium wilt. *Pl. Dis. Res.*, 9: 153–154.

Albert, W.B. (1946). The effect of certain nutrients upon the resistance of cotton to *Fusarium vasinfectum*. *Phytopath.*, 36: 703–710.

Alois, A.B. and Mace, M.E. (1981). Biochemistry and physiology of resistance. In: *Fungal Wilt Disease of Plants*, Mace, New York, pp. 447–449.

Anonymous (1987). Note on stem rot and wilt of chilli in Haryana (India). *Pl. Prot. Bull.*, 39(4): 35.

Anonymous (1989). *Annual Progress Report.* Division of Plant Pathology, SKUAST-K, Shalimar, Srinagar 30 pp.

Anonymous (1993). *Annual Progress Report.* Division of Plant Pathology, SKUAST–K, Shalimar, Srinagar 44 pp.

Anonymous (2000). *Kharif Final Forecast Figures.* Bulletin of Financial Commissioner Revenue, J&K Govt. Srinagar, 27 pp.

Babayan, D.N. and Shakhnubaryan, S.T. (1969). Presence of infection by fungal wilt in the seed of tomato, pepper and egg plant under conditions of the Ararat plain, the Armanian SSR. *Uchen–Zap. Erevan, Univ.*, 1(110): 136–147.

Baker, K.F. and Cook, R.J. (1974). *Biological Control of Plant Pathogens.* W.H. Freeman and Company, San Francisco, 433 pp.

Baris, M., Gulsoy, E., Guneu, M., Maden, S., Sagir, A., Senyurek, M., Ulukus, I., Yalcin, O. and Zengin, H. (1986). Investigations on sources of primary inoculum of crown blight (*Phytophthora capsici* L.) of capsicum and control measures against the disease. *Bilki Koruma Bullteni*, 26(3–4): 59–95.

Basak, A.B., Fakir, G.A. and Maridha, M.A.U. (1996). Relation of seed-borne infection to different infection grades in fruit rot disease of chilli. *Seed Research*, 24(1): 30–69.

Beattie, J.H., Doolittle, S.P., Beattie, W.R., Magruder, R. and Webster, R.E. (1944). Production of peppers. *Leaf Literature U.S.D.A.*, 140: 7.

Beckman, C.H. (1987). *Nature of Wilt Diseases.* American Phytopathological Society, St. Paul, 174 pp.

Behera, B., Narian, A. and Swain, N.C. (1989). Physiology and chemical control of *Fusarium pallidoroseum* affecting ground nut. *Orissa J. Agri. Res.*, 2:120–26.

Benlloch, M. and Dominguez, F. (1934). Leanfermedad de Los Pimentales en Aldeanueva Del Camino. Biology, pathology, vegetable and entomology. *Agri. Res.*, 7: 1–20.

Brown, J.G. and Gibson, F. (1926). *Plant Pathology.* 34th Annual Report Arizona Agriculture Experimentation Station for the year ending June 30th 1923, pp. 448–506.

Byrde, R.J.W. (1991). Slide effects of fungicides on crops and ecosystems. *Inter. J. Tropical Pl. Dis.*, 9: 180–193.

Café-Filho, A.C. and Duniway, J.M. (1995). Effects of furrow irrigation schedules and host genotypes on phytophthora root rot of pepper. *Pl. Dis.*, 79: 39–43.

Camino, V., Despestre, T. and Espinosa, J. (1987). Search for *Capsicum annum* susceptibility to *Fusarium. Capsicum Newsletter*, 6: 70.

Chakrabarti, D.K. and Basu-Chaudhary, K.C. (1978). Incidence of wilt of safflower caused by *Fusarium oxysporum* f.sp. *carthami* and its relationship with the age of the host, soil and environmental factors. *Pl. Dis. Rep.*, 62: 776–778.

Chakrabarti, S.K. and Sen, B. (1991). Suppression of Fusarium wilt of muskmelon by organic soil amendments. *Indian Phytopath.*, 44: 476–479.

Chaudry, M.N.A., Akhtar, A.S. and Khan, R.A.A. (1995). Phytophthora problem on chillies and its control. *Capsicum and Eggplant Newsletter*, 14: 62–64.

Chaves, B.A. (1947). Principais doencas das plantas en nordeste. *Brazil Bollettino Agriculture Pernambuco* 14(1): 5–46.

Chupp, C. and Sherf, A.F. (1960). *Vegetable Disease and their Control*. John Wiley and Sons, New York, USA, 693 pp.

Ciferri, R. (1926). Primer Informe, Annual de la Estacion Agron. U. Col. De. Agric. de Haina. Republica Dominicana 1: 27–36.

Clayton, E.E. (1923). The relation of soil moisture to the Fusarium wilt of tomato. *American Journal of Botany*, 10: 133–147.

Crawford, R.F. (1934). The etiology and control of chilli wilt produced by *Fusarium aannuum*. *Technical Bulletin, New Mexico Agriculture Experimental Station*, 223: 1–20 (c.f. 'RAM' XIV: 7).

Dar, G.M. and Mir, N.A. (1995a). Pepper wilt causes and prevalence in Kashmir. In: *Paper Presented at the National Symposium on Recent Trends in the Management of Biotic and Abiotic Stresses in Plants*, Held on 2–3 Nov. at HPKVV, Palampur (India).

Dar, G.M. and Mir, N.A. (1995b). Studies on the management of chilli wilt in Kashmir. : *Paper Presented at the National Symposium on Recent Trends in the Management of Biotic and Abiotic Stresses in Plants*, Held on 2–3 Nov. at HPKVV, Palampur (India).

Dimond, A.E. (1970). Biophysics and biochemistry of the vascular wilt syndrome. *Annual Rev. Phytopath.*, 8: 301–322.

Dwivedi, R.S. and Pathak, S.P. (1981). Effect of certain chemicals on population dynamics of *Fusarium oxysporum* f.sp. *lycopersici* in tomato field soil. In: *Proc. Indian Nat. Sci.Academy*–B, 47(5): 751–755.

Fernandez, M.C. (1983). *Phytophthora capsici:* Casual agent of wilt of *Capsicum aannuum* in chilli. *Agriicultura Technica*, 43: 91–93.

Fournet, J. and Jacqua, G. (1978). Note on attack of Fusarium solani on egg plant (*Solanum melongena*). *Farm Research*, 74: 165–168.

Garcia, F. (1933). Reduction of chilli wilt by cultural method. *New Mexico Agriculture Experimental Station Bulletin*, 216: 1–5 (c.f. 'RAM' XII: 420).

Garofalo, F. (1957). Preliminary experiments on soil sterilization with live steam and with vapam in plantings of egg plant and capsicum. Bollettino Laboratory Sperimentale. *Fitopatholgia*, Torino (Italy) N.S. 20: 59–78.

Griffin, G.J. (1969). *Fusarium oxysporum* and *Aspergillus flavus* spore germination in the rhizosphere of peanut. *Phytopath.*, 50: 1214–1218.

Grover, R.K. and Singh, G. (1970). Pathology of wilt of Okra (*Abelmoschus esculentus* L.) Moench caused by *Fusarium oxysporum* f.sp. *vasinfectum* (Atk.) Synder and Hansen, its host range and histopathology. *Indian J. Agri. Scie.*, 40: 989–996.

Hansford, C.G. (1940). *Report of the Senior Plant Pathologist*. Department of Agriculture, Uganda, Part–II 1938–39, pp 28–29.

Harender, R. and Kapoor, I.J. (1997). Possible management of fusarium wilt of tomato by soil amendments with compost. *Indian Phytopath.*,50: 387–395.

Hashmi, M.H. (1989). Seed-borne mycoflora of *Capsicum aannuum* L. *Pakistan J. Botany*, 21: 302–308.

Hoitink, H.A., Heer, L.J. and Schmitthenner, A.F. (1976). Survival of some plant pathogens during composting of hard tree bark. *Phytopath.*,66: 1369–1372.

Huber, D.M. (1981). The use of fertilizers and organic amendments in the control of plant disease. In: *Handbook of Pest Management in Agriculture, Vol. 1,* (Ed.) D. Pimental. CRC Press, Boca Raton, Florida, pp. 357–394.

Ibrahimillari, L. (1987). Some data on wilt organisms of pepper in the district Tirane of Albania. *Bulletini i Shkencave Bujgexre*, 26(3): 94–100.

Jimenez, J.M., Bustamante, E., Bermudez, W. and Gamboa, A. (1990). Identification and evaluation of sweet pepper lines resistant to fungal blight in Costa Rica. *Turrialba,* 40: 228–234.

Jones, J.P. and Overman, A.J. (1978). Evaluation of chemicals for the control of Verticillium and Fusarium wilt of tomato. *Pl. Dis. Repor.*, 62: 451–455.

Jones, J.P. and Woltz, S.S. (1970). Fusarium wilt of tomato: Interaction of soil liming and micro-nutrient amendments on disease development. *Phytopath.*, 60: 812–813.

Jones, J.P. and Woltz, S.S. (1972). Effect of liming and nitrogen source on Fusarium wilt of cucumber and watermelon. *Proce. Florida State Horti. Soc.*, 88: 200–203.

Jones, J.P. and Woltz, S.S. (1981). Fusarium-incited diseases of tomato and potato and their control. In: *Fusarium Diseases, Biology and Taxonomy,* (Eds.) P.E. Nelson, T.A. Toussoun and R.J.Cooke. Pennisylvania State University Press, Pennisylvania, pp. 157–158.

Jones, M.M. and Black, L.L. (1992). Source of resistance among *Capsicum* spp. to fusarium wilt of pepper. *Capsicum Newsletter*, 11: 33–34.

Kalloo, G. and Paroda, R.S. (1995). Vegetable Research with special reference to hybrid technology in the Asia–Pacific region. Food and Agricultural Organization, Bangkok, Thailand, 117 pp.

Kannaiyan, J. and Nene, Y.L. (1979). Effect of crop rotation and planting time on the incidence of wilt in lentil. *Indian J. Pl. Prot.*,7(2): 114–118.

Kapoor, I.J. (1988). Fungi involved in tomato in tomato wilt syndrome in Delhi, Maharashtra and Tamil Nadu. *India Phytopath.*, 41: 208–213.

Kapoor, K.S. and Sharma, S.R. (1988). Soil application of fungicides against wilt of egg plant. *Capsicum Newsletter*, 7: 90–91.

Kato, K. and Tomita,I. (1981). Effect of successive applications of various soil amendments on tomato Fusarium wilt. *Research Bulletin, Airchi Agriculture Research Center*, 13: 199–208.

Kaur, S. (1993). Fusarium wilt: A cause of chilli crop failure in Punjab. *Pl. Dis. Res.*, 8: 181–183.

Kaur, S., Hundal, J.S., Khurana, D.S. and Jindal, S. (1999). Evaluation and utilization of chilli (*Capsicum annum*) germplasm for improving resistance to wilt disease. In: *National Symp. on Plant Disease Scenario under Changing Agroecosystems*, HPKVV, Palampur.

Kaur, S., Kaur, R., Kaur, P. and Singh, D. (1985). Studies on wilt and fruit rot of brinjal caused by *Fusarium semitectum*. *Indian Phytopath.*, 38: 736–737.

Kelman, A. and Cook, R.J. (1977). Plant pathology in the peoples Republic of China. *Annual Rev. Phytopath.*, 17: 409–429.

Kim, K.Y., Park, S.K., Shin, Y.A. and Lee, E.J. (1990). Survey on the actual condition of cultural methods and continuous cultivation injury of red pepper in field. *Acta Horti.*, 32(1):1–10.

Koleva,K. and Vitanov, M. (1990). *Fusarium* species related to root rot of pepper. *Rasteniev dni Nauki*, 27(6): 61–63.

Kozlowska,C. (1964). Investigation on the biology of *Fusarium oxysporum* and attempts to control it. *Proce. Inst. Badaw*. Lesn., 246: 3–91.

Kumar, A., Aulakh, K.S., Grewal, R.K. and Kumar, A. (1986). Incidence of fungal fruit rots of brinjal in Punjab. *Indian Phytopath.*, 39: 482–485.

Kumar, T.S. and Lokesh, S. (1999). Evaluation of seed mycoflora of tomato and their management *in vitro*. *Seed Research*, 27: 181–184.

Liang, I.Z. (1990). Seed borne Fusarium of chilli and their pathogenic significance. *Acta Phytopathologica Sinica*, 20(2): 117–121.

Lukas, J. and Sjarka, J. (1988). Fusarium wilt of capsicum, A paprika Fusarium hervadasa. *Zoldse gtermeszlesi kutato interzet bulletine*, 21: 95–99.

Marchionatto, J.B. (1935). Species of Fusarium existing in Argentina. *International Bulletin of Plant protection*, 9(6): 125–128.

McRae, N. (1932). *Science Report for 1930–31*. Imperial Institution of Agriculture, Research, Pusa, 84 pp.

Movens, M. and Benaicha, B. (1990). Control of pepper wilt in Tunisia. *Parasitica*, 46(4): 103–109.

Najar, A.G. (2001). Cause and management of chilli wilt in Kashmir. *Ph.D. Thesis*, SKUAST–K, Shalimar, Srinagar, India, p. 15–159.

Nash, S.M., Christou, T. and Synder, W.C. (1961). Existence of *Fusarium solani* f.sp. *phaseoli* as chlamydospores in soil. *Phytopath.*, 51: 308–312.

Nasreen, S., Khan, S.A.J. and Khanzada, A.K. (1988). A new fusarium wilt of okra in Pakistan. *Pak. J. Scie. Indu. Res.*, 31(8): 577–578.

Nayeema, J., Ahmad, N., Tanki, M.I. and Dar, G.M. (1995). Screening of hot pepper germplasm for resistance to Fusarium wilt (*Fusarium pallidoroseum* [(Cooke) Sacc]. *Capsicum and Eggplant Newsletter*, 14: 68–71.

Nedumaran, S. and Vidhyasekaran, P. (1981) Control of *Fusarium semitectum* infection in tomato seed. *Seed Research*, 9(1): 28–31.

Nedumaran, S. and Vidhyasekaran, P. (1982) damage caused by *Fusarium semitectum* in tomato. *Indian Phytopath.*, 35: 322–323.

Nelson, P.E. (1981). Life cycle and epidemiology of *Fusarium oxysporum*. In: *Fungal Diseases of Plant Species*, 51–78, (Eds.) M.E. Mace, A.A. Bell and C.H. Becman. Academic Press, New York, 640 pp.

Nikolaeva, V. (1978). Results of trials of chemical preparations in the control of the pathogen of Fusarium wilt of tomato in the glass house. *Lozarska Nauka*, 15(1):66–71.

Nirwanto, H. and Djajati (1994). Study on the use of soil amendment with leguminous foliage and liming in effort to control fusarium wilt disease of tomato. *Acta Horticulturae*, 36: 155–162.

Pandey, U. (1976). Fungi associated with seeds of chillies grown in kumaon hills. *Indian Phytopath.*, 29: 472.

Pontis, R.E. (1940). El marchitonatto del Pimiento (*Capsicum annum*) en la provincial de Mendoza. *Rev. Argentina Agro.*, 7: 113–127.

Ristaino, J.B. (1991). Influence of rainfall, drip irrigation and inoculum density on development of Phytophthora crown and root rot epidemics and yield in bell pepper. *Phytopath.*, 81: 922–929.

Rivelli, V. and Black, L.L. (1992). Pathogenicity of *Fusarium oxysporum* f.sp. *capsici* to *Capsicum* spp. And the effect of temperature and seedling age on disease severity. *Plant Disease*, 76: 340–346.

Saharan, G.S. and Gupta,V.K. (1973). Pod rot and collar rot of soybean caused by *Fusarium semitectum*. *Pl. Dis. Repor.*, 52: 1330.

Sanford, G.B. (1926) Somr factors affecting pathogenicity of Actinomyces scabies. *Phytopath.*, 16: 525–527.

Sarhan, A.R.T. (1982). The influence of soil pH on the severity of tomato fusarium wilt infection. *Acta Phytopathologica Academiae Scientiarum Hungaricae*, 17: 292–294.

Sarhan, A.R.T and Sharif, F.M. (1986). Integrated control of Fusarium wilt of pepper. *Acta Phytopathologica et Entomologica Hungarica*, 21: 123–126.

Saxena, S.K. (1989). Soil moisture regimes in relation to Fusarium wilt of potato. *Proc. Nat. Symp. Epide. Fore.* Pl. Dis. at IARI, New Delhi.

Sen, B. (1986). Cultural management of soil borne diseases. In: *Vistas in Plant Pathology*, (Eds.) A. Verma and J.P. Verma. Malhotra Publishing House New Delhi, pp. 367–381

Sen, B. and Kapoor, I.J. (1974). Chemical control of wilt of tomato. *Pesticides*, 8: 40–42.

Shamsher, K., Kassim, M.Y., Abou Heilah, A.N. and Sheir, H.M. (1983). Effect of soil treatment with some fungicides on fusarium wilt of tomato. *Inter.J. Trop. Pl. Dis.*, 1(1): 61–64.

Singh, A. and Singh, A.K. (1995). Field performance of chilli genotypes to Fusarium wilt under natural infection. *Nat. Symp.*, HPKVV, Palampur.

Singh, R.S. (1983). Organic amendments for root disease control through amendment of soil microbial and the host. *Indian J. Mycol. Pl. Pathol.*, 13: 1–16.

Singh, S.P. (1989). *Production Technology of Vegetable Crops*. Agriculture Research Communication Centre, Karnal, India, pp. 2.

Swain, N.C., Narain, A., Behera, B. and Sahoo, K.C. (1989). Wilt of black gram caused by *Fusarium pallidoroseum*: A new record. *Indian J. Mycol. Pl. Pathol.*, 19: 172–177.

Thind, T.S. and Jhooty, J.S. (1985). Realative prevalence of fungal diseases of chilli fruits in Punjab. *Indian J. Mycol. Pl. Pathol.*, 15: 305–307.

Thomas, K.M. (1938). Detailed administration report of the Government of Mycologist, Madras, for the year ending 1937–39, pp. 21.

Thurston,H.D. (1990). Plant disease management practices of traditional farmers. *Pl. Dis.*, 74: 96–102.

Vaughn, E.K., Roberts, A.N. and Mellenthin, W.M. (1954). The influence of Douglas fir saw dust and certain fertilizer elements on the incidence of red stele disease of strawberry. *Phytopath.*, 44: 601–603.

Verma, B.R. and Sharma, B.K. (1995). Mangement of chilli wilt. *Nat. Symp.*, HPKVV, Plampur.

Vidhyasekaran, P. and Thiagarajan, C.P. (1981). Seed borne transmission of *Fusarium oxysporum* in chilli. *Indian Phytopath.*, 34: 211–213.

Vishwakarma, S.N. and Basu-chaudhary, K.C. (1982). Studies on survival of *Fusarium solani*, incitant of gram wilt. *Indian Phytopath.*, 35: 624–627.

Vitanov, M. (1989). Movement of *Phytophthora capsici* in soil and infection of pepper. *Rasteniev dni Nauki*, 26(4): 60–66.

Wang, M. and Wang, L.L. (1989). Studies on the resistance to Fusarium wilt of pepper (*Capsicum annum*). In: *Eucarpia* VIIth meeting, Kraqujevac, Yugosalavia, Institute for vegetable planka, Yugoosalavia, pp. 159–164.

Wani, M.A. (1994). Studies on fungal wilt of chilli (*Capsicum annum* L.). *M.Sc. Ag. Thesis*, Submitted to the Faculty of Post Graduate Studies, SKUAST–K, Srinagar (J&K), pp. 74.

Plant Diseases Management in Horticultural Crops (2011) *Pages* **120–134**
Editors: **Shahid Ahamad, Ali Anwar and P.K. Sharma**
Published by: **DAYA PUBLISHING HOUSE, NEW DELHI**

Chapter 10

Resistance in Plants: Ecofriendly Disease Management

Nasreen Fatima[1], Saba Banday[1], Sabiha Ashraf[1], S. Asifa Bukhari[2] and Shaheeda Iqbal[2]

[1]Division of Plant Pathology, [2]Division of Plant Breeding and Genetics, Sher-e-Kashmir University of Agricultural Sciences and Technology of Kashmir, Shalimar, Srinagar – 191 121

There is a serious concern for food security especially in developing countries like India, because of increasing food demand for rapidly expanding population, declining productivity and increasing vulnerability of agriculture to biotic and abiotic stresses, among the management strategies available for their management, the chemical strategies so for dominated to farmer community. Over emphases on chemical control of biotic and abiotic stresses caused serious imbalance in our agro-system. In addition, their high cost makes them inaccessible to many farmers and extended application of agrochemical reduce their effectiveness of development of resistance to chemical biotic stresses. Thus the problem of increasing production and productivity of agricultural produce unsustainable manner but not at the cost of human health and the loss to farmers is being solved to some extent by introduction of resistance in plants through breeding techniques whether conventional or non-conventional, employed to evolve the resistant varieties by the plant breeders.

Resistance Necessity

Biotic Stresses Causing Huge Losses

Biotic stresses reduce biomass (dry matter) of the crop in one or more of the following ways :

- ☆ Killing of plants
- ☆ Killing of branches
- ☆ General stunting

- ✰ Damage to the leaf tissues
- ✰ Damage to the reproductive organs including fruits and seeds.

Loss due to disease is caused by abiotic stresses may be ranged from a few to 20-30 per cent, in cases of severe epidemic.

Disappearance of Otherwise Excellent but Susceptible Varieties

This is the most striking effect of the plant diseases lending to the disappearance from cultivation of otherwise excellent but susceptible varieties *i.e.*, wheat variety Kalyansona dominated the Indian agriculture for about a decade, but had to abandoned as it become susceptible of leaf rust. Similarly, another wheat variety, Janak was eliminated from cultivation due to its susceptibility to karnal bunt.

Disease Scenario in Crop Plants keeps on Changing

For each crop species the importance of different diseases keeps on changing. Diseases which were minor in the past become important due to changes in crop varieties and agricultural practices *e.g.* Helminthosporum leaf blight in maize and Tungro virus in rice.

Evolution of New Pathotypes

New pathotypes have been emerged in most of the pathogens due to selection pressure. Thus, the resistance breeding is a continuous process and the task of the breeders not only to breed for the existing pathotypes but also for them which are likely to emerge in near future (Singh, 2002).

Sources of Resistance in Plants

The basis requirement for breeding disease resistant plants in sources of disease resistance and methodology for combining resistance with commercially or aesthetically acceptable plant types. To produce a new cultivars, there must be continual search for germplasm with disease resistant traits that meet the changing needs. Primary sources of disease resistance are being included :

A Known Variety

Disease reactions of most of the cultivar varieties are documented, and a breeder may find the resistance. He needs in a cultivated variety resistant plants were isolated from commercial varieties in the cases of cabbage yellows (causal agent, *Fusarium, oxysporum* f. sp. *conglutinans*) in cabbage (*B. oleracea*), only top resistance in susceptible (*B. vulgaris*) etc. These plants provided the basis for new resistant varieties of these crops.

Germplasm Collection

Often resistance to a disease may not be present in the varieties of the concerned crop species. In such cases, it would be necessary to transfer resistance genes from related species through interspecific hybridization. Resistance to yellow mosaic has been transferred from the wild species, *Abelmoschas monihot* and a yellow mosaic resistant variety of bhindi (*Abelmoschas esculentus*), Parbhani Kranti, has been evolved and achieved by such programme.

Related Species

When resistance is not present in the concerned varieties of the crop species, then, transfer of disease resistance genes from related wild species is done by inter-specific hybridization. This method has been used successfully and extensively in number of cases *e.g.* resistance to all three rusts in wheat and resistance to yellow mosaic in bhindi from wild species *Abelmaschas monihot* (Singh, 2002).

Mutation

Resistance to some disease may be obtained through mutations arising spontaneously, in some cases, or induced through mutagen treatment. Resistance to Victoria blight caused by *Helminthosporium victoriae* in oats (*A. sativa*) was induced by irradiation with X-rays or thermal neutrons, resistant mutants were also isolated spontaneously in low frequencies.

Somoclonal Variation

Genetic varieties excess in plant cell cultivated *in vitro* and in plants regenerated from such cells and in the progeny of such plants. Somoclonal variants resistant to various diseases have been isolated in many crops and in some of these have been released as varieties.

Unrelated Organism

Efforts are being made to utilize genes from other organisms to produce disease resistance in plants. These genes may be obtained from the pathogens themselves *e.g.* coat protein gene of pathogenic virus for varies resistance of from other plant example genes for novel phytoalexins.

Conventional Approaches

Selection

Selection is the primary aspect of isolating sources of disease resistance. But, the major drawback of selection is that we can choose only those individuals, which have vertical resistance as it can be easily overcome by the appearance of undetected races of the pathogen.

Plant Introduction

Collection of related materials from other countries particularly from areas where the pathogen and host species have co-evolved.

Induced Mutation

It has been succeeded by inducing the resistance in barley against common areas of *Erysiphe graminis hordei.*

Hybridisation

It is the most common method of breeding for disease resistance. It serve the following two major purposes:

1. Transfer of resistance from an agronomically undesirable variety in high yielding commercial backcross method.
2. Combining disease resistance and some other desirable characters of one variety with the superior characters of another variety by pedigree method.

Non-Conventional Approaches

Somoclonal Variation

Disease resistant somoclonal varieties can be obtained in the following two ways :

1. Plant regenerated from cultural cells or their progeny are subjected to disease test and resistant plants are isolated.

2. Cultured cells are selected for resistance to the toxin or culture filtrate produced by the pathogen and plants are regenerated from the selected cells.

Genetic Engineering

Genes expected to confer disease resistance are isolated, cloned and transferred into the plant. A plant in which a gene has been transferred through genetic engineering is called transgenic plants. Transgenic plant can be obtained by two methods:

1. Vector mediated
2. Vector less

By the use of vector mediated or vector less techniques to product transgenies.

Conventional Method

Among the conventional methods following two types of methods can be used :

Backcross Method

By the use of this method we can transfer a desirable trait directly or by the production of

1. Alien addition lines
2. Alien substitution lines
3. Pedigree method

Non-Conventional Method

1. Protoplast fusion
2. Embryo rescue
3. Genetic engineering

Conventional method has been described by Singh 2002 as follows :

Backcross Method

Backcross is a cross between a hybrid (F_1 or segregating generation) and one of its parents. In backcross method, the hybrid and the progenies in the subsequent generation are repeatedly backcrosses to one of their parents. As a result the genotype of backcross progeny would be increasingly similar to the parent of which the backcross are made :

Objectives of a Backcross

To transfer 1 or 2 desirable in variety which is well adapted by the farmers for specific region to the prevalent biotic stresses.

1. A suitable recurrent parent receipt parent
2. A suitable donor parent (non-recurrent parent)
3. Character to be transfer should be highly heritable and should be determined by one or few genes.
4. A sufficient number of backcrosses should be made so that genotype of the recurrent parent should be made. Generally 4-6 backcrosses are made.

The plant of transfer of gene by backcross method is depending upon whether the gene to be transferred is dominant or recessive (Figure 10.1).

Table 10.1: Successful Examples of Transfer of Resistance by Backcross

Cultivated Plant	*Donor*	*Pathogen*	*Reference*
Wheat (*Triticum aestivum*)	*Agropyron elongatum*	*Puccinia graminis tritici*	Roeffs (1988)
Rice (*Oryza sativa*)	*Oryza minata*	*Xanthomonas oryzae py.oryzae* and *Magnaporthae grisea*	Bordeas *et al.* (1992)
Bean (*Phaseolus vulgaris*)	*Phaseolus* (runner bean)	*Xanthamonas axanopodis*	Kumar *et al.* (2005)

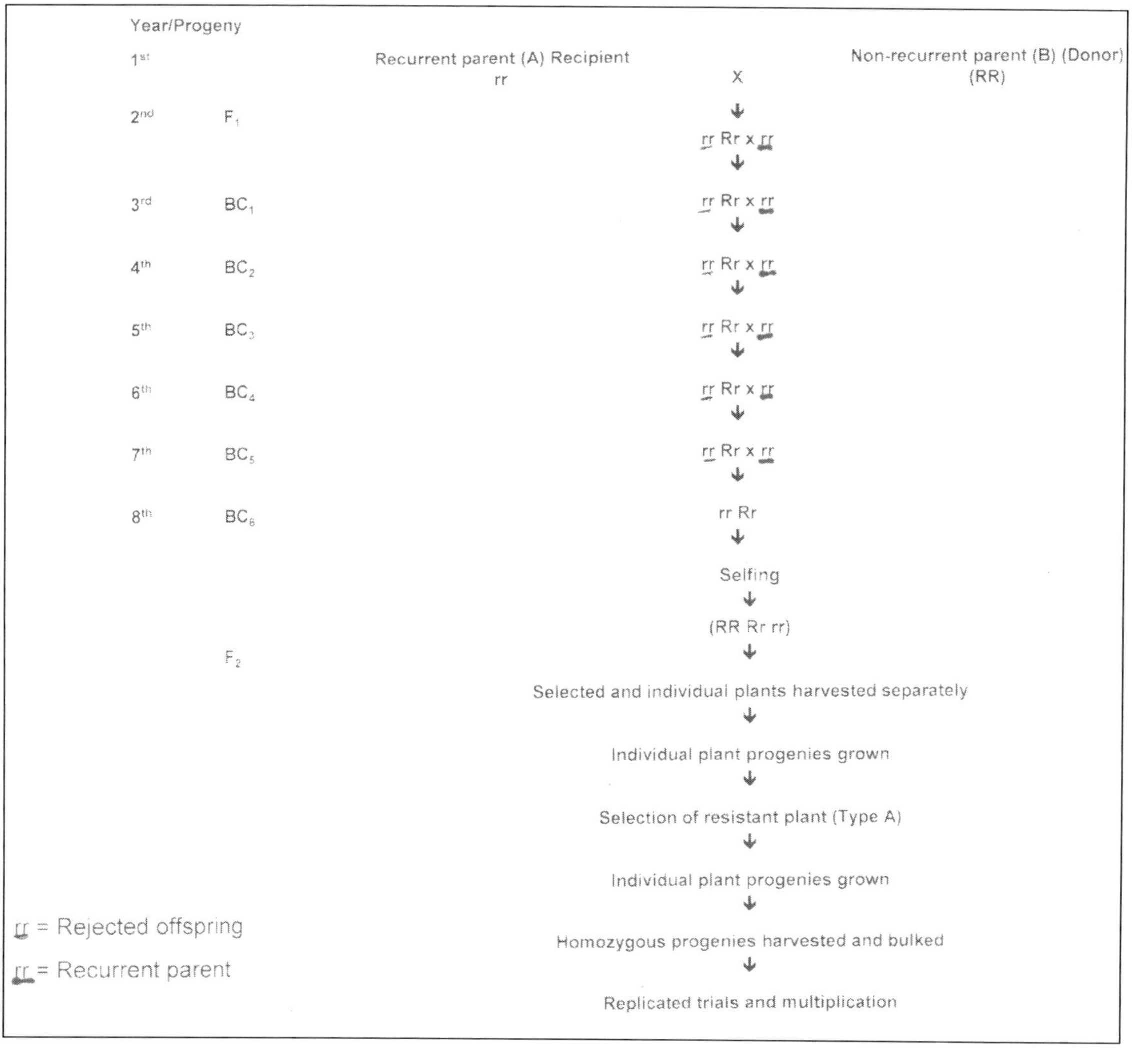

Figure 10.1: Transfer of Dominant Gene

Production of Alien Addition and Substitution Lines by Backcross Method

By the use of the backcross method the desired genes from the wild and alien source can be transferred by the production of alien addition lines or alien substitution lines Figure 10.2.

Alien Addition Line

It is a line which carries one chromosome pair from the donor parent in addition to the whole chromosome complement of the parent plant.

Figure 10.2: Transfer of a Recessive Gene by Backcross Method

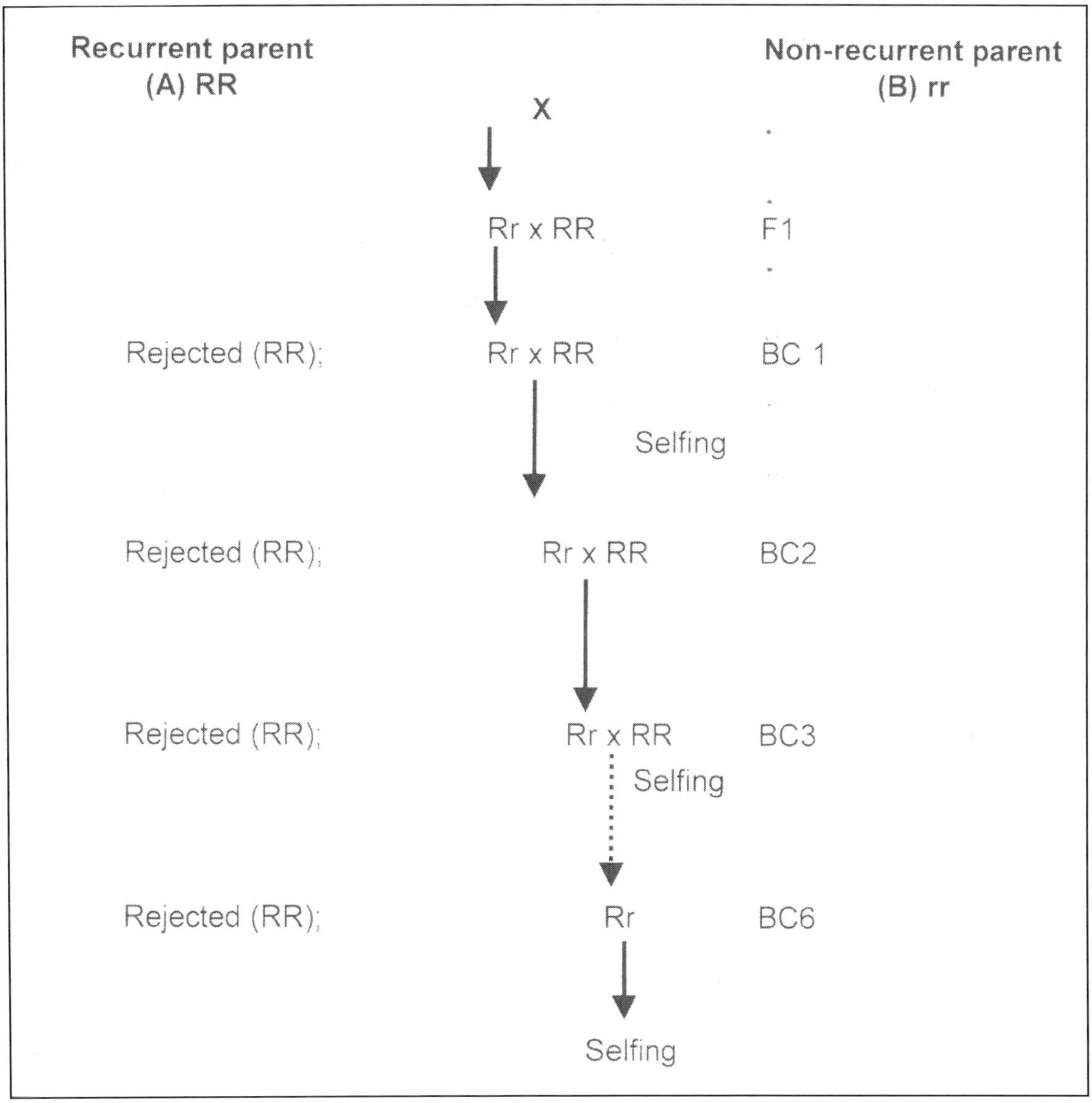

Contd...

Figure 10.2–Contd...

Selected and individual plants are harvested separately

↓

Individual plant progenies are grown

↓

Selection for resistant plant (Type A)

↓

Individual plant progenies are grown

↓

Homozygous progenies harvested and bulked

↓

Replicated trials and seed multiplication

Alien Substitution Line

Carries one chromosome pair from a different species in place of the chromosome pair of parent plant.

Table 10.2

Cultivated Plant	*Donor*	*Pathogen*	*Reference*
Wheat (*Triticum aestivum*)	*Aegilops triuncalis*	*Puccinia recondita*	Sarberzch *et al.* (2002)
	Aegilops geniculata	*P. striformis*	-do-
	Thinopyrum touschii	*Fusarium graminearum*	Oliver *et al.* (2005)

Pedigree Method

It has reported and promoted this method to the stable resistant genotypes against diseases. In this method individual plants are selected from F_2 and the subsequent segregating generation and their progenies are tested. During the entire operation a record of all the parent offspring relationship is considered to get relevant and desirable trait of crop. Individual plant selection is contributed till the progenies become homozygous and they do not the further segregation at this stage. Selection is done among the progenies because it should not genetic variation within the progenies (Figure 10.3).

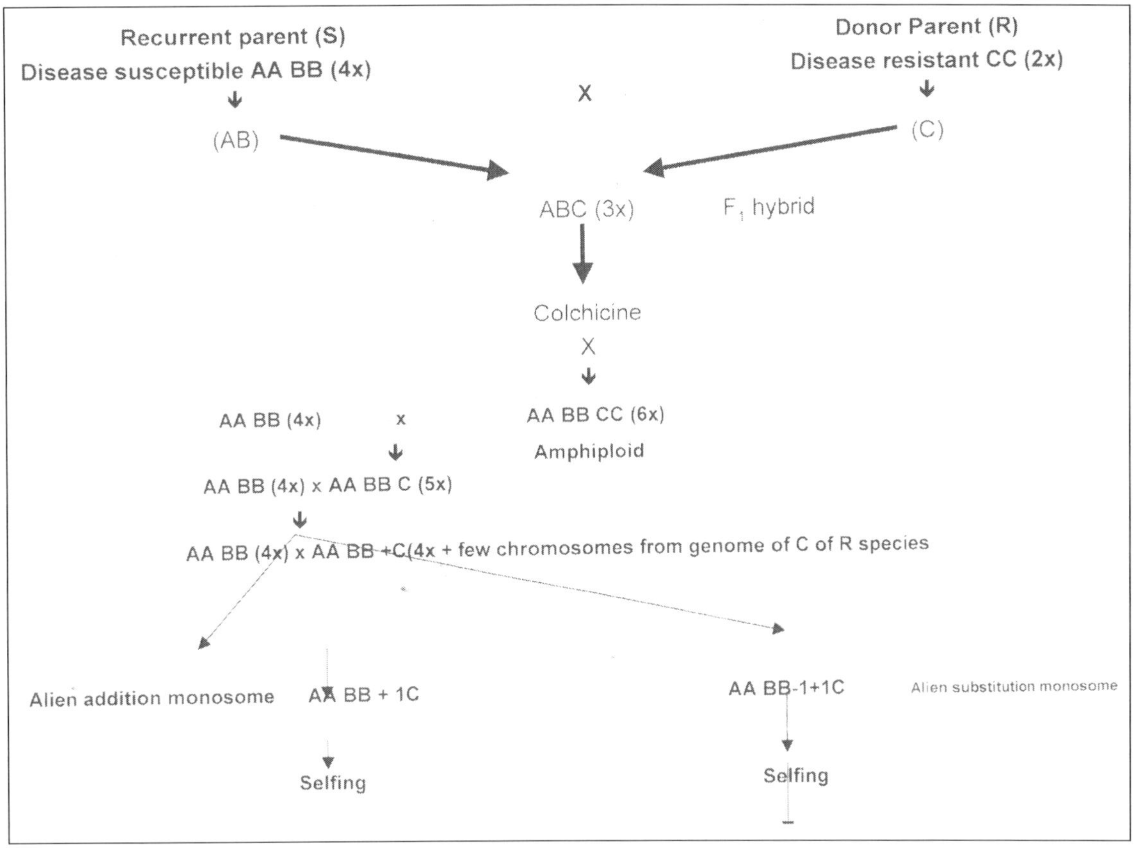

Figure 10.3: Production of Alien-Addition and Alien-Substitution Lines

Non-conventional Approaches

Gupta (2000) described the non-conventional method as follows:

Somatic Hybridization

Somatic hybridization is one of the most important uses of protoplast culture. This is particularly significant for hybridization between species or genera, which can not be made to cross by conventional method of several hybridization.

It involved 2 steps :

1. Protoplast isolation
2. Protoplast fusion

Disease Free Seed Produce through the Pedigree Method

The pedigree may be defined as a description of the ancestors of an individual and it generally goes back to some distant ancestors or ancestors in the past.

Table 10.3: Pedigree Method

	Selected parents planted in a crossing blocks and crosses made
F_1	10-30 seeds pace planted, harvested in bulk
F_2	i) 2000-10,000 plants pace planted
	ii) 100-150 superior plants selected
F_3	i) Individual plant progenies pace planted
	ii) Superior plant selected
F_4	As in-i-ii in F3
F_5	i) Individual plant progenies planted in multirow plots
	ii) Superior plants selected from superior progenies
F_6	i) As in I and ii in F_5
F_7	i) Primary yield trials
	ii) Quality test
F_8-F_{10}	i) Coordinated yield trials
	ii) Disease and quality tests
F_{11}	i) Seed increase for distribution begins

Protoplast isolation

By treating the cell with a mixture of cell wall degrading enzymes (usually a mixture of pectinase (0.1-1 per cent) and cellulose (1-2 per cent). Osmotic concentration of the mixture is increased by the addition of 500-800 m mol^{-1} sorbitol or mannitol. This is done for few to several hours. Naked cells/ protoplasts will come in the enzyme mixture. These protoplasts are washed with a washing medium in order to remove the enzymes and the debris. Then these protoplasts recultured on a suitable media and they generate cell wall followed by mitosis and numbers of colonies are produced. From here whole plant can be regenerated.

Protoplast Fusion

Various techniques are available for protoplast fusion as given below (Figure 10.4).

- High pH high Ca^{++} treatment
- Polyethylene glycol (PEG) method
- Electrofusion technique

*High pH high Ca** Treatment*

Protoplasts of different species are mixed in equal proportion. It involves spinning protoplasts in a fusion inducing solution (0.05 m (add $2H_2O$ in 0.4 M manitol at pH 10.5) for 30 minutes at 50 g, after which the tubes leads to water both (37 °C) for 40-50 minutes. This leads to fusion of 20-250 per cent of the protoplasts.

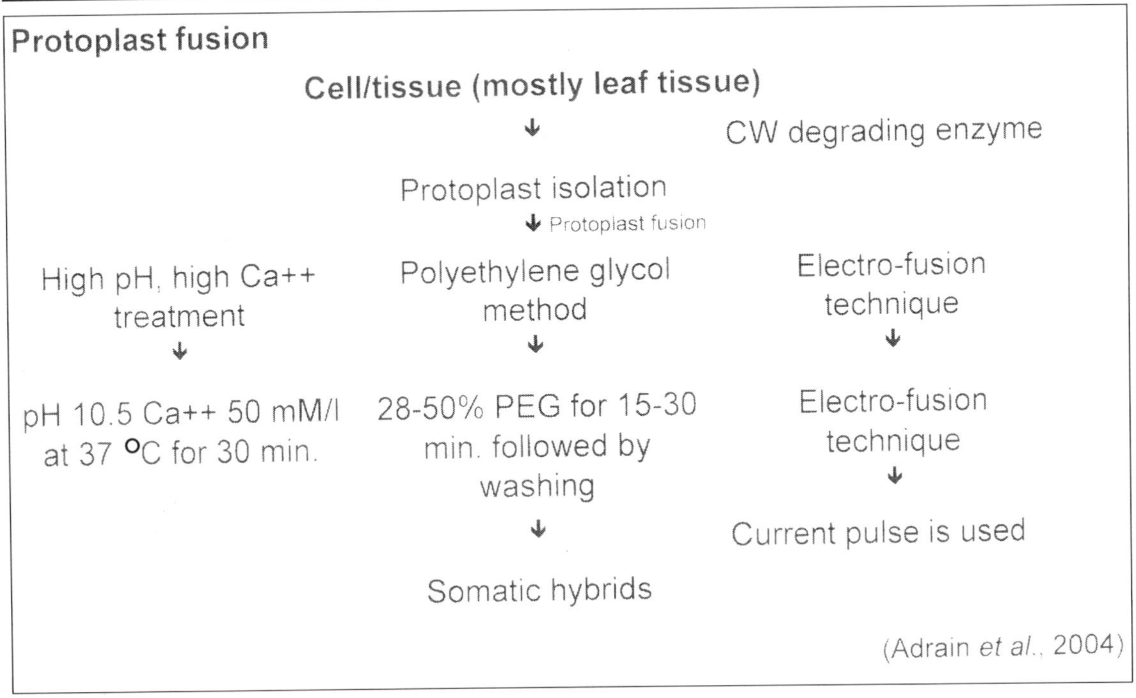

Figure 10.4: Non-conventional Approaches

Polyethylene Glycol (PEG) Treatment

The agglutination of protoplast, during PEG-treatment, can be brought about by the following two different methods :

1. When protoplasts are available in sufficient quantities 1 ml of culture medium with suspended protoplasts is added to 1 ml of 56 per cent solution of PEG and the tube shaken for 5 seconds. The protoplasts are allowed to sediment for 10 minutes, were led with growth medium and examined for successful agglutination and fusion.
2. If protoplasts are available in micro quantities drop cultures can be used. Two types of protoplasts are mixed in equal quantities and 4-6 micro-drops (100 µl each) are placed in a small petriplate and allowed to settle for 5-10 minutes at room temperature. Two to three micro drops (50 µl each) of PEG are added from the periphery in each petriplate which is incubated for 30 minutes at room temperature (24 °C). This levels to agglutination of protoplasts.

After PEG treatment, protoplasts are gradually washed and during the process, most of the fusion is achieved. PEG is then replaced by culture medium to allow growth of fusion protoplast.

Electrofusion Technique

Here current pulse is used for protoplast fusion. Protoplasts are arranged in a row and passed through a narrow gap between the electrodes. When the two protoplast come near to each other which are to be used. They orient them opposite to each other in between the electrodes. Following this a short pulse of high voltage is passed which leads to the protoplast fusion (Figure 10.4).

Table 10.4: Successful Examples of Somatic Hybrids

Cultivated Plant	*Donor*	*Pathogen*	*Reference*
Brassica (*B. rapa*)	*B. oleracea*	*Erwinia coratovora* subsp. *corotovora*	Kaloo (1988)
Brassica (*B. napus*)	*B. nigra*	*Phoma lingam*	Singh (2002)

Embryo Rescue Technique

In case of endosperm abortion hybrid plants may be raised by culturing the hybrid embryos on a suitable culture medium. In this case the embryos are removed from young seeds before endosperm abortion takes place. This called embryo rescue.

General Approach

Isolation of immature heart shaped embryos from the seed

↓

Culturing on artificial medium

↓

Transfer to rooting medium

↓

Hardening and regeneration of the plant

Genetic Engineering

General approach

Designing gene construct

↓

Introduction of gene of interest

↓

Integration of gene in genome of plant

↓

Expression of transgene

↓

Regeneration of whole plant

↓

Transmission of the transferred gene to sexual progeny of these plants

Methods of gene transfer

Vector mediated method

↓

Ti or Ri plasmid by *Agrobacterium* spp.

Caulimovisu or Gemini virus vectors

Vector less method

↓

Particle gun method

PEG induced DNA uptake electroportion

Direct DNA uptake by cells

Microinjection

Lipoinfection

Ribosome mediated DNA uptake

Vector Mediated Gene Transfer Methods

The Ti plasmid or Ri plasmid of *Agrobacterium tumefaciens or A. rhizogenes* is used mostly. In addition DNA viruses such as caulimovirus or geminiviruses can be used as vector for gene amplification because viruses multiply in plant cells. But, the virus genome is neither integrated into the plant cell nor they are transmitted to the seed. Therefore, these cannot be used for stable gene transfer.

Therefore, mostly the Ti plasmid is used for gene transfer. Fi plasmid have five regions *viz.,*

T-DNA region : These region transferred into the host genome

Vir region : Virulence region

Ori region

Occ region

Tra region

- ☆ T-DNA is separated from other regions by left and right repeat border sequences. Here tmsI and tms2 serve for IAA production and tmr for cytokinins production.
- ☆ Vir region mediates the transfer of I-DNA. Therefore, it is essential for vigilance. Here vir A acts as a receptor site for acetosytringone. Vir G acts a DNA binding protein and forms a dimmer after phosphorylation by vir A and induces expression of all vir operations. For the transfer of desired gene through Ti-plasmid is called disarmed plasmid. In place of this T-DNA of the Ti plasmid the desired DNA is incorporated (Figure 10.5).

Vectorless Gene Transfer

Particle Gun Method or Biolistic Method of DNA Delivery

In this method DNA is delivered by high velocity metal particles. Tungstem or gold particle (microprojectile) coated with DNA to be used for transformation are accelerated to high velocity (1400 ft/sec.) velocity which enable their entry into plant cell.

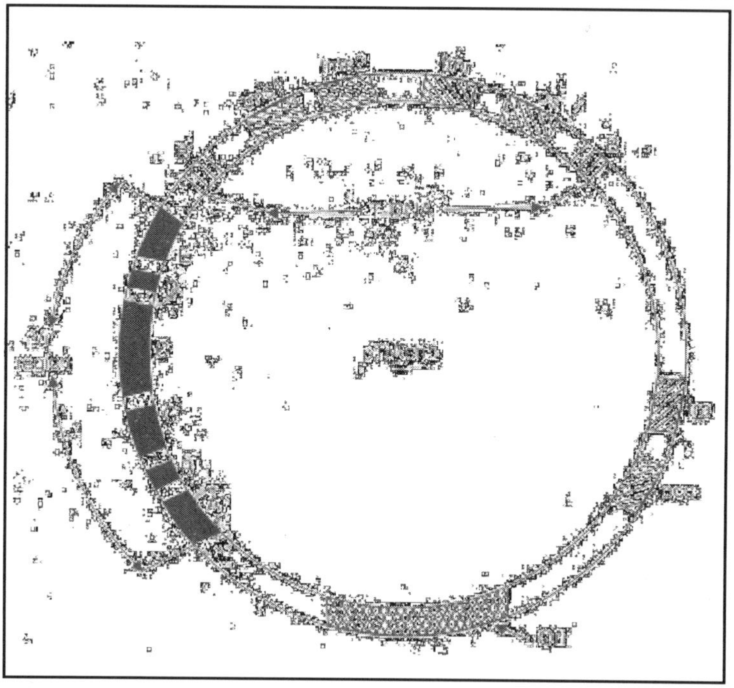

Figure 10.5: A schematic representation of the Ti plasmid of *A. tumefacients*

The T-region (*T*-DNA) is borded on both the sides by a direct repeat sequence of 25 bp (B); *T*-DNA contents genes *tms* 1 (now, *iaam*), *tms* 2 (now, *iaah*), *tmr* (now, *ipr*) and *os*. The *vir* region has at least six operons; *vir D* operon codes for an endonuclease responsible for the excision of a single stranded copy of *T*-DNA from the Ti plasmid and then guiding this DNA into the plant cell nucleus.

PEG Induced DNA Uptake

Polyethylene glycol enhance the uptake of DNA by protoplast plant protoplast are suspended in transformation medium rich in Ca ion. The plasmid DNA containing the gene constructs is added to the protoplast suspension followed PEG is added the pH is adjusted to 8.0

Electroporation

Induction of DNA into cell by exposing them for very brief period to a high voltage electric pulses which induce transient pore in the plasmalemma is known as electroportion.

Direct DNA Uptake by Cell

DNA is taken up directly plant cell, plant protoplast, pollen grain, mature embryos and even whole seed, in many cases a transient expression of foreign gene has also been reported.

Microinjection

In case of microinjection the DNA solution is injected directly inside the cell using micropipette with the help of micro-syringe. It is easier to use protoplast than cell since cell was interface with the process of microinjection.

Liposome–Mediated DNA Uptake

The DNA is enclosed within artificial lipid vesicles, which are taken up by plant protoplast through fusion.

Table 10.5

Cultivated Plant	Donor	Pathogen	Gene Transferred	Reference
Potato (*Solanum tubersum*)	*Bacillus ancyloliquifaciens*	*Phytophthora infestans*	Barnase	Singh (2002)
Wheat (*Triticum aestivum*)	*Tricicum dicoccum*	*Puccinia graminis tritici*	Sr_2	Gabrield and Rolfe (1990)
	Aegilops	–do–	Sr_{26}	–do–
	Sacale cereale	–do–	Sr_{31}	–do–

Conclusion

Disease resistance is still an exciting and highly promoting area of research because advancement in the knowledge of this field will directly help to boost up our economy. Chemicals means of controlling disease are inadequate and give adverse effect on our ecosystem. Wild and alien plants are always a source of potential resistance genes for both conventional plant breeding as well as for producing disease resistant transgenic plants. Transgenic not only gear up the speed for production of disease resistant plants but are also ecofriendly which is the need of hour.

References

Adrian, S., Nigel, S. and Mark, F. (2004). *Plant Biotechnology*. Oxford University Press, pp. 35–177.

Agrios, G.N. (2004). *Plant Pathology*. Academic Press, London, pp. 416.

Ashok, J., Fakruddin, B., Kuruvinashetti, M.S. and Kullaiswami, B.Y. (2005). Regeneration of stem and pod resistant groundnut plants from *Sclerotium rolfssi* Sacc. culture filtrate treated callus. *Indian J. Genetics*, 64(3): 221–224.

Bahadur, P., Singh, D., Kumar, A. and Charan, R. (2002). Genetics of resistance to leaf rust in wheat. *Indian Phytopath.*, 55: 438–443.

Bordeos, A.A., Sitch, L.A., Nelson, R., Dalmacio, R.D., Oliva, N.P., Aswiddinnoor, H. and Leung, H. (1992). Transfer of bacterial blight and blast resistance from tetraploid wild rice *Oryza minuta* to be cultivated rice. *Oryza sativa. Theoretical and Applied Genetics*, 84: 345–354.

Dantre, R.K., Rathi, Y.P.S. and Sinha, A.P. (2003). Elicitation of resistance in rice plants against sheath blight with non-conventional technique. *Indian Phytopath.*, 56(3): 283–285.

Gabriel, D. and Rolfe, B. (1990). Working models four specific recognition in plant microbe interactions. *Ann. Revi. Phytopath.*, 218: 365–391.

Gupta, P.K. (2004). *Elements of Biotechnology*. Rastogi Publications, pp. 277–346.

Kaloo, G. (1988). *Vegetable Breeding*, 3 Vols. Panima Education Book Agency, New Delhi, p. 284.

Kumar, A., Kaushal, R.P. and Sharma, R. (2005). Transfer of resistance to floury spot and bacterial blight in French bean from alien germplasm. *Indian Phytopath.*, 54(2): 136–142.

Lewis, R.S., MIlla, S.R. and Levin, J.S. (2005). Molecular and genetic characterization of *N. glutinosa* L. Chromosome segment in Tobacco. Mosaic virus resistant accessions. *Crop Science*, 45: 2355–2362.

McIntosh, R.A., Willings, C.R. and Park, R.F. (1995). Wheat rusts. *An Atlas of Resistance Gene* (SIRO Australia), pp. 200.

Moghaddm, R.A., Willings, C.R. and Park, R.F. (2005). Genetics of resistance in bread wheat cultivars to leaf rust. *Indian Phytopath.*, 57(4): 422–426.

Oliver, R.E., Cai, X., Zu, S.S., Chen, X. and Stack, R.W. (2005). Wheat aliend species derivatives: A novel source of resistance to Fusarium head blight in wheat. *Crop Science*, 45: 1353–1360.

Raman, M. and Hanneman, R.E. (2002). Introgression of resistance to late blight (*Phytophthora infestans*) from *Solani pimatisectum* into *Solanm tuberosum* using embryo rescue and double pollination. *Euphytica*, 127: 421–452.

Roelfs, A.P. (1988). Resistance to leaf and stem rusts in wheat. *Ann. Rev. Plant Path.*, 144: 10–22.

Sarbazeh, A.M., Ferrabi, M., Singh, H., Friebe, B., Gill, B.S. and Dhaliwal, H.S. (2002). pH induced transfer of leaf and stripe rust resistance genes from *Aegilops triuncalin* and *A. geniculata* to bread wheat. *Euphytica*, 127(3): 377–382.

Schoelz, J.E., Niggins, B.F., Wintermantal, W.M. and Russ, K. (2006). Genetics and resistance. *Phytopath.*, 96: 453–359.

Singh, B.D. (2002). *Plant Breeding*. Kalyani Publisher, New Delhi, pp. 65–76.

Singh, R.S. (2004). *Introduction to Principle of Plant Pathology*. Oxford and IBH Publishing Co. Pvt. Ltd., New Delhi, pp. 4–5.

Plant Diseases Management in Horticultural Crops (2011) *Pages* **135–143**
Editors: **Shahid Ahamad, Ali Anwar and P.K. Sharma**
Published by: **DAYA PUBLISHING HOUSE, NEW DELHI**

Chapter 11

Ecofriendly Management of Viral Diseases in Some Vegetables

Ravindra Kumar and J.P. Tewari

Plant Pathology Laboratory, Department of Postgraduate Studies and Research in Botany, M.L.K. (P.G.) College, Balrampur, U.P.

Healthy plants look good, grow well, and are productive. Plants remain healthy as long as conditions favour normal plant growth and development. Sometimes plants are unhealthy, due to irritation caused by biotic and abiotic factors the irritation may be somewhat continuous acting over extended period, or it may occur nearly instantaneously. Continuous irritation cause disease; instantaneous irritation causes injury. Thus disease is an exception rather than rule in nature that frequently expressed by production of symptoms.

Common agents that cause plant disease include nonliving (abiotic) and living (biotic) factors. Non-living factors that cause disease include harmful temperature, moisture, light, nutrients, pH, air pollutants, and pesticides. Living pathogens that cause disease include fungi, bacteria, viruses, viroids, mycoplasma and nematodes. Most pathogens are microscopic and derive their food by growing in or on the host plant. Viruses are unique among all pathogens being ultramicroscopic in size and obligate intracellular parasite in nature. Pathogen-host interaction in viral infections is more direct than in other diseases and could be characterized as genomic interaction.

Thus, disease is an abnormal condition in the plant produced by biotic or abiotic factors. The development of disease in plants commonly depends on the following factors:

1. Host genotype
2. Pathogen genotype
3. The environment

These factors are commonly called as disease triangle. Adding time to disease triangle as a *fourth dimension* gives a disease pyramid. If the four components of the disease pyramid could be quantified, its volume would be proportional to the amount of disease on a plant or in plant population. Reducing any component of disease pyramid through management (the host susceptibility, the virulence of pathogen, degree of favourableness of time or of the environment) will decrease the amount of disease in a plant. Disease development in cultivated plants is also greatly influenced by a *fifth component*; humans, human affect disease development in various ways. They affect the kind of plants, grown in a given area, their level of resistance, the number of plants planted, time of planting and density of plants. They also modify the environment of disease development in various ways.

In management of disease incidence one cannot exercise control on pathogen genotype because it changes continuously with time. We can modify environment and exercise some control on host genotype because in nature such adaptation in form of resistance is available, with certain limitations. It is very tedious to control or manage these limitations, which are not permanent (genotype). In such condition, chemicals that modify environment can be useful. These chemicals, pesticides and resistance genes played important role in success of green revolution.

When organic pesticides were discovered they were hailed for their positive contributions to increase food production and the reduction of serious human diseases such as malaria. These products today are considered by many to be environmental poison and health hazards. Thus on global level environmental issue is more important that cause increasing pressure for the reduction and eventual elimination of pesticides.

The complexity of many virus diseases has led to the development of a large number of approaches for their control. Among these use of leaf extracts of some higher plants have been reported to induce resistance against infection of few viruses in hypersensitive hosts. (Verma *et al.*, 1982, 1984 and 1998; Verma and Prasad, 1983) and non-hypersensitive hosts (Verma and Prasad, 1983). Many workers from time to time have investigated reduction of plant virus disease by application of several methods and found increased yield (Griffing, 1956). Simons (1960) have studied effects of some insecticides and physical barrier method on field spread of pepper vein banding mosaic virus. He also studied the effect of foliar spray of cytovirin on susceptibility and transmissibility of potato virus Y in pepper. Cook 1960 has studied use of resistant varieties against two virus diseases of chilli. In Israel, a significant reduction (40-90 per cent) of PVY and CMV incidence was obtained by oil spraying on bell peppers (Lobenstein *et al.*, 1970). Oil sprays have proved to be useful to prevent the spread of non-persistent viruses (Hein, 1971).

Black and Rolston (1972) have worked out that aphids were repelled and virus diseases were reduced in pepper plants by use of aluminum foil mulching. Trapping and yellow polythene sheet have been observed as an effective tool for the inhibition of aphid population and incidence of chilli mosaic viruses. (Cohen and Marco, 1973) has studied the transmission of pepper mottle virus from susceptible and resistant pepper cultivars. Soh *et al.* (1977) have also studied the inheritance of resistance to pepper veinal mottle virus in chilli. The use of a specially formulated mineral oil, JMS Stylet-oil, results in the control of virus diseases on such sensitive plants as crookneck squash and tomato without indication of phytotoxicity. This oil has been used in Florida on peppers, squash and tomato. The oil sprays were useful with the inhibitory effect of incidence of insect borne viruses and was also helpful to increase the yield (Asjes, 1979). A significant reduction of chilli mosaic virus and PVY was also reported by oil sparaying on pepper (Walkey and Dance, 1979; Lobenstein and Raccah, 1980). Singh and Thakur (1980) have studied the reaction of some hot peppers (*Capsicum annum L.*) lines to cucumber mosaic viruses. Villalon (1981) has studied the breeding of peppers to resist virus diseases.

The forms of resistance reported to control aphid borne viruses include immunity to PVY, pepper mottle and pepper vicinal mottle viruses (Rast, 1982; Ortega and Altega, 1988).

Murthy *et al.* (1981) reported the effect of prophylactic sprays of leaf extract on the infection of tobacco by tobacco mosaic virus.

Peshney and Moghem (1989) have reported the inhibition of tobacco mosaic viruses on chilli by different plant extracts. Cheema *et al.* (1991) have evaluated various plant products, insecticides and chemicals against cucumber mosaic virus on sweet pepper under green house conditions. Ravi (1991) has observed aluminum surfaced plastic film as an effective tool for the inhibition of vector population and reducing mosaic incidence of chilli. Green and Kim (1991) have reported the predominant role of non-additive gene action in controlling the expression of some traits. The genetic of resistance of CMV in *Capsicum annum* L. have been studied by Bansal *et al.* (1992). They have reported that perennial and Punjab *gucchedar* var. *Capsicum* were highly resistant to CMV in Punjab, India.

Duriat (1966) reported that CMV is one of the most prevalent and widespread virus infecting peppers. Paroda and Chadha (1996) have reported that some of the low yield of chilli in India (0.9t./ha.) in comparison to other varieties used were susceptible to various pests and diseases. Ariyaratne *et al.* (1996) worked out the resistance of *Capsicum* sp. genotypes to tobacco etch poty virus isolates from the Western hemisphere. Singh and Singh (1998) have reported that Punjab lal and perennial variety of *C. annum* L. were highly resistant to CMV. They have also reported that analysis of variance revealed significant difference for green fruit yield and CMV incidence of chilli indicating wide genetic variability among different genotypes. They have also studied the field reaction of some chilli genotypes for leaf curl virus in the Chhattisgarh of India.

Hobb *et al.* (2000) have reported many viruses among Solanaceous weeds and among these, potato virus Y (PVY), chilli veinal mottle virus (CVMV), tobacco etch virus (TEV), pepper vein banding virus (PVBV) and pepper veinal mottle virus (PVMV) belong to potyvirus group and cucumber mosaic virus belong to cucumo virus groups, are economically most important virus having greater host range. Predominant role of non-additive gene action in controlling the expressions of these traits, were reported by Lohithaswa *et al.* (2000), and Grube *et al.* (2000).

Alegbejo (2002) have studied reaction of pepper cultivars to pepper veinal mottle potyvirus. Forty-two pepper (*Capsicum annum* L. and *C. frutescence* L.) cultivars were screened for resistance to pepper veinal mottle poty virus over a two-year period (1997 to 1998) during wet seasons at Samaru, Northern Nigeria. The cultivar 'Tea-14' was immune, the cultivars 'Kenba' and 'Hashin burgu' were moderately resistant. Thirty-three cultivars were moderately susceptible and six cultivars were highly susceptible. The immune and moderately resistance cultivars has characteristics that are highly valued as producers and consumers, high yield, bright yellow colour when ripe, high pungency and good aroma.

A wide variety of coloured plastic mulches, including red, yellow, silver, blue, grey, and orange have been used among various efforts to achieve specific production goals. Coloured plastic mulches have been used in many parts of the United States of America to enhance the yield. These mulches, however, have produced mixed results (Green and Dole, 2003).

Srivastava *et al.* (2004) studied on the management of mosaic disease of pumpkin by using leaf extracts and plant derivatives. Shukla *et al.* (2006) reported that effect of mulching on the spread of aphid transmitted watermelon mosaic virus in cucumber (*Cucumis sativus*). Ramana *et al.* (2005) have developed F_1 hybrids of chilli (*Capsicum annum*) with high yield and resistance to cucumber mosaic virus. He studied combining ability of resistance to CMV and green fruit yield in chilli, in full diallel crosses between eight parents involving six resistant and two susceptible genotypes.

Ekbote and Patil (2005) have studied the management of chilli virus complex disease in Karnataka and Dharward. It was controlled by using plant extracts of sorghum, coconut, neem and chilli leaves at 10 per cent concentration combined with insecticides monocrotophos (0.2 per cent) and dicofol (0.25 per cent). Totally four spray starting from nursery to fruiting stages were given sorghum leaf extracts (10 per cent) in combination with monocrotophos (0.2 per cent) and dicofol (25 per cent) and dicofol (0.25 per cent) treatment gave maximum disase control over other treatments and gave the highest dry chilli yield (7.57 q/ha). Shukla *et al.* (2006) studied the integrated management of viral disease of *Luffa acutangula* Roxb.

Mulches have been commonly, used for commercial vegetable production for more than 30 years. Excellent sources of information on specific characteristics and application of mulches are available on the word-wide-web. Most mulch materials are made of either high or low density polythene ranging from 0.3 to 0.76 inches (7.7 to 20.2 mm) in thickness, are 5 to 6 feet (1.5 m to 1.8 m) wide and are available on rolls 555 to 1,338 yards (506 to 1233 m.) long, depending on the thickness of the mulch (Lamor, 2001). The colour of mulch is an important determinant of the microclimate around a crop plant. A wide variety of other coloured plastic mulches, including red, yellow, silver, blue, grey and orange have been used in various efforts to achieve specific production goals. Each of these colours has distinct spectral reflectivity characteristics and thus modifies the radiation balance in and below a crop canopy. These coloured mulches affect not only the microclimate around a crop but have also been shown to influence insect behavior (Zitter, 1975, 1978).

The use of polythene sheet and yellow polythene mulches approach to decrease the spread of insect born virus diseases and to get virus free planting material are being practiced in vegetable crops (Vani, 1987; Vani *et al.*, 1989; Black and Rolston, 1972; Black, 1980). Plastic mulches have been reported to be helpful to reduce aphid population, delay of virus infection and reduction in virus incidence (Chen and Chen, 1983; Stapleton and Summers, 2002). Straw mulch plots also had less disease incidence and increased the yield than the un-mulched plots. The effect of straw mulching on the spread of TLCV and the influence of straw on the behavior of the vector whitefly was studied in Israel (Nitzany *et al.*, 1964; Cohen and Madjar, 1974; Vani *et al.*, 1989; Shukla *et al.*, 2005). Similar results have also been recorded by several workers; Lobenstein and Raccah, 1980; Shukla *et al.*, 2005). Some workers also used combination of the two treatments and obtained the decrease in the incidence of virus disease (Basky, 1984). The mulches not only decreased the incidence of virus disease in various crop but also increased their yield (Zatyko, 1979; Eulitz, 1977; Ravi, 1991; Whyman *et al.*, 1979; Hassan *et al.*, 1995; Tripathi and Varma, 2002).

Advances in modern biology, especially biotechnology, offers possible solution to previously intractable problems and difficult targets of disease resistance. Transgenic crops raised through genetic engineering appear to be more efficient and durable. Transformed plants with either plant derived genes or derived from pathogens usually exhibit good degree of resistance against pathogens. In India the first transgenic crop, Bt. cotton developed for resistance to bollworm, was commercially permitted in 2002, from a small coverage of 30,000 hectare in 2002, the area under *Bt*-cotton has increased to 3.5 million hectare in 2006. As a result, the average yield increased from 280 to 460 kg.lint/ha. during 2002 to 2006.

Similarly, plants transformed with the virus coat protein gene or viral genes (other than coat protein) also show high level of resistance. Stark and Beachy, (1989), produced transgenic tobacco pants carrying a gene encoding coat protein of soybean mosaic virus (SBMV) a non-pathogen of tobacco. These transgenic plants showed resistance against two serologically unrelated poty viruses *viz.* potato virus-y and tobacco etch virus both are pathogens of tobacco. Thus, it could be utilized for induction of resistance in susceptible cultivars. Beachy *et al.* (1997) observed in some transgenic plants that accumulate the coat protein or a defective movement protein of TMV exhibit degree of resistance to infection by TMV that are dependent on accumulation of the protein.

Genetically resistant (naturally resistance) plants (cultivated or wild) are always a source of potential resistance genes for conventional plant breeding as well as for producing disease resistant transgenic plants. These resistant plants are more efficient, long lasting and reliable. Exploitation of non-host resistance/immunity is undoubtedly an excellent approach to plant disease management, but it is not an easy task to identify specific genes in non-host plant, which confer immunity to a large range of unrelated microbes.

Among all the measures available for disease management, no one is ultimate and permanent thus a new approach integrated disease management is more common utilizing all the possible ways of disease management.

Chemicals especially organic compounds that were popular in green revolution became hazardous now. Among all these circumstances certain botanicals or plant products or extracts of plants in different dilution alone or in combination with others. These plant or animal-based chemicals are biodegradable and offer no harm to ecosystem. Extracts of neem leaf together with cow urine is very popular in tea plantation due to no residual effect and effective in control of insects and microbes.

Table 11.1: Effect of Different Botanicals on Incidence of Watermelon Mosaic Viral Diseases and Yield of Ridge Gourd

Sl.No.	*Treatment*	*Average Incidence of WMV MM + WMV VB+ WMVC*	*Yield (q/ha)*	*Per cent Yield Increase Over Control*
1.	*Clerodendrum aculeatum*	8.31	6.45	49.30
2.	*Dillenia indica*	9.12	5.90	36.57
3.	*Grewia tenax*	9.42	5.40	25.00
4.	*Ocimum sanctum*	8.52	6.12	41.66
5.	*Phyllanthus emblica*	7.08	7.30	68.98
6.	*Solanum surratense*	9.18	5.80	34.25
7.	*Strychnos nox vomica*	8.71	6.00	38.88
8.	*Terminalia arjuna*	8.12	7.10	64.35
9.	*Vitex negundo*	9.35	5.70	31.94
10.	*Withania sominifera*	9.97	4.90	13.42
11.	Control	15.44	4.32	–

WMV MM: Water Melon mosaic virus-mild mosaic; WMV VB: Water Melon mosaic virus-vein banding; WMVC: Water Melon mosaic virus-chlorotic.

The data presented in Table 11.1 shows that the average of total incidence of the WMV MM, WMV VB and WMV C diseases were lower in *Phyllanthus* (7.08 per cent), *Terminalia* (8.12 per cent), and *Clerodendrum* (8.31 per cent) as compared to other leaf extract spraying treatments. Higher yield and least disease incidence were recorded in *Phyllanthus* and *Terminalia* (7.30q/ha and 7.10q/ha and 7.08

per cent and 8.12 per cent) followed by *Clerodendrum* (6.45q/ha and 8.31 per cent). *Ocimum* (6.12q/ha and 8.52 per cent), *Strychnos* (6.0 q/ha and 8.71 per cent), *Dillenia* (5.90 q/ha and 9.12 per cent) and *Solanum* leaf extract spray (5.80 q/ha and 9.81 per cent) were next best treatments followed by *Vitex* (5.70 q/ha and 9.35 per cent), *Grewia* (5.40 q/ha and 9.42 per cent) and *Withania* leaf extract spray (4.90 q/ha and 9.97 per cent) as compared to control. Table 11.1 indicated the highest yield increase over control was recorded in *Phyllanthus* (68.98 per cent), *Terminalia* 64.35 per cent) and *Clerodendrum* (49.30 per cent) leaf extract spray treatment as compared to control

Conclusion

Plant diseases have been of great concern to the Indian agriculture of these; viral diseases are a major constraint in improving crop production, particularly in tropics and sub-tropics that provides ideal conditions for the perpetuation of the viruses and their vectors. Although plant viruses are simple in composition, the management of their diseases is much more difficult owing to complex disease cycles, efficient system of transmission and above all, non-availability of viricides. Efficient management of viruses require complete understanding of their relationship with the host the environment and the vector.

Indiscriminate use of the non-biodegradable compounds created certain problems, which are hazardous to environment and human health. In order to overcome these problems alternative remedies where sought from plants in form of biopesticides and non a day it is becoming a success story all over the world.

References

Asjes, C.J. (1978). Mineral olien op lelies on virus verspreideng tegen te gaan. *Bloembollencultior*, 88: 1046–1047.

Ariyaratne, I., Hobbs, H.A., Valverde, R.A., Black, L.L. and Dufresne, D.J. (1996) Resistance of *Capsicum* sp. Genotypes to tobacco etch poty virus isolates from the Western Hemisphere. *Plant Dis.*, 80: 1257–1261.

Alegbejo, M.D. (2002). Reaction of pepper cultivars to pepper veinal mottle potyvirus. *J. Vegetable Crop Production*, 8: 19–26.

Black, L. L. and Rolston, L. H. (1972). Aphids repelled and virus diseases reduced in peppers planted on aluminum foil mulch. *Phytopathol.* 62: 747(Abstr.).

Basky, Z. (1984). Effect of reflective mulches and oil sprays on mosaic virus incidence in cucumber, *Ecology*, 7: 243–248.

Bansal, R.D., Aulakh, R.K. and Hundal, J.S. (1992). Reaction of different genotypes of pepper (*Capsicum annum* L.) to cucumber mosaic virus. In: *8th Meetting of Genetic and Breeding on Capsicum and Egg Plant*, Sept. 7–10, Rome, Italy, pp. 131–137.

Beachy, R.N., Bendahmane, M., Heinlein, M., Padgett, H., Kahn, T., Reichel, C. and Lim, C.O. (1997). Pathogen derived resistance to TMV in transgenic plants. Indian Phytopathalogical Society. Golden Jublee, IARI, New Delhi, Abstract.

Cook, A.A. (1960). Genetics of resistance in *Capsicum annum* to two virus diseases. *Phytopathol.*, 50: 364–367.

Cohen, S. and Marco. S. (1973). Reducing the spread of aphid-transmitted viruses in peppers by trapping the aphids on sticky yellow polythene sheets. *Phytopathol.*, 63: 1207–1209.

Cohen, S. and Madjar, V.M. (1974). Prevention of the spread of tomato yellow leaf curl virus transmitted by *Bemisia tabacci* in Israel. *Bull. Ent. Res.*, 64: 193–197.

Chen, C.M. and Chen, C.C. (1983). Plastic cloth mulching for the control of virus in peppers. In: *Abstracts of Experimental Report on Agriculture*, Department of Agriculture and Forestry, Taiwan Provincial Government, R.O.C, pp. 106.

Cheema, S.S., Chahal, A.S., Bansal, R.S. and Kapoor, S.P. (1991). Efficacy of various byproducts and chemical against cucumber mosaic virus on sweet pepper under green house conditions. *Indian J. Virol.*, 7: 169–175.

Duriat, A.S. (1996). Management of pepper viruses in Indonesia: Problems and progress. *IARD Journal*, 18: 45–50.

Eulitz, E.G. (1977). Aluminum foil for the control of water melon mosaic in vegetable marrow.

Ekbote, S.D. and Patil, M.S. (2005). Management of Chilli viral complex disease. *J. Mycol. Pl. Pathol.*, 35: 177–178.

Griffing, B. (1956). Concept of general combining ability in relation to diallel crossing system. *Aust. J. Biol. Sci.*, 9: 463–493.

Green, S.K. and Kim, J.S. (1991). Characteristics and control of viruses infecting peppers a literature review. *AVRDC Technical Bulletin*, 18: 60.

Grube, R.C., Zhang, Y., Murphy, J.E. Lowza Figuera, F., Lackney, V.K., Provvidenti, R. and John, M.K. (2000). New source of resistance to cucumber mosaic virus in *Capsicum frutescens*. *Plant Dis.*, 84: 885–891.

Green, L. and Dole, J.M. (2003). Aluminum foil, aluminum painted plastic, and degradable mulches increase yields decrease insect-vectored viral diseases of vegetables. *Hort. Tech.*, 13: 176–184.

Hassan, S.A., Zainal, A.R. and Ramlan, M.E. (1995). Growth and yield of chilli (*Capsicum annum* L.) in response to mulching and potassium fertilization. *Pertanika J. Trop. Agric. Sci.*, 1995: 113–117.

Hobb, H.A.D.M., Eastburn, C.T., Darey, J.D., Kindhart, J.B., Masiuan, D.J., Voeglin, R.A., Weingier and Coppin, N.K.Mc (2000). Solonaceous weeds as possible source of cucumber mosaic virus in Southern Illinois for aphid transmission to pepper. *Plant Dis.*, 84(1): 1221–1224.

Lobenstiein, G., Alper, M. and Levy, S. (1970). Field-tests with oil sprays for prevention of aphid-spread viruses in peppers. *Phytopathol.*, 6: 212–215.

Lobenstein, G. and Raccah, B. (1980). Control of non-persistently transmitted aphid borne viruses. *Phytoparasitica*, 8: 221–235.

Lohithaswa, H.C., Kulkurni, R.S. and Manyunath, A. (2000). Combining ability analysis for yield. Capsaisin and other quantitative traits in chilli (*Capsicum annum* L.) over environments. *Indian J. Genet. Plant Breed.*, 60: 511–518.

Lamor, W. L. (2001). Vegetable production using plasticulture food and fertilizer technology center. http: II www agnet.org/library/article/eb 476html.

Murthy, P.N., Nagrajan, K. and Sastry, A.B. (1981). Effect of prophylactic sprays of leaf extract on the infection of tobacco-by-tobacco mosaic virus. *Indian J. Agri. Sci.*, 51: 791–795.

Nitazany, F.E., Geisenberg, H. and Kock, B. (1964). Test for the protection of cucumbers from a whitefly borne virus. *Phytopathol.*, 54: 1059–1061.

Ortega, G.R. and Altega, L.M. (1988). Response of pepper to two Spanish isolates of CMV. *Capsicum News*, 7: 65–66.

Peshney, N. and Maghem, D.G. (1989). Inhibition of tobacco mosaic virus on chilli by different plant extracts. *PKV. Res. J.*, 13: 199–200.

Paroda, R.S. and Chadha, K.L. (1996). *50 Years of Crop Science Research in India*. ICAR, New Delhi, pp. 574–592.

Ravi, K.S. (1991). Studies on pepper vein banding virus and other components of Murda Syndrome in chilli (*Capsicum annum* L.). *Ph.D. Thesis*, University of Agricultural Sciences, Hebbal Bangalore, Karnataka, India.

Rast, A.T.B. (1992). Resistant of *Capsicum* species to tobacco, tomato and pepper strains of tobacco mosaic virus. Netherlands. *J. Plant Pathol.*, 88: 163–170.

Ramana, C.V., Reddy, K.M., Reddy, M.K. and Sadashiva, A.T. (2005). Development of F_1 hybrid in chilli (*Capsicum annum*) with high yield and resistance to cucumber mosaic virus. *J. Mycol.Pl. Pathol.*, 35: 179–182.

Simons, J.N. (1960). Effects of foliar sprays of Cytovirin on susceptibility and transmissibility of potato virus Y in pepper. *Phytopathol.*, 50: 109–111.

Soh, A.C., Yap, T.C. and Graham, K.M. (1977). Inheritance of resistance to pepper veinal mottle virus in chilli. *Phytopathol.*, 67: 115–117.

Singh, S.J. and Thakur, M.R. (1980). Reaction of some hot pepper (*Capsicum annum* L.) lines to cucumber mosaic virus. *Indian J. Mycol. Pl. Pathol.*, 9: 276.

Singh, M.J. and Singh, J. (1998). Source of resistance to cucumber mosaic virus in chillies. *Plant Dis. Res.*, 13(2): 184.

Stark, D.M. and Beachy, R.N. (1989). Protection against potyvirus infection in transgenic plants evidence for broad spectrum resistance. *Biotechnology*, 7: 1257–1262.

Stapleton, J.J. and Summers, C.G. (2002). Reflective mulches for management of aphids and aphid-borne virus diseases in late seasons cantaloupe (*Cucumis melo var. cantalupensis*). *Crop Protec.*, 21: 891–898.

Srivastava, G., Shukla, D., Srivastava, S. and Tewari, J.P. (2004). Management of a mosaic disease of pumpkin by using plant extracts and plant derivatives. *Vegetos*, 17: 1–6.

Shukla, D., Ansari, N.A., Srivastava, G., Shukla, S., Chaturvedi, R.C., Srivastava, A.K., Pandey, A., Yadav, A.K. and Tewari, J.P. (2006). Integrated management of a viral disease of *Luffa acutangula* (Roxb.) caused by watermelon mosaic virus I. *J. Liv. World*, 13: 44–47.

Tripathi, S. and Varma, A. (2002). Ecofriendly management of leaf curl disease of tomato. *Indian Phytopath.*, 55: 473–478.

Villalon, B. (1981). Breeding peppers to resist virus diseases. *Plant Dis.*, 65: 557–561.

Verma, H.N., Awasthi, L.P. and Mukherjee, K. (1982). In: *Advancing Frontiers of Mycology and Plant Pathology*, (Eds.) K.S. Bilgrami, P.S. Misra and R.S. Misra. Today and Tomorrow Printers and Publishers, New Delhi, pp. 255–264.

Verma, H.N. and Prasad, V. (1983). In: *Recent Advances in Plant Pathology*, (Eds.) A. Husain, K. Singh, B.P. Singh and V.P. Agnihotri. 312, Print House (India), Lucknow.

Verma, H.N., Chowdhury, B. and Rastogi, P. (1984). Antiviral activity of leaf extracts of different Clerodendrum L. species. Z. Pflanzenks. *Pflanzenschutz*, 91: 34–41.

Vani, S. (1987). Studies on viral diseases of muskmelon and watermelon. *Ph.D. Thesis*, PG School, IARI, New Delhi.

Vani, S., Verma, A., More, T.A. and Srivastava, K.P. (1989). Use of mulches for the management of mosaic disease of muskmelon. *Indian Phytopathol.*, 42: 227–235.

Verma, H.N., Baranwal, V.K. and Srivastava, S. (1998). Antiviral substances of plant origin. In: *Plant Virus Diseases Control*, (Eds.) A. Hadidi, R.K. Khetarpal and H. Koganezawa. APS Press, The American Phytopathological Society, St. Paul, Minnesota.

Walky, D.G.A. and Donce, M.C. (1979). The effect of oil spray on aphid transmission of turnip mosic and yellow bean common mosaic virus. *Plant Disease Reporter*, 63: 877–881.

Whyman, J.A., Toscano, N.C., Kido, K., Johnson, H. and Mayberry, K.S. (1979). Effects of mulching on the spread of aphid transmitted watermelon mosaic virus to summer squash. *J. Econ. Entomol.*, 72: 139–143.

Zitter, T.A. (1975). Transmission of pepper virus from susceptible and resistant pepper cultivars. *Phytopathol.*, 65: 110–114.

Zatyko, L. (1979). Melon mulching with coloured plastic. *Kertgaz da sag*, 11: 11–17.

Plant Diseases Management in Horticultural Crops (2011) *Pages* **144–154**
Editors: **Shahid Ahamad, Ali Anwar and P.K. Sharma**
Published by: **DAYA PUBLISHING HOUSE, NEW DELHI**

Chapter 12

Virus Diseases of Chilli (*Capsicum annuum* L.) Commonly Occurring in Jammu Division

Ranbir Singh, Sachin Gupta and V.K. Razdan

Division of Plant Pathology, Faculty of Agriculture,
Main Campus Chatha of Sher-e-Kashmir University of Agricultural Sciences and Technology, Jammu

Chilli (*Capsicum annuum* L.) is one of the most important cash crop among the spices and grown widely round the year. Chilli is used both as a green vegetable and is one of the most important spices belongs to family Solanaceae and has number of varieties differing in plant and fruit characteristics. Chillies with higher pungency are mostly grown in tropical countries and the type with mild pungency known as paprika are cultivated mainly in European countries. Red peppers or chillies are cultivated mainly in tropical and subtropical countries like Africa, India, Pakistan, Bangladesh, Malaysia, Indonesia, Japan, Mexico, Turkey, USA. etc.

Chilli can be grown successfully in both cold and warm climatic conditions. For rainy season crop seeds are sown in nursery during May-June and for summer crop in December-February. India is one of the major chilli producing countries in the world. Introduction of chilli in India is believed to be in 17Th century through Portuguese. In India,chilli is cultivated over an area of 965 thousand hectares with an annual production of 1075 thousand tones (Samnotra *et al.*, 2006). There is an immense possibility for export of dry chilli and its derivatives especially that have low pungency and high colour (Mathew *et al.*, 2000). In Jammu Division, chilli is grown over an area of 760 hectares with the production of 6380 metric tones (Anonymous 2005).

Inspite of availability of good varieties and technologies for high production the main constraint for low productivity or quality are due to attack of various viral diseases. which are very commonly found in all types of chilli cultivars, irrespective of seasons or geographical locations.

In Jammu Division various diseases which generally affect the production of chilli as well as its quality are commonly found from seedling to maturity stage of the crop. The major viral diseases are leaf curl and mosaic. These diseases are discussed as under:

Chilli Mosaic Virus

Chilli mosaic virus (CMV) is an important disease of chilli and is responsible for the considerable loss if infection initiate at early stage of plant growth. The disease was first reported from India by Kulkarni (1924) and McRae (1924) and subsequently by Raychaudhuri and Jha (1954) recorded it from Indian Agricultural Research Institute, New Delhi. Besides India, the occurrence of mosaic disease was reported from other countries like Trinidad (Dale, 1954), Argentina (Gracia *et al.*, 1968) and Nigeria (Lana *et al.*, 1975).

Bidari and Reddy (1990) reported that 50 per cent of commercial chilli growing areas in Karnataka were infected by chilli mosaic during 1979-80 and the isolates were identified as PVY, Pepper vein-banding virus, Pepper veinal mottle virus, Tobacco etch poty virus, Tobacco mosaic tobamo virus, Cucumber mosaic cucumo virus, Tobacco ring spot nepovirus and Tomato spotted wilt tospo virus. Jeyarajan and Ramakrishnan (1961) recorded the higher per cent age of reduction in the length of chilli leaf lamina, shoot weight, root weight and root length due to mosaic disease and consequently that plants yield less than the healthy plants.

Symptoms

As chilli mosaic disease is caused by number of virus, that contribute different symptoms. At the beginning Kulkarni (1924) reported mottling symptoms mostly pronounced in young leaves besides, blistering and distortion of leaves. Later Doolittle and Walker (1925) described in detail the mosaic symptoms caused by Cucumber mosaic virus in chillies. The symptoms were initiated with downward curling of young leaves along the midrib, light green basal portion of leaves and later such leaves developed mottling. Moreover, narrow leaves with filiform tips and shortened internodes were also observed.

Dale (1954) observed that infection with Pepper vein-banding virus caused irregular and discontinuous dark green chlorotic areas merged into dark interveinal tissue and mottling, accompanied by wrinkling and reduction in the size of lamina.

Raychaudhuri and Jha (1954) observed under field conditions, the earliest symptoms were vein-clearing of the younger leaves followed by severe mottling with patches of light and dark green island scattered all over the leaf surface. In some cases slight curling, marginal rolling and occasional smalling of leaves were also observed. General stunting of the aerial parts appeared to be a common feature with few flowers and fruits on the affected plants.

Causal Viruses

The causal viruses of mosaic symptoms are associated with at least eight viruses in India. Although in symptomatology, transmissibility, host range, these viruses are more or less similar but they differ in physical and genomic properties. The viruses responsible for causing chilli mosaic include, Potato virus X (Rishi and Dhawan, 1988), Potato Virus Y (Jeyarajan and Ramakrishnan, 1969), Pepper vein-banding virus (Prasadarao and Yaraguntaiah, 1979), Pepper veinal mottle virus; Tobacco mosaic virus (Ramakrishnan, 1961), Tobacco etch virus (Bidari and Reddy, 1991), Tobacco ring spot virus, Tomato spotted wilt virus (Bidari and Reddy, 1984) and Cucumber mosaic virus (Kandaswamy *et al.*, 1963).

Transmission

Maduewsi (1967) reported that CMV was mechanically transmitted and produced characteristic symptoms on *Solanum nigrum*, *Nicotiana glutinosa* and *Capsicum annuum*. Gahukar and Nariani (1982) reported that eight isolates of chilli mosaic disease were sap transmissible on *N. tabacum* var. White burley.

Nariani and Sastry (1958) reported two vectors namely *M. persicae* Sulz. and *Aphis evonymi* Fabr. of CMV and further they reported that a single viruliferous aphid (*A. gossypii*) could transmit CMV after feeding 30 seconds on a CMV infected plant and obtained maximum infection after 5 minutes of inoculation feeding (Nariani and Sastry, 1962).

Epidemiology

The severity of chilli mosaic disease depends on the activity of the vectors and the source of primary inoculums (Infected weeds, solanaceous and other hosts) available in and around the field and environmental conditions. The temperature effects both vector population as well as virus multiplication. Maximum population of *Myzus persicae* and *Aphis craccivora* in chilli fields was recorded at temperature range of 26-60°C along with the corresponding increase in the incidence of mosaic disease (Sharma *et al.*, 1991).

The plant debris can harbor the infectious particles of chilli mosaic virus for many years. These are also carried on infected seed. Secondary spread within a field or glasshouse occurs by contact between plants, which is frequently mediated by workers handling plants or using virus contaminated tools during staking.

Host Range

Host range of chilli mosaic virus is very wide and distributed within the cultivated crops and weed hosts from where they could perpetuate into chilli fields. (Gahukar and Nariani, 1982). Besides pepper, major crops susceptible to chilli mosaic virus are tomato, cucumber, melon and squash.

Management

Cultural Practices

Cultural practices like destruction of previous year's plant debris, removal of weeds that can harbor viruses, use of disease free seed and distancing of pepper plants from potato, tomato and other solanaceous crops are the best ways to limit viral infection.

Barrier crops, which are taller than the pepper, can impede aphid entry into the fields and proved quiet useful in checking CMV and PVY which are the causal viruses for chilli mosaic virus disease. Significant reduction in the primary spread of mosaic disease in chillies has been observed by planting sunflower, snap bean, sesame, sorghum, maize and pearl millet as a barrier crop (Ragupathi and Veeraragavathatham, 2002). Growing of marigold with pepper has also been found useful in reducing the incidence of viruses infection pepper in Mexico (Chew- Medinaveitia *et al.*, 1995). Intercropping of chilli with onion resulted in the lowest aphid population and disease incidence (Basavarajappa and Rajsekhar, 2001a). Mulching the field with aluminium foil was also found effective in reducing the aphid population as well as mosaic incidence in chilli (Basavarajappa and Rajsekhar, 2001b).

Varietal Screening

Anand *et al.* (1961) showed five chilli varieties *viz.* Puri red, Puri orange, Kondiverum, G-2 and a local variety proved to be resistant to CMV. Khatri and Sekhon (1974) tested few chilli varieties by feeding the viruliferous aphids (*A. gossypii*) and observed that varietal reaction of chilli varieties NP-46, Anaheim and 129-12 exhibited severe mottling and leaf deformities, while AC-2207, Malagache, Agronomico No.8, Long red-I and II showed faint mottling, P-II was hypersensitive and K-1 was immune.

Chowfla and Sharma (1990) screened the germplasms of bell pepper (*C. annuum* L.) against mosaic complex caused by viruses belonging to poty and cucumo virus groups in Himachal Pradesh. They showed pepper varieties Punjab Lal, Gauhati black, belonging to pungent group, were found to be highly resistant to both the groups of viruses. Bansal *et al.* (1992) tested twenty five genotypes under both field and artificial condition during 1989-90 for resistance to cucumber mosaic virus (CMV) on different genotypes of pepper (*C. annuum* L.) and they found the varieties like Punjab Lal, Indonesian selection and MS-13 were completely free from CMV and rated as resistant, while CA-586, ELS-1, ELS-2, Jawahar-218, KCI-159, Laichi 4-4, MF 41-1-2, Ornamental berry, Ornamental long, Pant C-1, TC-2 and 851201 were moderately resistant. Dhawan *et al.* (1996) evaluated forty six accessions of chilli for resistance to Cucumber mosaic cucumo virus (CMV) and potato virus Y in the field during 1992-94. The eight genotypes were found to be highly resistant under both field and artificial inoculation: and they were HCI-1, HC-15, HC-22, HC-28, HC-69, HC-226, Pusa Sadabahar and virus free-1.

Pinaki and Acharyya (1999) screened twenty five genotypes of *C. annuum* L. against strains of Cucumber mosaic virus during 1995-96 under both field and artificial conditions. The genotypes Pusa Sadabahar and Punjab Lal were rated as resistant while Utkal Ragini and HC-44 were rated as moderately resistant.

Chemical Control

There are numbers of aphids involved in transmitting CMV disease of cultivated crops in both, subtropical and tropical countries. In the plains of Jammu Division the most common aphids that are considered include *Aphis gossypii, Myzus persicae, A. craccivora* etc. Considering their importance one of the best method to control chilli mosaic disease under field condition is to control such vectors by chemical insecticides.

Spraying with systemic insecticides will be effective in reducing mosaic incidence that have been tested in various aphid transmitted viruses (Devi and Reddy, 1995) including Pepper vein-banding virus (PVBV) and Cucumber mosaic virus (CMV) on chilli. They tested 12 insecticides namely Quinolphos, Malathion, Methyl parathion, Chloropyriphos, Monocrotophos, Phosphomidon, Cypermethrin, Dimethoate and most of them were effective to reduce the aphid population. Basavarajappa and Patil (1999) reported that chilli mosaic caused by several viruses can be managed by using plant extracts as well as by use of insecticides.

Chilli Leaf Curl Virus

Chilli leaf curl virus (CLCV) is one of the economically important disease of chilli causing great loss to the crop. CLCV belongs family of Gemini-viridiae and transmitted by whitefly (*Bemisia tabaci* Genn.) in circulative non propagative manner. The disease was first reported in India by Hussain (1932) and subsequently by Vasudeva (1954), Mishra *et al.* (1963) and Dhanraj and Seth (1968).Sastry and Singh (1973) reported that the plants infected by this virus, within 20 days of transplanting remain stunted in growth and produce fewer leaves and fruits than those infected 35 to 50 days following transplanting. Early infection resulted in 92.3 per cent yield reduction as compared with 74.1 to 28.9 per cent yield reduction in plant infected 35 to 50 days after transplanting. Singh *et al.* (1979) reported that the disease caused heavy loss in yield and quality of fruits. If the plants gets infected within 20-25 days after transplanting the loss in yield goes up to 80-90 per cent, but in case of later infection the loss is comparatively less. Saikia and Muniyappa (1986) reported that the disease incidence in Karnataka varied from 6.4 to 52.2 per cent during summer season. The plants infected at 2, 4 and 6 weeks after planting were severely stunted with corresponding yield loss of about 94.9, 90.0, 78.0 per cent respectively. However, when the plants were infected 10 weeks after planting the loss in yield was 10.8 per cent. Rishi and Dhawan (1988) showed that the incidence of chilli leaf curl disease varied from 5.21 to 71.4 per cent. Naitam *et al.* (1990) found that low yield of chilli in Vidarbha region of Maharashtra was due to CLCV. Prasad *et al.* (1992) reported that capsicum leaf curl virus incidence was increased and yield decreased by delaying the date of transplanting of 6 weeks old seedlings from early December to early January possibly related to difference in population of the vector *B. tabaci*.

Symptoms

The symptoms of CLCV consists of abaxial and adaxial curling of leaves accompanied by puckering and blistering of interveinal areas and thickening and swelling of the veins. In advance stages, auxiliary buds are stimulated to produce clusters of leaves, which are reduced in size having malformed or bushy appearance with stunted growth. Flowers and fruits are few being curled at the end. Saha and Singh (1988) described the symptoms of the disease as downward curling of leaves, small sized leaves, shortened internodes and general stunting of the plants giving witches broom appearance. The affected plants stop bearing of fruits and if fruits that are formed they are small and deformed.

Virus Particle

The disease is caused by tobacco leaf curl virus(TLCV), which belongs to bigemini virus group. The virus consists of geminate isometric particles measuring 18-20 nm in diameter.

Transmission

Chilli leaf curl virus is only transmitted by whitefly (*Bemisia tabaci*) in circulative non propagative manner (Shalaby *et al.*, 1997 and Kumar *et al.*, 1999).

Epidemiology

The severity of CLCV is greatly influenced by number of factors such as meteorological parameters, vector population, cropping season and varietal susceptibility.

Varma (1959) and Nair (1971) reported that whitefly vector transmit the virus in circulative non-propagative manner and transmit virus over a considerable periods. Minimum acquisition and inoculation feeding period varied from 15 minutes to 4 hours and 30 minutes to 3 hours respectively. However, Costa (1976) and Muniyappa *et al.* (2000) found that vector has definite incubation period that may range from 6-7 hours to 20-21 hours. The rate of virus transmission however found to increased with the increase of incubation period Shivanathan (1982) reported that in Sri Lanka, tobacco leaf curl virus occurs in all agro-climatic regions of the country and observed marked difference in the rate of spread in four capsicum cultivars. He suggested that slower spread during wet weather was due to reduce mobility of the vector (*B. tabaci*) than to inadequate inoculum. Besides, environmental factor, spread of CLCV also varied with the host resistance. Slow spread in cultivars Wanni and TBS was associated with poor host receptivity to the vector, reduced infection potential, lack of generation of secondary inoculum and larger time taken by the vectors for acquisition and inoculation of the disease agent. While high epidemic in cultivar "Skantha" was associated with good host receptivity to the vector, generation of secondary inoculum and shorter acquisition and inoculation feeding period resulted in efficient dissemination and transmission of disease.

Saikia and Muniyappa (1986) found that tomato leaf curl virus (TLCV) infection was more in summer (upto 100 per cent) than any other season of the year. During July to November infection ranged from 17-53 per cent. A strong correlation was also obtained between the per cent age incidence of TLCV and *B. tabaci* population.

Singh *et al.* (1994) reported that increasing population of *B. tabaci* during July to September was due to prevalence of high humidity and moderate temperature (22 °C – 28 °C), which may be attributed to the chilli leaf curl virus disease in epidemic form. Alegbejo (1999) reported that pepper leaf curl virus caused more yield losses in dry season.

Host Range

Chilli leaf curl virus have a very wide host range that includes both cultivated crop and weed hosts (Pruthi and Samuel, 1942). Mishra *et al.* (1963) reported that CLCV exhibit typical symptoms on *Nicotiana tabacum, N. glutinosa, Lycopersicon esculentum, C. annuum, C. frutescens, C. sinensis, C. pubescens, C. pendulum* and *Crotalaria juncea*. Tsering and Patel (1990) described host range of tobacco leaf curl virus and they reported that out of 99 plant species belonging to 24 families, 28 were susceptible to tobacco leaf curl virus when inoculated artificially by *B. tabaci*. Considering the close relationship of tobacco leaf curl virus and CLCV it is presumed that, this chilli virus may also survived in number of host plants.

Management

Varietal Screening

The use of resistant variety is by far the most accepted method of disease management due to its hazardless effect upon environment. In India varietal screening against leaf curl virus have been carried by different workers.

Dhanraj *et al.* (1968) reported that tobacco leaf curl virus severely attack all important crosses and exotic collection of Capsicum. On inoculation studies all the 23 mutants of *C. annuum* variety NP 46-A were infected by the enation strain of CLCV that developed typical leaf curl symptoms. Capsicum varieties like Puri red and Puri orange reported as resistant to tobacco leaf curl were also found equally susceptible to enation strain. Bhalla *et al.* (1983) have screened 64 lines/cultivars of chilli against leaf curl virus under natural conditions in Madhya Pradesh. Among them only eight cultivars namely Karanja, Pant C-1, Sel. 46-1, IC-18235, IC-18885, JCA-196, Cross-218 and E.C.-121490 showed less than 30 per cent leaf curl incidence. Mename *et al.* (1987) tested 64 chilli varieties under natural field conditions for resistance to chilli leaf curl complex and the lowest disease incidence was observed in Pant C-1 (40.22 per cent), the varieties LIC-45 and N-146 were found as moderately resistant. Sanger *et al.* (1988) screened 120 varieties of *C. annuum* for resistance to Tobacco mosaic virus and Tobacco leaf curl virus under field condition and observed the different degree of susceptibility with the varieties like JCA-248, JCA-218, Pant C-1, NP 46-1, Selection 3, JCA-15 and Pandura. Brar *et al.* (1989) selected lines from 33 indigenous Capsicum genotypes for selfing and the progeny were exposed to infection by leaf curl and mosaic viruses, only six lines were found to be tolerant to both the diseases. Singh *et al.* (1990) screened 148 varieties of capsicum for resistance to leaf curl disease of which 122 lines were susceptible, while 26 were disease free. Sixty six genotypes of chilli were screened by Munshi and Sharma (1996) for resistance to leaf curl complex, thought to be caused by a combination of Tobacco leaf curl gemini virus, Cucumber mosaic cucumo virus, thrips and mites. Out of the genotypes tested only six were resistant and 31 were highly susceptible to the disease. Arora *et al.* (1996) reported that varieties like Hisar Vijay and Hisar shakti released in Haryana were found resistant to mosaic and leaf curl virus. Roy *et al.* (1997) assessed 24 genotypes of red pepper for incidence of tobacco leaf curl virus. Among them the genotypes D.C-18 was resistant and D.C-17 and D.C-16 were moderately resistant to tobacco leaf curl virus. Rest of the cultivars were moderately to highly susceptible. Singh *et al.* (1998) screened a number of cultivars against chilli leaf curl virus and Pusa Sadabahar, Pant C-2 and Jawahar Mirch-2 were found to be promising by exhibiting fewer infection and high yield. Jadhav *et al.* (2000) reported that 'Phule Sai" (GCH-8), a new rainfed chilli variety yielded more than the high yielding standard variety NP-46A and was moderately resistant to leaf curl virus under field conditions. Das *et al.* (2004) screened nine diverse genotypes of chilli against chilli leaf curl virus (CLCV). None of the entries were free from leaf curl disease, however minimum disease incidence was recorded in KS_1 (3.13 per cent) followed by HC-28 (5 per cent), CA-219 (5.39 per cent), Pusa Sadabahar (5.84 per cent) and KS_3 (6.66 per cent). While, the cultivars KS_2 and Bhutan oval shaped showed moderate disease reaction.

Chemical Control

The successful development of chemical control requires knowledge on the biology of the pest species as well as about host plants. However, the population dynamics of *B. tabaci* plays an important role in determining the severity and timing of initial infestation in a crop. The spatial distribution of the pest in plant canopy is also a factor for consideration in chemical control. Singh *et al.* (1979) pointed out that one soil application of Carbofuran or Disulfoton @ 1.5 kg a.i. ha^{-1} or four sprays of 1 per cent mineral oil at 10 days interval reduced *B. tabaci* population and tobacco leaf curl virus incidence on *C. annuum* to a great extent which increased the yield by 2-3 times over the untreated control. Dimethoate and oxydemeton methyl (@ 0.05 per cent) were also found effective for the control of *B. tabaci*.Monocrotophos (Nuvacron 0.03 per cent) applied four times at an interval of 10 days gave best control among eight insecticides against the vector *Bemisia tabaci* (Datar, 1980). Krishna *et al.* (1983) reported that Thimet (Phorate) at the rate of 0.1 per cent applied,after the seedling became established at 15 days interval gave the best control of insect vector followed by Furadon (Carbofuran) at the rate of 0.03 per cent. Nandihalli and Thontadarya (1986) tested Phorate (0.75, 1.5 and 2.25 kg a.i. ha^{-1}), Carbofuran (0.3, 0.6 and 0.75 kg a.i. ha^{-1}), Oxydemeton-methyl (0.313 kg a.i. ha^{-1}), Monocrotophos (0.45 kg a.i. ha^{-1}), Endosulfan (0.438 kg a.i. ha^{-1}), Phosalone (0.438 kg a.i. ha^{-1}) and Malathion (0.625 kg a.i. ha^{-1}) for the control of leaf curl of chillies in Karnataka, during 1987. Phorate at 2.25 kg a.i. ha^{-1} and Phosalone at 0.438 kg a.i. ha^{-1} gave the best result. Phosalone at 0.438 kg a.i. ha^{-1} had the highest cost benefit ratio. Rustamani *et al.* (1994) evaluated four insecticides namely Thiodan 35 EC (Endosulfan), Anthio 25 EC (Formothion) and Curacron 35 EC (Profenofos) against *B. tabaci* on chilli and found that Formothion was significantly more toxic against *B. tabaci*. Patnaik and Mahapatra (1997) observed that substantial reduction of chilli leaf curl virus could be achieved by the application of Methomyl @ 0.5 kg a.i. ha^{-1} at 50 days after planting and Monocrotophos @ 0.5 kg a.i. ha^{-1} at 75 days after planting. Venkatesh *et al.* (1998) observed that alternate spraying of Triazophos 40 EC (1.5 ml $litre^{-1}$) and Neemark (5 ml/litre) both in nursery and in the field at 15 days interval reduced leaf curl incidence (8.31 per cent) and increased yield of green chilli compared to control (80.23 per cent). Spraying Triazophos alone both in nursery and main field was effective in reducing leaf curl incidence (7.86 per cent).Panickar and Patel (2001) studied the efficacy of seven insecticides against pepper leaf curl virus and *S. dorsalis* and observed that Triazophos at 0.04 per cent was the most effective for managing population of *S. dorsalis* and preventing incidence of leaf curl in chilli.

References

Anonymous (2005). *Area and Production Data of Vegetables Crop in Jammu Division*. Directorate of Agriculture, Talab Tillo, Jammu, p. 6.

Alegbejo, M.D. (1999). Screening of pepper cultivars for resistance of pepper leaf curl virus. *Capsicum and Egg Plant Newsletter*, 18: 69–72.

Anand, G.P.S., Mishra, M.D. and Singh, A. (1961). Resistance to mosaic in certain chilli varieties. *Indian Phytopath.* 14: 113–114.

Arora, S. K., Pandita, M. L., Pratap, P. S., Malik, Y. S., Rakesh, M., Poonam, D., Gandhi, S. K., Mehra, R. and Dhawan, P. (1996). Hisar Vijay and Hisar Shakti–two new varieties of chilli. *Haryana Agril. Univ. J. of Res.*, 26(4): 227–233.

Bansal, R.D., Aulakh, R.K. and Hundal, J.S. (1992). Reaction of different genotypes of pepper (*Capsicum annuum* L.) to cucumber mosaic virus. *Capsicum Newsletter*, p. 132–137

Basavarajappa, M.P. and Patil, M.S. (1999). Management of chilli mosaic by using plant extract and insecticides. *Indian J. Pl. Pathol.*, 17(1 and 2): 70–72.

Basavarajappa, M.P. and Rajsekhar, D.W. (2001a). Effect of intercropping in the management of chilli mosaic transmitted by insect vector. *Insect Environ.*, 7: 63–64.

Basavarajappa, M.P. and Rajsekhar, D.W. (2001b). Effect of mulching on chilli mosaic transmitted by insect vector. *Insect Environ.*, 7: 64–65.

Bhalla, P.L., Baghel, B.S., Krishna, A., Muthukrishnan, C.R., Muthuswamy, S. and Arumugam, R. (1983). Screening of chilli cultivars/lines against leaf curl and anthracnose. *Proceedings of National Seminar on the Production Technology of Tomato, Chillies,* held at Coimbatore, Tamil Nadu, pp. 147.

Bidari, V.B. and Reddy, H.R. (1990). Identification of naturally occurring viruses on commercial cultivars of chilli. *Mysore J. Agric. Sci.*, 24(1): 45–51.

Bidari, V.B. and Reddy, H.R. (1991). Incidence of chilli mosaic and its distribution in the commercial field of Karnataka. *J. Plantation Crops*, 19(1): 21–25.

Chowfla, S.C. and Sharma, P.N. (1990). Management of bell pepper mosaic disease complex in Himachal Pradesh. *Indian Phytopath.*, 43(3): 349–353

Chew-medinaveitia, Y.I., Zavaleia-Mejia, E., Delgadillo-Sanchez, F., Valfivis-Alcala, R., Pena- Martinez, M.R. and Cardenas-Suniano, E. (1995). Evaluation of control strategies for virus disease of pepper. *Fitopatologia*, 30: 74–78.

Costa, A.S. (1976). Whitefly transmitted plant diseases. *Ann. Rev. Phytopathol.*, 14: 429–449.

Dale, W.T. (1954). Sap transmissible mosaic diseases of solanaceous crops in Trinidad. *Annals of Applied Biology*, 41: 240–247.

Das, S., Sil, S. and Kabir, J. (2004). Disease reaction and its progress on different chilli cultivars (*Capsicum annuum* L.) against anthracnose and leaf curl virus under field condition. *J. Mycopathol. Res.*, 42(1): 49–52.

Datar, V.V. (1980). Chemical control of chilli leaf curl complex in Maharashtra. *Pesticides*, 14(9): 19–20.

Devi, P.H.S. and Reddy, H.R. (1995). Effect of antibiotics on aphid transmission of pepper vein banding virus and cucumber mosaic virus in chilli. *Indian J. Hill Farming*, 8(1): 42–46.

Dhanraj, K.S. and Seth, M.L. (1968). Enations in *Capsicum annuum* L. (Chilli) caused by a new strain of leaf curl virus. *Indian J. Horticulture*, 25(1–2): 70–71.

Dhanraj, K.S., Seth, M.L. and Bansal, H.C. (1968). Reaction of certain chilli mutants and varieties to leaf curl virus. *Indian Phytopath.*, 21(3): 342–343.

Dhawan, P., Dang, J.K., Sangwan, M.S. and Arora, S.K. (1996). Screening of chilli cultivars and accessions for resistance to cucumber mosaic virus and potato virus Y. *Capsicum and Egg plant Newsletter*, 15: 55–57.

Doolittle, S.P. and Walker, M.M. (1925). Further studies on over wintering and dissemination of cucurbit mosaic. *J. Agric. Res.*, 31(1): 58.

Gahukar, K. . and Nariani, T.K. (1982). Studies on an aphid borne mosaic disease of chilli. *Indian Phytopath.*, 35(1): 75–79.

Gracia, O., Feldman, J.M., Pontis, R.S. and Boninsegna, J. (1968). Some viruses affecting tomatoes and peppers in Argentina. *Plant Dis. Reptr.*, 52: 674–676.

Jadhav, M.G., Dhumal, S.A., Burli, A.V. and Moro, S.M. (2000). "Phule Sai" (GCH–8): A new rainfed chilli variety. *Journal of Maharashtra Agricultural University*, 25(1): 110–112.

Jeyarajan, R. and Ramakrishnan, K. (1961). Studies on a virus disease of chilli (*Capsicum* sp.). *South Indian Hort.*, 9: 1–12.

Jeyarajan, R. and Ramakrishnan, K. (1969). Potato virus Y on chilli (*Capsicum annuum* L.) in Tamil Nadu. *Madras Agric. J.*, 56(12): 761–766.

Kandaswamy, T.K., Janaki, I.P., Ramakrishnan, K., Thangamani, G., Subba Raja, K.T. and Sellamal, S. (1963). A preliminary note on virus diseases of chilli in Madras State. *Madras Agric. J.*, 50: 110–111.

Krishna, A., Neema, A.G., Nadkarni, P.G. and Bhalla, P.C. (1983). Evaluation of chemical for control of leaf curl of chillies. In: *Proceedings of National Seminar on the Production Technology of Tomato and Chillies,* (Eds.) C. R. Muthukrishnan, S. Muthuswamy and R.J.Arumugam. Tamil Nadu Agricultural University, Coimbatore, March 22–23, pp. 167–169.

Kulkarni, G.S. (1924). Mosaic and other related diseases of crops in Bombay Presidency. *Agric. College Mag.*, 16: 6–12.

Kumar R., Rai, N. and Lakpale, N. (1999). Field reaction of some chilli genotypes for leaf curl virus in Chattisgarh region of India. *Orissa Journal of Horticulture,* 27(1): 100–102.

Lana, A.O., Glimer, R.M., Wilson, G.F. and Shoyinka, S.A. (1975). An unusual new virus, possibly of the poty virus group from pepper in Nigeria. *Phytopathol.*, 65(12): 1329–1332.

Maduewsi, J.N.C. (1967). Studies on the mosaic disease of pepper in Nigeria: Transmission, symptomatology and physical properties. *Nigerian Agric. J.*, 4(91): 28–32.

Mathew, P.A., Peter, K.V. and John, Z.T. (2000). Production and export potential of paprika. *Spice India*, 13: 13–16.

McRae, W. (1924). Economic Botany, Part–III. *Mycology. Ann. Rept. Board. Sci. Advice, India*, 1922–23, 31–35.

Mename, S.A., Joi, M.B. and Kale, P.N. (1987). Screening of chilli cultivars against leaf curl complex. *Current Res. Reporter*, 3(1): 98–99.

Mishra, M.D., Raychaudhury, S. P. and Ashrafi, J. (1963). Virus causing leaf curl of chilli (*Capsicum annuum* L.). *Indian J. Microbiology*, 3: 73–76.

Muniyappa, V., Venkatesh, H.M., Ramappa, H.K., Kulkarni, R.S., Zeidan, M., Tarba, C.Y., Ghanim, M. and Czosnek, H. (2000). Tomato leaf curl virus from Bangalore. Sequence comparison with Indian to LCV isolates, detection in plants and insects and vector relationships. *Archives of Virology*, 145(8): 1583–1598.

Munshi, A.D. and Sharma, R.K. (1996). Field screening of chilli germplasms against leaf curl complex. *Annals of Pl. Prot. Sciences*, 4(1): 87–89.

Nair, N.G. (1971). Studies on the yellow mosaic of urdbean caused by mungbean yellow mosaic virus. *Ph.D. Thesis*, U.P. Agric. Univ. Pantnagar, U.P., India, pp. 1–66.

Naitam, N.R., Patangrao, D.A. and Deshmukh, S.D. (1990). Resistance response of chilli cultivars to leaf curl. *P.K.V. Res. Journal*, 14(2): 206–207

Nariani, T.K. and Sastri, K.S.M. (1958). Two additional vectors of chilli mosaic virus. *Indian Phytopath.*, 11: 193–194.

Nariani, T.K. and Sastry, K.S.M. (1962). Studies on the relationship of chilli mosaic virus and its vector, *Aphis gossypii* Glover, *Indian Phytopath.*, 15: 173–183.

Panickar, B.K. and Patel, J.R. (2001a). Population to dynamics of different species of thrips on chilli, cotton and pigeon pea. *Indian J. Entomology*, 63(2): 170–175.

Patnaik, H.P. and Mahapatra, S. (1997). Effect of fertilizers on the incidence of thrips and leaf curl in chilli under protected and unprotected conditions. *Mysore J. Agric. Sciences*, 31(2): 159–169.

Pinaki, A. and Acharya, P. (1999). Screening of chilli germplasms against strains of cucumber mosaic virus. *Environment and Ecology*, 17(2): 484–487.

Prasad, S.M., Kudada, N., Mishra, B. and Dhar, V. (1992). Date of transplanting and incidence of leaf curl in chillies. *J. Res. Birsa Agri. Univ.*, 4(1): 81–82.

Prasadrao, R.D.V.J. and Yaraguntaiah, R.C. (1979). The occurrence of pepper veinal mottle virus on chilli in India. *Mysore J. Agric. Sci.*, 13(4): 445–448.

Ramakrishnan, K. (1961). Potato virus X on chilli (*Capsicum* spp.). *South Indian Hort.*, 7: 41–42.

Raychaudhuri, S.P. and Jha, A. (1954). Chilli mosaic occurring in Delhi. *Proc. 41st Indian Sci. Congr. Assoc.*, pp. 256–259.

Rishi, N. and Dhawan, P. (1988). Field reaction of chilli germplasms to different viral diseases. *Pl. Dis. Res.*, 3: 69–70.

Ragupathi, N. and Veeraragavathatham, D. (2002). Management of chilli mosaic virus disease using insecticides, botanicals and barrier crops. *South Indian Hortic.*, 50: 273–275.

Roy, A., Bordoloi, D.K. and Paul, S.R. (1997). Screening of resistant genotypes of red pepper (*Capsicum annuum* L.) against leaf curl infection. *P.K.V. Res. Journal*, 21(2): 175–176.

Rustamani, M.A., Hussain, T., Baloch, H.B., Talpur, M.A. and Mal, K. (1994). Comparative effectiveness of different insecticides in controlling whitefly and mosaic disease of chillies. *Proceedings of Pakistan Congress of Zoology*, 14: 61–64.

Saha, L.R. and Singh, H.B. (1988). Diseases of chilli and their management. *Int. J. Tropical Plant Diseases*, 6: 135–143.

Saikia, A.K. and Muniyappa, V. (1986). Epidemiology and control of tomato leaf curl virus in India. In: *Proceedings of Workshop on Epidemiology of Plant Virus Diseases*, Orlando, Florida, USA, August 6–8, pp. 18–19.

Samnotra, R.K., Sidhu, A.S. and Khurana, D.S., 2006. Stability studies for quality traits in chilli (*Capsicum annum* L.). *Environment and Ecology*, 245: 570–574.

Sastry, K.S. and Singh, S.J. (1973). Assessment of losses in tomato by tomato leaf curl virus. *Indian J. Mycol. and Pl. Pathol.*, 3: 50–54.

Sharma, P.N., Chowfla, S. C. and Andotra, P.S. (1991). Epidemiology of mosaic disease complex of bell pepper in Himachal Pradesh. *Indian Phytopath.*, 44: 112–114

Shalaby, A.A., Nakhla, M.K., Shafie, M.S., Mazyad, H.M. and Maxwell, D.P. (1997). Molecular characterization of tomato yellow leaf curl gemini virus isolated from pepper collected in Egypt. *Annals of Agricultural Science Moshtohor*, 35(2): 819–831.

Shivanathan, P. (1982). Epidemiology of chilli leaf curl disease in Sri Lanka. *Tropical Agriculturist*, 138: 99–109.

Singh, H.B., Upadhyay, D.N. and Saha, L.R. (1994). Observation of some important viral diseases of crop plants in Nagaland, India. In: *Virology in the Tropics*, (Eds.) N. Rishi, K.L. Ahuja and B.P. Singh. Malhotra Publishing House, New Delhi, p. 316.

Singh, S.J., Sastry, K.S. and Sastry, K.S.M. (1979). Combating leaf curl virus in Chilli. *Indian Horticulture*, 24: 9–10.

Singh, U.C., Singh, R. and Nagaich, K.N. (1998). Reaction of some promising chilli varieties against major insect pest and leaf curl diseases. *Indian J. Entomology*, 60(2): 181–183.

Tsering, K. and Patel, B.N. (1990). Host range of tobacco leaf curl virus. *Tobacco Res.*, 16(2): 79–82.

Varma, P.M. (1959). Tomato leaf curl. In: *ICAR Proc. Seminar on Diseases of Host Plants*, Simla, pp. 182–200.

Vasudeva, R.S. (1954). *Report of the Division of Mycology and Plant Pathology*. Sci. Rept., Indian Agric. Res. Inst., New Delhi, 1952–53: 79–89.

Venkatesh, H.M., Muniyappa, V., Ravi, K.S., Krishnaprasad, P.R., Reddy, P.P. Kumar, N.K.K. and Verghese, A. (1998). Advances in IPM for horticultural crops. In: *Proceedings of First National Symposium on Pest Management in Horticultural Crops: Environment Implications and Thrusts*, Bangalore, India, October, 15–17, 1997, pp. 111–117.

Plant Diseases Management in Horticultural Crops (2011) *Pages* **155–169**
Editors: **Shahid Ahamad, Ali Anwar and P.K. Sharma**
Published by: **DAYA PUBLISHING HOUSE, NEW DELHI**

Chapter 13

Diseases of Strawberry and their Management

M.K. Pandey[1], P. Mishra[1], J.N. Srivastava[2] and Anwar Ali[3]

[1]*Division of Plant Pathology, SKAUST-Jammu, J&K*
[2]*Regional Horticultural Research Sub-station, Bhaderwah, SKUAST, Jammu*
[3]*Division of Plant Pathology, SKAUST-K., Srinagar, J&K*

The strawberry (*Fragaria ananassa* Duch.) is the most important soft fruit crop, which is cultivated throughout the world. Diseases caused by fungi, bacteria, nematode, viruses and virus like organisms are affecting the cultivation of strawberry, which reduces the yield and quality of fruits considerably. The fungal pathogens are main problem in all area of cultivation and occurring on all parts of infected plant.

Powdery Mildew

This is the most important foliar disease of strawberry caused by fungus *Sphaerotheca macularis* (Wall ex. Fr.) Jazc., and reported from all strawberry grown areas in world. The disease recently occurred in epidemic form in Taiwan and caused severe loss to crop. The symptoms occurr at all aerial parts including flower calyx and fruits. Initially fungal infection starts from underside of leaves which become developed into white powdery growth. These leaves curve upward cup like and show reddish blotches. In severe infection leaves become burnt at margins to a greater extent. The white powdery mass also develops on infected flower stalk, flower and fruits. Delayed reddening is reported in infected fruits (Wilhelm, 1961). The pathogen *Sphaerotheca macularis* (Wall. ex. Fr.) Jacz f.sp. *fragariae* produces conidia from conidiophores and ascospores. The pathogen is air borne and overwinters through mycelium in old infected plants. The optimum temperature and relative humidity ranges from 16–18°C and 80 – 100 per cent for germination of conidia but maximum germination occurs at 20°C temperature. It was reported that the free water has lethal effect on conidial germination and rain has deleterious effect on spore dispersal (Peris, 1962; Jhooty and Mc Kean, 1965). Cleistothecia as sexual fruiting bodies are formed in nature (Haward and Alberts, 1973), while the role of cleistothecia as primary source of inoculum is not important because this stage is rarely formed.

Management

The disease can manage by the use of many fungicides *i.e.* flusilazole, mycobutonil, tridemifon, fenarimol and buprimate were found effective. The sulphure containing fungicides and benomyl were found most effective but they cause phytotoxicity, which burn the leaves at above 27°C temperature. Two to three sprays of buprimate or tridemifon at 15 days interval give good result at nurshery stage. The fungicides give effective control when they applied after transplanting and just before flowering. Cultivars like ARKING and REAL are reported to be resistant against disease (Moore, 1981).

Anthracnose

Anthracnose disease of strawberries caused by *Colletrotichum fragariae* Brooks was first reported from Florida (1926–29) by Brooks (1931). Dai *et al.* (2006) reported anthracnose fruit rot of strawberry caused by *Colletotrichum acutatum* in China. In India it was first reported by Singh (1974) from Banglore. The disease is fairly wide spread in strawberry growing regions. Strawberry anthracnose disease is a complex of three *Colletotricum* species, *C. acutatum*, *C. fragariae* and *C. gloeosporioides*, and causes a variety of symptoms that make anthracnose the most important disease of strawberry. *C. acutatum* causes an anthracnose fruit rot and black leaf spots (Inada and Yamaguchi, 2006), whereas *C. gleosporiods* Penz and *C. fragariae* Brooks infect primarily the crown of the plants and causes crown rot and wilt.

Symptoms may appear as sunken, dark brown, circular lesions on petioles, stolon (runners), which may then be girdled, resulting in wilting and death of the leaf or of the daughter plant beyond the lesion on the stolon. The fungus often spreads into the crowns of young plants, which it rots, and the plants then die in the nursery or after they are transplanted in the field.

On fruits, rot is initially restricted to only ripe fruits and characterized by one or more light brown soaked lesions. Two or more lesions collapse and occasionally covered by buff colored spores (Haward, 1972). On the leaf, minute dull, violet-black or black spots surrounded by yellow region appear. In later stages spots coalesce and cause defoliation. Other symptoms include bud rot and flower blight. The conidia of *C. acutatum* were hyaline, straight and fusiform, and produced melanized, ellipsoid, ball- or pear-shaped appressoria at the end of germ tubes (Dai *et al.*, 2006).

The fungus mostly survived on infected or contaminated transplants, stolons in the soil, on weed hosts, and possibly in plant debris in the soil. The fungus was able to survive in buried strawberry tissues for a month (Eastburm and Gubler, 1990). *Cassia obtusifolia*, a weed can serve as source of primary inoculum of *C. fragariae* (Howard and Albregts, 1973). Warm humid weather (>21°C), during flowering and fruiting favoured the disease but extensive losses are being caused after rainy periods (Howard, 1972). The fungus produces fruiting body like acervuli in lesions which contain numerous spores *i.e.* conidia borne in a mucilioginous matrix, these spores are splashed by irrigation or rain to nearby plants or are carried by insects, animals and humans moving among the plants.

Disease Management

Use of straw mulch (Maiden *et al.*, 1993) and use of planting stock obtained from disease free nurseries may provide effective control of the disease. Growing nursery plants with little fertilizer and removing all infected fruits at each harvest helps in controlling disease (Agrios, 2005). Using resistant cultivars like Sequoia, Florida balle, Doner (Delp, 1981) and FL 79 – 1126 (Chandler *et al.*, 1988). Hot water treatment (for 7 minutes at 35°C, followed by 2 or 3 minutes at 50°C) was successful in reducing the proportion of cuttings infected with *C. acutatum* (Johnson *et al.*, 2006). The infections by *C. acutatum* can be completely managed by the fumigation of soil and planting material with methyl bromide and chloropicrin. Use of soil disinfectants like chloropicrin and dazomet effectively control the disease in nursery bed (Matsuo *et al.*, 1994).

Twelve resistant varieties of strawberry against anthracnose were reported by Denoyes and Guerin (1996). 'Honeoye' and 'Pandora' cultivars showed a useful level of resistance, even against the most aggressive isolates (Simpson *et al.*, 2006).. 'Sweet Charlie', 'Carmine', and 'Earlibrite' were the most resistant cultivars. 'Strawberry Festival' has intermediate susceptibility; while 'Camarosa' and 'Treasure' were highly susceptible (Chandler *et al.*, 2006).

Prochloraz and difenoconazole were most effective fungicides for control of strawberry black flower (Domingues *et al.*, 2001). In 1984 Cheah and Soterosi reported that the use of benomyl with captan gave good results. Ergosterol biosynthesis inhibitors (EBI) fungicide reduced infection of *C. fragariae* as compared to other fungicides (Smith and Black, 1993). OctaveReg. (462 g/kg prochloraz as the $MnCl_2$ complex) was highly effective in reducing the incidence of colletotrichum crown and stolon rot in runner production (Hutton, 2006).

Verticillium Wilt

Verticillium wilt or vascular wilt occurs worldwide but is most important in temperate zone of the world. In United States and England, this disease has gained much importance due to its heavy losses (Wilhelm and Paulus, 1980; Simpson *et al.*, 1994). Verticillium wilt disease is caused by fungus *Verticillium* spp. In England, *V. dahliae* Kleb. and *V. alboatrum* both are reported as pathogen (Keyworth and Benett, 1951) but in United States only *V. alboatrum* is reported as pathogen (Wilhelm, 1952).

The symptoms of Verticillium wilts are almost identical to those of Fusarium wilt but Verticillium induces wilt at lower temperatures than Fusarium. Verticillium wilt develops primarily in seedlings, which usually die shortly after infection. More common are late infections, which cause upper leaves to drop and other leaves to develop irregular chlorotic patches that become necrotic. Older plants infected with Verticillium are usually stunted and their vascular tissues show characteristic brownish discoloration. Verticillium infection may result in defoliation, gradual wilting and death of successive branches, or abrupt collapse and death of the entire plant. From India, Mitra (1993) reported that the symptoms of Verticillium wilt in strawberry are characterized by browning of wood at the base of crown, wilting decreases runner production and causes reddening of petiole and stolon late in season. This disease causes total collapse of stawberry plants during peak of the first year's growth. Plants showed stunted growth.

Both species of Verticillium, *V. dahliae* and *V. alboatrum* produce short lived conidia. *V. alboatrum* produces dark, thick-walled mycelium but not microsclerotia. Fungus overwinters in the soil as mycelium or as microsclerotia and can survive upto 15 years. *V. alboatrum* grows best at 20 to 25°C, whereas *V. dahliae* prefers slightly higher temperatures (25-28°C) and sometime is more common in warmer regions (Agrios, 2005). The conidia, however, play an important role in spread of disease by wind or by water. Penetration mainly occurs through root but sometime also through stem at base line of soil. The fungus enters in vascular system and by its growth and gum deposition check the uptake of water which result in wilting of plants. The temperature is the most important factor for disease development. It was reported that the low temperature and cool weather favors the disease development (Anderson, 1956), it has been found that the population of nematode (*Paratylenchus penetrance*) enhances the infection of *V. dahliae* in soil (Mc Kinley and Tolboys, 1979).

Management

The disease can be controlled by adoption of long crop rotation, use of disease free planting material in disease free soil, use of soil disinfectant and resistant cultivars to combat the disease (Ercole, 1976 and Agrios, 2005). The thermal inactivation via soil solarization is useful for the control of Verticillium wilt in regions with high summer temperature and low rainfall. Use black mulch with low dose of nitrogenous fertilizers. Best control of disease was achieved by the use of chloropicrin (8 kg/ha), where 2.5 times more yield than land fumigated with ethyl dibromide was obtained (Wilhelm and Paulus, 1980). In addition to chloropicrin, dozomat was also found effective against *V. dahliae* infection (Harris 1980a). The pre – planting treatment of soil with dozomat and chloropicrin followed by covering the land with polyethylene sheet or nylon laminate is very effective(Harris 1980b). Total fumigation by chloropropene followed by strip fumigation with methyle bromide controls the disease effectively (Meesters, *et al.*, 1996). Soil drenching with fungicides *i.e.* benomyl and thiophanate M @ 0.1 per cent in the holes and soil solarization during hot months followed by fumigation with chloropicrin reduce the infection of Verticillium in strawberry (Hartz *et al.*, 1993). Resistant cultivars like Etna (Faedi *et al.*, 1986), Calypso, Tang and Eros (Simpson and Blanke, 1996), Pegasus (Simpson *et al.*, 1994) have been reported as resistant against the disease.

Black Root Rot

Black root rot is a serious and common problem of strawberries. The term 'Black root rot' is the general name for several root disorders that produce similar symptoms. These disorders are generally referred to as a root rot complex disease of strawberry and it is reported to be incited by a number of soil inhabiting fungi *i.e. Laptosphaeria coniothyruim* (Fckl). Sacc. *Fusarium ortocereras* App. and Wollenw, *Pezizella lytheri* Shear and Dodge, *Ramullaria* species, *Corticium vagum* Berk and Cut., *Packybasidium cadium* Sacc., *Pythium* species Trow, *P. irrigulare* Buis, *Pyrenochaeta* spp., *Cylindrocarpon destrctans* and *Ceratobasidium* species. (Bhardwas and Sharma, 1999). Besides the fungal pathogens certain nematodes, winter injury, fertilizer burn, soil compaction, drought and excess salt, water or improper soil pH are also involved in this disease complex.

The affected plants show an uneven patchy appearance due to dwarfing and death of affected plants. The root system is much smaller than normal. The main roots develop brown areas and small, fibrous rootlets are killed. Lesions extend and the entire root is blackened. Leaves remain small with short petioles and turn yellowish, old leaves wither. Berries, if formed, remain small or wither before ripening. The disease is more prevalent in high clay content and waterlogged soil.

Management

Avoid continuous cultivation of strawberries on the same land, planting material should be free from diseases and avoid water logged or poor drained areas, for plantation. Soil treatment with chemicals can reduce the population of undesirable nematodes and fungi in the soil. Soil fumigation with methyl bromide and chloropicrin reduce the population of fungi involved in the black root rot disease (Drozdowski, 1987).

Red Stele

Red stele also called as red core, red root rot, brown stele and black stele. This is the most important disease of strawberry caused by the soil-borne fungus *Phytophthora fragariae* Hickman. The disease was noticed in Scotland in 1920 (Alcock *et al.*, 1930) and the was fungus isolated and reported by Hickman in 1940 for the first time. Fifteen physiological races of *P. fragariae* have been reported from worldwide (Mass, 1984).

Infected plants are stunted, lose their shiny green luster and produce few runners. Younger leaves often have a metallic bluish-green cast. Small sized fruits with poor quality are formed on infected plants, sometimes fruits are not formed. Older leaves turn prematurely yellow or red and with the first hot, dry weather of early summer, diseased plants wilt rapidly and die. Diseased plants have very few new roots compared to healthy plants that have thick, bushy white roots with many secondary feeder roots. Infected strawberry roots usually appear gray, while the new roots of a healthy plant are yellowish-white. The inside or central portion of the root of wilted plant stele becomes pink to brick red or brownish red and on this basis disease is referred to as red stele disease while stele of normal plants is yellowish-white. This red color may show near the tip of the root or it may be extend the length of the root. The red stele disease is best seen in the spring up to the time of fruiting.

Heavy rain fall, frequent irrigation and water logged conditions help in libration of the zoospores. Oospores survive in soil and cause root infection. Germinated oospores produce sporangia which germinate indirectly to produce many zoospores. The zoospore germinate on the host roots and oospore development in the host tissues occurred at 15°C to 20°C temperature. Root exudates attract the zoospores which encyst around root tips and produces germ tube which penetrates epidermis (Anderson and Colby, 1942). Fungus reaches the phloem and pericycle, within three days of infection (George and Milholland, 1986, 1987).

Management

The disease can be controlled by use of disease free planting material taken from disease free areas. Acidic soil should be avoided for the planting. Soil fumigation with ethyl bromide and chloropicrin significantly reduces inoculum potential (Joffens, 1957). Drenching with systemic fungicide metalaxyl effectively manages the disease (Mcintyre and Walton, 1981).

The disease can be controlled by use of resistant varieties but due to pressure of physiological races, it is not easy to maintain resistance. The cultivars of strawberry *viz.* loke Delmarvel, Mohawk, Primetime and Erose have been found resistant against *P. fragariae* (Igarashi *et al.*, 1994; Gallete *et el.*, 1995a, 1995b; Simpson and Blanke 1996).

Fruit Rot Diseases

Rhizopus Rot

It is a fungal disease caused by *Rhizopus stolonifer* (Ehrenb. Fr.), Vuill. Infected fruits turn light brown, soft, watery and give exudates juice even at slight pressure. The whole fluffy mycelial growth with black sporangia are formed on severely infected fruit during transportation and storage. The pathogen is a wound parasite and can be prevented by pre – cooling of fruits before shipment.

Grey Mould Rot

Grey mould rot disease is caused by *Botrytis cinerea*. It appears as fruit rot under humid conditions, fungus produces a noticeable grey – mould fruiting layer on the affected tissues (Agrios, 2005). In high humidity, white cottony mycelial growth may be formed (Bhardwas and sharma 1999). The fungus overwinters as mycelium in senescent petals, stamens and calyces of marketable fruits (Powelson, 1960). Mycelium produces 90 – 99 per cent of inoculum in the laminae and petioles of dead strawberry leaves (Braun and Sutton, 1987). Humidity is the main determining factor for occurrence of grey mould rot, while frequent rains induced higher disease incidence. The optimum temperature 15 – 20°C and more than 90 per cent relative humidity are most suitable for disease development in field (Devaux, 1978). The fungus requires cool (18 – 23°C) weather for best growth, sporulation, germination and establishment of infection (Agrios, 2005). Hundred per cent infection caused by 24 hours of wetness with 20°C optimum temperature was reported in the case of flower infection whereas flower and fruit infection greatly reduced bellow 15°C and 25°C temperature for all wetness duration (Bulger, *et al.*, 1987).

Management

Removal of infected and infested debris from the field and storage rooms by providing conditions for proper aeration and quick drying of plants and plants products controls the disease (Agrios, 2005). Proper drainage and aeration in field also reduces the incidence of grey mould rot (Daubeny and Pepin, 1977; Mass, 1978 and Barritt 1980). In green house humidity should be reduced by ventilation and heating. Storage organs of plants are wrapped in sealed polyvinyl chloride wrappers for better maintenance and protected by keeping them at 31 – 50°C for 2 – 4 days to remove excess moisture and then kept at 5°C or bellow or a combination of 10 – 15 Per cent CO_2 in the controlled atmosphere with 10°C below temperature (Sommer *et al.*, 1973). The persistent off – flavours developed when strawberries are stored in atmosphere containing 20 or 30 per cent CO_2 (Harris and Harvey, 1973).

Biological control of Botrytis gray mould is achieved by spraying the flowers and fruits with spore suspensions of certain antagonistic fungi and bacteria. *Trichoderma harzianum* inhibit the growth of *B. cinerea in vitro* and reduced infection on fruits *in vivo* under greenhouse and field conditions (EL – Zayat *et al.*, 1993).

Management of *Botrytis* through chemicals has been partially successful. Regular spray of fungicides like captan or thiram and benomyl or thiophanate methyl was found effective against gray mould in strawberry (Horn, 1961; Borecka *et al.*, 1973 and Gourley, 1974). Spray of vincolozolin was found most effective (Paulus *et al.*, 1978). The fruits decay due to grey mould can reduced by fumigation with acetaldehyde vapour (Prasad and Stadelbacher, 1974).

Leather Rot

Leather rot in strawberry was reported by Rose (1926) in 1924 from USA. This disease is caused by a fungi *Phytophthora cactorum* (Lebert and Cohn) Schrot. Bhardwaj and Sharma (1999) observed that this disease causes heavy losses (20-30 per cent) due to attack on the strawberry in Himanchal Pradesh during 1997. This disease may be more severe due to heavy rainfall in the month of April – May and shade of poplar trees. The disease has also been reported from Europe and Asia (Kao and Leu 1979; Seemuller and Schmidle, 1979). Crown rot and wilt of stawberry caused by *P. cactorum* was reported from Europe and significant loss by this disease was recorded by Ellis and Grove (1983) from USA.

The symptoms of leather rot are associated with ripe as well as unripe fruits, but the infection on green fruits is more. Dark brown or natural green disease area surrounded by light brown margin on newly green fruits is formed. The rotted area spreads over fruits, whole fruits become brown, leathery with rough texture. The pathogen causes slight change in color of pulpy mature fruits which becomes brown to dark purple. The vascular tissues of each seed become dark in severely infected fruit. Under high moisture fine white mycelial growth is observed around surface of infected fruits. Both infected green and mature fruits dry to form hard, shrivelled mummies. Infected fruits have an unpleasant odour and taste.

Excessive moisture (dew or rain) and temperature are most important factors for the infection of *P. cactorum* on fruits of strawberry and disease incidence increases along with wetness duration (0.5 hrs) at all tested temperatures (Grove *et el.*, 1985c). The disease incidence increased after optimum temperature of about 21°C and declined at each wetness duration. Huge sporulation and sporangia production increased when wetness duration increased from 3 – 24 hours at 12.5°C–27°C temperature. The temperature between 15°C to 25°C is best for production of sporangia and 21°C is optimum for the same (Grove *et el.*, 1985a and 1985b).

Leaf Diseases

Leaf Spot

Mycosphaerella Leaf Spot

White centered purple margined leaf spot is a fungal disease caused by *Mycosphaerella fragariae* Tul. (= *Ramularia tulasnii* Sacc, imperfect stage). In India this disease was first reported from Niglar, Bhowadi hill of Uttar Pradesh during October 1952 (Bosh, 1970). In initial stage, circular purple scattered spots appear on upper surface of young leaves with an average diameter of 2-6 mm. The spots are most frequent on blades of leaflets but also appear on petioles, fruits and fruit stems. The leaf spots enlarge, then turn to white and are surrounded by dark purple margins, rendering birds eye effect. On lower surface of leaves, prominent veins touching the spots become reddish purple and entire leaf may die in advanced stages. The lesion on stem, fruit stalks are elongated as well as circular. On berries infection is not common however, the pulp of berry becomes discolored and render the fruits unmarketable under high pressure of disease (Fulton, 1958).

The fungus belongs to class – ascomycetes. The ascomata are usually very small, mostly emerg in the host tissues in leaves. The single septate hyaline ascospores measure 11-45x2-3μm dimensions, and are usually formed on the upper surface of the leaf, sometimes formed on petioles and calyx also. The sclerotia on dead leaves may also produce conidia.

The fungus overwinters through perithecia and sclerotia. Perithecia are produced at the edge of leaf in autumn season, whereas, sclerotia are produce on infected stolons and petioles. The prolonged wet period with low temperature in winter and spring favours the disease development. Continuous rainfall in month of April is mainly responsible for the occurrence of epiphytotics (Dale and fulton, 1957). The disease spreads in a very short period of time when strawberry plants are grown during warm days and cold nights (Nemee, 1972). The optimum temperature for the conidial germination ranges from 13 °C to 21°C (Fall *et al.*, 1957). Six virulence groups on nine strawberry cultivars (Bolton, 1962) and four races of fungus (Namee, 1971) were noted. Lower temperature delayed the maturation of young leaflets, thus, extending their susceptible period.

Management

Cultural practices to reduce disease incidence include planting the strawberry plants in well drained soil, keeping out the weeds and maintaining proper spacing. Avoid excessive use of nitrogenous fertilizers (Harnendo and Casada, 1976). During sprouting in spring and budding, the spray of micronutrients like manganese, copper and boron can reduce the infection and enhance the yield upto 13.20 Per cent (Dorozhkin and Grisanovich, 1972). Chemically this disease is managed by spraying of Bordeaux mixture (0.8 per cent), Zineb, Ferbam, Captan, Cuprex (0.2 per cent) and by benomyl (Stubbs, 1956). Spray of ditianon and Euparen M is effective against the disease (Agnello, 1975; Rebandel, 1975) and Vincolozolin and Iprodione have also been found effective (Alofs and Bryon, 1980). The application of soil sterilants *i.e.*, methyl bromide, Chloropicrin or Basamide can completely control the disease caused by sclerotia. The use of benomyl and thiophanate – M as dips to nursery plants or as spray after planting renders best control (Paulus *et al.*, 1974). The varieties like,

Combridge (Festic and Delkic, 1977), Takane Kurumal 103, Horida's wonder, Hukuda and Benzuru (Kim *et al.*, 1978), Arking (Moore, 1981), Premier, Covaliar, Dilpasand and Albitron (Sindhan and roy, 1981), Elista and Tioga (Saharan and Badiyala, 1985), Tarda vicoda and Marmalada (Zurawicz and Daminikowski, 1995), Joliette (Khanizadeh *et al.*, 1996) were reported as resistant against disease.

Pestaliopsis Leaf Spot

This is a common leaf spot disease of strawberry caused by fungus *Pestaliopsis disseminate* (Theum) stey. The disease is prevalent in all strawberry growing areas in India and was for the first time recorded by Singh *et al.* (1975) from Bangalore on cv. Pusa early dwarf.

Symptoms are characterized by circular, dark brown to chocolate coloured spots surrounded by reddish brown or yellowish margins. Spots may coalesce and shows patchy appearance. The infection normally starts from the leaf margin and extends towards midrib and covers the whole leaf. The infected areas become brittle and get detached from the healthy ones. Severely infected leaves get defoliated. Dot like fruiting bodies (acervuli) of the pathogenic fungus appeared on the necrotic areas of the infected leaf. The acervuli are amphiginous, scattered, distinct sometime gregarious, globose to centricular, conic or pyramidal, puntiform, erumpent rupturing the epidermis or diplacing it completely, exposing the black conidial mass. Short conidiophores slightly broader at the base with tapering end emerge from the basal cells, Conidia are elliptical or clavate, fusiform, tampering at the base, straight or slightly constricted at septa, 5 celled, measuring 18.5 x 26 x 5.5 μm (Singh *et al.*, 1975).

Yellow Leaf Spot

The disease is caused by the fungus *Dendrophoma obscurans* (Ell. And Ev.) Anderson, and is prevalent in strawberry growing areas. In India it has been reported from hills of Uttar Pradesh and Himachal Pradesh (Bosh, 1970). It has been found in moderate to severe form from 1,000 to 24,00 meters above sea level. The first symptoms appear on leaves with onset of the rains and the maximum damage occurs, during July to September. In initial stage of infection minute, circular, reddish to purple spots appear on the upper surface of leaves and calyces. The spots rapidly increase in size with yellow or yellow brown centre surrounded by a purple margin. The spots may coalesce and form irregular or oblong patches, covering more than half of leaflets surface. The large spots are distinctly divided in three; outer purple, inner light brown and central dark brown zones, and spots often extend towards the edge of lamina, which become "V" or fan shaped. The affected areas with disease may dry up and becomes puckered or crumpled. Sometimes, shot holes are formed. Pycnidia found in necrotic area of spot, are deeply embedded in host tissue with a distinct neck extending through the epidermis and act as a source of primary infection (Anderson, 1956; Bose, 1970).

The disease can be controlled by the use of Phenyl mercury acetate or captan. Spraying captan @ 0.2 per cent with the onset of rains. Albitron, Premier, Cavalier and Diplasand varieties are reported as resistant against disease (Sindhan and Roy, 1981).

Angular Leaf Spot

This is a most important bacterial disease of strawberry and has been reported from Carolina, Euthopia, Argentina and Australlia (Mc Gechan and Fahy, 1976; Alipi *et al.*, 1989; Navarthan *et al.*, 1991 and Ritchie *et al.*, 1993). Angular Leaf spot disease is caused by *Xanthomonas fragariae.*

The water soaked spots appear on upper surface of leaves in initial stage, which later on become dark brown and angular in shape. In severe condition of disease, the affected plants may be die.

Disease can be effectively controlled by spraying of streptomycin or streptocycline, when the symptoms begin to appear (Alipi *et al.*, 1989).

Hainesia Leaf Spot

Hainesia Leaf spot disease is caused by *Hainesia lythri* (Desm.) Hohn., and perfect stage of this pathogen is *Discochainesia ocrietherol* (Crok and Ellis) Nanuf. The perfect stage of this fungus has not been reported from India. This disease was first reported from Delhi (Pathak and Payak, 1965) and from Kumaon hills of U.P. (Bose, 1970).

Spherical to oval spots with light centers appeared on older leaves, which increased in size and covered larger areas of leaf lamina. Minute fruiting bodies pycnidia developed in the infected portion of leaves with blighted appearance.

Pestalotia Leaf Spot

This disease was reported by Bose (1970) and is caused by fungus *Pestalotia jeolikotensis* sp. nov. Bose. The fungus produces grey or dirty white spots on the upper surface of leaves, sometimes they coalesced and formed larger spot. The fungus produces dot – like black structure, acervuli in the form of concentric rings over spots. Later on shot – holes were formed in the center of spots.

Phyllosticta Leaf Spot

This is a fungal disease caused by *Phyllosticta fragaricola* Desim. et Rob. (Anon., 1960). Minute, round to irregular spots with grayish center and surrounded by dark brown margins are formed on upper leaf surface. The spots coalesce to form bigger spots. In advanced stage numerous pycnidia are formed on the grayish region of the spot.

Leaf Blight

This disease was first reported by Lele and Pathak (1965) from India. The disease is caused by *Rhizoctonia bataticola* (Taub). Bull. [= *Macrophomina phaseolina* (Tassi.) Goid].

Symptoms are produced on all plant parts during September to October months. More or less circular to spreading spots with ash grey centre and purplish dark margins were found on leaves. In advanced stage these spots become oval to irregular in shape and lesion started spreading from margin of the leaves towards centre. Infected runners and stalks turned dark brown or black with irregular lesions. Due to infection in plants new growth was very much suppressed. The poor development in new root and shredded appearance in older ones at their distal portion was found, but no irregular rot was seen.

Leaf Scorch

Leaf scorch disease was first reported by Wolf (1924) and in India it was reported by Bose (1970) from Kumaon hills of Uttar Pradesh. This disease is caused by fungus *Marssonina fragariae* (Sacc.) Kleb [(*Diplocarpon carliana*) (E and E) Wolf]. The disease mostly occurred in month of April or May and symptoms appeared on leaves, petioles, peduncle and runners. Irregular, oval or angular and purple colored symptoms surrounded by reddish or pinkish halo are found on leaves. Elongated, reddish brown or purple streaks developed on petiole and peduncle. Similar symptoms appear on runners which become discolored and girdled (Bose, 1970). Disease can be effectively managed by spraying of carbendazim.

References

Agnello, A.V. (1975). Chemical control trials against leaf spot of strawberry. *Annali dell Instituto Sperimentale per la Pathologia Vegetable* 4: 47–49.

Agrios, G.N. (2005). *Plant Pathology*, 5th edn Elsevier Academic Press, UK, pp. 922.

Alcock N.L., Howells, D.V. and Foister, E.C. (1930). Strawberry disease in Lankashire. *Scott. J. Agric.* 13: 242–251.

Alipi, A.M., Ronco, B.L. and Carrainza, M.R. (1989). Angular leaf spot of strawberry, a new disease in Argentina. Comparative control with antibiotics and fungicides. *Advances Hortic. Sci.* 1: 3–6.

Alofs, W. and Bryon, J. De. (1980). Control of Botrytis fruit rot of strawberry. *Fruittelet*, 70: 340–341.

Anderson, H.W. (1956). *Diseases of Fruit Crops*. McGraw Hill Book Company Inc., New York, Toronto, London, 501pp.

Anderson, H.W. and Colby, A.S. (1942). Red stele of strawberry on Path finder and Aberdeen. *Plant Dis. Rep.* 26: 291–292.

Anonymous. (1960). Index of plant diseases in the United States. U.S. Dep. Agric. Handb. No. 165. Washington DC. 531 pp.

Barritt, B.H. (1980). Resistance of strawberry clones to Botrytis fruit rot. J. Am. Soc. *Hortic. Sci.* 105: 160–164.

Bhardwaj, L.N. and Sharma, S.K. (1999). Diseases of strawberry, Gooseberry and raspberry. In: *Diseases of Horticultural Crops: Fruits*. Indus Pub. Com., New Delhi, p. 316–336.

Borecka, H., Borecki, Z. and Millikan, D.F. (1973). Recent investigations on the control of grey mould of strawberries in Poland. *Plant Dis. Rep.* 57: 31–33.

Bose, S.K. (1970). Diseases of valley fruits in Kumaon (III) leaf spot disease of strawberry. *Prog. Hortic.* 2: 33–53.

Braun, P.G. and Sutton, J.C. (1987). Inoculum source of *Botrytis cinerea* in fruit rot of strawberries in Ontario. *Can. J. Plant Pathol.* 9: 1–5.

Brooks, A.N. (1931). Anthracnose of strawberry caused by *Colletotrichum fragariae*, N. Sp. *Phytopathol.* 21: 739–744.

Bulger, M.A., Ellio, M.A. and Madden, L.V. (1987). Influence of temperature and wetness on infection of strawberry flowers by *Botrytis cinerea* and disease incidence of fruit originating from infected flowers. *Phytopathol.* 77: 1225–1230.

Chandler, C.K., Albregts, E.E and Howard, C.M. (1988). Evaluation of strawberry cultivars and advanced selections at Dover Florida, 1986–88. *Adv. Strawberry Prod.*, 8: 19–22.

Chandler, C.K., Mertely, J.C. and Peres, N. (2006). Resistance of selected strawberry cultivars to anthracnose fruit rot and Botrytis fruit rot. *Acta–Horticulturae*. 708: 123–126

Cheah, L.H. and Soterosi, J.J. (1984). Control of black fruit rot of strawberry. *Proc. 37th N.Z. Weed Pest Control Conf.*, 37: 160–162.

Dai, F.M., Ren, X.J. and Lu, J.P. (2006). First report of anthracnose fruit rot of strawberry caused by *Colletotrichum acutatum* in China. *Plant Disease*. 90(11): 1460

Dale, J.L. and Fulton, J.P. (1957). Severe loss from strawberry leaf spot in Arkansas in 1957. *Plant Dis. Rep.*, 41: 681–682.

Daubeny, H.A. and Pepin, H.S. (1977). Evaluation of strawberry clones for fruit rot resistance. J. Am. Soc. *Hortic. Sci.*, 102: 431–435.

Delp, C.J. (1981). Pesticide resistance–its impact on pest and disease control. *Proc. 1981. Br. Crop Prot. Conf.*, 3: 865–871.

Denoyes, R.B. and Guerin, G. (1996). Comparison of six inoculation techniques with *Colletotrichum acutatum* on cold stored strawberry plants and screening for resistance to this fungus in French strawberry collections. *Eur. J. Plant Pathol.*, 102: 615–621.

Devaux, A. (1978). Epidemiological studies on grey mould of strawberry and control trials. *Phytoprotection*, 59: 19–27.

Domingues, R.J., Tofoli, J-G., Oliveira, S.H.F. and Garcia-Junior, O. (2001). Chemical control of strawberry black flower rot (*Colletotrichum acutatum* Simmonds) under field conditions. *Arquivos-do-Instituto-Biologico, Sao-Paulo*, 68(2): 37–42

Dorozhkin, N.A. and Grisanovich, A.K. (1972). Effect of microelements on susceptibility of strawberry to white spot and grey rot. *Khimiyav-Sel'Shom-khozyaistve*, 10: 50–51.

Drozdowski, J. (1987). A survey of fungi associated with strawberry black root rot in commercial fields. *Adv. Strawberry Prod., N. America Strawberry Growers Assoc.*, 6: 47–49.

Eastburm, D.M. and Gubler, W.D. (1990). The detection and survival of *Colletotrichum acutatum*, a causal agent of strawberry anthracnose in soil. *Plant Dis.*, 74: 161–163.

El-Zayat, M.M., Okashakha, K.H.A., El-Tobshy, Z.M., El-Kohli, M.M.A. and El-neshuary, S.M. (1993). Efficiency of *Trichoderma harziamum* as a biocontrol agent against gray mould rot (*Botrytis cinerea*) of strawberry fruits. *Ann. Agric. Sci. Cairo.* 38: 283–290.

Ellis, M.A. and Grove, G.G. (1983. Leather rot in Ohio strawberries. *Plant Dis.*, 97: 549.

Ercole, N. (1976). Verticillium wilt of strawberry, a difficult disease to control. *Informattore Agrario*, 32: 925–926.

Faedi, W., Rosati, P. and Ercole, N. (1986). Etna, Ferrarra and Gea: There new strawberry cultivars. *Annali dell Instituto Sperimental per Ia Fritticoltura, Italy*, 27: 180.

Festic, H. and Delkic, I. (1977). The reactions of some strawberry varieties to *Mycosphaerella fragariae* (Schw.) Lind at Bosanska Krajina. *Jugoslovonsko-Vocarstovo*, 10: 633–637.

Fulton, R.H. (1958). Studies on strawberry leaf spot in Michigan. *Michigan Expt. Sta Quart. Bull.* 40: 581–588.

Galletta, G.J., Mass, J.L., Enns, J.M., Draper, A.D., Fiola, J.A., Scheerens, J.C., Archbold, D.D. and Ballington, J.R. (1995b). Delmarvel strawberry. *Hort. Sci.*, 30, 1099–1103.

Galletta, G.J., Mass, J.L., Enns, J.M., Draper, A.D., Dale, A. and Swartz, H.J. (1995a). Mohawk strawberry. *Hort. Sci.*, 30, 631–634.

George, S.W. and Milholland, R.D. (1986). Inoculation and evaluation of strawberry plants with *Phytophthora fragariae. Plant Dis.*, 70: 371–375.

George, S.W. and Milholland, R.D. (1987). Seasonal occurrence and recovery of *Phytophthora fragariae* in infected strawberry roots. *Hort. Sci.*, 22: 1254–1255.

Gourley, C.O. (1974). A comparison of benomyl, thiophanate methyl and captan for control of strawberry fruit rot. *Can. Plant Dis. Surv.*, 54: 27–30.

Grove, G.G., Maddan, L.V. and Ellis, M.A. (1985a). Influence of temperature and wetness duration on sporulation of *Phytophthora cactorum* on infected strawberry fruit. *Phytopathol.*, 75: 700–703.

Grove, G.G., Maddan, L.V. and Ellis, M.A. (1985b). Splash dispersal of *Phytophthora cactorum* from infected strawberry fruit. *Phytopathol.*, 75: 611–615.

Grove, G.G., Maddan, L.V., Ellis, M.A. and Schmitthenner, A.F. (1985c). Influence of temperature and wetness duration on infection of immature strawberry fruit by *Phytophthora cactorum. Phytopathol.*, 75: 165–169.

Harnendo, V. and Casada, M. (1976). The relationship between *Mycosphaerella fragariae* infection and the nutritional status of strawberry plants. *Anales de Edafologia-y-Agrobiologia*, 35: 1–2.

Harris, C.M. and Harvey, J.M. (1973). Quality and decay of California strawberries stored in CO_2 enriched atmospheres. *Plant Dis. Rep.*, 57: 44–46.

Harris, D.C. (1989a). The use of soil sterilants for controlling Verticillium wilt of strawberry in Britain. *Acta Hortic.*, 255: 100.

Harris, D.C. (1989b). Control of Verticillium wilt of strawberry in Britain by chemical soil disinfestations. *J. Hortic. Sci.*, 64: 683–686.

Hartz, T.K., DeVay, J.E. and Elmore, C.L. (1993). Solarization is an effective disinfestations technique for strawberry production. *Hort. Sci.*, 28: 104–106.

Haward, C.M. (1972). A strawberry fruit rot caused by *Colletotrichum fragariae. Phytopathology*, 62: 600–602.

Horn, N.L. (1961). Control of Botrytis rot of strawberries. *Plant Dis. Rep.* 45:818–822.

Howard, C.M. and Alberts, E.E. (1973). Cassia obtusifolia, a possible reservoir for inoculation of *Colletotrichum fragariae. Phytopathology*, 63: 533–534.

Hutton, D. (2006). Successful management of Colletotrichum crown and stolon rot in runner production in sub-tropical Australia. *Acta Hort.*, 708: 293–298

Igarashi, I., Monma. S., Fujino, M., Okimura, M., Okitsu, S., Takada, K. and Nii, T. (1994). Breeding of new everbearing strawberry variety 'Ever Berry'. Bull. Natl. Res. Inst. Veg. Ornamental plant and Ta Series A. Vegetables and Ornamental Plants No. 9: 69–84.

Inada, M. and Yamaguchi, J.I. (2006). Occurrence of anthracnose fruit rot caused by *Colletotrichum acutatum* and *Colletotrichum gloeosporioides* on strawberry in forcing culture. *Kyushu-Plant-Protection-Research*, 52: 11–17.

Jhooty, J.S. and Mc Kean, W.E. (1965). Studies on powdery mildew of strawberry caused by *Sphaerotheca macularis. Phytopathol.*, 55: 281–285.

Joffens, W.F. (1957). Soil treatments for control of the red stele disease of strawberries. *Plant Dis. Rep.*, 41: 415–418.

Johnson, A.W., Simpson, D.W. and Berrie, A. (2006). Hot water treatment to eliminate *Colletotrichum acutatum* from strawberry runner cuttings. *Acta Horticulturae*, 708: 255–258

Kao, C.W. and Lu, L.S. (1979). Strawberry fruit rot caused by *Phytophthora cactorum* and *Phytophthora citrophthora. Plant Prot. Bull. Taipei*, 21: 239–243.

Keyworth, W. and Benett, M. (1951). Verticillium wilt of the strawberry. *J. Hortic. Sci.*, 26: 304–316.

Khanizadeh, S., Buszard, D., Carrise, O. and Thiobdean, P.O. (1996). Joliette strawberry. *Hortic. Sci.*, 31: 1036–1037.

Kim, J.G., Cho, C.T., Bia, T.U., Han, H.S. and Mun, B.J. (1978). Varietal resistance to strawberry to two strains of *Mycosphaerella fragariae* and chemical control of the pathogen. *Korean J. Plant Prot.*, 17: 149–154.

Lele, V.C. and Pathak, H.C. (1965). Leaf blight and dry stalk rot of strawberry Caused by *Rhizoctonia bataticola. Indian Phytopath.*, 18: 38–42.

Madden, L. V., Wilson, L.L and Ellis, M.A. (1993). Field spread of anthracnose fruit rot of strawberry in relation to ground cover and ambient weather conditions. *Plant Dis.*, 77: 861–866.

Mass, J.L. (1978). Screening for resistance to fruit rot in strawberries and red raspberries: A review. *Hort. Science*, 13: 423–426.

Mass, J.L. (1984). *Compendium of Strawberry Diseases*. Am. Phytopathol. Soc. St. Paul, Minn.

Matsuo, K., Suga, Y. and Jan, M. (1994). Chemical control of strawberry anthracnose caused by *Colletotrichum acutatum. Proc. Assoc. Plant Prot. Kyushu*, 40: 17–21.

McGechan, J.K. and Fahy, P.C. (1976). Angular leaf spot of strawberry, *Xanthomonas fragariae*: first record of its occurrence in Australia and attempts to eradicate the disease. *APPS–Newsletter*, 5: 57–59.

Mcintyre, J.L. and Walton, G.S. (1981). Control of strawberry anthracnose caused by *Colletotrichum acutatum. Proc. Assoc. Plant Prot. Kyushu*, 40: 17–21.

McKinley, R.T. and Tolboys, P.W. (1979). Effect of *Pratylenchus penetrans* on the development of strawberry wilt caused by *V. dahliae. Ann. Appl. Biol.*, 92: 347–351.

Meesters, P. Meurrens, F. and Brugmans, W. (1996). Strawberry cultivation. Total soil fumigation and fumigation in strips for growing *Elsanta. Fruit Belge.*, 64:65–69.

Mitra,S.K. (1993). Strawberries. In : *Temperate Fruits*, (Eds.) S.K. Mitra, T.K. Bose and D.S. Rathore. Alert and Allied Publishers 27/3 Chakraberia Lane, Kolkata, pp. 549–596.

Moore, J.N. (1981). 'Arking', a late season, disease resistant strawberry variety. *Arkansas Farm Res.*, 30: 6.

Namec, S. (1971). Studies on resistance of strawberry varieties and selections to *Mycosphaerella fragariae* in southern Illinois. *Plant Dis. Rep.*, 55: 573–576.

Navarthan, S.J., Degefu, T. and haite, M. (1991). Outbreak and new records. Ethopia: Angular leaf spot of strawberry in Ethopia. *FAO Plant Prot. Bull.*, 39: 116.

Nemec, S. (1972). Studies on resistance of strawberry varieties and selection to *Mycosphaerella fragariae* in southern Illinois. *Plant Dis. Rep.*, 55: 573–576.

Pathak, H.C. and Payak, M.M. (1965). *Hainesia lythri* on strawberry: Anew record from India. *Indian Phytopath.*, 18: 237–239.

Paulus, A.O., Voth, V., Nelson, J. and Bowen, H. (1978). Control of Botrytis fruit rot of strawberry. *Calif. Agric.*, 32: 9.

Paulus, A.O., Voth, V., Bringhurst, R.S. and Welch, N. (1973). Comparison of fungicides and methods of application for the control of Ramularia leaf spot of strawberry. *Proc. 7th Br. Insecti. Fungic. Conf.*, 2: 471–476.

Peris, O.S. (1962). Studies on strawberry mildew, caused by *Sphaerotheca macularis* (Wallr. Ex Fries) Jaczcwski II: Host parasite relationships on foliage on strawberry varieties. *Ann. Appl. Biol.*, 50: 225–233.

Powelson, R.L. (1960). Initiation of strawberry fruit rot caused by *Botrytis cinerea*. *Phytopathol.*, 50: 491–494.

Prasad, K. and Stadelbacher, G.J. (1974). Control of postharvest decay of fresh raspberries by acctaldehyde vapor. *Phytopathol.*, 64: 948–951.

Rebandel, Z. (1975a). Leaf white spot of strawberry. *Sad. Mowoczesny*, 6: 16–18.

Ritchie, P.F., Averre, C.W. and Milholland, R.D. (1993). First report of angular leaf spot, caused by *Xanthomonas fragariae*, on strawberry in north Carolina. *Plant Dis.*, 77: 1263.

Rose, D.H. (1926). Relation of strawberry fruits rots to weather conditions in the field. *Phytopathol.*, 16: 229–232.

Saharan, G. S. and Badiyala, S.D. (1985). Progress of Mycosphaerella leaf spot on strawberry cultivars in relation to environment. *Indian Phytopathol.*, 38: 139–141.

Seemuller, F. and Schmidle, A. (1979). Influence of the origin of *Phytophthora cactorum* isolates on their virulence on apple bark, strawberry rhizomes and strawberry fruits. *Phytopathol.*, 94: 218–225.

Simpson, D. and Blanke, M. (1996). Calypso and Tango perpetual flowering strawberry varieties from Horticultural Research International East Malling. *Erwerbsobstbau*, 38: 55–57.

Simpson, D., Bell, J.A., Harris, D.C., Schmidt, H. and Kellaerhols, M. (1994). Breeding to fungal diseases in strawberry. Proc. Eucarpia Fruit Breeding to fungal diseases in strawberry. *Proc. Eucarpia Fruit Breeding Section Meeting*, Wadenswil. Einsiedeln. Switzerland 30 August to 3 September, 1993. pp. 63–66.

Simpson, D., Hammond, K., Lesemann, S. and Whitehouse, A. (2006). Pathogenicity of UK isolates of *Colletotrichum acutatum* and relative resistance among a range of strawberry cultivars. *Acta Horticulturae*, 708: 281–285

Sindhan, G.S. and Roy, A.J. (1981). Screening of different varieties and selections of strawberry against leaf diseases and their control. *Indian Phytopathol.*, 34: 304–306.

Singh, S.J. (1974). A ripe fruit rot of strawberry caused by *Colletotrichum fragariae*. *Indian Phytopath.*, 27: 433–434.

Singh, S.J., Sastry, K.S.M. and Sastry, K.S. (1975). Investigations on mosaic disease of cape gooseberry. *Curr. Sci.*, 44: 95–96.

Smith, B. J. and Black, L.L. (1993). *In vitro* activity of fungicides against *Colletotrichum fragariae*. *Acta Hortic.*, 348: 509–512.

Sommer, N.F., Fortlage, R.J., Mitchell, F.C. and Maxi, E.C. (1973). Reduction of post-harvest losses of strawberry fruits from grey mould. *J. Am. Soc. Hortic. Sci.*, 98: 285–288.

Stubbs, L.L. (1956). Diseases of strawberries. *J. Agric. (Victoria)*, 54: 232–236.

Wilhelm, M. (1952). Verticillium wilt and black root rot of strawberry. *Calif. Agric.*, 6: 8–9.

Wilhelm, M. (1961). Diseases of strawberry. A guide for the commercial growers. *Univ. Calif. Expt. Sta. Circular*, 494 pp.

Wilhelm, S. and Paulus, A.O. (1980). How soil fumigation benefits the California strawberry industry. *Plant Dis.*, 64: 264–269.

Zurawicz, E. and Daminikowski, J. (1995). Preliminary evaluation of productivity of several new strawberry (Fragaria x ananassa Duch.) clone and cultivars in central Poland. *Zesszyty Naukouse Instytute-Sadowhictwa-i-Kwiaciarstwa-w-Skiniewicach*, 2: 5–11.

Plant Diseases Management in Horticultural Crops (2011) *Pages* ***170–186***
Editors: **Shahid Ahamad, Ali Anwar and P.K. Sharma**
Published by: **DAYA PUBLISHING HOUSE, NEW DELHI**

Chapter 14

Soil Borne Diseases of Apple and their Management

Farah Rasool and Shahzad Ahmad

Division of Plant Pathology,
S.K.University of Agricultural Sciences and Technology of Kashmir,
Shalimar, Srinagar – 191 121, J&K

Apple is one of the most important fruit crops of the temperate regions of the world. It is reported to have originated in the temperate regions of Western Asia. The commercial cultivation of apple is practiced in the north-western Himalayan states of J&K, Himachal Pradesh and Uttarakhand besides some hilly areas of the north-western part of the India. In addition to providing healthy and nutritious food, it has also increased the socio-economic status of farmers. Diseases and pests have always acted as bottlenecks in its qualitative and quantitative production. Among various diseases affecting apple, soil-borne diseases such as White root rot (*Dermatophora necatrix*), collar rot (*Phytophthora cactorum*), seedling blight (*Sclerotium rolfsii*), Crown gall (*Agrobacterium tumefaciens*) and Hairy root (*Agrobacterium rhizogenes*) cause significant losses amounting to around 40 per cent to apple both in the nurseries as well as in orchards (Gupta and Sharma, 1999, Sharma and Sharma 2001).

The persistence of soil borne pathogens in the soil and immobility of chemicals in soil makes management of soil borne diseases cumbersome. The problem is more complex in perennial trees due to continuous contact of the host with the pathogen (Sharma *et al.*, 1999). Soil-borne diseases of apple have however been managed successfully by integrating various methods like cultural, chemical, biological and use of resistant varieties.

White Root–Rot

It is one of the most important soil borne disease of apple having a host range of about 158 plant species belonging to over 45 families (Ito and Nakamura, 1984), comprising of fruit plants, vegetables, forest trees and field crops. The disease was first reported in India in Uttarakhand Hills in1929(Bose and Sindhan, 1976) and latter on in Himachal Pradesh (Agarwal, 1961). Agarwala and Sharma (1966) have been recorded 8-33 per cent incidence of white root rot in different apple growing areas of Himachal Pradesh.

Symptoms

The above ground symptoms are expressed as bronzing of leaves, stunted tree growth, resulting in the progressive decline in the vigour of the tree as whole or certain branches. It is followed by wilting, defoliation and death of the tree (Jain, 1961 Agarwala and Sharma, 1966 Gupta 1977, Sztejnberg *et al.*, 1987).

In the below ground expression of the disease the lateral roots turn dark brown and are covered with a greenish grey or white mycelial mat having a flocculent web of whitish strands or ribbons during monsoon season. The cortical cells are ruptured resulting in disruption of the plant system leading to the death of the tree and hence expressing disease on above ground plant parts (Nattrass (1972, Agarwala and Sharma, 1966) Agarwala and Sharma (1966) reported that the fungus does not infect xylem tissues of apple roots but the bark is completely damaged.

Causal Organism

The disease is caused by *Rosellinia necatrix* Berl.ex Prill (Anamorph: *Dermatophora necatrix* Hartig).However the perfect stage of the fungus has not been reported from India. The fungus is distinguished on the basis of pear shaped swellings at each septum of the fungal hyphae (Gupta and Sharma, 1999: Sharma and Sharma, 2004). Two types of mycelial characteristics represented by small black microsclerotia, which unite to form microsclerotial sheets.

Management

White root rot of apple is controlled by practicing preventive as well as curative measures consisting of cultural, biological, chemical methods and resistant root stocks.

Cultural Practices

It includes hot water treatment of infected seedlings at 45°C for one hour before plantation, digging isolation trenches and removal of rotten roots followed by application of disinfectant paste. Burning and heat drying of infected roots have also been reported to increase the life of the trees (Agarwala and Sharma, 1975, Jain, 1961). Soil solarization has also been found effective in reducing the fungus inoculum in soil (Freeman *et al.*, 1986). Soil amendment with neem cake and deodar needles has been reported to reduce the incidence of the disease (Gupta and Jindal, 1989; Sharma and Gupta, 1996 b.)

Chemical Control

Initially, broad spectrum chemicals such as carbon bisulphide, choropicrin, calcium cynamide, formaldehyde etc. were recommended for checking root rot of apple. With the advent of systemic fungicides, such a benomyl, aureofungin, carbendazim etc, the earlier recommendations were replaced. The fungicide should be applied through deep holes (15-20 cm) made in the basin of tree at 30 cm distance from each other and the tree basin is brought to a soil moisture level of 30 to 40 per cent prior to the application. The treatment is done 3 to 4 times every year during rainy season.

Resistant Rootstocks

Some degree of resistance against the pathogen in three rootstocks namely MM109,M16 and MM104 has been observed, but until and unless, a rootstock with better resistance is developed, the root rot problem will persist. Modern biotechnological approaches are required to be exploited for combating this pernicious soil borne disease of apple.

Biocontrol

Various antagonists have been explored for the management of disease such as *Trichoderma viride, T. harzianum* (Mercer and Krik, 1984, Freeman *et al.*, 1986a; Sztejnberg *et al.*, 1980, *Enterobacter aerogenes* (Gupta and Jindal, 1989) have been found to be protected the plants from *D. necatrix*. These antagonists inhibited the growth of *R. necartix* and covered the pathogen culture with different degrees of sporulation. Bacterial antagonists like *Enterobacter aerogenes* and *Bacillus subtilis* have also been found to inhibit *Dematophora necatrix* and reduce disease up to 45 days of inoculation (Gupta and Jindal, 1989). The efficacy of the antagonists was increased due to their repeated applications Sharma and Gupta (1996c)

Collar Rot

Symptoms

The infection starts from the collar region and spread mostly to the underground parts. Bark at the soil level becomes slimy and rots resulting in cankered areas. The wounds are irregular in outline but usually roughly oval which extend rapidly, often resulting in girdling the tree. The attacked trees are recognized by chlorotic foliage with red colouration of veins and margins. The above ground symptoms of collar rot are often confused with white root rot, however, examination of the underground parts reveals the differences

Causal Organism

Collar rot is mainly caused by *Phytophthora cactorum* (Lebert-Cohn) Schroete, which is semi aquatic, soil borne homothallic fungus with a complex life cycle consisting of an asexual phase in which motile zoospores are produced from sporangia and a sexual phase resulting in the formation of oospores. The optimum temperature for mycelial growth in culture is shown variously as 25 to 27°C and 30 to 34°C. The sporangia are papillate, borne on sympodialoly branched sporangiophores.

Management

The management of collar rot of apple like other soil borne diseases involve the integration of various cultural, chemical and biological measures

Chemical Control

The fungicides are applied around the affected trees or by painting the wounds with fungicidal paints. The disease can also be controlled by drenching with Dithane M-45 (0.03 per cent) or Blitox/ Fytolan/Blue copper (0.5-1 per cent) in 30 cm radius around the tree trunk (Agarwala, 1970; Rana and Gupta, 1981). Systemic fungicides like Ridomil MZ and Aliette are also effective both as soil drench and paint (Gupta, 1990; Bleicher, 1994) Fosetyl-aluminium applied as foliar spray completely control the disease and increase the growth and fruit yield (Utkhede and Smith, 1995).

Resistant Rootstocks

In soil-borne diseases use of resistant root-stock is most effective and desirable approach. Root-stocks such as M2,M4, MM110,MM114,MM115 and Crab apple have been reported to posses high degree of resistance against *P. cactorum* as well as *Pythium ultimum* under Indian conditions (Grag and

Gupta, 1989;Sharma and Gupta, 1989). Utkhede and Smith. (1994) reported that rootstock MM111 is highly resistant to crown rot which remained disease free for more than ten years.

Biocontrol

Bacillus subtilis and *Enterbacter aerogenes* have been identified as potential antagonists against the disease which need commercial exploitation (Utkhede, 1986;Orlikowski and Schmidle, 1985).Smith *et al.* (1990) identified The isolates of *Trichoderma* and *Glicocladium* were identified as potential biocontrol agents for *P. cactorum* and reported that both these antagonists reduced root damage and increased the plant growth.

Seedling Blight

The appearance of the disease occurs in nurseries as well as 2 to 3 years old seedlings. In India, the disease was first reported on apple seedlings from Mysore state (Anon, 1937) and later in Himachal Pradesh by Jain (1962). Bhardwaj and Agarwala (1986a) recorded 40 per cent losses in registered nurseries of Himachal Pradesh.

Symptoms

The preliminary symptoms subsequent to root infection include wilting of the leaves showing a characteristic reddish or grayish purple discoloration as well as brightening of foliage of the infected seedlings. The above ground symptoms are generally confused with white root rot but its identification can be confirmed by examination of roots, where mustard seed size sclerotia are seen in the vicinity of dead seedlings. The disease is aggravated by root injuries. A soil temperature of 30 to 33 °C, pH 6.0 and 38 per cent or above soil moisture has been found to be most suitable for disease development (Bhardwaj and Agarwala, 1986a). It has also been reported that the mortality of seedlings is higher in sandy loam soil than in loamy sand and clay loam soil.

Causal Organism

The disease is caused by *Sclerotium rolfsii* Sacc. The fungus grows on a wide range of media and produce mustard seed size, tan to reddish brown to dark brown sclerotia. The asexual spores are not produced. *Corticium rolfsii* is the sclerotial state of *S. rolfsii*, which has now been transferred to Athelia. The pathogen is plurivorous, unspecialized but important pathogen and more frequent in warm moist conditions. The sclerotia have a long but variable survival period and germinate eruptively to infect plants directly.

Management

Affected seedlings in the nursery beds are treated with thiram, brassicol and aureofungin to keep this disease under control (Gupta and Agarwala, 1973; Bhardwaj and Agarwala, 1986b. As the disease is aggravated by high soil moisture so water logging in nursery beds should be avoided and loam soil should be selected for raising nursery. The nursery site must be rotated every 4 to 5 years. Maize should be planted continuously for 4 to 5 years before site is again selected for nursery production.

Crown Gall and Hairy Root

Crown gall and hairy root are serious problems in pome and stone fruits. The disease is prevalent in all the apple growing countries of the world and causes heavy losses in nursery because the diseased seedlings are rendered unfit for transplanting. It also occurs on grown up trees, which develop tumors at or near the ground line. The disease was first of all reported from California in 1892

(Wickson and Woodworth, 1892). Smith and Townsend (1907) first reported the bacterial nature of crown gall.

Symptoms

The symptoms of hairy root and crown gall are characterized by extensive proliferation of adventitious roots, singly or in clusters. On the basis of tissue morphology airy root is divided into simple hairy root wherein large number of small roots appear from stem without any associated callus or tumor tissues; wooly knot wherein fibrous roots arise from graft overgrowth or tumors on young tissues; broom root wherein fine roots develop from the tops of roots, which themselves originate from tumor tissues. The characteristic of these disorders is te transformation of the cells into autonomously proliferating tumor cells, resulting from unregulated cell division giving rise to clearly visible galls. These galls are visible at or near the graft union, surrounding the stem or root or connected to the host.

Causal Organism

Agrobacterium tumefaciens (Smith and Townsend)Conn. and *A. rhizogenes* (Lee *et al.*, 1935) conn, belonging to Rhizobiaceae family are responsible for crown gall and hairy root, respectively. These bacteria are rod shaped (0.6-1.0x1.5-3.0μm), aerobic and contain peritrichous flagella. The most characteristic feature of bacterium is its ability to transform normal plant cell to tumor cells. The pathogenic bacteria carrying Ri and Ti plasmids are attracted by the phenolic substances produced by the wounded cells, subsequently, the bacteria produce growth and the bacteria multiply in the watery intercellular spaces that cuts the Ti plasmid at specific sites releasing the segment of T-DNA. Several copies of this T-DNA from Ri or Ti-plasmid get integrated into plant cell chromosomes. The T-DNA expression results in overproduction of plant hormones and synthesis of opines, resulting in uncontrolled cell division and growth, producing a tumor.

Management

Cultural Practices

Destruction of the infected plant material by uprooting and burning and rotation in the nursery site is helpful in preventing the disease to some extent, budding instead of grafting also helps in preventing the disease spread as grafting leads to the spread of the infection.

Chemical Control

To avoid the dissemination of pathogen, the entire root system of apparently healthy grafted plants should be dipped in 1 per cent copper sulphate solution for ½ hour prior to transplanting.

Biological Control

Biological control of crown gall was the first commercial deployment of specific microorganism to control a soil borne pathogen. Use of antagonistic bacterium against crown gall, irrespective of host species has been exploited; inoculation of seeds and roots of peach with a non pathogenic strain of *A radiobacter* has been reported to give 99 per cent disease control. Cooley (1986) demonstrated that ineffectiveness of strain 84 against crown gall on pome fruits was due to the fact that root exudates affected sensitivity of strain of pathogen to agrocin -84.

References

Agarwal, R.K. (1961). Problem of root rot in Himachal Pradesh and prospects of its control with antibiotics. *Himachal Hortic.*, 2: 171–178.

Agarwal, R.K. (1970). Relative importance of the control methods of phytophthora collar rot disease of apple. In: *Plant Disease Problems*, (Eds.) S.P. Raychaudhuri *et al.* IPS, IARI, New Delhi, pp. 632–638.

Agarwal, R.K. and Sharma, V.C. (1966). White root rot disease of apple in Himachal Pradesh. *Indian Phytopathol.*, 19: 82–86.

Agarwal, S.C., Khare, M.N. and Agarwal, P.S. (1977). Biological conrol of *Sclerotium rolfsii* causing collar rot of lentil. *Indian Phytopathol.*, 30: 176–179

Anonoymous (1937). Some diseases of apples. *Mysore Agric.*, Cal, 25 p.

Baker, A. and Khan, A.A. (1981). Effect of nitrogenous amendments on the incidence of Sclerotium wilt of potato. *Potato Res.*, 24: 363–365.

Bhardwaj, L.N. and Agarwala, R.K. (1986). Seedling blight of apple and its control. In: *Advances in Research on Temperate Fruits*, (Eds.) T.R. Chadha, V.P. Bhutani and J.L. Kaul. UHF, Nauni, pp. 397–401.

Braun, H. and Nienhaus, F. (1959). Further studies on collar rot of apple (*Phytophthora cactorum*) *in vitro* with *Enterbobacter aerogenes*. *N.Z. J. Crop Hortic. Sci.*, 25: 9–18

Brown, E.A. and Hendrix, H. (1980). Efficacy and *in vitro* activity of selected fungicides for control of *Phytophthora cactorum*, collar rot of apple. *Phytopathology*, 78: 846–851.

Burr, T.J. and Reid, C.L. (1995). Survival and tumorigenicity of *Agrobacterium vitis* in living and decaying grape roots and canes in soil. *Plant Dis.*, 79: 677–682

Cooley, J.S. (1936). *Sclerotium rolfsii* as a disease of nursery apple trees.*Phytopathology*, 26: 1081–1083

De Sousa, A.J.T., Guillaumin, J.J., Sharples, G.P. and Whalley, A.J.S. (1995). *Rosellinia necatrix* and White root rot of fruit trees and other plants in Portugal and nearby regions. *Mycologist*, 9: 31–33

Elad, Y., Chet, I. and Katan, J. (1980). *Trichoderma barzianum*: A biocontrol agent effective against *Sclerotium rolfsii* and *Rhizoctonia solani*. *Phytopathology*, 70: 119–121.

Ellis, M.A., Grover, G.G. and Ferree, D.C. (1982). Effect of Metalaxyl on *Phtytophthora cactorum*. *Phytopathology*, 72: 1431–1433.

Fitzpatrick, R.E., Mellor, F.C. and Welsh, M.F. (1944). Crown rot of apple trees in British Columbia rootstocks and scion resistance trials. *Sci. Agric.*, 24: 533–544.

Freeman, S., Sztehnberg, A. and Chet, I. (1986). Evaluation of *Trichoderma* as a biocontrol agent for *Rosellinia necatrix*. *Plant Soil*, 94: 163–170.

Garg, R.C. and Gupta, V.K. (1989). Reaction of apple rootstocks to *Phytophthora cactorum*. *Indian Phtopathol.*, 42: 580–581

Gupta, G.K. and Agarwala, R.K. (1973). Field testing of soil fungicides for the control of seedling blight of apple. *Indian J. Mycol. Plant Pathol.*, 3: 109–111.

Gupta, V.K. and Mir, N.M., 1983. Field testing of apple rootstocks against *Phytophthora cactorum* A new technique. *J. Tree Sci.*, 2: 81–83

Gupta, V.K. and Rana, K.S. and Mir, N.M., 1985. Variability in *Phytophthora cactorum* in India. In: *Ecology and Management of Soil Borne Plant Pathogens*, (Eds.) C.A. Parker, A.D. Rovira, A.J. Moore, P.T.W. Wong and J.F. Koilmongen. American Phytopathological Society, USA, pp. 167–171.

Gupta, V.K. and Sharma, R.C. (1999). Concepts and practices in the management of soil borne diseases. In: *Modern Approaches and Innovations in Soil Management*, (Eds.) D.J. Bagyaraj, A.M. Verma, K.K. Khanna and H.K. Kehri. Rastogi Publications, Meerut, pp. 333–334.

Gupta, V.K. and Sharma, S.K. (1999). Soil borne diseases of apple and their management. In: *Diseases of Horticultural Crops: Fruits*, (Eds.) L.R. Verma and R.C. Shama. Indus Publishing Co., New Delhi, pp. 89–104.

Gupta, V.K. and Verma, K.D. (1978). Comparative susceptibility of apple rootstocks to *Dematophora necatrix*. *Indian Phytopathol.*, 31: 377–378

Gupta, V.K. and Rana, K.S. (1982). Effect of soil factors and amendments on the viability of *Phytophthora cactorum* sporangia and oospores in apple orchard soil. *J. Tree Sci.*, 1: 64–70.

Handea, H., Matsuzaki, K. and Mitsueda, T. (1985).Organic materials application to the soil on root infecting fungi. *Annual Report of the Society of a Plant Protection of North Japan*, 36: 157–159.

Hartig, R. (1883). Untermechengen aus dem forest rot amschen. *Institute Zu Mmunchen*, 3: 95–141.

Ito, S.I. and Nakamura, N. (1984). An outbreak of white root rot and its environmental conditions in the experimental arboretum. *J. IPM For. Sci.*, 66: 262–267.

Jain, S.S. (1962). Studies on apple seedling blight Sclerotium rolfsii Sacc. In: *Himachal Pradesh Proc. 49th Indian Sci. Cong.*, 3: 247.

Keyser, H.A. and Ferrierira, J.H.S. (1988). Chemical and biological control of *Sclerotium rolfsii* in grapevine nurseries. *S. Afr. J. Enol. Vitic.*, 9: 43–44.

Laha, G.S., Verma, J.P. and Singh, R.P. (1996). Effectiveness of fluorescent pseduomonads in the management of sclerotial wilt of cotton. *Indian Phytopathol.*, 49: 3–8.

Lee, S., Chong, B., Jang, H., Kim, K.H. and Choi, Y. (1935). Incidence of soil borne diseases in apple orchards in Korea. *J. Plant Pathol.*, 11: 32–138.

Maiti, S. and Sen, C. (1984). Population density of *Sclerotium rolfsii* in amended Soil. *Indian J. Plant Pathol.*, 2: 83–89.

Mathur, S.N. and Sarbhoy, A.K. (1978). Biological control of *Sclerotium* root rot of sugarbeet. *Indian Phytopathol.*, 31: 365–367.

Mcintosh, D.L. and Mellor, F.C. (1953). Crown rot of fruit trees in British Columbia II. Rootstock and scion resistance trials of apple, pear and some stone fruits. *Cn. J. Agric. Sci.*, 33: 615–619.

Mcintosh, D.L. (1975). Proceedings of the 1974 APDW Workshop on crown rot caused by *Phytophthora cactocum* in planting of apple trees aged 1 to 7 years. *Plant Dis. Rep.*, 59: 539–541.

Mclntosh, D.L. (1970). Dilution plates used to evaluate initial and residual toxicity of fungicides in soil to zoospores of *Phytophthora cactorum*, the cause of crown rot of apple trees. *Plant Dis. Rep.*, 55: 213–216.

Nair, P.K.U.M., Ammoottty, K.P., Sasikumaran, S. and Pillai, V.S. (1993). *Phytophthora* root rot of black pepper (*Piper nigrum* L.)A management study with organic soil amendments. *Indian Cocoa, Arecanut and Spices J.*, 17: 1–2.

Papavizas, G.C. and Lewis, J.A. (1989). Effect of *Gliocladium* and *Trichoderma* on damping-off and blight of snap bean caused by *Sclerotium rolfsii* in the greenhouse. *Plant Pathol.*, 38: 278–286.

Podile, A.R., Kumar, B.S.D. and Dube, H.C. (1988). Antibiosis of rhizobacteria against some plant pathogens. *Indian J .Microbiol.*, 28: 108–111.

Prasad, R.D., Rangeshwaran, R. and Kumar, P.S., 1999. Biological control of root and collar rot of sunflower. *J. Mycol. Plant Pathol.*, 29: 184–188.

Sharma, M. and Sharma, S.K. (2002). Control of white root rot of apple caused by *Dermatophora necatrix* with *Bacillus* sp. *Plant Dis. Res.*, 17: 308–312.

Sharma, M. (2000). Non-chemical methods for the management of white root rot of apple. *Ph.D. Thesis*, UHF, Solan, 96p.

Sharma, R.C., Sharma, S. and Sharma, J.N. (1999). Present concepts and strategies of disease management in pome and stone fruits. *Int. J. Trop. Plant Dis.*, 17: 67–80

Sharma, S.K. and Gupta, V.K. (1995). Management of root rot of apple through soil amendments with plant materials. *Plant Dis. Res.*, 10: 164–167.

Sharma, S.K. (1993). Studies on management of white root rot of apple. *Ph.D. Thesis*, UHF, Solan, 146 p.

Singh, R.S. and Pandey, K.R. (1996). Effect of green and mature plant residues and compost on population of *Pythium aphanidermatum* in soil. *Indian Phytopathol.*, 19: 367–372.

Sztejnberg, A., Greeman, S., Chet, I. and Katan, J. (1987). Control of *Rosellinia necatrix* in soil and in apple orchard by solarization and *Trichoderma harzianum*. *Plant Dis.*, 77: 365–369.

Viala, P. (1891). Monograph Duepourridie (Dematophora) des Vigens, Paris.

Wazir, F.K., Meladul, K., Qureshi, J.A., Barech, A.R. and Kakar, K.M. (2000). Effect of physical and biological control measures on Colt rootstock of cherry grown on crown gall infested soil. *Sarhad J. Agric.*, 16: 49–51.

Chapter 15

Disease Management of Horticultural Crops through Organics

***Shripad Kulkarni*[1]*, V.I. Benagi*[1] *and Jameel Akhtar*[2]**

[1]*Department of Plant Pathology, UAS, Dharwad, Karnataka*
[2]*Department of Plant Pathology, BAU, Ranchi, Jharkhand*

Shift of Indian agriculture from a state of food deficiency to food sufficiency through introduction of fast growing, high yielding hybrid varieties, usage of high dose of chemical fertilizers, pesticides and weedicides. No doubt green revolution was yielded rich dividends but at the same time gifted serious pest problems and degradation of environment followed by deleterious effects on environment.

World over it has been estimated that more than 67,000 different pest species attack crop. To protect them, many chemical pesticides are used which have unsafe environment impact, and hence there is a pressure for decreased reliance on such agents and greater regulatory control of their use. Besides many of the pathogens developed many fold of resistance to fungicide. (Krishna *et al.*, 2005; Gupta and Thind, 2006; Ray *et al.*, 2004 and Singh, 2005)

Most of Horticultural crops are eaten fresh or used for health care: hence any contamination in the form of pesticide/chemical residue may lead to health hazards; hence, organic horticulture offers a better possibility of producing healthy food. Plant diseases caused by different groups of organisms belonging fungi,bacteria, viruses, rickettsia, spiroplasma, nematodes and few others have remained important in causing significant losses in different crops indicating the urgent need of their integrated management. The continuous and indiscriminate use of chemical pesticides has posed several serious problems such as pesticide residue, development of resistant strains, environmental pollution and adverse effect on beneficial microorganisms and created a greater concern over global food safety and security.

Organic farming relies on crop protection, crop residues, animal manures, legumes, green manures, off farm organic wastes, cultural practices,mineral bearing rocks and aspects of biological pest control to maintain soil productivity and to supply plant nutrients and to control diseases,insects, weeds and other pests.

Cultural Methods

By practicing the following methods we can regulate/modify the pest and disease incidence effectively.

Selection of Adopted and Resistant Varieties

By choosing varieties which are well adopted to the local environmental conditions (such as temperature, nutrient supply, pests and disease resistance) by which crop is allowed to grow healthier and stronger against attack of pests and pathogens.

Defense Mechanisms Influencing Diseases

1. *Phenols*: Host enzymes like plyphenol oxidase and peroxidase oxidize phenolics to quinines and the quinines are more fungitoxic than phenolics
2. *Phytoalexins*: Phytoalexins are mostly isoflavonoids, terpenoids and poly acetylene compounds and synthesized *de novo* on infection by the pathogens. Phenylalanine and acetic acid may be involved in the biosynthesis of Phytoalexins and Phenylalanine ammonialyse (PAL) has been considered to be the key enzyme.
3. *Lignin*: Phenylalanine and cinnamic acid are the important precursors. Lignin may act as physical barrier to the pathogens
4. *Callose*: It is the substance found in the sieve tubes and may prevent the leakage of sieve tube sap or water in the cell walls. Penetration of incompatible pathogens into the host tissues results in the production of papillae and the papillae may mostly contain callose.
5. *Sugars*: Sugars are precursors of synthesis of phenolics, Phytoalexins, lignin and callose
6. *Amino acids*: Aminoacids are the corner stones for synthesis of proteins and some of them are essential for the synthesis of phenolics, Phytoalexins and lignin.

Cropping System

Cropping system in a particular agricultural ecosystem plays major role. By adopting a suitable cropping system in right time, we can avoid most of the harmful pests and diseases. Some of the practical practices are as follows.

1. Activity in the soil and enhance the presence of beneficial organism. For instance plants colonized by arbuscular mycorrhizal may increase pests also. So careful selection of proper green manure is essential.
2. Crop rotation: Cauliflower – paddy – cauliflower rotation is highly effective in controlling Stalk rot of Cauliflower disease caused by *Sclerotinia sclerotiorum* and it reduces infection by >60 per cent and > 161 per cent increase in seed yield has been observed.
3. By adopting mixed cropping system pest and disease incidence can be minimized since pest has less host plants to feed on.
4. By following crop rotation practices we can increase the soil fertility and reduce the chances of soil born diseases.
5. By cultivating green manure cover crops like Horse gram, Cow pea, Sun hemp, Sesbania, Dhaincha, Glyrisidia we can increase the biological activity in the soil and enhance the

presence of beneficial organism. For instance plants colonized by arbuscular mycorrhizal may increase pests also. So careful selection of proper green manure is essential.

Selection of Clean Seed and Planting Materials

Seeds and planting materials are the primary sources of diseses and hence selection and use of disease free seeds after inspection for pathogens and weeds is very much essential. Further, it is advised to get seeds and planting materials from the reliable safe sources only. Use of healthy seed and it's hot water treatment at 50°C for 25 minutes or for sodium phosphate 90 gram/Lit. for 20 minute is effective for tomato mosaic which is most dangerous disease of Tomato.

Selection of Optimum Planting/Sowing Time and Spacing

Most of the pests or diseases attack the crops only in a certain life stage. By adopting sufficient spacing between plants we can reduce the spread of a disease as well as allows good sunlight to the plants which facilitates less moisture on the leaves leading to hinder and of pathogen development and infection. In the same way more sunlight allows plants to do more photosynthesis. This practice not only avoids disease and pests in cropping system but also increase the crop productivity.

Balance Organic Nutrition

Gradual and steady growth makes plants less vulnerable to infection. So this steady growth could be achieved by applying organic fertilizers timely and moderately because, excess and indiscriminate use of fertilizers often results in damaging the roots. This damage facilitate to secondary infection. To overcome this problem we can adopt integrated nutrient management system with organic manures like FYM, compost, nutrients slowly when the plant needs. Further, by using liquid Bio – fertilizers like Potash mobilizers namely *Frateuria aurentia* along with organic manures provides balanced potassium and contributes to the prevention of fungi and bacterial infections:

Addition of more and more Organic Matter

Organic content of the soil is directly related to density and activities of microorganisms in the soil there by pathogenic and soil borne fungal population can be reduced. Besides this, organic matter provides.

- ☆ All the nutrients that are required by the plants.
- ☆ Corrects C:N ratio in the soil.
- ☆ Good physical chemical and biological support to soil
- ☆ More water holding capacity to the soil.
- ☆ Cover from evaporation losses of the moisture from the soil

Ultimately, organic matter supplies substances which strengthen plants with their own protection mechanisms.

Soil Amendments

The decomposition of organic matter helps in alteration of the physical, chemical and biological conditions of the soil and the altered conditions reduce the inoculums potential of a soil borne pathogens. In addition, the practice improves soil structures, which promotes root growth of the host. Various biochemical substances like antibiotics and phenols are released during decomposition, which in turn induce resistance in the root system. Soil amendments like sunflower, rape seed cakes, mustard cake, gypsum and been straw can be used.

Figure 15.1: Banana Cariety Against Sigatoka Disease

Healthy Turmeric Crop in Soil Solarized Plot

Sakkarebale (Highly resistant)

Figure 15.1–Contd...

Soil Solarization Method

Rajapuri (Highly susceptible)

Rhizoctonia solani causal organism of damping off in nursery stage and wire stem, bottom rot and head rot after transplanting plants in crucifiers. Soil amended with neem cake 1 kg/m^2 and solarized by covering with white polythene sheet for two weeks and treatment of seeds with *Trichoderma harzianum* proved to be effective in the management of damping off of vegetables crops. Soil amended with cellulose powder was also found effective in reducing the disease incidence, which may be due to increase in C:N ratio in soil. This increase results in the decrease of fungal population but actinomycetes and bacterial populations increased.

Organic amendment is used here to mean organic material incorporation into the soil that comes from external sources such as processing residues or industrial waste products. Organic material added as fresh crop residue and grown in the field in rotation – break, cover, trap, antagonistic or green manure crops- are discussed below. Incorporation into the soil of large amounts of any organic material will reduce nematode densities. Oil cakes, coffee husks, paper wastes, crustacean skeletons, sawdust and chicken manure, amongst others, have been used with some success. Nematode control may be due to any one or more of the following mechanisms:

- ✰ Toxic and non-toxic compounds present in the organic material;
- ✰ Toxic metabolites produced during microbial degradation; or
- ✰ Enhancement of the soil antagonistic potential.

Chitin amendments have received much interest in the past as an organic amendment in that they stimulate the antagonistic potential in soil towards nematodes. Organic amendments have also been combined with various biocontrol agents with reports of enhanced levels of control. The use of organic amendments is often limited by availability and, in some cases, by the large quantities needed. In addition to their effects on nematode density, organic amendments also improve soil structure and water-holding capacity, reduce diseases and limit weed growth, all of which ultimately lead to a stronger plant and improved tolerance to nematode attack.

Biofumigation

This term normally refers to suppression of soil-borne pests and pathogens by biocidal compounds, principally isothiocyanates, released in soil; when glucosinolates in cruciferous crop residues are hydrolysed. Soil amended with fresh or dried cruciferous residues at 38°C day and 27°C night temeperatures reduced *Meloidogyne incognita* galling by 95-100 per cent after 7 days' incubation in controlled environment tests. It should noted here, however, that many cruciferous plants are good hosts of some important species of *Meloidogyne.*

The term biofumigation is now used more freely whenever volatile substances are produced through microbial degradation of organic amendments that result in significant toxic activity toward a nematode or disease. The release of toxic compounds already present in antagonistic plants used as amendments, *e.g.* neem, marigold and castor, or the production of toxic compounds due to microbial fermentation of nutrient-rich organic amendments, *e.g.* velvet bean, sunnhemp or elephant grass, lead to significant levels of nematode control.

Biofumigation under these circumstances is greatest when there is an optimum combination of organic matter, high soil temperature and adequate moisture to promote microbial activity leading to toxin production. In tropical and subtropical production systems, plastic mulch and drip irrigation improve effectiveness of biofumigation. Transporting organic amendments to the field or incorporating cover crops that produce large amounts of biomass into the soil, together with plastic mulch and drip irrigation, should significantly increase the level of control attained.

Biofumigation using fresh marigold as an amendment is used effectively in root knot management in protected cultivation in Morocco. Tagetes is grown in the raised beds prior to the planting of susceptible horticultural crops. The crop is then incorporated into the soil after 2-3 months. The beds are fitted with drip irrigation and covered with plastic mulch. The soil in the bed is then biofumigated under conditions of the high temperature and optimum soil moisture.

Control due to any form of biofumigation is probably the result of multifaceted mechanisms including:

1. Non-host or trap cropping depending on the host status of the plant used.
2. Lethal temperature due to solarization.
3. Nematicidal action of toxic by-products produced during organic matter degradation.
4. Stimulation of antagonists in the soil after biofumigation.

Water Management

By practicing good water management water logging in the field and stress on plants can be avoided otherwise pathogens take chance and infect the crop. Further, sprinkling water on foliage shall be avoided as it increase diseases by giving chance to pathogenic fungal spores to germinate.

Use of Proper Sanitation Measures

"Pull and Burn" is the best method to control disease and removal of infected plant parts (leaves, fruits) from the ground to prevent the disease from spreading. Eliminate residues of infected plants after harvesting.

Soil Solarization

Soil solarization for 4 weeks during summer months coupled with application of neem cake @ 400 g/m2 proved effective against damping-off (76.9 per cent) in nursery and resulted in significant, higher number of healthy transplants.

Table 15.1: Successful Control of Plant Diseases Using *Trichoderma* Species

Chick pea wilt	Carbendazim seed treatment (ST) + *T. harzianum* or neem cake + *T. harzianum*
Redgram wilt	ICP8863 + ST with *T. harzianum*
Foot rot of paper	Neem cake + *T. harzianum* (soil solarization and application of Metalaxyl MZ (two spray)
Cotton root rot	*T. harzianum* + Benomyl
Sclerotium wilt of sugarbeet	Captan + *T. harzianum*
Damping off of tobacco, tomato, egg plant	Metalaxyl MZ + *T. harzianum*
Sclerotium rolfsii in tomato, beans, groundnut, sugar beet and chick pea	Furrow application of 130-160 kg/ha with 4 g/kg of seed treatment of (*T. harzianum, T. viride, T. hamatum*)
Bean root rot *Macrophomina phaseolina*	Furrow application of *T. harzianum* (130-160 kg/ha)
Fusarium oxysporum in bean, chrysanthemum, cotton, melon and redgram	Soil application of *T. harzianum* at 130-160 kg/ha
Different wood rotting fungi in wood trees	Dusting *T. viride* on wounds
Grey mold, *Botrytis cinerea* in grapes	Aerial spray with *T. viride* (106 – 108 cfu/ml)
Stalk rot of rabi sorghum	*T. harzianum* kg/ha seed)

Mechanical Methods

1. By removing and burning Clipping of lower leaves upto 20 cm and weading to reduce the alternaria blight in tomato
2. This is the best method, before reaching the loss beyond economic level. To save crop from fungal and bacterial diseases pull and burn method is most effective.

Botanicals

Plants during their long evolution, have synthesized a diverse array of chemicals to prevent the colonization by pathogens. They produce secondary metobolites like terpenoids alkaloids, flavonoids, phenolic compounds. These secondary metabolites are having disease suppressing properties. In India several plants such as nicotinoids, natural pyrethrins, rotenoids, neem products have been used in the past for suppression of diseases. Among the botanical pesticides neem occupies very important place in the pest and disease control. Different parts of neem tree can affect more than 200 insects and diseases, some of them are effective against nematodes, fungi, bacteria and viruses.

Biological Control

Management of soil borne pathogens is difficult because of non-availability of desired level of resistance against major soil borne diseases caused by species of *Fusarium, Sclerotium, Macrophomina, Pythium, Phytophthora* and few others forced to search or resorting to new approaches to manage the diseases. Therefore, biocontrol agents or antagonists are means of plant disease control has gained importance in the recent years. The biocontrol agents multiply in soil and remain near root zone of the plants and offer protection even at later stages of crop growth.

Mechanisms of Biocontrol

- ✰ Competition
- ✰ Mycoparasitism
- ✰ Antibiosis and Enzymes
- ✰ Cross protection
- ✰ Induced systemic resistance (ISR)

Methods of Application

1. Broad cast application (125-250 kg/ha)
2. Furrow application (130-160 kg/ha)
3. Root zone application (Formulation mixed in soil @ 1 kg/plant)
4. Seed treatment (4g/kg seeds)
5. Wound application (Applied to wounds in peach, plum etc)
6. Spraying (106 – 108 cfu/ml)

Advantages of Biopesticides

1. Avoids environmental pollution (soil, air and water).
2. Avoids adverse an effect on beneficial organisms *i.e.* maintains healthy biological control balance.
3. Less expensive thou pesticides and avoids problems of resistance.
4. Biopesticides are self maintaining in simple application and fungicide needs repeated application.
5. Biopesticides are very effective for soil borne pathogen where fungicidal approach is not feasible.
6. Biopesticides ecofriendly, durable and long lasting.

7. Very high control potential by integrating fungicide resistant antagonist.
8. Biopesticides help in induced system resistance among the crop species.

Mycoparasites

Mycoparasites *Ampelomyces quisqualis* for management of anthracnose, downy mildew and powdery mildew of grape. It is commercially available as AQ 10 (USA), Bio – Dewcon (India) 1000ml has to be mixed with 500 ml Neem Oil in 250 Liters of water.This also useful in managing powdery mildew of cucurbits caused by *Sphaerotheca fuliginea*.

Use of Bleaching Powder

12 kg/ha found to reduce bacterial blight by 80 per cent when applied in furrows at the time of planting.

References

Chandra, Krishna, Greep, S. and Srivatha, R.S. H. (2005). *Biocontrol Agents and Biopesticides*. Regional Center of Organic Farming, Hebbal, Bangalore.

Gupta, S.K. and Thind, T.S. (2006). *Disease Problems in Vegetables Production*. Scientific Publishers (India), Jodhpur.

Ray, A.B., Sarma, B.K. and Singh, U.P. (2004). *Medicinal Properties of Plants: Antifungal, Antibacterial and Antiviral Activities*. International Book Distributing Co., Lucknow, U.P.

Singh, R.S. (2005). *Plant Diseases*, 8th edn. Oxford and IBH Publishing Co. Pvt. Ltd., New Delhi.

Plant Diseases Management in Horticultural Crops (2011) *Pages* ***187–208***
Editors: **Shahid Ahamad, Ali Anwar and P.K. Sharma**
Published by: **DAYA PUBLISHING HOUSE, NEW DELHI**

Chapter 16

Saffron Corm Rot Disease and its Biological Management

G.H. Mir[1], V. Manoj Kumar[2], Sobita Devi[2], S. Ahmad[2] and P. Williams[2]

[2]Division of Plant Pathology,
S.K.University of Agricultural Sciences and Technology of Kashmir,
Shalimar, Srinagar – 191 121
[2]Department of Plant Protection, Allahabad Agricultural Institute-Deemed University,
Allahabad, U.P.

Saffron (*Crocus sativus* L.) is highly prized and one of the best natural food flavouring and colouring substance having general panacea properties. In Kashmir, it is the legendry crop of the well-drained plateau of Pampore, where it is being grown since ancient times. The recorded account of saffron cultivation in Kashmir dates to 550 A.D, nearly four centuries earlier than its recorded cultivation in Spain by Arabs around 961 A.D. The plant, native of Iran and Asia Minor, was earlier grown widely in Europe and Asia, however, presently major producing countries are Spain, Iran and India. It is considered as the costiest crop in the world and is also known as "Royal Flower". The flower has six lilac petals encircling three yellow stamens and three red styles. It is the three fine red premium strands that form the true saffron or "Shahi Zaffaran", once they are minutely separated from the petals and stamens and dried in the mid-October sun.

Being of ancient origins, saffron has remained synonymous with Kashmir. So close and connected have saffron and Kashmir become through the ages the flower has become the "Symbol of Kashmir". It is famous for its bouquet and is in great demand as a condiment and as a pigment for the forehead marks of the Hindus.

Importance of Crop

Saffron has a variety of uses; particularly it has got immense medicinal properties. Dried stigmas

Figure 16.1: *Crocus sativus* (Saffron Plant)

and tops of the styles of *Crocus sativus* (saffron plant) are used. It is generally used as a condiment for its aromatic odour and beautiful colouring matter.

Medicinally, it is used in small doses in fevers, melancholia, enlargement of the liver, and in spasmodic cough and asthma, and in catarrhal afflictions of children. It is also given in anemia, chlorosis and seminal debility.

Chinese Traditional Medicine

Saffron has been employed in many medicinal remedies against numerous conditions. In Chinese traditional medicine, saffron has been widely used for it anodyne, tranquiling and emetic properties. It has also been used in the treatment of menstrual disturbances, thrombus diseases and some other diseases related to high blood viscosity. It has found applications in nervous disorders to allay fears, cure trances and in the treatment of some disorders of the central nervous system (Suzhou New Medical College 1977, Zhou *et al.*, 1987). Its medical value was recorded in Yi-Lin-Li-Yao, a traditional Chinese medical book composed during the ming Dynasty (16th Century); Notable among the effects described was the promotion of blood circulation to remove blood stasis. The book Yinshanzhengyao ("The Importance of Diet") (Circa 1550) contains 136 recipes which include saffron for treating a variety of conditions, saffron also appears in several traditional Chinese pharmaceutical compendia (Ni 1992).

Traditional Indian and Azerbaijani Medicine

It has been used in traditional Indian and Azerbaijani medicine to treat various diseases including cancer, cardiac disease, eye disease, blood disease, muscle paralysis, enlargement of liver, dysentery, urological and gall bladder infections, sedative, antispasmodic, aphrodisiac properties, irregular menstruation problems, stimulating and reduce hangovers caused by wine (Kasumov 1970, Pfander and Witwer 1975, Nadkarni 1976, Damirov *et al.*, 1988, Zhang *et al.*, 1994). Saffron is now also used as natural dye in Nematodes for identifying their taxonomy (Gaur and Chandel, 1998). Saffron (*C. sativus* L.) a product of high commercial value obtained from the stigmas. The stigmas of the saffron flower contain many chemical substances such as carbohydrates, minerals, mucilages, vitamins (especially riboflavin and carotene, lycopene, zigzanthin).

Externally, saffron is used in headache in the form of paste, also applied to bruises and superficial sores. Saffron is used in snake-bite also, and is useful in chronic diarrhea.

Unani System of Medicine

According to Unani System of Medicine, saffron is used to reduce inflammation, in the treatment of enlarged liver and infection of urinary bladder and kidney, acts as refrigerant for brain and gives strength to heart, is useful in menstrual disorders, acts diuretic if soaked overnight in water and used with honey and when pounded with clarified butter, it is used for treating diabetic patients. (C.C.R.A.S., Ministry of Health and Family Welfare). Most "Mughlai" preparations including the well-known "Biryani" get their colour and smell from the expensive saffron. Most North Indian flavoured dishes including famous Kashmiri preparations, pudding, pastry and confectioneries will be dull without saffron.

Saffron in Kashmir

Saffron is mentioned by many old Sanskrit writers, Sushrata, the great Indian Surgeon-Physician, who flourished in 800 B.C., has described the medicinal value of saffron. It was known as "Kumkum" of which its Kashmiri name "Kwang" appears to have derived. Among other names by which saffron is known in Sanskrit literature is "Kashmirja" (born in Kashmir), which suggests that it was a monopoly of Kashmir. (Sir Walter Lawrence "The Valley of Kashmir").

It grows in very few countries in the world. Among them being France, Spain, Sicily, England, Iran and India. And in India, it grows only in the state of Jammu and Kashmir. Saffron is not a native of Pampore (Kashmir) but hails from the Mediterranean. Waverly root in his masterly book "Dictionary of Foods" credits Asia Minor and Greece as the probable areas of origin. He has unearthed all manners of information on its antiquity. It seems saffron grew in Sicily and in Cilicia, where the locals perfumed themselves with it. It also grew in Canaan and all over Greece as early as the 4th century B.C. Apparently, Homer has written that Zeus slept on a bed of saffron, lotus and hyacinth leaves and flowers.

Saffron grew wild in Italy from its earliest days, when Italian ladies, envying the blend locks of Northern lands, would dye their hair with saffron. According to Waverly Root, it did not come to China until the Mongols invaded the country in early 13th century. The use of saffron in Europe was common during the Roman Empire. Emperor Nero is said to have had the streets of Rome sprinkled with saffron for his official entry into the city. The high and the mighty were all charmed by it.

It declined, however, after the fall of Roman Empire and did not reappear in culinary history until the advent of the Moorish Arab invaders in the 10th century. They reintroduced it to Spanish cuisine, and it is when "Crocus" became "Zaffaran". At once time during the crusades in Europe, that precious spice was so much rage that Venice, the chief port of entry for goods from the Middle east, opened a special office of saffron.

As mentioned in the "Rajatarangini", the ancient chronicle of Kashmir composed by Pandit Kalhana in the 12th century, "saffron" is one of the five objects for which Kashmir is famous and which are difficult to secure in paradise". Kalhana further states that in the rule of King Lalitaditya (695-732 A.D.), there lived a famous physician, Waghbhatta in Padampur, a city founded by one of the Lalitaditya's minister, Wazir Padam. A nag, snake diety, Takshaka Naga, or water-god fell sick of an eye complaint and went to Waghbhatta, who tried in vain to cure him. Baffled, the physician at last asked Takshaka (water-god) whether he was a man, and on finding out that he was actually a Nag, the physician at once realized that the remedies applied to the Nag's eyes were nullified by the poisonous vapours issuing from the water god's mouth. He bound Takshaka Nag's eyes with a piece of cloth and the diety was restored to health. In his gratitude, the Nag gave the physician a bulb of saffron, and thus its cultivation sprang up at Padampur, now known as Pampur (Sir Walter Lawrence "The Valley of Kashmir").

Table 16.1: Production and Productivity of Saffron in J&K (India)

Sl.No.	*Year*	*Area (ha)*	*Production (q)*	*Productivity (kg/ha.)*
1.	1980-1981	3405	200.44	7.86
2.	1990-91	4050	130	3.21
3.	1991-92	4050	130	3.21
4.	1992-93	4050	130	3.21
5.	1993-94	4496	130.65	2.90
6.	1994-95	4496	134.25	2.98
7.	1995-96	4496	141.28	3.14
8.	1996-97	5707	159.52	2.79
9.	1997-98	5361	173.82	3.24
10.	1998-99	4116	128.80	3.03
11.	1999-2000	3997	75.54	1.89
12.	2000-2001	3827	48.60	1.27
13.	2001-2002	3827	35.90	0.94
14.	2002-2003	3830	50.32	1.30

Source: (1) Financial Commissioner's Office, Sgr. Kashmir.

Agriculture office, Lalmandi, Sgr. Kashmir.

G.M. Mir (Saffron Agronomy in Kashmir Valley)

Table 16.2: Production and Productivity of Saffron in Abroad

Country	*Year*	*Area (ha)*	*Production (q)*	*Productivity (kg/ha.)*
Iran	2003	43000	176300	4.10
Italy	2003	—-	—-	10.00
Spain	2003	—-	—-	8.00

Source: Malcolm Douglas, June 2004.

Takshaka Naga is worshipped to this day in a large pool of limpid waters in that very place. An annual fair is also organized on the full moon night of Kartik Poornima when the saffron flowers are in full majestic bloom.

Matta, besides Pampur, is the only other area where saffron is grown in Jammu and Kashmir. The village claims to have been established by a descendent of King Vikramaditya of Ujjain way back in 6th century. Matta saffron is said to have been granted by none else than God Indra. Legend has it that the diety was once afflicted by some serious disease which could not be cured even by the celestial physicians. Ultimately, a clever medicine man of this region succeeded in curing Lord Indra. Pleased by his talents, Indra asked the Matta (Kishtwar) physician to desire any boon. The patriotic physician asked Lord Indra to grant something that would make his land famous. So Lord granted Matta the fascinating saffron (The Kashmir Times, Nov. 2000).

Interestingly, the famous poet Kalidas has referred to saffron in some of his immortal works. He gives such elaborate details of this flower that some persons took its as an evidence that the poet must have hailed from Kashmir since only an inhabitant of the valley could have described the bewitching flower so mightyly.

Mughal Emperor Akbar conquered Kashmir in 1586 A.D. and in "Ain-e-Akbari", Abul Fazal points out that "Saffron fields in blossom afford a prospect that would enhance those who were most difficult to please". So also has the famous Chinese traveler Heun Tsang praised the exquisite charm and beauty of Kashmir saffron in his travelogue (Sir Walter Lawrence "The valley of Kashmir").

The saffron growing farmers of Kashmir valley under their own traditional practices of cultivation, due to non-availability of modern technology, the yield as well as quality of crop is drastically reducing day by day.

Biotic Limiting Factors Saffron in India

The corm rot of saffron is likely to pose serious threat to saffron cultivation in Kashmir. In 1980 the yield per hectare was 7.86 kg/ha (Mir, 2003) and now its present productivity is 1.3 kg per ha, which is the lowest in the world as compared to 10 kg per ha in other regions like Spain and Italy, due to widespread fungal infection of saffron corms this crop has faced a virtual threat from corm rot diseases caused by complex of *Fusarium oxysporum* f. sp. *gladioli, Fusarium oxysporum, Fusarium solani, Rhizoctonia solani, Phytopthora* sp. and sterile Basidiomycetes fungus limiting the production of saffron corm and pistil (style and stamen) economical part of saffron.

The control measures, such as rotation of cultures, use of resistant varieties and treatment of seeds and/or soil with fungicides, become unsuitable or not effective, mainly due to the perennial nature of the crop remaining in soil and yielding upto 15 years, non varietal crop, genetic variability presented by the pathogen to the hosts range, capacity to survive in the soil and in the corms for long time. During the last few decades the yield of the saffron crop decreased year after year by these major fungal pathogens and due to lack of plant protection and agronomical practices, post nutritional management are a major problems for the production of saffron crop of the region resulted in heavy mortality of saffron corms upto 87 per cent (Mir and Devi 2003, 2004).

Status of Corm Rot Disease

Corm rot disease cused by several fungal pathogens *viz., Fusarium* spp., *Phytophthora* sp., *Penicillium* spp., *Rhizoctonia* spp., *Sclerotium* spp., *Macrophomina phaseolina, Phoma crocophila*, sterile Basidiomycetes fungus (Cappelli, 1994, 1991; Nannizzi, 1941; Carta *et al.*, 1982; Francesconi, 1983; Mir and Devi,

2003, 2004) in India and abroad. As a matter of fact the saffron is cultivated in suitable climatic conditions and in specific soil types in few pockets of Globe having the optimum conditions for cultivation of this precious season bound crop. So the present attempt is to highlight the available research under suitable heads in detail.

Symptom and Causal Organisms

Mir and Devi (2003) have been reported that saffron crop was affected by severe wilting of shoots producing from corms due to *Fusarium* sp., *Penicillium* sp., *Rhizoctonia* sp. resulting in reduction in yield. Primary symptoms during flowering shows yellowing and wilting of shoots due to basal stem rot and white rounded spots on the corm. Black powdery appearance develops beneath the outer tunic layer of the corm in the severe infestation. Cappelli (1994) reported that saffron crop has been affected by a severe blight and growers have suffered a serious reduction in the yield and in some cases a complete loss of the crop in Italy. Affected saffron plants show either drooping and damping off, basal stem rot or yellowing and wilting of shoots as well as corm rot. Corms used by farmers for planting have dark lesions beneath the other tunic and sometimes patches of blue-green mould on the surface. Cappelli *et al.* (1991) reported that infested saffron plants manifested damping off basal stem rot, drooping and wilting of shoots. Tammaro (1990) reported that an unidentified species of Fusarium from infected corms shows drooping, basal stem rot, wilting of shoots and corm rots, resulted in serious reduction in the yield.

Five different genera of fungal pathogen were isolated from the rotted corms of saffron *viz., Fusarium* sp., *Penicillium* sp., *Rhizoctonia* sp., *Phytophthora* sp. and sterile Basidiomycetes fungus and *Criconematoide nematode* (Mir, 2006). The symptoms with taxonomy of the pathogens are caused diseases corms have been carefully noted by Mir (2006) as given below.

Fusarium spp.

Sub-division – Deuteromycotina, Class – Hyphomycetes, Order – Moniliales, Family–Tuberculariaceae

Infected corms shows minute black dots distributed or scattered on the corm below the tunic layer which later turns into blackish brown irregular spots, coalescing with each other leaving the entire corm dry and reducing the size of the corms as compared to healthy ones.

Fusarium oxysporum Schlechtendahl, stroma brownish white to violet, smooth, medium high aerial mycelium, later forming sporodochia more seldom pionnotes with three-septate spindle – sickle–shaped conidia curved or almost straight, rarely pedicellate. Smaller conidia, one or two celled, oval to reniform are numerous in the aerial mycelium but are lacking in the typical fruiting layers of the macro conidia.

0-septate, 6 -13 x 2 –4 m, 1 septate 12-23 x 2-3.5 m, 3 septate 21-43 x 2.5-5m, 5 septate 32-51 x 3.5-4.5 m. Chlamydospores terminal and intercalary globose smooth or wrinkled and conidia, 6-14 m long, in mycelium

Fusarium solani (Martius) Appel and Wollenweber, conidia scattered in false heads, sporodochia or pionnotes, in mass brownish-white to loamy-yellow. Larger conidia strongly twisted spindle form, slightly curved, both ends rounded to tenpin-like base, three to five septate: 0-septate 11 x 3.5µ, one septate 20 x 4.1µ, three septate 21-48 x 3.8-6.7µ, five septate 33-58 x 4.3-6.5µ. Chlamydospores terminal and intercalary, single, globose to pear-shaped, one celled 8 x 8µ, two celled 10-15 x 6.7-9.5µ.

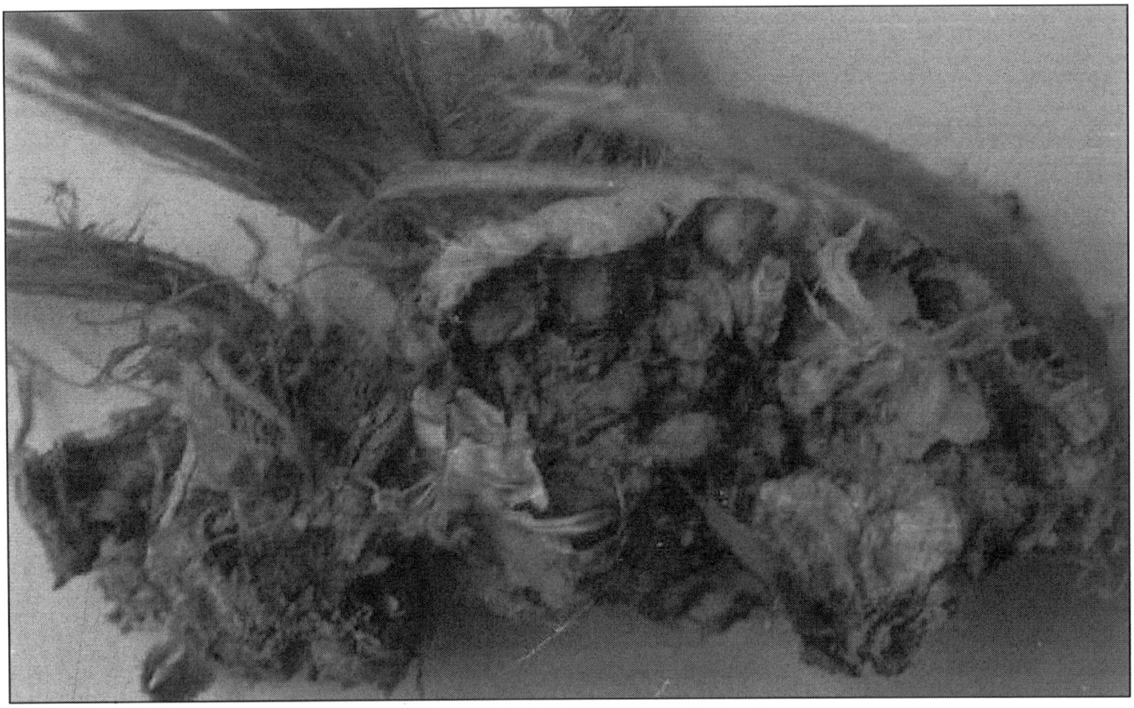

Figure 16.2: *Fusarium oxysporum* Infected Saffron Corms (Photo by G. Hassan Mir, 2003)

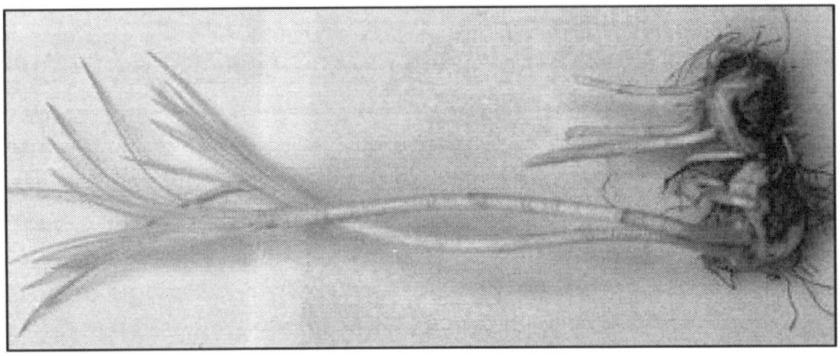

Figure 16.3: *Fusarium solani* Infected Saffron Corms (Photo by G. Hassan Mir, 2003)

Fusarium oxysporum f. sp. *gladioli* Massey and Hansen, mycelium is white, abundant, time textured and 3-5 mm deep. Substratum is colourless to pale buff, lilac.

Microconidia are abundant, mostly non-septate, hyaline, 90 per cent avoid, kidney sharper or ovate measuring 3-14 x 2.2 – 2.0m, 1-septate measuring 14.0-2.5 x 3.0 – 5.0m.

Macro conidia are scattered in the aerial mycelium along with the micro-Conidia, straight to slightly curved; slightly pedicellate; ends pointed and curved; typically 3-septate 31.5 x 4.3, range (16.8-54.6 x 2.5-5.6) m 4-septate rare, 39.0 x 4.7 range (25.5-54.0) x 4.2-5.6) m, 0-1 and 2-septate, rare in mature cultures, 10.0 – 40.0 x 2.5-4.5m, average diameter 3.6m.

Figure 16.4a: ***Fusarium solani*** **Infected Saffron Corms**
(Photo by G. Hassan Mir, 2003)

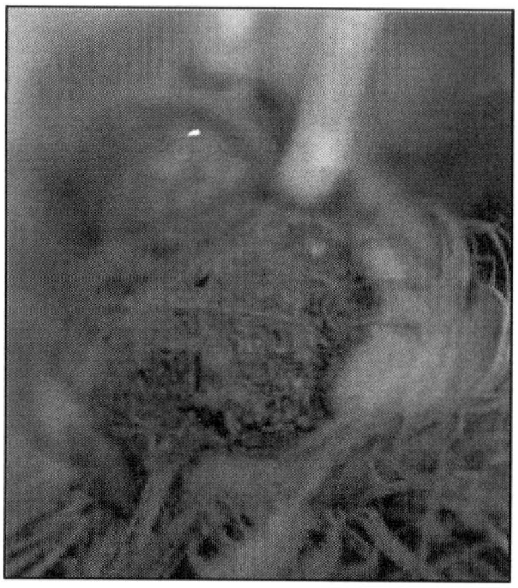

Figure 16.4b: ***Fusarium oxysporum*** **f. sp.** ***gladioli*** **Infected Saffron Corms**
(Photo by G. Hassan Mir, 2003)

Penicillium spp.

Sub-division – Deuteromycotina, Class – Hyphomycetes, Order – Moniliales, Family – Moniliaceae.

Initial infection shows minute dots on corm below the tunic layer which later spread and slightly irregularly white fluffy mycelium appears which later turns greenish to yellow and finally brownish/dark and appears on the tunic layer, sunken irregular pits distributed on the corm. In advance stages the entire corm may turn into dark powdery mass.

Penicillium purpurogenum Stoll, colonies on Czapek's agar slowly spreading very closely floccose at almost velvety, white at first, becoming yellow to pinkish shades and finally light grey-green. Conidiophores arise from aerial mycelium upto 100 m or 300 m long. Conidial fructifications consist of long, divergent chains, upto 100 m long, in two stages; metulae 10-16 x 2-2.5 m; phialides 11-12 x 2.5m, conidia elliptical 3.4-3.8 x 2-2.5 m, smooth pale green.

Penicillium corymbiferum Westling, colonies on Czapek agar growing rapidly fasiculate, margins deep, white mycelium bright yellow. Conidiogenesis heavy, dull green to jade green, axudate abundant, deep –brown to maroon, deep brown, reverse reddish-brown. Conidial heads large, conidiophores rough 0.1-0.2mm when in margins, upto 1.0mm when in central fasciculate areas. Metulae rough, appressed, 10-15 mm long. Phalides smooth, ampulliform, 9-12 mm long conidia globose, smooth to finally rough end, 2.5-4.0 mm diameter.

Sterile Basidiomycetes Fungus

The disease first appears on the surface and predominantly on the buds (eyes) on which sunken white spots appear having whitish mycelial covering on tunic layer of the buds/eyes. Irregular patches, slowly deep sunken, due to this rotting the tissues get disintegrated resulting in the spongy nature of the corm. The infection from the buds penetrates into the corm and make deep hole inside the corm. In advance stages white slightly mycelium appears on the tunic layer and nature of the pathogen is all the daughter corms were rotted and were spongy. The colour of spongy tissue were dirty white to slightly grey. On PDA it suppresses other soil fungi.

Figure 16.5: ***Penicillium purpurogenum*** **Infected Saffron Corms (Photo by G. Hassan Mir, 2003)**

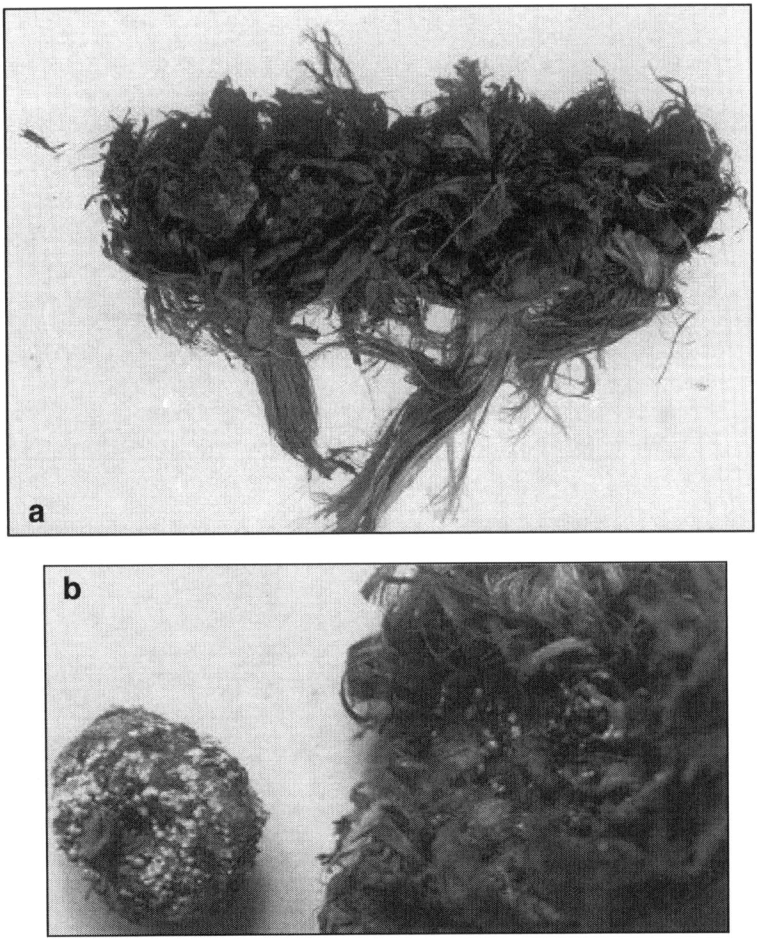

Figure 16.6a and b: ***Penicillium corymbiferum*** **Infected Saffron Corms (Photo by G. Hassan Mir, 2003)**

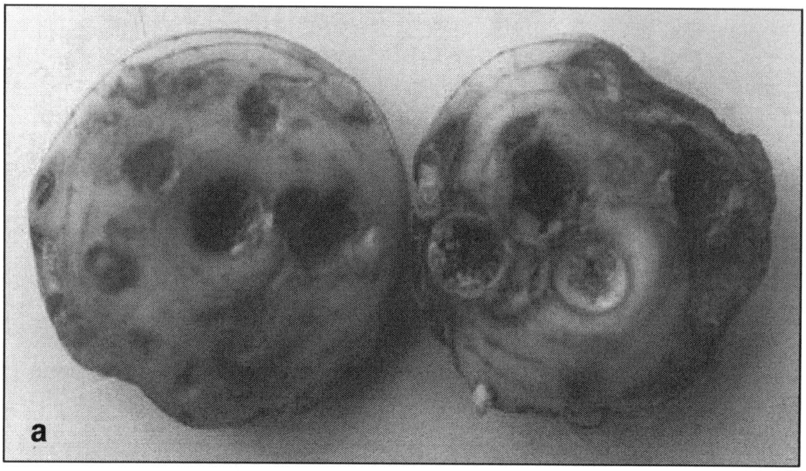

Figure 16.7a and b: Sterile Basidiomycetes Fungus Infected Saffron Corms (Photo by G. Hassan Mir, 2003)

Figure 16.8: Saffron Field Highly Infested with Sterile Basidiomycetes Fungus (Photo by G. Hassan Mir, 2003)

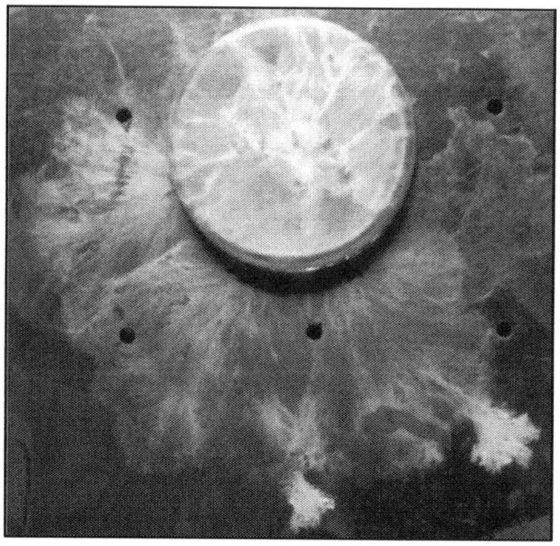

Figure 16.9: Saffron Field Highly Infested with Sterile Basidiomycetes Fungus (Photo by G. Hassan Mir, 2003)

White mycelium appears on Czapek agar suppressed later slightly fluffy, quick growth, mycelium comes out from petriplate and overlaps the petriplate inside and outside, spreads on the base of the shelf of BOD. Clamp connections were found.

Rhizoctonia solani khun

Sclerotial stage: Sub-division – Deuteromycotina, Class – Deuteromycetes, Order – Agonomycetales, Genus – Rhizoctonia, Species – solani (Kuhn),Basidial stage : Sub division – Basidiomycotina, Class – Basidiomycetes, Sub class – Holobasidomycitidae (Hymenomycetes), Order – Tulasnellales, Family – Ceratobasidiaceae, Genus -Thanetephorus, Species – cucumeris (Frank) Donk

Initially brownish lesions appear which are irregular having sparse growth of mycelium on the central portions of the lesions. Gradually these lesions are black in colour. Sclerotium observed after 15 days when the infected corm is cut, the symptoms appeared to develop all over the corm. The symptoms observed inside the corm were same as that of outside ones. One the periphery of the lesions dark brownish colour was found. The nature of the rotting is dry one. In severe infection whole of the corm transformed into dry powdery mass.

The basic colour of the colony was initially dirty white, later grayish brown, the young branches arise at right angles to the main hyphae but they later bend toward the direction of the growth of the main filaments, cells varied in size at different ages and on different diameter of hyphae as 4-6m. Monilioid cells form on branches that arise from long hyphae, size 20-22 x 30-35m. Sclerotia- colour grayish brown to cinnamon brown, irregular and flat, tending to elongate, size range from pinhead to 5-6mm in diameter.

Phytophthora sp.

Sub-division – Mastigomycotina, Class – Phycomycetes, Order – Peronosporales, Family – Pythiaceae.

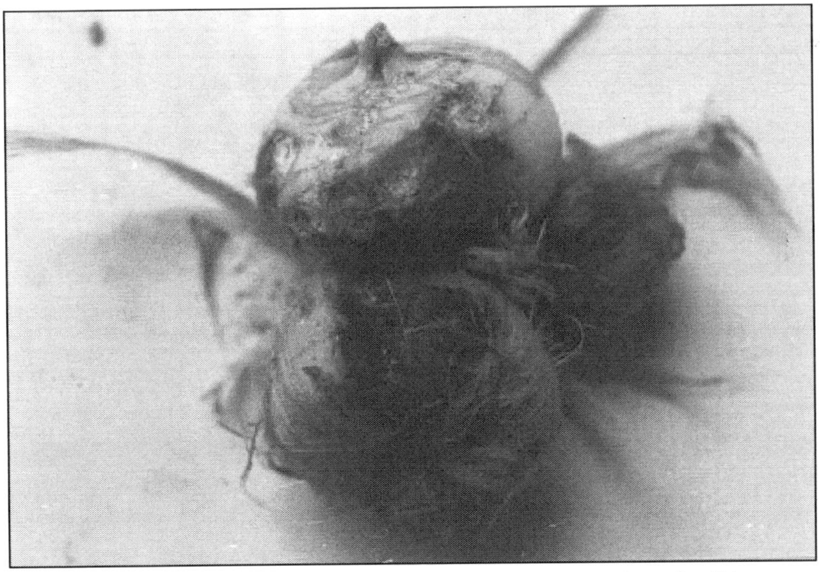

Figure 16.10: Saffron Corms Infected by *Rhizoctonia solani* khun (Photo by G. Hassan Mir, 2003)

PDA, the culture morphology of all the isolates was found to be similar. *Trichoderma viride*, *Trichoderma harzianum*, *Pseudomonas fluorescence* were inhibitive to the corm rot pathogens of saffron *in vitro* and in field Paecilomyces lilacinus along with above mentioned bioagents was also effective individually and in combination on corm rot diseases of saffron *Fusarium oxysporum* f. sp. *gladioli* and sterile Basidiomycetes fungus, it was found that the combination of these biocontrol agents was highly effective in reducing the pathogens *Fusarium oxysporum* f. sp. *gladioli* and sterile Basidiomycetes fungus and maximizing corm production and yield.

The effect of organic manures *viz.*, FYM, poultry and sheep manure @ 10, 15 and 20 tons per hectare each individually was studied, sheep manure @ 20 tons per hectare was highly effective in reducing the corm rot pathogens *Fusarium oxysporum* f. sp. *gladioli* and sterile Basidiomycetes fungus in saffron with maximum corm production and yield.

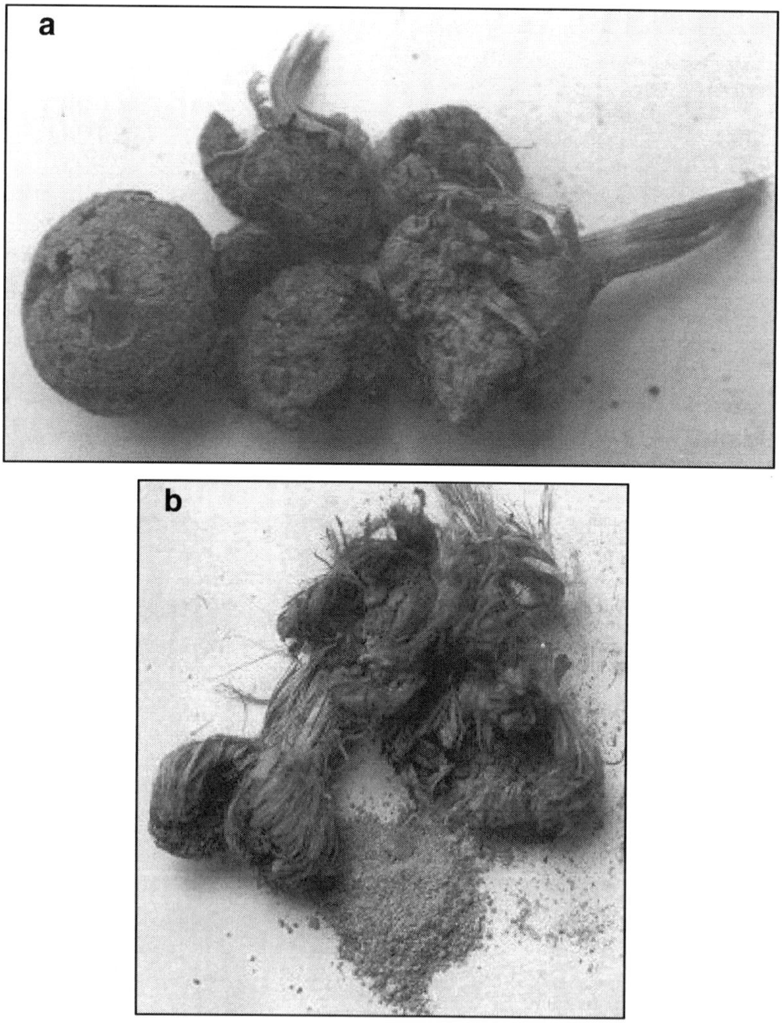

Figure 16.11a and b: *Phytophthora* sp. Infected Saffron Corms (Photo by G. Hassan Mir, 2003)

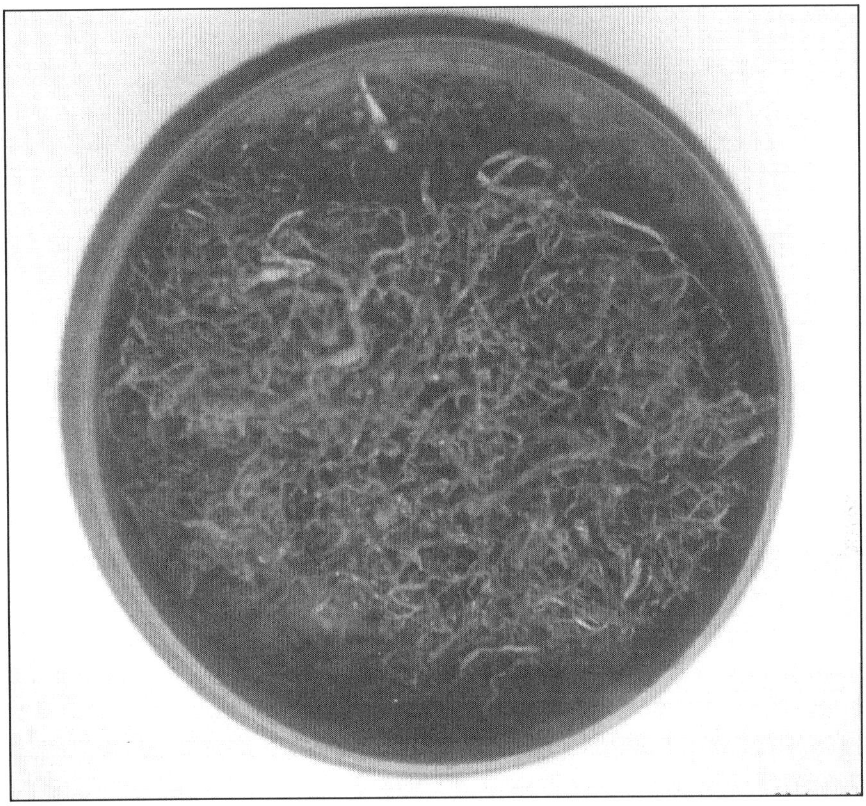

Figure 16.12: Postharvest Diseases of Saffron Stigmas by *Aspergillus* sp. (Photo by G. Hassan Mir, 2003)

The first sign of corm infection is a brown to purple discoloration of the surface followed by a brownish dry rot which extents gradually inside the corm. When the disease progresses the entire affected corm may be rapidly decayed and when the infected corm is pressed or squeezed it appears spongy due to disintegration of tissues.

Colonies on Czapek agar were initially white colored, slow growth, later the dense mycelium becomes purple colored and fluffy. The systemic position of the sclerotial and basidial stages are as follows :

Weeds

There are number of weeds associated with the saffron crop, the saffron farmers are weeding after the full growth of the weeds and the importance of the weeds are very useful for cattle's. During my research two weeds were found namely *Cyprus rotundus* and *Sorghum helepensis*. *Cyprus rotundus* produces number of dent like structures and are parasitic to the saffron corms while as *Sorghum helepensis* is deep rooted and roots grow quickly they cover the space and spread faster and suppress the corm production.

Nematodes

The plant parasitic nematodes were found in association with saffron crop *viz., Criconematiode* sp., *Hemicriconemaides* sp., *Xiphinema* sp., *Pratylenchus* sp and *Aphelenchus* sp.

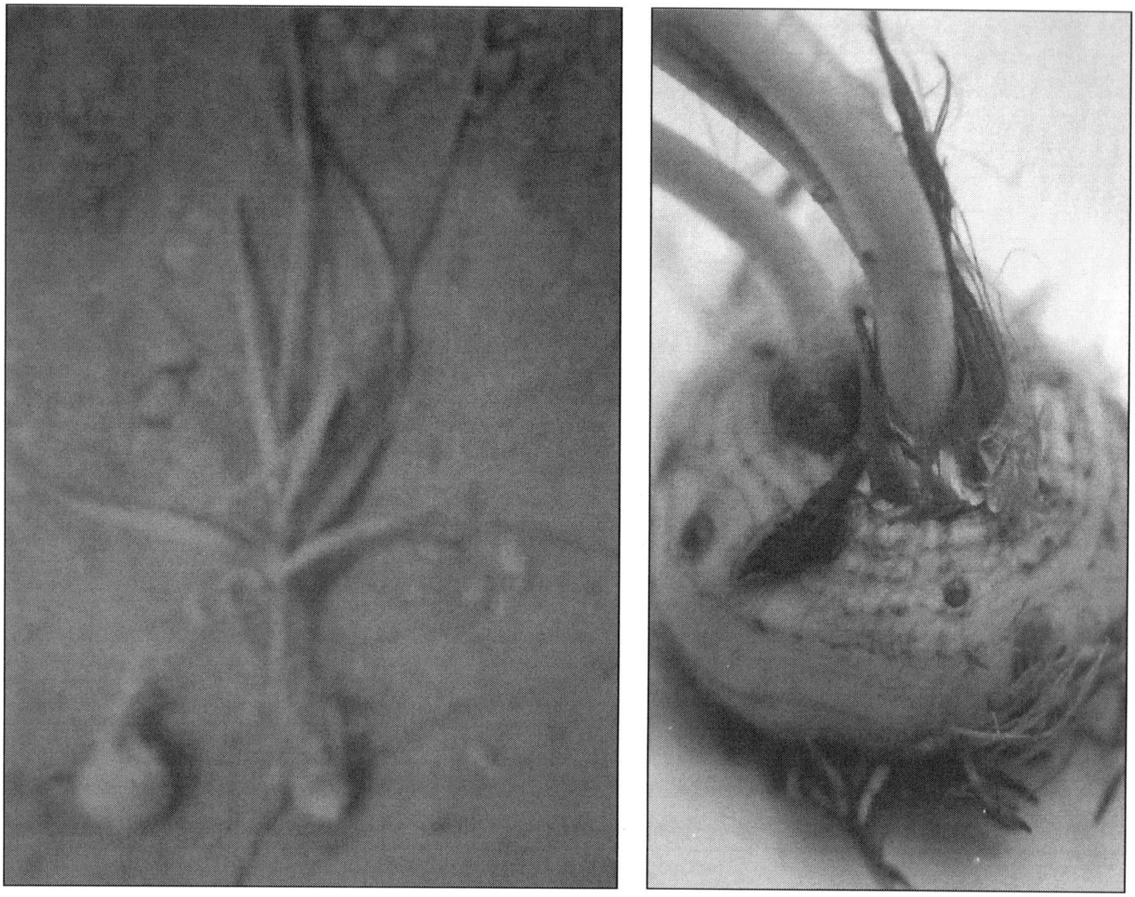

Figure 16.13: *Cyprus rotundus* Parasitizes the Saffron Corms (Photo by G. Hassan Mir, 2003)

Biological control

This involved selecting an antagonist with high antagonistic potential against corm rot pathogens. The target antagonist was *Trichoderma harzianum and Trichoderma viride* of which fifteen isolates were collected from different horticultural fields of Kashmir valley (Figure 16.15).

Investigation was undertaken to screen the potential native isolate of *Trichoderma* spp. for bio suppression of corm rot pathogen complex, as *Trichoderma* spp. are the most successful and widely used biocontrol agents, they can be applied to soil, seed, foliage or roots. Inspite of their high efficacy under controlled conditions, their performance in farmers field is not consistent. Unlike chemicals these Biocontrol agents need support even after their application to get established in targeted niche. Taking the advantage and constraints of *Trichoderma* spp. into consideration, efforts were made to encourage the native isolate against corm rot pathogens. Fifteen isolates of *Trichoderma* (*T. harzianum* and *T. viride*) were isolated from soils of different orchard plantations of Kashmir valley on modified Trichoderma specific medium (TSM). The isolates were studied for their antagonistic potential against six newly recorded major fungal pathogens of saffron *viz.* sterile Basidiomycetes fungus, *Rhizoctonia solani, Phytophthora* sp., *Fusarium oxysporum* f. sp. *gladioli*, *F. oxysporum* and *F. solani* individually on

Figure 16.14: Pathogencity of *Helminthosporium* sp. a–f: *Sorghum helepensis* and g–i: Saffron corms (Photo by G. Hassan Mir, 2010)

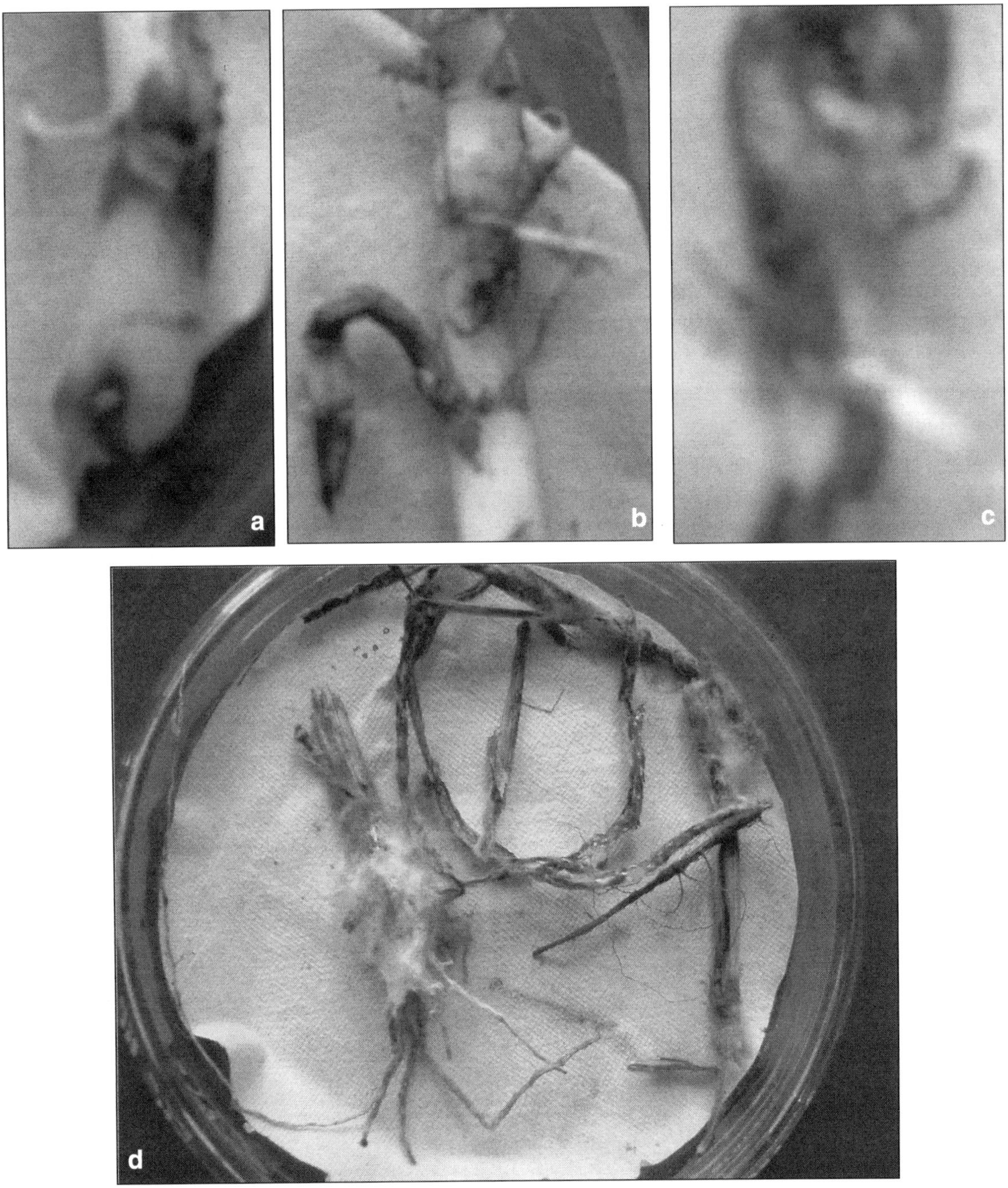

Contd..

Figure 16.14–Contd...

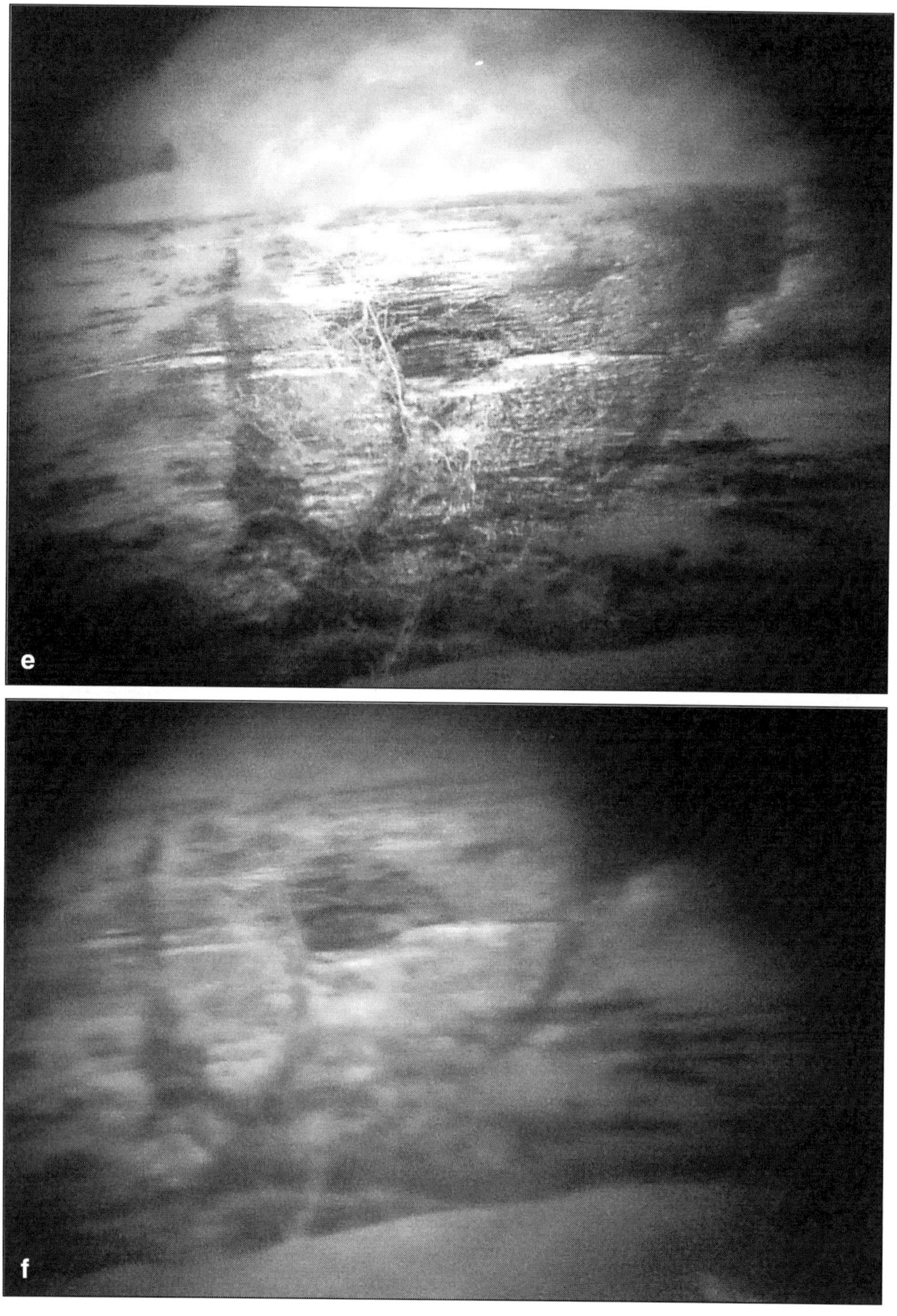

Contd..

Figure 16.14–Contd...

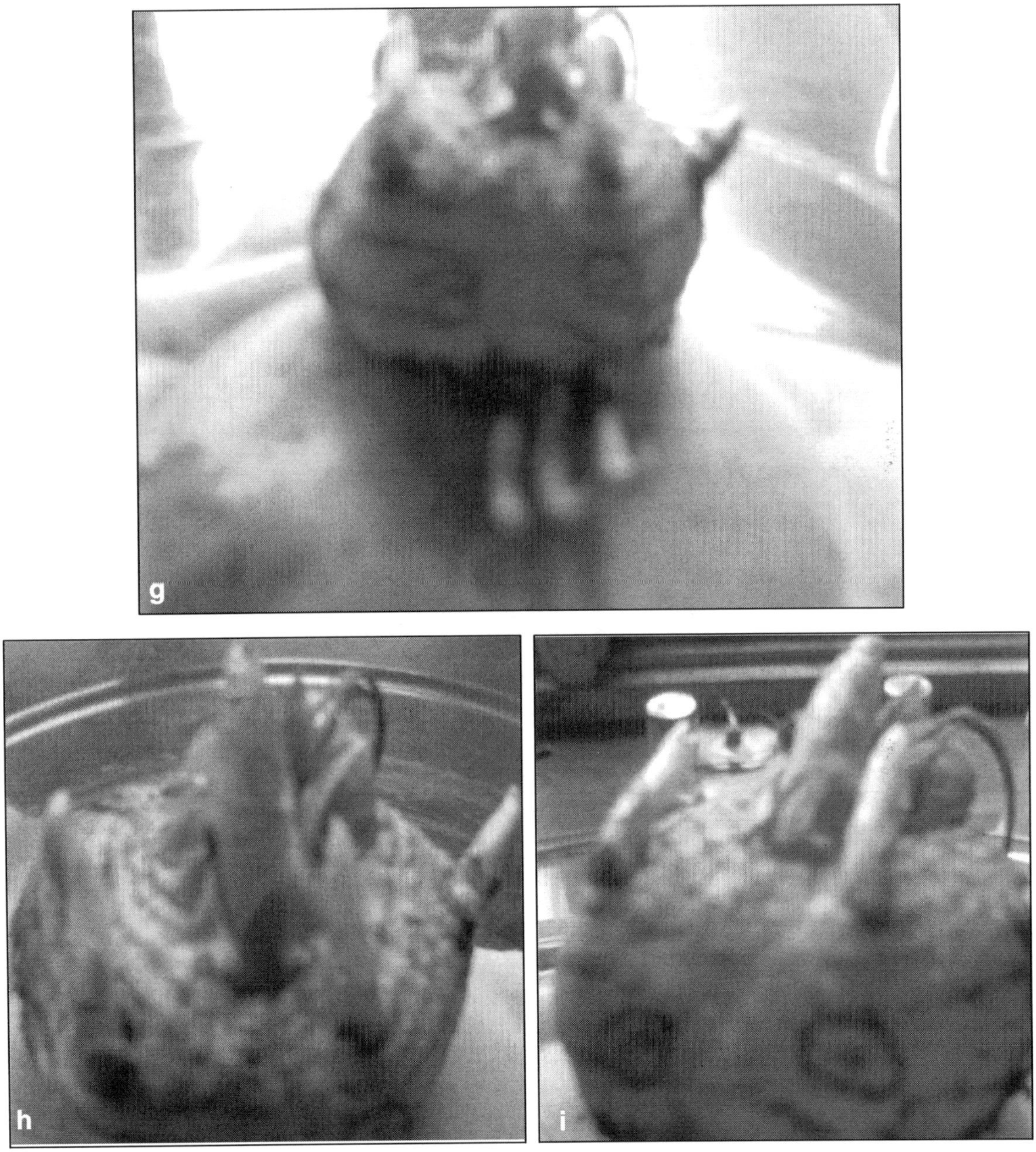

The effect of different commercially neem products *viz.*, Nemola, Nemoria, Neem Gold, Neem Churn, were found least effective *in vitro* studies but Nemoria, Neem Churn favoured the radial growth *of Rhizoctonia solani* and *Phytophthora* sp. respectively. In field studies, none was found to be effective.

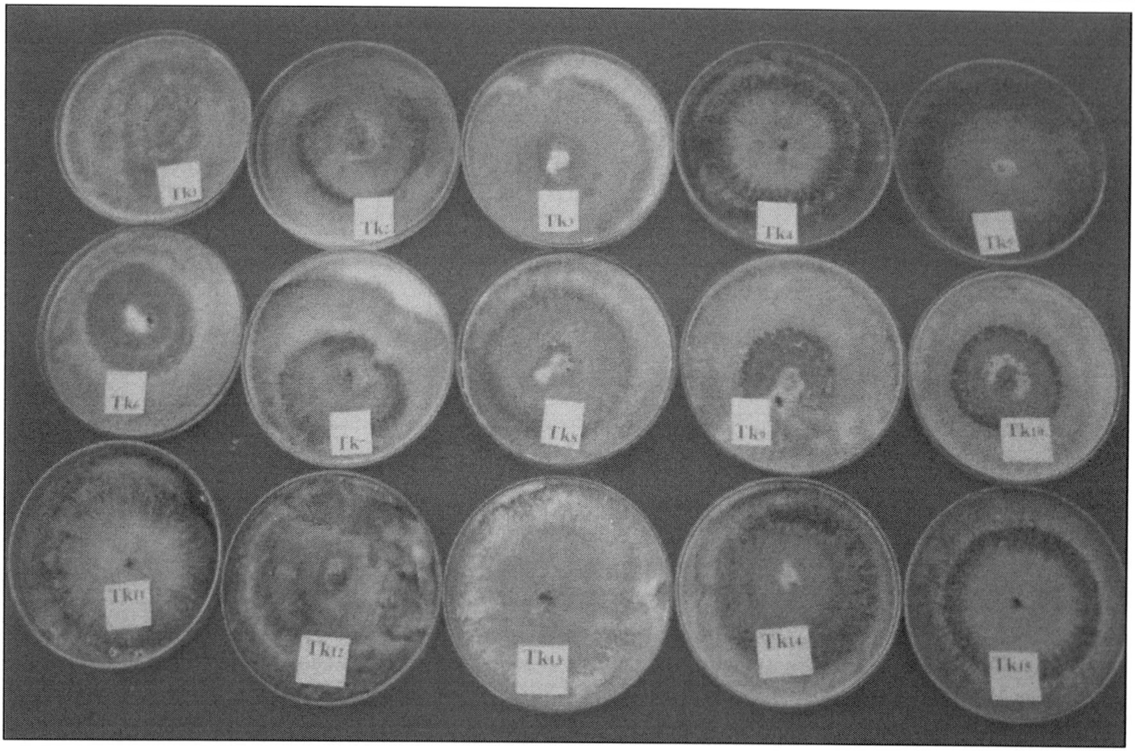

Figure 16.15: Native Isolates of *Trichoderma viride* and *Trichoderma harzianum*

From the results of the present investigation it can be concluded that the chemical pesticides Carbendazim 50 per cent WP, Zineb 75 per cent WP, Captan 50 per cent WP, Dithianon 75 per cent WP and biocontrol agents *Trichoderma viride, Trichoderma harzianum, Paecilomyces lilacinus, Pseudomonas fluorescens* and organic manures *viz.*, sheep manure 20 tons per hectare are highly effective in reducing the corm rot diseases incidence caused by *Fusarium oxysporum* f. sp. *gladioli* and sterile Basidiomycetes fungus in saffron crop and improves the yield. But there is certain limitation of chemical pesticides on this valuable medicinal crop, which may causes pesticidal residue in saffron.

Hence, biological and organic farming is well suitable for this medicinally important crop to protect from pesticidal residue. Therefore, use of *Trichoderma viride, Trichoderma harzianum, Pseudomonas fluorescens* and *Paecilomyces lilacinus* and sheep and poultry manure as suitable manure, ecofriendly alternative to the use of chemicals in the management of corm rot and enhancement of crop production and yield.

Since biocontrol and organic farming technology is not popular among the saffron growers, all success with bioagents should be demonstrated and published in the local language to encourage efforts to register and commercialize the product and to increase the familiarity with biocontrol agents among the target groups. Farmers should be provided incentives or subsidy on the bioproducts preparation, as the large scale use of biocontrol agents will significantly enhance, the environment and agricultural productivity.

Many chemical pesticides were tested for the efficacy against these four pathogens in vitro thus, bioassay was conducted with seven different fungicides Carbendazim 50 per cent WP, Sulphur 80 per cent WP, Triadimefen 25 per cent WP, Zineb 75 per cent WP, Captan 50 per cent WP, Bitertanol 25 per cent WP and Dithianon 75 per cent WP against four pathogens namely *Rhizoctonia solani*, *Phytopthora* sp., *Fusarium oxysporum* f. sp. *gladioli* and sterile Basidiomycetes fungus.

The results showed that Carbendazim 50 per cent WP was of great promise against *Phytopthora* sp. followed by Captan 50 per cent WP, Zineb 75 per cent WP and Dithianon 75 per cent WP towards inhibition of the fungus growth, while the Sulphur improve the radial growth of sterile Basidiomycetes fungus and *Fusarium oxysporum* f. sp. *gladioli*. When tested in the field, these fungicides were applied as corm treatment in standard doses against two pathogens *Fusarium oxysporum* f. sp. *gladioli* and sterile Basidiomycetes fungus in the saffron infected fields. Yield attributing characters and corm rot prevention was observed by Carbendazim 50 per cent WP, Zineb 75 per cent WP, Captan 50 per cent WP, Dithianon 75 per cent WP in both the infected fields.

Trichoderma viride, Trichoderma harzianum, Pseudomonas fluorescence were inhibitive to the corm rot pathogens of saffron *in vitro* and in field Paecilomyces lilacinus along with above mentioned bioagents was also effective individually and in combination on corm rot diseases of saffron *Fusarium oxysporum* f. sp. *gladioli* and sterile Basidiomycetes fungus, it was found that the combination of these biocontrol agents was highly effective in reducing the pathogens *Fusarium oxysporum* f. sp. *gladioli* and sterile Basidiomycetes fungus and maximizing corm production and yield.

The effect of organic manures *viz.*, FYM, poultry and sheep manure @ 10, 15 and 20 tons per hectare each individually was studied, sheep manure @ 20 tons per hectare was highly effective in reducing the corm rot pathogens *Fusarium oxysporum* f. sp. *gladioli* and sterile Basidiomycetes fungus in saffron with maximum corm production and yield.

The effect of different commercially neem products *viz.*, Nemola, Nemoria, Neem Gold, Neem Churn, were found least effective in vitro studies but Nemoria, Neem Churn favoured the radial growth of Rhizoctonia solani and Phytophthora sp. respectively. In field studies, none was found to be effective.

From the results of the present investigation it can be concluded that the chemical pesticides Carbendazim 50 per cent WP, Zineb 75 per cent WP, Captan 50 per cent WP, Dithianon 75 per cent WP and biocontrol agents *Trichoderma viride, Trichoderma harzianum, Paecilomyces lilacinus, Pseudomonas fluorescens* and organic manures *viz.*, sheep manure 20 tons per hectare are highly effective in reducing the corm rot diseases incidence caused by *Fusarium oxysporum* f. sp. *gladioli* and sterile Basidiomycetes fungus in saffron crop and improves the yield. But there is certain limitation of chemical pesticides on this valuable medicinal crop, which may causes pesticidal residue in saffron.

Hence, biological and organic farming is well suitable for this medicinally important crop to protect from pesticidal residue. Therefore, use of *Trichoderma viride, Trichoderma harzianum, Pseudomonas fluorescens* and *Paecilomyces lilacinus* and sheep and poultry manure as suitable manure, ecofriendly alternative to the use of chemicals in the management of corm rot and enhancement of crop production and yield.

Since biocontrol and organic farming technology is not popular among the saffron growers, all success with bioagents should be demonstrated and published in the local language to encourage efforts to register and commercialize the product and to increase the familiarity with biocontrol agents among the target groups. Farmers should be provided incentives or subsidy on the bioproducts preparation, as the large scale use of biocontrol agents will significantly enhance, the environment and agricultural productivity.

Constraints

Some articles and papers have been written in the country but all these attempts are independent, isolated and fragmented piece meal not much related with the present problem. Few of the foreign scientists have made efforts of pure scientific nature but having the least relevance for our country.

Problems

- ✰ The mortality rate of seed/corm lets has increased.
- ✰ Ninety per cent growers complained that Agriculture Department has so far paid no attention to saffron cultivation
- ✰ No proper extension facilities have been made available to the growers.
- ✰ Adulteration of saffron done by certain big growers and dealers/wholesalers has brought bad name to Kashmir saffron in the national and international markets
- ✰ Saffron is cultivated as a rainfed crop.

References

Cappalli, C. (1994). Recent observations on saffron diseases in Italy. In: *Proceedings of the Congress of the Mediterranean Phytopathological Union*, Sep. 18–24, Kusadari-aydin, Turkiye, p. 497.

Cappelli, C. (1994). Occurrence of *Fusarium oxysporum* f. sp. *gladioli* on saffron in Italy. *Phytopathologia Mediterranea*, 33(1): 93–94.

Cappelli, C. and Di Donato, G. (1998). Attacchi di *Fusarium oxysporum* schhecht. In: *Coltivazioni Di Zafferano in Abruzgo. L. informatore Agrario*, 25: 75–76.

Cappelli, C. and Di Minco, G. (1998). Control of *Fusarium oxysporum* f. sp. *gladioli* based on the production of pathogen fee saffron corms. *Journal of Plant Pathology*, 80(3): 253.

Cappelli, C., Buonaurio, R. and Polverari, A. (1991). Occurrence of *Penicillium corymbiferum* on saffron in Italy. *Plant Pathology*, 40(1): 148–149.

Cappelli, C. and Di Minco, G. (1999). Risultati di un Triennio di studi sulle malattie dello Zafferano riscontrate nele principali aree di coltivazione dell' Abruzzo. *Informatore Filopatologico*, 49(9): 27–32.

Cappelli, C., di Minco, G. and G. Minco Di. (1999). Instituto di Patologia Vegetale, Universita di Perugia, Italy. *Informatore Fitopatologico*, 49(9): 27–32.

Cappelli, C., Donato, G. and Di Donato, G. (1994). Attacks of Fusarium oxysporum Schlecht. in crops of saffron in Abruzzo. *Informatore-Agrario*, 50(25): 75–76.

Carta, C., Fiori, M. and Franceschini, A. (1982). II marciume Carborioso dei bulbi delo zafferano (*Crocus sativus* L.). *Studi Sassaresi*, III, 29: 193–197.

Damirov, I.A., Prilipko, L.I., Shukurov, D.Z. and Kerimov, J.B. (1988). *Remedy Plants of Azerbaijan.* Maarif, Baku, Azerbaijan, pp. 90–93. Anonymous, Directorate of Agriculture, Lalmandi, Srinagar, J&K.

Kalhana (12th Century). *Rajtarangini*. An ancient cronical of Kashmir.

Kasumov, F.J. (1970). *The Extract of Saffron Flowers*. Gos Plan Press, Baku, Azerbaijan.

Mir, G.H. and Devi, L.S. (2003). Corm rot diseases of saffron in Kashmir valley. *Indian Phytopathology*, 56(1): 122.

Mir, G.H. and Devi, L.S. (2004). Saffron corm rot and their management. In: *National Symposium on Detection and Management of Plant Diseases Using Conventional and Modern Tools*, IPS Zonal Meeting (MEZ) Dec. 31, p. 20.

Mir, G.M. (1992). *Saffron Agronomy in Kashmir: A Study in Habitat, Economy and Society*. Gulshan Publishers. Srinagar.

Mir,G.H. and Devi, L.S. (2004). Studies of diseases associated with saffron (*Crocus sativus*). In: *National Symposium of Indian Phytopathological Society on Crop Surveillance: Disease Forcasting and Management*, Feb. 19–21, p. 63–64.

Mir Gulam Hassan and Devi, L.S. (2003). Corm rot diseases of saffron in Kashmir valley: A new record. *Indian Phytopath.*, 56(1): 122.

Mir Gulam Hassan and Devi, L.S. (2004). Antagonistic potential of *Trichoderma harzianum* isolates against major corm rot pathogens of saffron (*Accepted by Phytopathology*).

Mir Gulam Hassan and Devi, L.S. (2004). Studies of diseases associated with saffron. In: *Indian Phytopathological Society National Symposium on Crop Surveillance Disease Forecasting and Management*, 19–21 Feb. at IARI New Delhi.

Mir Gulam Hassan and Devi, L.S. (2004). Saffron corm rot diseases and their management. In: *National Symposium of Indian Phytopathology on Detection and Management of Plant Disease Using Conventional and Modern Tools*, December 31 at CISH, Lucknow.

Primo, P. di. and Cappelli, C. (2000). Indagine preliminare sui gruppi di compatilita vegetative di isolati di *Fusarium oxysporum* f. sp. *gladioli* ottenuti da zafferano in Italia. *Informatore Fitopatologico*, 11: 35–38.

Primo, P. di., Cappelli, C., Katan, T. and Primo, P. Di (2002). Vegetative compatibility grouping of *Fusarium oxysporum* f. sp. *gladioli* from saffron. *European Journal of Plant Pathology*, 108(9) : 869–875.

Sud, A.K., Pual, Y.S. and Thakur, B.R. (1999). Corm rot of saffron and its management. *Journal of Mycology and Plant Pathology*, 29(3): 380–382.

Sutton, M.W. and Wale, S.J. (1985). The control of *Penecillium corymbiferum* on Crocus and its effect on corm production. *Plant Pahtology*, 34: 566–570.

Suzhou New Medical College (1977). *Dictionary of Traditional Chinese Medicine*. Zhong Yao Da Zi Dianj, Vol. 2. Shanghai People's Publication House, Shanghai, pp. 2622–2623.

Szita, E. (1987). Wild about saffron a contemporary guide to an ancient spice. *Saffron Rose*, Daly City. C.A.

Tammaro, F. (1990). *Crocus sativus* L.c.v. Piano di Navdli– L' Aquila (Zafferano doll' Aquila): ambiente, coltinazione, caratteristiche morformatriche dello Zafferano (*Crocus sativus* L.) L' Aquile instituto Teen. *Propag. Agr., Roma*, p. 1–20.

Tammaro, F. and Di. Francesco, L. (1978). Lo Zafferano de L'Aquila. 1st.Tecn. Propag. Agr., Roma.

Yamamoto, W., Omatsu, T. and Takami, K. (1954). Studies on the corm rots of *Crocus sativus* L. I. On saprophytic propagation of *Sclerotinia gladioli* and *Fusarium oxysporum* f. sp. gladioli on various plants and soils. In: Scientific reports of the Hyogo University of Agriculture, 1, 64–70, Rev. appl. Mycol., 35, 327.

Zhang, Y., Shoyama, Y., Sugiura, M., and Saito, H. (1994). Effects of *Crocus sativus* L. on the ethanol induced impairment of passive avoidance performances in mice. *Biol. Pharm. Bail.*, 17: 217–221.

Zhou, Q., Sun, Y. and Zhang, X. (1987). Saffron, *Crocus sativus* L. *J. Traditional Chinease Med.*, 28 : 59–61.

Plant Diseases Management in Horticultural Crops (2011) *Pages* **209–224**
Editors: **Shahid Ahamad, Ali Anwar and P.K. Sharma**
Published by: **DAYA PUBLISHING HOUSE, NEW DELHI**

Chapter 17

Ecofriendly Management of Fusarial Wilts of Vegetable and Fruit Crops

V.S. Verma, V.K. Singh and Sonika Jamwal

Dryland Research Sub Station, SKUAST-J, Dhiansar, Jammu, J&K

Soil is the favourable habitat and provides home for a variety of microorganisms such as bacteria, fungi, actinomycetes, algae, protozoa and viruses wherein bacteria and fungi are most prevalent. Soil-borne diseases result from a reduction of biodiversity of friendly microorganisms in soil. Restoring beneficial organisms that attack, repel or antagonize disease-causing pathogens renders a soil disease-suppressive. Plants growing in disease-suppressive soil resist diseases much better than in soils low in biological diversity. The higher the diversity, the more stable is the soil biological system. These beneficial organisms suppress disease through competition, antibiosis, antagonism and direct feeding on pathogenic fungi, bacteria, and nematodes. Plant diseases may occur in natural environments, but they rarely go out of control and cause major problems unless a susceptible host and a disease-causing pathogen meet in a favourable environment. If any one of these conditions is not met, there is no disease. As plants and soils have become sick, growers have to respond with new and more potent chemicals in an effort to get rid of the problem pathogens. Soil-borne plant pathogens are responsible for many acute and chronic diseases of crop plants that can result in severe losses to growers. Economic losses to soil-borne pathogens are estimated at 50-75 per cent of the achievable yield in many crops. Yield failures resulting from acute diseases such as vascular wilts, take-all of cereals, Phymatotrichum root rot, *Verticillium* and *Phytophthora* may be even more severe. About 90 per cent of the major diseases of the principle crops are caused by soil-borne plant pathogens (Lewis and Papavizas, 1991). The annual monetary losses due to soil-borne diseases have been estimated to be more than 4 billion US dollars (Stanghellini, 2008).

While it may seem the reasonable course of action, chemical intervention only serves to make things worse over time. Endemic soil-borne plant diseases frequently inflict huge and recurring economic losses to growers and are usually difficult to manage. Many intervention practices such as the use of fungicides indiscriminately eradicate soil microorganisms including the beneficial ones, thereby reducing the disease suppressive features of the soil. The alternative ecofriendly methods have proved effective in managing many soil-borne diseases involving manageable areas, especially in fruit plant nurseries and vegetable crops.

Organic Soil Amendments

Organic soil amendments play a major role in changing the soil ecology and their role has been studied in connection with various soil borne diseases. These amendments act both on host or well as on pathogen and enrich the soil with microflora potentially competitive or antagonistic to pathogens by inhibitory substances during decomposition.

Oilseed Cakes

Effect on Soil Microflora

Soil amendment with oilseed cakes is one of the important methods of altering soil environments, thereby affecting plant growth and soil microflora. Oilseed cakes have been used as soil amendments for controlling various soil borne plant pathogens. Singh and Pandey (1966) reported that low quantity (0.1 per cent) of groundnut and castor seed cakes enhanced the growth of *Pythium aphanidermatum*, but in the presence of higher quantities of these cakes, the growth of the fungus in soil was inhibited.

Oilseed cakes of neem (*Azadirachta indica*), castor (*Ricinus communis*) and linseed (*Linum robusta*) were stimulatory to soil microflora but sawdust was inhibitory (Khanna and Singh, 1975). They further reported that microbial counts of amended non-rhizosphere soil were generally higher than in unnamended soil. Singh and Singh (1980) studied the effect of oil cakes on *F. oxysporum* f.sp. *udum* and observed that neem seed cake, rice husk or saw dust with or without supplemental nitrogen in un-autoclaved soil greatly enhanced its lytic effect on the fungus whereas the lytic effect was mostly absent in unamended soil, due to absence of normal soil microflora responsible for the lytic effect in soil. Un-autoclaved soil contained its natural soil microflora and hence showed better lytic effect than autoclaved soil.

While studying the quantitative and qualitative changes in the microflora in oilseed cake amended soil, Khalis and Manoharachary (1985) recorded increase in the fungal population within 20-45 days in soil amended with neem, groundnut and safflower oilseed cakes. They isolated the fungal species like *Aspergillus, Curvularia, Drechslera, Penicillium, Rhizopus* and *Trichoderma* from the natural soil and reported that *Fusarium solani* disappeared from the soil which was amended with neem, groundnut and safflower oilseed cakes.

Effect on Disease control

Khanna and Singh (1974) observed that neem seed cake caused inhibition of sporulation of *Fusarium* spp. in rhizosphere soil. Out of the nine organic amendments tested against Fusarium wilt of tomato in pot and field, Homma *et al.* (1979) observed that soybean and rapeseed cakes reduced the disease severity in the pot experiments. Chakrabarti *et al.* (1991) studied the effect of oilseed cakes on wilt of muskmelon caused by *Fusarium solani* and reported that neem cake reduced wilt incidence by 80 per cent, mustard seed cake by 65 per cent and groundnut seed cake by 37 per cent. Rai and Singh (1996) reported neem, mustard and mahua oilseed cakes as most effective in reducing the growth of *F. udum* on pigeonpea. Neem cake was most effective in controlling wilt of pigeonpea caused by *F. udum*. Neem seed cake (250 kg/ha) has been found to be highly effective in reducing collar rot disease of groundnut and the reduction in disease was attributed to the increase in population of antagonistic organisms by the application of neem seed cake and farm yard manure (Karthikeyan, 1996). Groundnut and mustard seed cakes each at 2 per cent concentration (w/w) effectively reduced the population of tomato wilt pathogen *F. oxysporum* f.sp. *lycopersici* by more than 70 per cent (Raj and Kapoor, 1996). The rhizosphere and non-rhizosphere population of fungi, bacteria and actinomycetes was higher in neem seed cake amended plots (Ushamalini *et al.*, 1997). *Cannabis* and *Lantana* leaves as well as wheat and barley straws when amended in soil at 1 and 2 per cent recorded up to 71 per cent seedling wilt of mango (Verma and Sharma, 2004).

Fresh and Dry Plant Materials

Effect on Microflora

Plant residues such as fresh leaves and green manures are important components of soil which are added to the soils by farmers, or by nature itself. The influence of these residues is felt, either directly or indirectly, by the magnitude of living components of that soil. An environment that favours the increase of organisms antagonistic to pathogens in soil is a desirable goal, and a hostile environment can be created by culturally favouring the production of aggressive compounds in soil (Linderman, 1970). Singh and Pandey (1966) studied the effect of finely ground mature dry straws of wheat, barley, oat, pea, dry maize cob powder, chopped leaves of Krishnaneel (*Anagallis arvensis*), bhang (*Cannabis sativa*), bakayan (*Melia azadirach*), neem (*Azadirachta indica*) and leaf composts of neem and bakayan each at 5 per cent (w/w) to study their effect on the population of *Pythium aphanidermatum*. It was reported that amendments with dry straws of wheat, dry maize cob powder as well as green leaves of neem, bhang and bakayan suppressed the development of *Pythium* in soil.

Adams *et al.* (1968) reported a poor germination of clamydospores of *F. solani* f.sp. *phaseoli* when the soils were amended with cellulose and oat straw at the rate of 1 per cent (w/w). Later Papavizas (1968) found a reduction in inoculum density of *Thielaviopsis basicola*, causing bean root rot, from 30×10^3-40×10^3 propagules per gram of soil to fewer than 0.5×10^3 propagules in soil amended with cellulose, chitin, barley straw, corn stover and alfalfa hay. Similarly, germination of conidia, micro-sclerotia and mycelial growth of *Verticillium dahliae* in soil was inhibited by the addition of chitin, limanarin, wheat straw and oven dried green clover. The population of bacteria and actinomycetes in the rhizosphere of plants in amended soil was more than the natural soil (Jordan *et al.*, 1972).

Effect of fresh and dry plant materials on the microbial population of rhizosphere and non-rhizosphere soils has also been reported by adding a number of dry and green plant materials *viz.* wheat straw, sorghum straw and sugarcane straw as dry matter and green leaves of *Datura alba*, *Ipomea*, *Cornea*, *Calotropis procera* and entire green plants of *Lemma paucicostata*, *Eichornia crassipes* and *Salvinia natans* (Chandra *et al.*, 1981). Amendment of soil with dry and green plant materials suppressed the population of *Fusaria* and stimulated that of *Aspergilli* (Chandra *et al.*, 1981). All the amendments had marked stimulatory/inhibitory effect on the population of fungi and bacteria in the rhizosphere, but the magnitude varied with the amendment as well as variety and age of the plant. Significant suppression of *Fusarium* population in the rhizosphere was reported with wheat straw, *Calotropis*, *Salvinia* and *Lemma* whereas majority of amendments stimulated the population of Aspergilli in the rhizosphere. Raj and Kapoor (1997) studied the effect of composts of banana leaves, bagasse, synthetic mushroom compost, paddy straw and spent mushroom at 0.5, 1 and 2 per cent (w/w) and reported enhanced microbial activity (fungi and bacteria) resulting in reduction of population of tomato wilt pathogen *Fusarium oxysporum* f.sp. *lycopersici* by 44.74 per cent. Fresh leaves of *Cannabis sativa*, *Lantana camera*, mango (*Mangifera indica*), wheat and barley straw when amended in soil at 1 and 2 per cent (w/w) were effective against wilt of mango caused by *Fusarium solani* (Verma and Sharma, 2003).

Effect on Disease Control

Soil amendment with a mixture of barley straw and faba bean seed powder each at 1 per cent concentration effectively suppressed the infection of cowpea roots by *Fusarium solani* and *Rhizoctonia solani*, improved seedling stand and increased root and shoot weight in healthy cowpea (Barkat *et al.*, 1983). In a soil incubated with chlamydospores of *F. solani*, the infection rate decreased with garlic roots, maize stalks and barley straws (Son *et al.*, 1985). Ramirez and Munnecke (1988) amended dry straws of cruciferous plants, lucerne and wheat in pot soil and their effect was observed on cabbage yellows caused by *F. oxysporum* f.sp. *conglutinans* and reported a significant reduction in disease severity. They also observed a fungistatic effect when the fungus, growing on PDA, was suspended above the soil containing cabbage residues whereas a fungicidal effect was noticed when the infested soil was suspended above the decomposing residues.

Biological Control

An ecologically balanced soil is the best medium for growing healthy plants. Such soils can be maintained by following sound cultural practices. Though biological antagonists are subjected to numerous ecological limitations under field conditions, yet this method can be expected to become an important control measure against many soil borne diseases. As early as 1920, antagonistic fungi were introduced in forest nursery soils to reduce damping off of pine seedlings (Hartley, 1921). Reduction in disease incidence was recorded in some treatments, and it was concluded that competition of different fungi is a factor to be considered. The various mechanisms suggested to be involved in biocontrol by the antagonistic fungi are antibiosis, lysis, competition, mycoparasitism and promotion of plant growth (Baker, 1968; Henis, 1984; Papavizas, 1985; Chet, 1987; Lynch, 1990).

Antagonistic Activity of Bioagents

In vitro Efficacy

Sivan and Chet (1986) observed that germination of chlamydospores of *Fusarium* spp. decreased after the addition of conidia of *T. harzianum* in the soil. Dwivedi (1992) recorded a growth inhibition of *Fusarium oxysporum* f.sp. *psidii* by 71.4 per cent in *Trichoderma viride* while *T. harzianum* retarded the growth of *F. solani* and *F. longipes* by 60.0 and 60.4 per cent, respectively. Among seven antagonistic fungi isolated from soil, *T. viride* inhibited the growth of *F. oxysporum* f.sp. *lini* from flax by 68.2 per cent while *T. piluliferum* inhibited that of *F. oxysporum* f.sp. *lycopersici* in tomato by 64 per cent (Dwivedi *et al.*, 1993). Antagonistic activity of four isolates of *G. virens* was demonstrated towards four pathogenic fungi including *Fusarium solani*. Spore germination and radial growth of the test fungi were inhibited by the cell free culture filtrates of all the bioagents and the disease symptoms were restricted on pre-treated plants which were artificially parasitized by the test fungus (Mishra and Narain, 1994). High inhibitory effect of volatile toxic substances emitted by *Trichoderma* spp. on the radial growth of *Fusarium* spp. has also been reported by many workers (Padmodaya and Reddy, 1996; Verma and Sharma, 2007; Dubey *et al.*, 2007).

Trichoderma viride produced non-volatile substances which inhibited the growth of rhizome rot pathogen (*Pythium myriotylum*) and *F. solani* in ginger by 70 and 10 per cent, respectively when these organisms were grown on media plates previously used for *T. viride*. Species of *Trichoderma* are known to produce a number of volatile and non-volatile fungitoxic compounds like trichoviridin, trichodermadin, dermadin, viridin and metabolites containing an isocyanide group (Brian, 1946; Tamura, 1975; Ghiselbarti and Sivasithamparam, 1990). *G. virens* also produces antibiotics namely glioviridin, gliotoxin and viridin (Howell and Stipanovic, 1983). Hyphal interactions with strong antagonistic activity of *G. virens* and *T. hamatum* was observed by Cipriano *et al.* (1989) which inhibited the growth of *F. oxysporum* f.sp. *lycopersici*. Volatile substances produced by *T. viride* completely inhibited the growth of *P. myriotylum* but reduced the colony diameter of *F. solani* by only 3.4 per cent (Rathore *et al.*, 1992). A negative correlation between hyper-parasitic and antibiotic activities was noticed when *T. aureoviride, T. harzianum, T. viride* and *G. virens* isolates were tested against *F. solani, F. oxysporum, Pythium* sp., *Rhizoctonia solani* and *Sclerotinia sclorotiorum* in pea (Velikanov *et al.*, 1994).

Dennis and Webster (1971) reported that isolates of *Trichoderma* ageing 7-9 days were effective in producing volatile fungitoxic substances. Gaseous metabolites of *T. harzianum* have been reported by Mukherji *et al.* (1983) to be most inhibitory to *Sclerotium rolfsii* followed by *T. viride* and *G. deliquescens. Trichoderma harzianum* showed mycoparasitism whereas *T. viride* exhibited antibiosis and 80 and 60 per cent plants were observed healthy in sterilized as well as unsterilized soil inoculated with *F. oxysporum* f.sp. *ciceri* causing wilt of pea plants (Singh *et al.*, 1997). They further reported that *T. harzianum* proved to be highly antagonistic under field conditions and the disease incidence was reduced by 88.7 and 77.3 per cent, respectively.

Field Efficacy of Bioagents

Crown and root rot disease of tomato caused by Fusarium oxysporum f. sp. radicis lycopersici has effectively been controlled by *Trichoderma koningii, T. hamatum, T. harzianum, Gliocladium virens* (Sivan *et al.*, 1987; Cipriano *et al.*, 1989; Sivan and Chet, 1993). Seed treatment with *T. harzianum*, and *G. virens* gave complete control of *F. oxysporum* on 30 and 120 days old tomato plants (Parveen *et al.*, 1991). *T. harzianum* proved to be most effective when introduced to pots, and reduced the number of wilted plants by more than 70 per cent and also reduced disease severity in the non-wilted plants (Rettink, 1993). Dwivedi (1993) while working on wilt of guava reported that *Trichoderma* spp. were most effective in inhibiting the growth of almost all the species of *Fusarium* causing wilt of guava. *T. lignorum* inhibited the growth of *F. oxysporum* f.sp. *psidii* and *F. moniliforme* by 70 per cent while *T. viride* inhibited the growth of *F. oxysporum* f.sp. *psidii* by 71.4 per cent, *T. harzianum* reduced the growth of *F. solani* and *F. longipes* by 60.0 and 60.4 per cent, respectively. *Fusarium* infection of mustard was completely controlled by *T. viride* and *Bradyrhizobium japonicum* and also by the combined application of *B. japonicum* with *T. harzianum, T. viride* or *G. virens* in 30 day old seedlings (Haque and Ghaffar, 1991). Both these bioagents gave the same level of disease control as benomyl both under greenhouse and field conditions (Flori and Roberti, 1993). Seed treatment with mancozeb or carbendazim, in addition to biological control with *T. harzianum, T. hamatum* and *G. virens* gave excellent control of rhizome rot (*F. oxysporum* f.sp. *zingiberi*) of ginger (*Zingiber officinale*) and also increased yield (Dohroo, 1995). Ushamalini *et al.* (1997) reported 98.8 per cent seed germination of cowpea with *T. harzianum* and it was at par with the standard practice of seed treatment with fungicide carbendazim compared to the lower germination in control. *T. harzianum, T. viride, T. virens* and *T. hamatum* exhibited up to 89 per cent control of root rot/wilt of different crops caused by *Fusarium solani* (Rojo *et al.*, 2007; Verma and Sharma, 2007). An improvement in plant growth substanted by increase in shoot length (102 per cent), root length (221 per cent), shoot weight (196 per cent) and root weight (312 per cent) has been recorded by using *Trichoderma* spp as biocontrol agents against seedling wilt of mango caused by *Fusarium solani* (Verma and Sharma, 2007) and on other crops by different workers (Sharma and Dureja, 2004; Mukhopadhyay, 2005).

Soil Solarization

Soil solarization or solar heating of soil is a method whereby soil is disinfected by trapping solar irradiations in wet soil during hot summer months for obtaining better disease control. The method involves the use of low density transparent polyethylene tarps placed over wet soil during hot summer months to trap solar irradiation and build up enough heat to reduce population of soil-borne pests (Katan, 1981). The solar heating for disease control is similar, in principle, to that of artificial soil heating by steaming or other means except some biological and technical differences. In soil solarization there is no need to transport the heat from its source to the open field. Solar heating is carried out at relatively low temperature as compared to artificial heating, thus its effects on living and non-living components are likely to be less drastic. The negative side effects like phytotoxicity due to release of manganese or other toxic products and a rapid soil re-infestation due to creation of biological vacuum (Dawson *et al.*, 1965) have not yet been reported with solar heating which is nowadays becoming a popular method of disease control in soil borne diseases.

Various methods of using solar irradiation for controlling pests have been developed much before the development of plasticulture. These include heating of seeds, tubers or soil (by exposing to sunshine) for controlling plant pathogens. Plastics provided a new and effective tool for solar heating of soil. Polyethylene, popularly known as polythene, one of the most important materials used in agriculture, was first introduced on a commercial scale in 1939 (Byrdson, 1970). However, Jones *et al.* (1966) for the first time used polyethylene tarps to increase the soil temperature which proved lethal to soil borne pathogens of tomato. The experience of extension workers and growers of Israel over a period of time revealed that the intensive heat generated in mulched soil might be used for the control of plant diseases and suggested the use of polyethylene sheets for soil disinfestation in summer, prior to planting (Katan *et al.*, 1976). The method was also used in California for controlling wilt disease of cotton caused by *Verticillium dahliae* (Pullman and DeVay, 1977). Similar studies have also been carried out in Jordan, England, Italy, Greece, California and India (Al-Raddad, 1979; White and Buczacki, 1979; Tamietti and Garibaldi, 1980; Tjamos and Faridis, 1980; Stapleton and DeVay, 1982; Chauhan *et al.*, 1988) against fungal pathogenic diseases of various crops.

Among different polyethylene mulches transparent mulch was able to transmit most of the solar radiations and was quite effective for solar heating of the soil (Macfayden, 1967; Katan *et al.*, 1976). They also recommended that soil mulching should be carried out during periods of high temperatures and soil should be kept wet during mulching in order to increase thermal sensitivity of resting structures and also to improve heat conduction. Pullman *et al.* (1979) suggested that mulching period should be sufficiently extended, usually four weeks or longer, in order to achieve better pathogen control at all desired soil depths. Later, Horowitz and Regev (1980) showed that as thin as 25-30 um polyethylene tarp transmitted better radiation and suggested the use of thinnest possible polyethylene tarps for better results.

Microbial Population

Ben Yaphet *et al.* (1987) found that viability of propagules of *F. oxysporum* f.sp. *vasinfectum* burried at 30 cm soil depth, was reduced after 31 days of soil solarization by 97.5, 58 and 0 per cent under a double layer, single layer of polyethylene and in uncovered soil, respectively. Sterilization of soil by polyvinyl sheet in a greenhouse considerably increased the soil temperature and decreased the number of spores of *F. oxysporum* in soil, and the low level of spores was maintained for a period of up to 6 months (Kodama, 1989). Bourbos and Skoudridakis (1991) reported the reduction of pathogen population of *F. oxysporum* f.sp. *radicis lycopersici*, *F. solani* and *Colletotrichum coccodes* by 83.5-100 per cent after soil solarization for a period of 10 weeks, whereas Dwivedi (1991) reported a reduction of 70 per cent population of *F. oxysporum* in tomato at 5 cm soil depth after 30 days of solarization with transparent polyethylene sheets (25 um). Solarization with transparent polyethylene for 60 days eliminated the mango wilt pathogen (*Fusarium solani*) up to 30 cm soil depth both under irrigated and unirrigated conditions (Raj *et al.*, 1993) as well as the citrus wilt pathogen *F. oxysporum* (Raj and Gupta, 1995). A similar study on wilt of mango seedlings caused by *Fusarium solani* (Raj and Gupta, 1996) revealed that soil solarization for 45 days was effective in eliminating the pathogen up to 30 cm soil depth under irrigated conditions whereas the pathogen was still detected under unirrigated conditions even after 45 days at 30 cm soil depth. They found that soil solarization for 15 days was sufficient to decrease the soil population of the test fungus to zero level up to 15 cm depth both under irrigated and unirrigated conditions.

In addition to reduction of pathogen population, soil solarization encouraged the populations of saprophytic fungi such as *Trichoderma* spp. and better plant growth and yield (Hasan, 1989; Sivan and Chet, 1993). Soil solarization by covering moist soil with transparent polyethylene sheet achieved high soil temperatures and population of actinomycetes as well as thermo-tolerant and thermophilic fungi increased (Stapleton and DeVay, 1982). Hasan (1989) recorded the growth of Trichoderma spp. by solar heating of soil with transparent mulches. The use of transparent tarp for 10 weeks in a field in Jorden Valley reduced the populations of *F. oxysporum* and *F. solani* but increased the populations of saprophytic fungi like *Aspergillus*, *Penicillium* and *Trichoderma* (Triolo *et al.*, 1988; Abu Gharbieh *et al.*, 1991). Similarly, AbuBlan and Abu Gharbieh (1994) also recorded higher populations of *Aspergillus* and *Penicillium* species in soils after 12 weeks of solarization by transparent or black plastic tarpings in potato, cauliflower and cucumber. Besides controlling a variety of soil borne pathogens of plants, soil solarization has also been effective in checking the populations of nematodes in soil (Abu Gharbieh and AbuBlan, 1991; Patel *et al.*, 1995).

Effect on Soil Temperature

In late seventies and early eighties a number of workers studied the pattern of heating in mulched as well as unmulched soils. While studying the course of heating in unmulched and polythene mulched wet soils at different depths Mahrer (1979) observed that the maximal temperature (49°C) at the uppermost mulched layer was 9°C higher than the unmulched one. Pullman *et al.* (1979) found that temperatures of tarped soils were higher than untarped soils, sometimes reaching upto 60°C at a depth of 5 cm. Jacobson *et al.* (1980) observed that the maximal temperature decreased with increasing soil depths, but the peaks (highest temperatures) lasted longer and reached later in the day. The investigations carried out in various regions of Israel in July-August during different years also revealed similar observations; the temperature in mulched soil at 5 and 20 cm depths were found to be 55 and 45°C, respectively, whereas the temperature maxima of unmulched soil at depths of 5 and 20 cm were 38 and 36°C, respectively (Katan, 1976; Al-Raddad, 1979; Grinstein *et al.*, 1979; Jacobson *et al.*, 1980). The temperature maxima in solarized soils were 60, 50, 42 and 37°C at soil depths of 2, 10, 20 and 30 cm respectively (Martyn and Hartz, 1986). In a field covered with transparent polyethylene 100 fm thick sheeting for 6-8 weeks from April to May, soil temperature increased by 6-10°C in 0-20 cm soil (Chauhan *et al.*, 1988). Mohamed (1990) recorded an increase in soil temperature by 9.2 and 10.0°C at a depth of 10 cm after mulching wet soil with plastic sheets for 45 days. Raj and Gupta (1995) observed an increase in temperature of 13.3°C in soil mulched with transparent polyethylene tarping (25 μm). The highest mean temperatures recorded at 6 cm depth beneath the transparent polyethylene sheets (25 μm) both under irrigated and unirrigated conditions were 50.1 and 52.2°C which were 14.3 and 12.6°C higher than in the unsolarized plots (Raj *et al.*, 1997). The maximum temperature of 55°C was recorded in the mulched plot (Dwivedi, 1993). Solarization for 40 and 45 days with transparent polyethylene mulch under irrigated condition was more effective in eliminating the pathogen up to 30 cm but the pathogen was still detected under unirrigated plots (Raj and Gupta, 1996; Raj *et al.*, 1997).

Effect on Chemical Characteristics of Soil

In tomato wilt, caused by *Fusarium oxysporum* f.sp. *lycopersici*, a direct correlation between soil pH and the disease was reported by Chand and Thakur (1969). They observed that the per cent infection decreased from 94 to 24 with the increase in soil pH from 5.4 to 7.0. In a tomato field, having a history of infestation of *Fusarium oxysporum* f.sp. *lycopersici*, solarization with clear polyethylene film raised soil pH, controlled the pathogen and increased tomato yield by 26 per cent (Overman and Jones, 1986).

Nitrate nitrogen content has also been reported to be increased in solarized soil whereas phosphorus, calcium and magnesium contents decreased (Stapleton and DeVay, 1982). Similarly, an increase in the soil concentrations of NO_3- nitrogen as well as NH_4-nitrogen to six times of those in untreated soils, has also been recorded (Stapleton *et al.*, 1985). However, solarization did not consistently affect the available K^+, Fe^{3+}, Mn^{2-}, Cu^{2+} and Cl^- concentrations in the soil. Solarization of soil within plastic bags for 4 weeks also increased the availability of nutrients such as NH_4, N, PO_4 and K in comparison to bags kept in shade for the same period (Kaewruang *et al.*, 1989).

Effect on Weeds

Katan (1976, 1980) reported almost complete control of weeds by soil solarization whereas many weeds belonging to the graminae family including the hard to manage weed *Cyperus rotundus* were completely eradicated. A reduction of 60-100 per cent in weed population with soil solarization has been reported in different crops including phanerogamic plant pathogen, Orobanche by many workers (Katan *et al.*, 1983; Lodha *et al.*, 1991, Huang and Huang, 1993; Sharma *et al.*, 2003). As compared to twenty five weed species found growing in unsolarized field plots, Sharma *et al.* (2003) reported the occurrence of only 6 weed species such as *Fumaria parviflora*, *Medicago* spp., *Melilotus* spp., *Ranunculus muricatus*, *Stellaria media* and *Vicia* spp. in plots solarized under low density, transparent polyethylene tarps. They also reported that most of the weeds growing in unsolarized plots were not recorded in solarized plots, thereby recording 100 per cent elimination whereas some weeds, though found growing both under mulched and unsolarized plots, their population was reduced by 80-90.47 per cent in solarized plots, whereas the least reduction in population was recorded in weed species of *Medicago* (7.69 per cent) and *Melilotus* (1.59 per cent). Out of 28 species found in unsolarized plots, 21 were completely eliminated in solarized plots.

Disease Control and Plant Health

Katan *et al.* (1976) found that the incidence of *F. oxysporum* f.sp. *lycopersici* was reduced by 94 to 100 per cent at 5 cm, 68-100 per cent at 15 cm and 54-63 per cent at 25 cm depth, respectively. Likewise, the incidence of *Pyrenochaeta terrestris* in onion was also reduced by 90 per cent after 195 days of planting in solarized soil (Katan *et al.*, 1980). Mortality of sclerotia of *Sclerotium rolfsii* at 5 and 20 cm depths was 100 and 25 per cent, respectively after 19 days of mulching whereas substantially higher sclerotial mortality of 80 per cent could be achieved at 20 cm soil depth by extending the mulching period for another 21 days (Elad *et al.*, 1980). The effectiveness of solar heating in controlling cotton wilt pathogen (*F. oxysporum* f.sp. *vasinfectum*) could be seen even after three years of mulching (Katan *et al.*, 1980). Subsequently Katan *et al.* (1983) corroborated that soil solarization not only effectively reduced pathogen population in soil and decreased wilt incidence in cotton, but also improved plant growth, weed control and yield.

An increase of 25 per cent in the plant height and 42 per cent in fresh weight was recorded in peach seedlings whereas in walnut seedlings 26 and 58 per cent increase in height and fresh weight was observed (Stapleton and DeVay, 1982). Soil solarization for 30 or 60 days not only reduced the total disease incidence in a *Fusarium* susceptible Cv. Sugarbaby but the effects lasted over two growing seasons (Martyn and Hartz, 1986). Solar heating with transparent polyethylene mulching controlled *Fusarium* spp. and increased plant growth and yield in tomato (Hasan, 1989). Moens and Ben (1990) observed that soil solarization with plastic covering of the soil during the warmest months of the year reduced the incidence of wilt of pepper (*Capsicum*) caused by a complex of fungi including *F. solani* with significantly higher yields than unmulched control. The per cent age of broad bean plants infected by *F. solani* decreased and plant growth increased in moistened field soil covered with polyethylene sheets for six weeks as compared to unmulched control (Sarhan, 1991). Increase in total and quality yields of 29 and 57 per cent of onion, and 79 and 65 per cent yield increases of tomatoes, 500 and 498 per cent in aubergines and 21 and 10 per cent in cucumbers in soils solarized with transparent and black plastic respectively have been recorded (Satour *et al.*, 1989; Abu Gharbieh *et al.*, 1991). Soil solarization led to an increase in yields of more than twice, as compared to untreated soil, in watermelons, 31 and 29 per cent in potato, 48 and 10 per cent in cauliflower (Gonzalez *et al.*, 1993; Abu Blan and Abu Gharbieh, 1994). Plants raised in the solarized field soil recorded a higher shoot and root length as compared to plants raised in unsolarized soil (Vannacci *et al.*, 1993). Soil solarization with transparent polyethylene mulch (25 um) in irrigated and unirrigated plots infected with citrus wilt pathogen (*F. oxysporum*) during June-July recorded higher shoot and root lengths (Raj and Gupta, 1995) They further reported that improvement in growth of mango seedlings was maximum in irrigated plots solarized with transparent polyethylene mulch (25 µm) showing that solarization under irrigated condition was more effective whereas sapling mortality due to wilt (*Fusarium solani*) was 13.7 per cent (irrigated) and 17.5 per cent (unirrigated) in solarized plots in comparison to 68.7 and 64.9 per cent respectively in unsolarized irrigated and unirrigated plots (Raj and Gupta, 1996).

Integrated Disease Management

Combined use of *T. harzianum* and soil solarization under field conditions resulted in significant control of crown rot of tomato induced by *F. oxysporum* f. sp. *radicis-lycopersici*. Treatments that resulted in disease control also significantly increased yields, with the highest yield improvement of 105 per cent over control recorded in plots following application of the antagonist in combination with soil solarization (Sivan and Chet, 1993). Minuto *et al.* (1993) also studied the effect of soil solarization in combination with *T. harzianum* as seed inoculants. The application of polyethylene mulching alone allowed significant control of *F. oxysporum* f.sp. *basilicum* on basil (*Ocimum basilicum*). They, however, observed that the integration of solar heating and the antagonist *T. harzianum*, however, did not provide a significantly different level of control from solarization alone.

Remirez and Munnecke (1988) observed that soil amendment with 9 cruciferous species significantly reduced the population of *F. oxysporum* f. sp. *conglutinans*. Solar heating (full sunlight, polythene cover) of pot soil amended with cabbage residues practically eliminated the propagules of the pathogen and cabbage Yellows was undetected. Cuciferous amendments in conjunction with solar heating were far more effective than solar heating alone. Solarization in combination with oilseed organic residues significantly reduced the population of *F. oxysporum* f.sp. *ciceri* as well as the incidence of chickpea wilt in the field trials (Pandey *et al.*, 1996). They also observed that neem oilseed meal amended soil exhibited the maximum reduction in the population of the pathogen (99.2 per cent), completely eliminated the chickpea wilt and obtained a positive correlation between the reduction in the population of the pathogen and disease incidence in oilseed meal amended solarized plants. In a tobacco nursery, soil solarization and rabbing with bajra husk greatly reduced the population of microbes including total fungi and bacteria as compared to control while green manuring with sunhemp (*Crotolaria juncea*) and ekkad (*Sesbania bispinosa*) significantly increased the populations (Patel and Patel, 1997).

Though a variety of effective ecofriendly methods are available for the management of a variety of soil-borne plant diseases caused by *Fusarium* or other fungal and bacterial pathogens as well as nematodes can hold good promise in managing, yet a judicious and convenient combinations of these methods has proved more efficacious in managing a variety of plant diseases. Hence, a choice between the combinations of the available ecofriendly options hostile against the known pathogen should always be the desirable goal of a judicious plant protection strategy.

References

AbuBlan, H.A. and Abu Gharbieh, W.I. (1994). Effect of soil solarization on winter planting of potato, cauliflower and cucumber in the central Jordan valley. *Appl. Sci.* 21: 203–213.

Abu Gharbieh, W.I., Saleh, H. and AbuBlan, H. (1991). Use of black plastic for soil solarization and post plant mulching. F.A.O. Plant Prod. and Protec. Paper No. 109, 229–242.

Adams, P.B., Papavizas, G.C. and Lewis, J.A. (1968). Mechanism of control of *Fusarium* root rot of bean with spent coffee grounds. *Phytopathology*. 58: 883.

Al–Raddad, A.M.M. (1979). Soil disinfestation by plastic tarping. M.Sc. Thesis, Univ. Jordan, pp.95.

Baker, R. (1968). Mechanisms of biological control of soil borne pathogens. *Ann. Rev. Phytopathol.* 6: 263–294.

Baker, R. (1988). *Trichoderma* spp. as plant growth regulators. *CRC Critical Rev. Biotech.* 7: 97–106.

Barkat, F.M., Mohamed, H.A. and Habib, W.F. (1983). Effect of organic matter on soil and antibiosis against *Rhizoctonia solani* and *Fusarium solani* affecting cowpea roots. *Agril. Res. Rev.* 61: 113–128.

Bourbos, V.A. and Skoudridakis, M.T. (1991). Control of tomato brown rot in green house using soil solarization. *Bull. SROP* 14: 172–177.

Brian, P.W. (1951). Antibiotics produced by fungi. *Bot. Rev.* 17: 357–430.

Byrdson, J.A. (1970). *Plastic Materials*. Iliffe Books, London, 597p.

Chakrabarti, S.K., Sen, B. and Sen, B. (1991). Suppression of *Fusarium* wilt of muskmelon by organic soil amendments. *Indian Phytopath.* 44: 476–479.

Chand, J.N. and Thakur, D.P. (1969). Soil type in relation to incidence of *Fusarium* wilt of tomato in Bihar, *Indian J. Hort.* 26: 87–88.

Chandra, S., Raizada, M. and Khanna, K.K. (1981). Effect of organic soil amendments on the rhizosphere microflora of tomato. Proc. Indian Acad. Sci., *Pl. Sci.* 90: 189–197.

Chauhan, Y.S., Nene, Y.L., Johonson, C., Haware, M.P., Saxena, N.P., Singh, S., Sharma, S.B., Sahrawat, K.L., Burford, J.R., Rupela, O.P., Kumar, J.V., Rao, D.K., Sithananthan, S. and Singh, S. (1988). Effects of soil solarization on pigeonpea and chickpea. *Res. Bull. ICRISAT* No. 11, 16p.

Chet, I. (1987). *Trichoderma*–Application, mode of action and potential as a biocontrol agent of soil borne plant pathogenic fungi. In: Innovative Approaches to Plant Disease Control (I. Chet, ed.) John Wiley Inc., New York. pp. 137–160.

Cipriano, T., Cirviller, G. and Cartia, G. (1989). *In vitro activity* of antagonistic microorganisms against *Fusarium oxysporium* f.sp.*radicis–lycopersici,* the causal agent of tomato crown root rot. *Informatore Fitopatologica* 39: 46–48.

Cobb, N.A. (1918). Estimating the nematode population of the soil. *Agric. Tech. Circ. I. Bur. Pl. Indus USDA*, 41p.

Cook, R.J. (1982). Progress towards biological control of plant pathogens with special reference to take–all of wheat. *Agric. For. Bul.* 5: 22.

Dawson, J.R., Johnson, R.A., Adams, P. and Last, F.T. (1965). Influence of steam/air mixtures, when used for heating soil, on biological and chemical properties that affect seedling growth. *Ann. Rev. Appl. Biol.* 56: 243–251.

Dennis, C. and Webster,J. (1971). Antagonistic properties of species – groups of *Trichoderma.* II. Production of volatile antibiotics. *Trans. Br. Mycol. Soc.* 57: 41–48.

DeVay, C.B. and Paparizas, G.C. (1963). Saprophytic activity of *Rhizoctonia* as affected by carbon nitrogen balance of certain organic soil amendments. *Soil Sci. Soc. Am. Proc.* 27: 164–167.

Dohroo, N.P. (1995). Integrated management of yellow of ginger. *Indian Phytopath.* 48: 90–92.

Dubey, S. C., Suresh, M. and Singh, B. (2007). Evaluation of *Trichoderma* species against *Fusarium oxysporium* f.sp. *ciceris* for integrated management of chickpea wilt. *Biol. Control* 40: 118–127

Dwivedi, S.K. (1991). Studies on population dynamics of *Fusarium oxysporium* f. sp. *lycopersici* in solar heated soil. *Nat. Acad. Sci. Lett.* 14: 235–237.

Dwivedi, S.K. (1992). Effect of culture filtrates of soil microbes on pathogens inciting wilt disease of guava (*Psidium guajava* L.) under *in vitro* conditions. *Nat. Acad. Sci. Lett.* 15: 33–35.

Dwivedi, S.K. (1993). Soil solarization adversly affects fungal pathogen causing wilt disease of guava (*Psidium guajava* L.). *Soil Biol. Biochem.* 25: 1635–1636.

Elad, Y., Katan, J. and Chet, I. (1980). Physical, biological and chemical control integrated for soil borne diseases in potatoes. *Phytopathology* 70: 418–422.

Flori, P. and Roberti, R. (1993). Treatment of onion bulbs with antagonistic fungi for the control of *Fusarium oxysporium* f. sp. *cepae. Difessadelle Piante* 16: 5–12.

Ghisalberti, E.L. and Sivasithamparam, K. (1990). Secondary metabolites produced by biological control agents. *Proceedings of the fourth International Mycological Congress,* Regensburg, fed. Rep. Germany, 230p.

Gilbert, R.G., Menzies, J.D. and Gabriel, G.E. (1968). The influence of volatile substances from alfalfa on growth and survival of *Verticillium dahliae* in soil. *Phytopathology*. 58: 1051.

Gonzalez–Torris, R., Melero–Vara, J.M., Gomez–Vazquez, J. and Jimenez–Diaz, R.M. (1993).The effects of soil solarization and soil fumigation on *Fusarium* wilt of watermelon grown in plastic houses in South–Eastern Spain. *Pl. Pathol.* 42: 858–864.

Grinstein, A., Katan, J. Abdul Razik, A, Zeydan, O. and Elad, Y. (1979). Control of *Sclerotium rolfsii* and weeds in peanuts by solar heating of the soil. *Pl. Dis. Reptr.* 63: 1056–1059.

Haque, S.E., Ghaffar, A. and Ghaffar, A. (1992). Efficacy of *Trichoderma* spp. and *Rhizobium melilotiin* the control of root rot of fenugreek. *Pakistan J. Bot.* 24: 217–221.

Hartley, C. (1921). Damping–off in forest nurseries..*U. S. Dep. Agric. Bull.* 934: 1–99.

Hasan, M.S. (1989). Soil sterilization by solar heating in Iraq. *Arab J. Pl. Protect.*7: 122–125.

Homma, Y., Kubo, C., Ishi, M. and Ohata, K. (1979). The effect of organic soil amendments on the suppression of disease severity of *Fusarium* wilt of tomatoes. *Bull. Shikou Agril. Experi. Sta.* No. 34, pp. 89–101.

Horowitz, M. and Regev, Y. (1980). Mulching with plastic sheets as a method of weed control. *Hassadeh* 60: 395–399.

Howell, C.R. and Stipanovic, R.D. (1983). Glioviridin, a new antibiotic from *Gliocladium virens* and its role in the biological control of *Pythium ultimum*. *Can. J. Microbiol.* 29: 321–324.

Huang, S. and Huang, S.H. (1993). Application of solarization for controlling soil borne fungal disease. Sustainable Agriculture: Proceedings of a symposium held at Taichung District Agricultural Improvement Station, Taiwan, 8–10. April, 1993. *Special Publication, Taichung Distt. Agril Imp. Sta.*, 1993, No. 32, 247–256.

Jacobson, R., Greenberger, A., Katan, J., Levi, M. and Alon, H. (1980). Control of Egyptian broomrape (*Orobanche aegyptica*) and other weeds by means of solar heating of the soil by polyethylene mulching. *Weed Sci.* 28: 312–316.

Jones, J.P., Overman, A.J. and Geraldson, C.M. (1966). Effects of fumigants and plastic films on the control of several soilborne pathogens of tomato. *Phytopathology*. 56: 929–932.

Jordan, V.W.E., Sneh, B. and Eddy, B.P. (1972). Influence of organic soil amendements on *Verticillium dahliae* and on the microbial composition of strawberry rhizosphere. *Ann. Appl. Biol.* 70: 139–148.

Kaewruang, W., Sivasithamparam, K. and Hardy, G.E (1989). Effect of solarization of soil within plastic bags on root rot of gerbera (*Gerbera jamesonii* L.). *Plant and Soil* 120: 303–306.

Karthikeyan, A. (1996). Effect of organic amendements, antagonist *Trichoderma viridi* and fungicides on seed and collar rot of groundnut. *Pl. Dis. Res.* 11: 72–74.

Katan, J. (1980). Solar pasteurization of soils for disease control: status and prospects. *Pl. Dis.* 64: 450–454.

Katan, J. (1981). Solar heating (solarization) of soil for control of soil borne pests. *Ann. Rev. Phytopathol.* 19: 211–236.

Katan, J. (1983). Soil solarization. *Acta Hort.* 152: 227–236.

Katan, J. (1995). Soil solarization: A non–chemical tool in plant protection. *Indian J.Mycol. Pl. Pathol.* 25: 46–47.

Katan, J., Greenberger, A., Alon, H. and Grinstein, A. (1976). Solar heating by polyethylene mulching for the control of diseases caused by soil borne pathogens. *Phytopathology* 66: 683–688.

Katan, J., Fisher, G. and Grinstein, A. (1983). Short and long term effects of soil solarization and crop sequence on Fusarium wilt and yield of cotton in Israel. *Phytopathology*. 73: 1215–1219.

Katan, J., Rotem, I., Finkel, Y. and Daniel, J. (1980). Solar heating of the soil for the control of pink root and other soil borne diseases of onions. *Phytoparasitica*. 8: 39–50.

Khalis, N. and Manoharachary, C. (1985). Studies on the microflora changes in oil cake amended and unamended soils. *Indian Phytopath.* 38: 462–466.

Khanna, R.N. (1970). Effect of organic amendments of soil on rhizosphere microflora of certain crop plants. Ph.D. Thesis, G.B. Pantnagar University, Pantnagar (U.P.).

Khanna, R.N. and Singh, R.S. (1974). Rhizosphere populations of *Fusarium* species in amended soils. *Indian Phytopath.* 27: 331–339.

Khanna, R.N. and Singh, R.S. (1975). Microbial populations of pigeonpea rhizosphere in amended soils. *Indian J. Mycol. Pl. Pathol.*5: 131–138

Kodama, T. (1989). The use of heat (solar radiation) for disease control. *Agric. Hort.* 64: 183–188.

Linderman, R.G. (1970). Plant residue decomposition products and their effects on host roots and fungi pathogenic to roots. *Phytopathology* 60: 19–22.

Lodha, S. and Lodha, S. (1995). Soil solarization, summer irrigation and amendments for the control of *Fusarium oxysporum* f.sp. *cumini* and *Macrophomina phaseolina* in arid soils. *Crop Protec.* 14: 215–219.

Lodha, S., Singh, M., Sharma, B.M., Lodha, S. and Singh, M. (1991). Solarization brings down soil borne pathogens in arid lands. *Indian Farming* 40: 12–13.

Macfayden, A. (1967). Thermal energy as a factor in the biology of soils. In: Thermology (ed. Rose, A.H.), London and New York, Academic Press, 653p.

Martyn, R.D. and Hartz, T.K. (1986). Use of soil solarization to control *Fusarium* wilt of watermelon. *Pl. Dis.* 70: 762–766.

Mahrer, Y. (1979). Predictions of soil temperature of a soil mulched with transparent polyethylene. *J. Appl. Meteorol.* 18: 263–1267.

Moens, M.and Ben, A.B. (1990). Control of pepper wilt in Tunisia. *Parasitica* 46: 103–109.

Mohamed, M.S. (1990). Effect of soil solarization on incidence of *Fusarium* wilt of broadbean (*Vicia faba*). *Assuit J. Agric. Sci.* 21: 49–58.

Mukhopadhyay, A. N. (2005). *Trichoderma*– Promises and pitfalls. *J. Mycol. Pl. Pathol.* 35: 533–534.

Overman, A.J. and Jones, J.P. (1986). Soil solarization, reaction and fumigation effects on double–cropped tomato under full bed mulch. *Proc. Fl. Sta. Hort. Soc.*1986 (publ. 1987) 99: 315–318.

Pandey, A.K., Arora, D.K., Pandey, R.R. and Srivastava, A.K. (1996). Integrated control of *Fusarium* wilt of chickpea by solar heating of soil amended with oilseed meals and fungicides. *Indian Phytopath.* 49: 247–253.

Papavizas, G.C. (1968). Survival of root rot infecting fungi in soil. VI. Effect of amendments on bean root rot caused by *Thielaviopsis basicola* and on inoculum density of the causal organism. *Phytopathology* 58: 421–428.

Papavizas, G.C. (1985). *Trichoderma* and *Gliocladium*: biology, ecology and potential for biocontrol. *Ann. Rev. Phytopathol.* 23: 23–54.

Parveen, S., Gaffar, A. and Parveen, S. (1991). Effect of microbial antagonists in the control of root rot of tomato. *Pakistan J. Bot.* 22: 179–182.

Patel, B.K. and Patel, H.R. (1997). Effect of soil solarization, rabbing, nematicides and green manuring on soil microbes in bidi tobacco nursery. *J. Env. Toxicol.*7: 42–46.

Patel, H.R., Lakshminarayana, R. and Makwana, M.G. (1995). Effect of soil solarization using clear plastic alone and with sebuphos on nematode management in bidi tobacco nursery. *Indian J. Mycol. Pl. Pathol.* 25: 51.

Pullman, G.S. and DeVay, J.E. (1977). Control of *Verticillium dahliae* by plastic tarping. *Proc. Am. Phytopath. Soc.* 4: 210 (Abstr.)

Pullman, G.S., DeVay, J.E., Garber, R.H. and Weinhold, A.R. (1979). Control of soil borne pathogens by plastic tarping of soil. In: Soil Borne Plant Pathogens (Eds. Schippers, B. and Gams, W.), London, New York, San Francisco, Academic Press 686pp.

Pullman, G.S., DeVay, J.E., Garber, R.H. and Weinhold, A.R. (1981). Soil solarization for the control of Verticillium wilt of cotton and the reduction of soil–borne population of *Verticillium dahliae, Pythium* spp., *Rhizoctonia solani* and *Thielaviopsis basicola. Phytopathology*. 71: 954–959.

Rai, P.K. and Singh, K.P. (1996). Efficacy of certain oilcake amendments on *Heterodera cajani*, *Fusarium udum* and associated wilt of pigeonpea. *Int. J. Trop. Pl. Dis.* 14: 51–58.

Raj, H., Bhardwaj, M.L. and Sharma, N.K. (1997). Soil solarization for the control of damping–off of different vegetable crops in the nursery. *Indian Phytopath.* 50: 524–528.

Raj, H. and Kapoor, I.J. (1993). Soil solarization for the control of tomato wilt pathogen (*Fusarium oxysporium* Schl.). *J. Pl. Dis. Protec.* 100: 652–661.

Raj, H. and Kapoor, I.J. (1996). Effect of oilcake amendments of soil on tomato wilt caused by *Fusarium oxysporium* f. sp. *lycopersici. Indian Phytopath.* 49: 355–361.

Raj, H. and Kapoor, I.J. (1997). Possible managements of *Fusarium* wilt of tomato by soil amendments with composts. *Indian Phytopath.* 50: 387–395.

Raj, H., Sharma, N.K. and Gupta, V.K. (1993). Effect of soil solarization on population of *Fusarium solani. Bilogical and Cultural Tests for control of Plant Diseases.* 10: 55.

Raj, H. and Gupta, V.K. (1995). Soil solarization for the control of citrus wilt (*Fusarium oxysporium*) in the nursery. *Indian J. Mycol. Pl. Pathol.* 25: 50.

Raj, H. and Gupta, V.K. (1996). Soil solarization for controlling mango wilt. *Indian J. Agric. Sci.* 66: 258–262.

Remirez, V.J. and Munnecke, D.E. (1988). Effect of soil heating and soil amendments of cruciferous residues on *Fusarium oxysporum* f.sp. *conglutinans* and other organisms. *Phytopathology* 78: 289–295.

Rathore, V.R.S., Mathur, K., Lodha, B.C. and Mathur, K. (1992). Activity of volatile and nonvolatile substances produced by *Trichoderma viride* on ginger rhizome rot pathogens. *Indian Phytopath.* 45: 253–254.

Rattink, H. (1993).Biological control of *Fusarium* crown and root rot of tomato on a recirculation substrate system. *Mededelingen van de Faculteit Landbouwwetenschappen* 58: 1329–1336.

Rojo, F. G., Ryenoso, M. M., Fezez, M., Chulze, S. N. and Torres, A. M. (2007). Biological control by *Trichoderma* species of *Fusarium solani* causing peanut brown root rot under field conditions. *Crop Protec.* 26: 549–555.

Sarhan, A.R.T. (1991). Control of *Fusarium solani* in broad bean by solar heating of the soil in northern Iraq. FAO Plant Prod. Protec. Pap. No. 109, 108–117.

Satour, M.M., Abdel Rahim, M.F., El Yamani, T., Radian, A., Grinstein, A., Robinowitch, H.D. and Katan, J. (1989). Soil solarization in onion fields in Egypt and Israil: short and long term effects. *Acta Hort.* No. 255, 151–159.

Sharma, P. and Dureja, P. (2004) Evaluation of *Trichoderma harzianum* and *T. viride* at BCA pathogen interface. *J. Mycolol. Pl. Pathol.* 34: 47–55.

Singh, N. and Singh, R.S. (1980). Lysis of *Fusarium oxysporum* f.sp. *udum* caused by soil amendment with organic matter. *Indian J. Mycol. Pl. Pathol.* 10: 46–150.

Singh, N. and Singh, R.S. (1982). Effect of oilcake amended soil atmosphere on pigeonpea wilt pathogen. *Indian Phytopath.* 35: 300–305.

Singh, R.S. (1983). Organic amendments for root disease control through management of soil microbiota and the host. *Indian J. Mycol. Pl. Pathol.* 13: 1–16.

Singh, R.S., Singh, D. and Singh, H.V. (1997). Effect of fungal antagonists on the growth of chickpea plants and wilt caused by *Fusarium oxysporum* f. sp. *ciceri*. *Pl. Dis. Res.* 12: 103–107.

Singh, R. and Pandey, K.R. (1966). Effect of green and mature plant residues and compost on population of *Pythium aphanidermatum* in soil. *Indian Phytopath.* 19: 367–372.

Sivan, A. and Chet, I. (1986). Possible machanisms for control of *Fusarium* spp. by *Trichoderma harzianum*. British Crop Protection Conference, 1986. *Pests and Diseases* 2: 865–872.

Sivan, A. and Chet, I. (1993). Integrated control of *Fusarium* crown and root rot with *Trichoderma harzianum* in combination with methyl bromide or soil solarization. *Crop Protec.* 12: 380–386.

Sivan, A., Ucko, O. and Chet, I. (1987). Biological control of *Fusarium* crown rot of tomato by *Trichoderma harzianum* under field conditions. *Pl. Dis.* 71: 587–592.

Son, H.S., Shin, H.S. and Lew, M.W. (1985). Effects of amendments on ginseng root rot caused by *Fusarium solani*: Population changes of microorganisms in soil. *Korean J. Mycol.* 13: 41–47.

Staplaton, J.J. and DeVay, J.E. (1982). Effect of soil solarization on populations of selected soil borne microorganisms and growth of deciduous fruit tree seedlings. *Phytopathology* 72: 323–326.

Staplaton, J.J., Quick, J. and DeVay, J.E. (1985). Soil solarization: effects on soil properties, crop fertilization and plant growth. *Ann. Biol. Biochem.* 17: 369–373.

Tamietti, G. and Garibaldi, A. (1980). Control of corky root in tomato by solar heating of the soil in greenhouse of Riviera Li Gure. *Le defesesa delle Piente.*

Tamura, A., Kotani, H. and Naruto, S. (1975). Trichoviridin and dermadin from *Trichoderma* spp. TK–I *J. Antibiotics.* 28: 161–162.

Tjamos, E.C. and Faridis, A. (1980). Control of soil borne pathogens by solar heating in plastic houses. *Proc. 5th Congr. Mediterr. Phytopath. Union*, Hellenic Phytopathol. Soc. Athens, pp. 82–84.

Triolo, E., Vannacci, G. and Materazzi,A. (1988). Soil solarization in vegetable production. Part 2. Studies on possible mechanisms of the effect. *Colture Protelle* 17: 59–62.

Ushamalini, C., Rajappan, K. and Gangadharan, K. (1997). Control of cowpea wilt by non–chemical means. *Pl. Dis. Res.* 12: 122–129.

Vannacci, G., Panattoni, A., Materazzi, A. and Triolo, E. (1993). Experiments in soil solarization for the control of *Fusarium oxysporum* f.sp. *melonis* in protected cultivation. *Colture Protette* 22: 69–72.

Verma, V. S. and Sharma, S. K. (2007). Evaluation of biocontrol agents against *Fusarium solani* causing wilt of mango seedlings. *Journal of Mycology and Plant Pathology* 37: 393–397.

Verma, V.S. and Sharma, S. K. (2004). Effect of plant materials on the management of mango wilt caused by *Fusarium solani*. *Indian Phytopathology* 57: 368.

White, J.G. and Buczacki, S.T. (1979). Observation on suppression of clubroot by artificial or natural heating of the soil. *Trans. Br. Mycol. Soc.* 73: 271–275.

Plant Diseases Management in Horticultural Crops (2011) *Pages* 225–242
Editors: **Shahid Ahamad, Ali Anwar and P.K. Sharma**
Published by: **DAYA PUBLISHING HOUSE, NEW DELHI**

Chapter 18

Integrtated Diseases Management of Chilli

V.B. Nargund[1], *A.P. Shivakumar*[1], *A.S. Byadagi*[1] *and Jameel Akhtar*[2]

[1]*Department of Plant Pathology, University of Agricultural Sciences, Dharwad – 580 005, Karnataka*
[2]*Department of Plant Pathology, Birsa Agricultural University, Ranchi – 834 006*

Introduction

Chilli (*Capsicum annuum* L.) is the universal spice of India, belong to family solanaceae. Chillies are native of central and southern America and were introduced to India by Portuguese in the later part of sixteenth century. India is the major and highly erratic producer, consumer and exporter of chilli. The chilli fruits do not undergo any processing other than drying and grinding before being added to foods. They are often damaged by several fungi such as, *Colletotrichum spp., Alternaria spp.* and *Fusarium* spp. (Prabhavathy and Reddy, 1995). Several reports also indicate the deterioration of chilli seeds by seed borne fungi (Mridha and Chowdhuary, 1990). It is so often estimated that around 30 per cent of looses occur in horticultural produce of faulty handling and storage (Salunkhe, 1985, Angadi, 1999 and Shivakumar, 2006).

Anthracnose caused by *Colletotrichum* spp. is one of the most serious diseases of chilli worldwide. The fungus was responsible for seed rot, seedling decay, dieback and fruit rot (Shivakumar, 2006) lead to direct reduction in yield. While quiescent infection of fruits leads to post-harvest disease development that can render the fruits unfit for marketing and consumption. *Alternaria alternata* (Fr.) Keissler was found responsible for severe seed rot, seedling decay, tender twig tip drying and fruit rot in chilli (Shiva kumar, 2006) and black/green/yellow mold caused by *Aspergillus* spp. are the other important diseases in India.

In the present review, post harvest diseases of chilli, its occurrence, severity, pathogens associated, and management of the disease by screening of genotypes for resistance and use of chemicals, plant extracts, bio-agents, hot water and salt solutions are included.

Anthracnose or fruit rot caused by *Colletotrichum capsici* (Sydow.) Butler and Bisby. is the major pre and post harvest diseases of chilli (Azad, 1992; Sulochana *et al.*, 1992; Singh *et al.*, 1993; Basak, 1998; Roy *et al.*, 1998; Hegde and Kulkarni, 2001). Alternaria rot or black rot caused by *Alternaria alternata* (Fr.) Keissal, *Curvularia lunata* (Prabhavathy and Reddy, 1995), black mold rot caused by *Aspergillus niger* Van Tiegh, Rhizopus rot caused by *Rhyzopus stolonifer* (Prasad *et al.*, 2000) and Fusarium rot caused by *Fusarium semitectum* (Basak, 1994) were the other important diseases which contribute to the post harvest losses of chilli.

History and Distribution

Anthracnose

The fungus, *Colletotrichum capsici* was reported for the first time in India by Sydow on chillies from Coimbatore of Madras presidency in 1913. Since then it has been reported and described from several parts of the world (Butler, 1918; Sydow, 1919; Dastur, 1921; Seaver *et al.*, 1932; Marchionatto, 1935; Ling and Lin, 1944; Thind and Jhooty, 1985; Hegde and Kulkarni, 2001; Menugupta and Garg, 2002) Akthar (2007). Arx (1957) noted *C. capsici* as a synonym of *Colletotrichum dematium* (Pres. Ex. Fr.) Groove.

Choudhury (1957) reported that this disease as quite serious and wide spread in Assam.

Thind and Jhooty (1985) noticed *C. capsici* as predominant fungus causing fruit rot of chilli. The incidence varied between 66-84 per cent. Whereas, Mathew *et al.* (1995) made survey during 1989-91 in Vellanikkara, Trichur, Kerala and reported that dieback and fruit rot were serious problems during rainy season. Ripe fruit rot was more conspicuous as it caused 10-15 per cent loss to mature fruits during transit and storage (Bagri *et al.*, 2004).

Khodke and Gahukar (1995) isolated *C. gloeosporioides* (*Glomerella cingulata*) from ripe chilli fruits in pure culture and its pathogenicity was confirmed. This was considered to be the first record of this species occurring on chilli in Maharashtra, India.

Patil *et al.* (1993) studied on losses due to dieback and fruit rot disease on capsicum varieties. A variety with large fruit with thick pericarp was affected less and gave higher yield, compared with varieties with smaller fruits and thin pericarp.

Basak (1994) recorded the fruit rot disease on chilli fruits in farmers field. Disease reduced dry weight compared with healthy fruits. Fusarium rot caused the greatest reduction and *Alternaria* spp. the least.

Black rot disease on chilli fruits collected from various markets in the Warangal district, Andhra Pradesh, India was found to be caused by three species of fungi *Alternaria alternata*, *Curvularia lunata*, *C. pallescens* and other fungi for the first time Prabhavathy and Reddy (1995).

The data collected from some studies carried out in India on post harvest diseases of chilli are put to an extent of 10 to 15 per cent (Datar, 1995; Bagri *et al.*, 2004). Three to 31 per cent loss was recoded (Shivakumar, 2006) and Das *et al.*, 2004, recorded 10 to 60 per cent.

Ekbote (2002) surveyed on prevalent diseases of chilli (*Capsicum annuum*) in six taluks in the Haveri district of Karnataka, India. Fruit rot caused by *C. capsici* was most prevalent.

Symptomatology

Anthracnose

The appearance of a small black circular spot sharply defined but at times diffused. The disease spread was more in the direction of long axis of the fruit, so that original circular spot becomes

Symptom on Stalk of Fruit

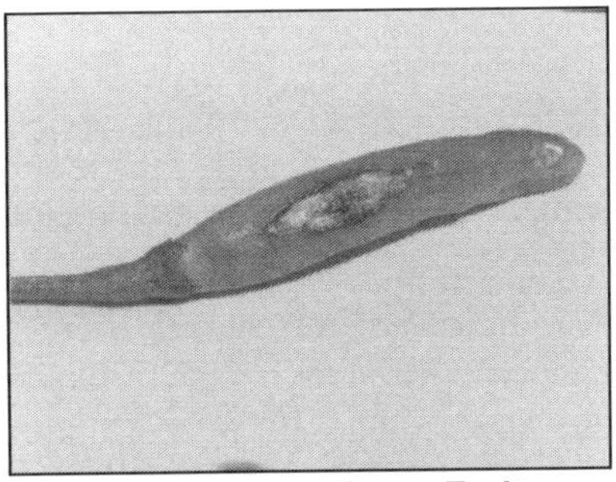

Symptom on Green Fruit

Die Back Initial Symptom

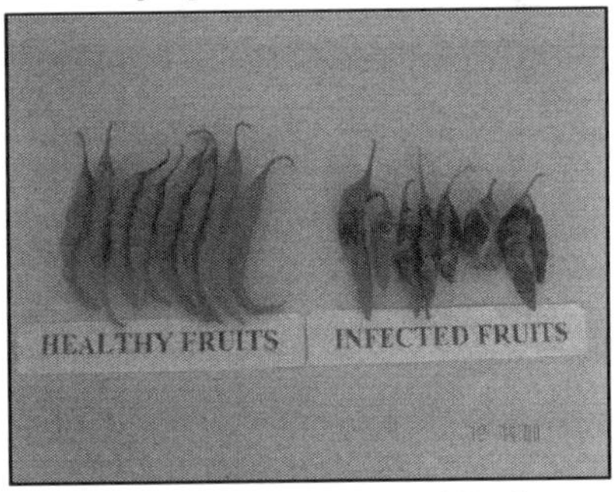

Symptom on Matured Fruits

Severe Die-Back Symptoms

Figure 18.1: Symptoms of Fruit Rot of Chilli

elliptical. As the infection progressed the spots showed either diffused black, greenish black or dirty gray in colour, markedly delimited by a thick sharp black outline enclosing a dark lighten black or straw coloured area. Two or more spots coalesced to form bigger spots. Diseased fruits lost their normal red colour and turned straw coloured or in some cases white. When diseased fruit was cut open the lower surface of the skin was found to be covered with minute, black, spherical elevations. In advanced cases, seeds were covered by a mat of mycelium (Choudhury, 1957).

Khodke and Gahukar (1995) described the disease symptoms of fruit rot of chilli (*Colletotrichum gloeosporioides*) as depressed sunken, discoloured, circular to irregular spots of varying sizes. Oh *et al.* (1998) observed that initial anthracnose symptoms were detected on some green fruits at two days after inoculation resulting in typical sunken necrosis within five days after inoculation.

Ripe fruit rot phase of the disease occurs when chilli fruits just start maturing (turning red). However, anthracnose symptoms do occur frequently on well developed matured green fruits as well. Initial development of symptom on fruits starts first after the formation of depressed lesions of variable sizes *i.e.* from mere dots to those covering most of the fruit surface areas (Figure 18.2). In anthracnose caused by *C. capsici*, numerous long black setae, infected fruits appear black (Smith and Crossan, 1958). Fully developed lesions are sunken and appear dark red to light tan with varying amount of visible dark acervuli of the fungus. The pathogen also colonizes frozen green fruits because of leakage of carbohydrates, proteins and electrolytes.

Park *et al.* (1989) observed that susceptibility of red fruit is due to higher amount of total carbohydrate, reducing sugars and total amino acids compared to their amounts in green fruits. Sucrose is the main carbohydrate in green fruits while glucose and fructose are dominant in red fruits.

Colletotrichum coccodes has been reported to cause anthracnose of capsicum seedlings. The initial spots that develop become sunken, coalescing to form large irregular lesions and finally showing blight symptom. Severe infection may lead to defoliation and lodging of the plants (Oh *et al.*, 1988).

The lower portion of infected fruit skin is invariably found covered with minute black, spherical elevated stromatic masses or sclerotia of the fungus. In advanced stages, the seeds are covered with felt

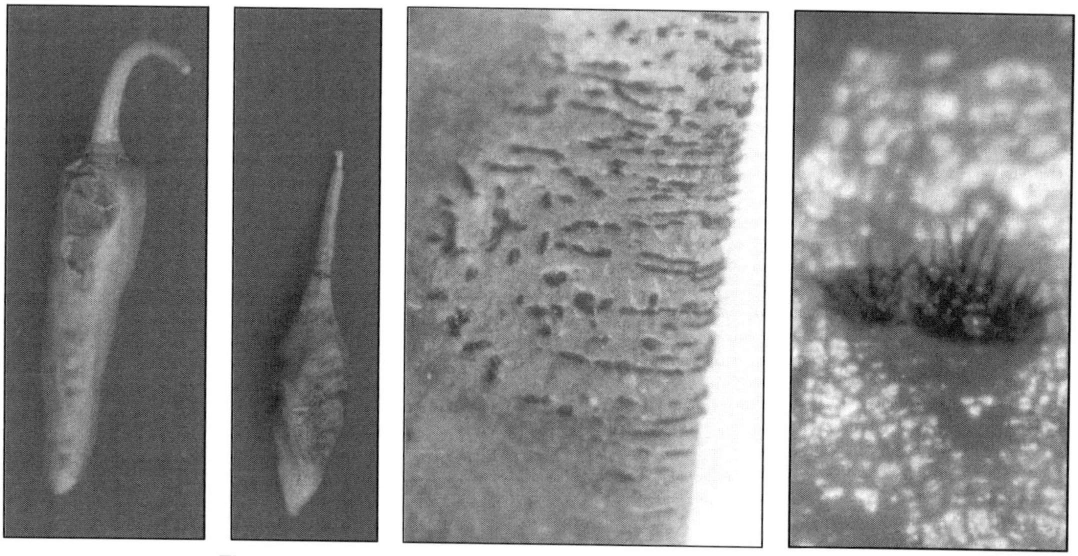

Figure 18.2: Fruit-rot and Anthracnose Symptoms on Chilli

of white mycelium in which are embedded a few black to grey green stromatic bodies. Infected seeds turn rusty in color. (Choudhary 1957; Rai and Chohan, 1966; Akhtar and Singh, 2007).

Alternaria Rot

Alternaria alternata (FR: FR) Keissal was found to be responsible for severe seed rot, seedling decay, tender twig tip drying and fruit rot in chilli. The seed borne nature of the fungus was also tested by standard blotter method. Among the sixteen fungi isolated *Alternaria alternata* was found and 78 per cent of harvested chilli fruits were infected by this mold and fungus also produced toxic metabolites such as alternariol and alternariol mono methyl ether (Vijayalakshmi and Rao, 2003).

Aspergillus Rot

Black or green or yellow mold caused by *Aspergillus* spp. on the surface of chilli fruits and in severe condition, the moldy growth covers the entire fruits and become seed borne. A global survey carried out in different parts of the world revealed that maize and groundnut are most affected by the mycotoxins followed by chilli (Hesseltine, 1986).

Stored chilli fruits exhibited varying degree of decay with respect to discolouration and breaking of the pericarp, detachment of the pedicel and spore dust formation within the fruits due to infection by *Aspergillus flavus, A. terreus, A. candidus* and *A. niger* (Prasad *et al.*, 2000).

Fusarium Rot

(Basak, 1994) obtained that Capsicum fruit rot caused by fungus, *Fusarium* spp. which causes greatest reduction in dry weight followed by *Alternaria* spp. Whereas, the tip rot caused by *Fusarium semitectum* was reported in chilli from Andhra Pradesh (Prabhavathy and Reddy, 1995).

Rhizopus Rot

Prabhavathy and Reddy (1995) conducted a survey on post harvest fungal diseases of fleshy (green and ripened) chilli fruits in various markets of Warangal district, Andhra Pradesh during 1990-92. Among the twenty four fungal species, Rhizopus rot caused by *Rhizopus stolonifer* on chilli fruits are recorded and inflicting heavy loss both in green and ripened chilli fruits.

Causal Agents

Colletotrichum capsici (Sydow.) Butler and Bisby. and *Colletotrichum gloeosporioides* (Penz) Penz and Sacc.

The fungus *Colletotrichum* was described by Corda (1831-32) under the name *Colletotrichum* with a single species. Later, Saccardo (1884) and Potebinia (1910) placed *Colletotrichum* in Melanconiales and Acervulales respectively.

Earlier reports revealed that *Colletotrichum gloeosporioides,* which has been considered a conidial stage of *Glomerella cingulata* (Small, 1924 and 1926) subsequently, Arx (1957) made a detailed study on the species of the genus *Colletotrichum* and assigned the ascogenous state of *Colletotrichum gloeosporioides* as *Glomerella cingulata*, and noted *Colletotrichum capsici* as a synonym of *Colletotrichum dematium* (Pres. Ex. Fr.) Groove.

Mesta (1996) studied the comparative measurement of acervuli, setae and conidia of *Colletotrichum capsici* obtained from host and potato dextrose agar (PDA) culture. Acervuli were gregarious, abundant, circular and saucer shaped and diameter of the acervuli measured 53.3-136.4 µm in host and 71.0-161.9 µm in culture. Setae were dark in colour but with paler at the apex, swollen at the base and tapering at the apex and more or less errect. Length of setae measured 48.3-140.0 µm in host and 69.3-172.4 µm in culture. Conidia were one celled, hyaline, smooth walled with a central oil globule, sickle shaped, tapering gradually at both the ends and with acute apex. Breadth of conidia measured 13.6-18.3 µm X 2-4 µm in host and 14.3-19.2 X 2.5-4.2 µm in culture.

Colletotrichum gloeosporioides fungus with grayish white to dark gray on Potato dextrose agar (PDA) and may produce aerial mycelium ranging from a thick mat to sparse or tufts of mycelium. Conidia are hyaline, unicellular and either cylindrical with obtuse ends or elliptical with a rounded apex and narrow, truncate base. They are 7-20 X 2.5-5 µm and were formed on hyaline brown conidiophore in acervuli that were irregular in shape and about 500 µm in diameter. Setae were 4-8 X 200 µm, one to four septate, brown and slightly swollen at the base and tapered at the apex. Orange slimy conidial masses can be formed as the acervuli matures.

Alternaria alternata (Fr.) Keissler.

The conidiophores were dark, simple, rather short or elongate, typically with both cross and longitudinal septa; variously shaped, obclavate to elliptical or ovoid, frequently borne acropetally in long chains, less often borne singly and having apical simple or branched appendages; parasitic or saprophytic (Dodd *et al.*, 1991).

Aspergillus niger V. Tieghem

Aspergillus produces large typical aspergilloid heads with long metulae and short phialides. The conidia were block and round 4.0-5.0 µm diameter: no sexual state was known (Dodd *et al.*, 1991).

Other Fungi Associated with Postharvest Diseases of Chilli

Basak (1994) reported the six major fruit rot on capsicum fruits caused by *Alternaria capsici* and *A. tenuis*, *Cercospora capsici*, *Colletotrichum capsici*, *C. gloeosporioides*, *Fusarium* spp. and *Periconia byssoides* from farmers' fields. Among them Fusarium rot caused the greater reduction in dry weight of fruits and it was maximum at the ripened stage.

Prabhavathy and Reddy (1995) isolated fungi causing black rot disease on chilli fruits *viz.*, *Alternaria alternata*, *Curvularia lunatus*, *C. pallescens*, *Pythyium butleri*, *Botryodiplodia theobromae*, *Phomopsis equiseti*, *Rhizopus stolonifer*, *Fusarium semitectum* and *Choanephora cucurbitarum* for the first time in Andhra Pradesh. Incidence and severity of *Alternaria alternata* on chilli fruits causing fruit rot is also a problem in recent years, its pathogenicity was confirmed on chilli fruit (Khodke and Gahukar, 1993). *Phoma sorgina* was reported from *Capsicum annuum* (Khare *et al.*, 1995).

Khodke and Gahukar (1995) isolated *Colletotrichum gloeosporioides* from ripe chilli fruits and its pathogenicity was confirmed. The pathogen *Aureobasidium pullulans*, which causes fruit rot in chilli, is recorded for the first time in India and the fungus was characterized by superficial black mycelium (Suryawanshi and Deokar, 2001).

Anthracnose of bell pepper caused by *C. capsici* incidence was observed from 3 to 15 per cent. The most severe disease occurred on ripened fruits and pathogenicity was confirmed by tooth prick inoculation method (Roy *et al.*, 1997).

Cultural Studies of *Colletotrichum* spp.

Misra and Mahmood (1960) observed that chilli leaf extract was found to be the best for germ tube growth of *Colletotrichum capsici* whereas Richard's medium for mycelial growth and Czapek's medium for sporulation.

The population of conidia of *Colletotrichum capsici* varied with the seasons, age of the host plants and characteristics of the substrate. The colonies of fungus was peak (8.7 colonies/cm^2) in the month of December with simultaneous decrease of temperature (18° C) and RH (75 per cent) and the population was more on fruit compared to leaves and flowers (Gupta *et al.*, 1983).

Ekbote *et al.* (1997) reported the maximum radial growth (90 mm) of *C. gloeosporioides* on Richard's, Brown's and Potato Dextrose Agar medium followed by Czapek's agar (89.6 mm). Richard's medium also supported good sporulation. The maximum dry mycelial weight of the fungus was recorded on the 12th day after seeding (426 mg) in potato dextrose broth.

Palarpawar and Ghurde (1997) reported that the peptone was remarkably good source of nitrogen for the growth of *C. capsici*. However, urea and ammonium oxalate proved to be the poorest sources of nitrogen for the growth of fungus. On contrary, urea was the best source of nitrogen, followed by ammonium carbonate and sodium nitrate to *C. curcumae* and excellent sporulation was also observed.

Nutritional Studies of the Pathogen

Maximum spore germination took place in neutral solution and in 1 per cent sucrose solution. However, leaf extract and leaf decoction media were most suitable substrates for the growth and sporulation (Mishra and Mahmood, 1960).

Akhtar and Singh (2007) observed colonies of *C. capsici* on five culture media. In general, the growth of mycelium production of conidia and setae on potato dextrose medium was better than on other media. Five monoconidial isolates of *C. capsici* differ in producing acervuli and spores on different culture media. This signifies the fact that different culture media do show differences in supporting growth and devlopment.

Akhtar and Singh. (2007) have also reported significant variations in the acervular density, conidial dimensions and germination among the five isolates of *C. capsici* derived from different geographical locations of Uttarakhand.

Two isolates of *Alternaria alternata* were grown on six different media, maximum growth and sporulation of both the isolates was recorded on Richard's medium followed by Czapek's Dox and modified Dox (Mathur and Sarbhoy, 1977).

Epidemiology of the Disease

Effect of Temperature and Relative Humidity (RH)

Choudhury (1957) recorded that, temperature of 28°C and relative humidity of 92 per cent was required for optimum growth of the pathogen under laboratory condition.

Padhi and Rath (1973) reported that the sporulation of *A. solani* was induced by exposing three day old culture to direct sunlight for ten minutes. Light affected the oxidative process of the fungus for inducing sporulation.

Mathur and Sarbhoy (1977) suggested that the optimum temperature for linear growth of two isolates of *A. alternata* at 30°C followed by at 25°C. However, excellent sporulation was recorded at 30° C of both isolates.

The incidence of *Aspergillus niger* was found to be high as seed rot and seedling emergence in hot climate. Seed rot and seedling emergence were usually severe at 30 and 35° C as compared to low temperature (Hayden and Maude, 1992).

The disease incidence varied with changes in climatic factors such as, temperature, humidity and rain fall and also with the stages of fruit development, location and growing years (Basak, 1994).Temperature affects the fruit rot disease of chilli. At 0 and 5°C, the fungal pathogen was completely arrested and at 10 and 15°C the disease development was slow but it increased at the temperature between 20 and 30°C (Datar, 1995). The temperature of 20 and 25°C were the most favorable for colony growth and sporulation of *Colletotrichum gloeosporioides*. Greater sporulation was obtained under incubation in constant light and at 25° C the greatest number of conidia and colony growth on oat medium was recorded (Mello *et al.*, 2004).

Extent of Seed Contamination

Surekha *et al.* (1990) screened capsicum seeds infected by *Colletotrichum dematium* and was mostly confined to seed coat and superficial layers of endosperm. Where as in heavily infected seeds inter and intracellular mycelium and acervulli were present throughout. It causes disintegration of paranchymatous layers of the seed coat, depletion of food materials in two endosperm and embryo, cell division in the palisade cells and lysigenous cavities in the embryo. Fruits with 76-100 per cent infected seeds but in those with only 1-25 per cent more than 92 per cent of the seed was still infected. In 85 per cent of infected seeds, the fungus was detected in the embryo region (Gahukar *et al.*, 1989). The incidence of *C. dematium* in chili seeds ranged from 0.5 to 69.5 per cent and it was found in all the seed components and the seed embryo (Banu *et al.*, 1990).

Lokesh and Shetty (1990) screened the seed samples of four hybrids and showed the presence of *C. dematium*. Both captan and thiram were effective in reducing the incidence of *C. dematium*.

Padaganur and Naik (1991) isolated the mycoflora from the seeds of diseased and apparently healthy fruits. Among the different fungi isolated most of them were eliminated by surface sterilization of the seed, the few exceptions including *C. capsici*, which is externally seed borne while *Fusarium* spp. appears to be internally seed borne to some extent. Further they studied, on the incidence of *C. capsici* (75.5 per cent) in chilli seeds collected from diseased fruits followed by *Fusarium* spp. (16.25 per cent) and *Alternaria* spp. (5 per cent). While, seeds collected from apparently healthy fruits yielded 31.75 per cent of *Alternaria* spp., 16.5 per cent of *Aspergillus flavus*, 14.5 per cent of *Fusarium* spp. and lower per cent of *Curvularia* spp., *Colletotrichum* spp., *Helminthosporium* spp. and *Cladosporium* spp.

Basak (1994) identified seventeen fungi (11 genera) from chilli seeds included *Alterrnaria capsici*, *A. solani*, *Colletotrichum gloeosporioides*, *Phomopsis capsici*, *Bipolaris* spp. and *Verticillium* sp. Five *Curvularia* species were detected more frequently in the seed coat, in decreased amounts in the endosperm and rarely in the seed embryo (Dhyani *et al.*, 1993). The different species of *Fusarium* was isolated by blotter method in chilli seed samples and also observed the seedling mortality after 30 days (Liang, 1990).

Jindal *et al.* (1994) observed the mixed infection in the seed samples of *Capsicum annuum* which included *Colletotrichum capsici*, *Phytophthora nicotianae* var. *nicotianae* and *Xanthomonas campestris* pv. *vesicatoria*. Lesser population of fungi and bacteria were recorded in seed samples from healthy fruits and showed the 71 per cent of germination as compared to seeds from diseased fruits, which was recorded highest fungi and bacterial population and lowest seed germination (17.3 per cent) (Krishna *et al.*, 1994).

Among the sixteen fungi, *Colletotrichum dematium* and *Alternaria alternata* associated with chilli seeds were found responsible for severe seed rot and seedling decay. Standard blotter method was better than agar plate method for their detection. Pre and post emergence mortality was also recorded due to fungi (Shivakumar, 2006).

The association of thirteen fungal species in bell pepper seeds by standard blotter and agar plate method, showed the incidence of seed borne fungi varied between 0-34 per cent. *A. flavus* and *C. capsici* were the major associated fungi resulting in poor seed germination and causing up to 77 per cent inhibition and low seedling vigour index (Nandita and Sunita, 2004).

Disease Management

Chemical Control

Jayasekar *et al.* (1987) studied the efficacy of different fungicides at varied concentrations to control chilli fruit rot caused by *Colletotrichum capsici* and *Alternaria solani*. Foltaf 0.2 per cent (captafol) gave the most effective control followed by Fytolan 0.25 per cent (copper oxychloride) and Bavistin (carbendazim) 0.1 per cent. The artificially inoculated *Colletotrichum capsici* to chilli seedlings was most effectively controlled by spraying with Dithane M- 45 (mancozeb) and was followed by Dithane Z- 78 (zineb) (Malraja and Narayanaswami, 1998).

The incidence of *Colletotrichum capsici* was reduced by Bavistin (carbendazim), Dithane M- 45 and Fenapanil. Among the different fungicides, fruit rot was least after seed treatment with thiram and spraying with mancozeb (Perane and Joi, 1989).

The four fungicides were tested for the control of fruit rot of chillies. Among the tested fungicides, Iprobenfos showed effectiveness in reducing the incidence at 0.05, 0.1 and 0.15 per cent. The residues were not found in dry chillies in any treatment of iprobenfos (Sharma and Thakore, 1999).

Hegde and Anahosur (2001) reported the efficacy of four each of non-systemic (mancozeb, chlorothalonil, copper oxychloride and iprodione at 0.3 per cent) and systemic fungicides (carbendazim, triadimefon, propiconazole and hexaconazole at 0.1 per cent).

Suryawanshi and Deokar (2001) reported on controlling the fruit rot of chillies under *in vitro* conditions. Carbendazim (0.1 per cent) and copper oxychloride (0.25 per cent) completely inhibited the growth and sporulation of *F. oxysporum* whereas carbendazim (0.1 per cent) and thiram completely inhibited the growth and sporulation of *F. moniliformae* and mancozeb (0.25 per cent) and carbendazim (0.1 per cent) gave the highest per cent inhibition of *Aureobasidium pullulans* and *Aspergillus niger* respectively.

The mycelial growth of the fungus was significantly inhibited by hexaconazole (85.29 per cent) followed by propiconazole (80.97 per cent) and triadimefon (79.67 per cent) (Hegde *et al.*, 2002). The tricyclazole was very effective under laboratory condition against *Colletotrichum gloeosporioides* among all tested fungicides (Venkataravanappa and Nargund, 2002).

Yenjarappa *et al.* (2002) studied the efficacy of thirteen fungicides and two biological control agents against fruit rot of chilli (*Colletotrichum capsici*). Among the different treatments hexaconazole (0.1 per cent) recorded the least incidence followed by benomyl (0.1 per cent), ziram (0.3 per cent) and propiconazole (0.1 per cent). These fungicides were on par with each other and showed significantly superior efficacy to other fungicides and biological control agents.

Ekbote (2003) evaluated different fungicides for the control of fruit rot and dieback of chilli (*Colletotrichum capsici*). Among the fungicides tested, prochloraz 45 EC at 0.125 per cent showed the least per cent disease index (PDI) for dieback (9.7) and fruit rot (12.1) with highest yield (6.2 q/ha).

Mandal and Beura (2003) recorded the reduced per cent disease index (PDI) of anthracnose (2.5) and ripe fruit rot (3.7) of chilli by spray with 0.05 per cent monocrotophos + 0.25 per cent mancozeb.

Hingole and Kurundkar (2004) reported that mancozeb + metalaxyl 72 WP (0.25 per cent) was effective against fruit rot caused by *Colletotrichum capsici*, which was most economical and profitable fungicide, giving a cost: benefit ratio of 1: 15.36.

Joi *et al.* (2004) employed different treatments for the control of anthracnose and fruit rot of chilli. Ekbote (2003)reported the significance of tebuconazole in the management of fruit rot of chilli.

Plant Products

In recent years, the increasing use of potentially hazardous fungicides in agriculture has been the subject of growing concern of both environmentalists and public health authorities. Therefore, hunt for renewable sources of alternative chemotherapeutants is highly desirable. Green plants appear to be the reservoir of effective chemotherapeutants and can provide renewable sources of useful biofungicides.Jeyalakshmi and Seetharaman (1998) reported that Palmarosa (*Cymbopogon martinii*) oil was most effective for reducing the fruit rot incidence followed by *Ocimum sanctum* leaf extract, Neem (*Azadirachta indica*) oil and *Saccharomyces cerevisiae*.

The leaf extracts of *Datura metal* (10 per cent), reduces the *C. capsici* incidence on chilli seedlings. The activities of peroxidase, polyphenol oxidase, polymethylgalacturonase and cellulase were reduced in treated seedlings (Asha and Kannabiran, 2001).

The ethanolic root extracts (10 per cent) of 18 different plant species were evaluated against anthracnose of chilli. Among the different ethanolic root extracts, *Abrus precatorius* (53.5 per cent) and *Rauwolfia tetraphylla* (42.9 per cent) showed significantly inhibitory effects on both the conidial germination and radial growth of *C. capsici* (Kumaran *et al.*, 2003).

The antimicrobial properties of Bitter Temru (*Diospyros cordifolia*), Datura (*Datura stramonium*), Amaltas (*Cassia fistula*), Brhati (*Solanum indicum*), Sandal (*Santalum album*), Mehandi (*Lawsonia inermis*) and Babool (*Acacia nilotica*) were tested for chilli fruit rot management. Disease severity was minimum with Bitter Temru extract (6.5 per cent) followed by Datura leaf extract. Disease severity was maximum with Babool seed extract (34.2 per cent) (Bagri *et al.*, 2004).

The botanicals *viz.*, NSKE, Nimbicidin, Rakshak and Pongamia oil at five per cent concentration were found effective in inhibiting the mycelial growth of aflatoxin fungus. However, in post harvest treatments, maximum control was achieved by treating fruits with cooking soda at 5 per cent concentration and also with hot water (Ajithkumar *et al.*, 2005).

Growth Regulators

Datar (1996) evaluated the effects of 4 growth regulators and fungicides on the chilli fruit rot pathogens. Fruit dipping in indole-3-butyric acid or naphthalene acetic acid at 200 μg/l for 30 minutes delayed the fruit rot caused by most of the pathogens by 6 days. However, 10 minutes dipping of chilli fruits in carbendazim (1000 μg/ml) solution effectively controlled all the pathogens.

Among the different growth regulators, GA_3 at 100 ppm spray at flowering stage on Byadagi Kaddi has recorded higher fruit and seed yield with better quality seeds (Natesh *et al.*, 2005).

Biological Control

There is an urgent need for increased crop production to feed the world's teeming million will force for quick results perhaps away from biological control. It has been now realized that lasting success cannot be achieved by poisoning the environment and is increasingly turning to "Natural control" by restoring a biological balance favorable to his crops.

Sanford (1926) was the first to realize that soil microorganisms themselves exert a natural biological control on diseases. Then onwards, there has been a dramatic increase in research reports and several books (Baker and Cook, 1974; Huffaker and Messenger, 1976; Cook and Baker, 1983) and review articles (Baker, 1968; Papavizas and Lumsden, 1980) have come up, covering the use of specific microbes, their mechanisms of active and commercial evaluation.

Trichoderma spp. represents the most thoroughly studied fungi amongst the antagonistics towards plant pathogens. The first of brilliant series of papers on parasitism of *T. viride* was that of Weindling (1932).

Kolte *et al.* (1994) reported the ecofriendly management practices of dieback and fruit rot of chillies by dipping the seedlings in a suspension of *Azospirillum* sp., gave the best control of shoot dieback compared with *Azotobacter* dip. Yield of green and red chillies were highest for mixed dip of both *Azotobacter* + *Azospirillum*. Ramamurthy and Samiyappan (2001) reported that *P. fluorescens* isolate Pf1 effectively inhibited the mycelial growth of the pathogen under *in vitro* conditions by producing maximum amount of indole acetic acid leads to decreased fruit rot incidence under green house conditions. Seed treatment plus soil application of talc-based formulation of *P. fluorescens* isolate Pf1 effectively reduced the disease incidence.

Thirteen plant growth promoting antagonistic rhizobacterial (PGPR) strains were evaluated against chilli fruit rot and dieback incited by *C. capsici*. Among them *Pseudomonas fluorescens* (Pf1) and *Bacillus subtilis* were found to be effective in seed germination and seedling vigour. The PGPR mixed formulation Pf1 + *B. subtilis* + neem cake + chitin was found to be the best for reducing fruit rot incidence besides increasing the plant growth and yield parameters under both green house and field conditions (Bharathi *et al.*, 2004).

A study of the fungal growth in dual cultures revealed that two *Trichoderma* spp. (*T. harzianum* and *T. viride*) inhibited the growth of *C. capsici* isolates. Effectiveness of *Pseudomonas fluorescens* against fruit rot causing pathogen of chilli has been reported by Shivakumar (2006) and Vinaya, (2007)

Seed Treatment

Setty *et al.* (1988) recorded the effective control of *Aspergillus*, *Colletotrichum* and *Rhizopus* spp. by treating the seeds with Emisan.

The infected chilli seeds were treated with benlate, manzate and vitavax- 200 in a mechanical shaker for 20 minutes for 48 hours. After 6 days of incubation the fungal infection was recorded by using the standard blotter method. All the treatments at 0.45 and 0.6 per cent of seed weight completely controlled *Alternaria alternata* and *Fusarium moniliformae*. While, infection by *Colletotrichum capsici* was reduced in all treatments (Mridha and Chowdhury, 1990).

Azad (1992) evaluated six antibiotics and four fungicides for control of *C. capsici* on chilli seeds. The seed treatment with agrimycin, cycloheximide, zineb and tridemorph recorded no contamination, while only 10 per cent contamination was observed on seeds treating with carbendazim and aureofungin as compared to other treatments. Treated seed showed highest per cent age germination than untreated seeds.

Karade *et al.* (1999) studied on possible use of medicinal plant extracts for seed treatment against pre-inoculated *Alternaria solani*, *Colletotrichum capsici* and *Macrophomina phaseolina* in chilli seeds. All the crude plant extracts and thiram (3 g/kg) significantly improved the germination per cent age and was recorded by blotter method. The complete inhibition of mycelial growth with maximum seedling growth was observed in acetone clove extract of *Allium sativum* in chilli seeds.

Rahman *et al.* (2004) suggested Bion (0.005 per cent) fungicide for seed treatment in inducing systemic resistance to anthracnose (*Colletotrichum capsici*) in chilli, and it was recorded reduced seedling mortality and dieback symptoms.

Varietal Screening

Singh *et al.* (1993) evaluated nineteen varieties of chilli against *C. capsici* under laboratory condition. Red ripe fruits were pinpricked and sprayed with 8000 conidia/ml of suspension of *C. capsici* and held in moist chamber at 28° C for 24 hours. The lesion spread in the first 6 days was lowest in varieties BG1 and Lorai (5.75 and 6.0 mm, respectively) and there were the only two varieties graded as resistant to fruit rot.

One hundred and fifty indigenous and exotic *Capsicum* collections were evaluated against multiple disease resistant in chilli. Among them, thirteen were free from the diseases namely leaf spot, powdery mildew and fruit rot (Muneem *et al.*, 1995).

Basak (1998) screened ten chilli cultivars against three major fruit rot fungi, *Colletotrichum capsici*, *C. gloeosporioides* and *Fusarium semitectum*. None of the cultivars were found to be immune however, except few remaining were rated as moderately resistant. Cultivars C-011 and C-045 were susceptible to *C. gloeosporioides*, *C. capsici*, C-123 to *C. capsici* and Chittagong local and Bogra local were susceptible to *F. semitectum* and highly susceptible to *Colletotrichum* spp.

Of fifty chilli cultivars/lines tested under artificial inoculation, five (Pusa Sada Bahar, 91-2, DC-9, DC-27 and Achar) were free from infection whereas six (86-5, Aparna, Kalyanpur Red, Sabour Anil, BG-1, Lorai and Perennial) were resistant and four (Pant C-1, DC-18, Suryamani and PS-1) were moderately resistant to *C. capsici* (Singh *et al.*, 1997).

In pot culture experiment, only one of the forty genotypes (CA87-4) showed highly resistant to *C. capsici* however, seven were categorized as resistant, 19 were moderately susceptioble and 13 were susceptible group (Jeyalakshmi and Seetharaman, 1998). Among the 24 chilli genotypes, none of the genotypes was rated resistant to *C. gloeosporioides* causing fruit rot disease of chilli. However, six were rated moderately resistant (Roy *et al.*, 1998).

Fifty one chilli cultivars were screened for field resistance to fruit rot fungus *Colletotrichum capsici*. Of the cultivars tested, none were tolerant, one was resistant, three were moderately resistant, five were moderately susceptible, seven were susceptible and nine were highly susceptible to the disease (Ekbote *et al.*, 2002).

Naik and Rawal (2002) used pinprick method of inoculation under laboratory condition to identify the resistant source against aflatoxin fungus and anthracnose pathogen. The observation on incidence of fruit rot was recorded at each picking. None of the cultivars was found to be immune to the disease. However, cultuivars namely Jwala, Phule Suryamukhi, Arka Lohit, KAC-86-25, Agni Rekha, AKC-BC-89-11, Parbhani tall, X-235, AKC-BC-89-8, G-4 and Surkta were found to be moderately resistant and seven genotypes were susceptible to disease.

Das *et al.* (2004) tested chilli genotypes against anthracnose (dieback and ripe fruit rot) disease under field condition. CA 219, KS1 were tolerant to dieback disease. In case of ripe fruit rot, lowest rotting was observed in KS1 and KS3.

References

Ajithkumar, K., Naik, M.K., Alloli, T.B. and Hosmani, R.M. (2005). Evaluation of genotypes, fungicides and plant extracts for the management of aflatoxin contamination in chilli caused by *Aspergillus flavus* Link, Fres. *Indian Journal of Plant Protection*, 33(1): 115–118.

Angadi, H.D. (1999).Studies on anthracnose of chilli (*Capsicum annum* L.) and its management. *M.Sc. (Agri.) Thesis,* University of Agricultural Sciences, Dharwad.

Akthar, J. (2007). Effect of fungicides on growth and sensitivity of the isolates of *Colletotrichum capsici* causing fruit rot of chilli. *J. Pl. Dis. Sci.,* 2(1): 59–62

Akhtar, J. and Singh, M.K. (2007). Studies on the variability in *Colletotrichum capsici* causing chilli anthracnose. *Indian Phytopath.* 60 (1): 63–67.

Akthar, J., Singh, M. K. and Chaube, H. S. (2008). Effect of nutrition on formation of acervuli, setae and sporulation of the isolates of *Colletotrichum capsici., Pantnagar Journal of Research* 6 (1): 110 –113.

Arx, J.A.V. (1957). Die Arten der Gattung Colletotrichum. Cda. *Phytopathologische, Zeitschrift,* 29: 413–468.

Asha, A.N. and Kannabiran, B. (2001). Effect of Datura metel leaf extract on hthe enzymatic and nucleic acid changes in the chilli seedlings infected with *Colletotrichum capsici. Indian Phytopathology,* 54(3): 373–375.

Azad, P. (1992). Efficacy of certain fungitoxicants against *Colletotrichum capsici* (Syd.) Butler and Bisby, the incitant of ripe rot of chilli. *Journal of the Assam Science Soceity,* 34(2): 34–39.

Bagri, R.K., Choudhary, S.L. and Rai, P.K. (2004).Management of fruit rot of chilli with different plant products. *Indian Phytopathology,* 57(1): 107–109.

Baker, K.F. and Cook, R.J. (1974). *Biological control of plant pathogens.* W. H. Freeman and Company. San Francisco, pp. 433.

Baker, R. (1968). Mechanisms of biological control of soil borne pathogens. *Annual Review of Phytopathology,* 6: 263–294.

Banu, I.S.K.F., Shivanna, M.B. and Shetty, H.S. (1990). Seed borne nature and transmission of *Colletotrichum dematium* in chilli. *Advances in Plant Sciences,* 3(2): 200–206.

Basak, A.B. (1994). Mycoflora associated with chilli seeds collected from Bogra District. *Chittagong University Studies, Science,* 18(1): 121–123.

Basak, A.B. (1998). Studies on the location of *Colletotrichum capsici* (Syd.) Butler and Bisby in the infected chilli seeds. *Seed Reaserch,* 26(1): 101–104.

Biswas, A. (1992). Efficacy of fungicides in control of anthracnose disease of chilli in Sundarban region of West Bengal. *Journal of Mycopatological Research* 30: 31–35.

Butler, E.J. (1918). *Fungi and Diseases in Plants.* Thacker, Spink and Co., Culcutta, VI, p. 547.

Choudhury, S. (1957). Studies on development and control of fruit rot of chillies. *Indian Phytopathology,* 10: 55–62.

Cook, R.J. And Baker, K.F. (1983). *The Nature and Practice of Biological Control of Plant Pathogens.* American Phytopathological Society, St. Poul, Minneosota, pp. 539.

Corda, A.C.I. (1831–32). *Srums Dentschlands flora nurnberg.*

Das, S., Somnathsil and Jahangir, K. (2004). Disease reaction and its progress on different chilli cultivars (*Capsicum annuum* L.) against anthracnose and leaf curl virus under field condition. *Journal of Mycological Research,* 42(1): 49–52.

Dastur, J.F., 1921, Dieback of chillies (*Capsicum* spp.) in Bihar. Memoirs. *Deportment of Agriculture in India. Botanical Series,* 35: 409–413.

Datar, V.V. (1995). Pathogenicity and effect of temperature on six fungi causing fruit rot of chilli. *Indian Journal of Mycology and Plant Pathology*, 25(3): 195–197.

Datar, V.V. (1996). Efficacy of growth regulators and fungitoxicants on fruit rot of chilli. *Indian Journal of Mycology and Plant Pathology*, 26(3): 239–242.

Dhyani, A.P., Sati, M.C. and Khulbe, R.D. (1993). Location and seed plant transmission of *Curvularia* spp. in chilli seeds in Kumaun Himalaya, India. *Madras Agricultural Journal*, 80(11): 645–648.

Dodd, J.C., Bugante, R., Koomen, I., Jefferies, P. and Jeger, M.J. (1991). Pre and post-harvest control of mango anthracnose in the Phillipines. *Plant Pathology*, 40(4): 576–583.

Ekbote, S.D. (2002). Bio–efficacy of Copper hydroxide (Coxid) against anthracnose of chilli. *Karnataka Journal of Agricultural Sciences*, 15(4): 729–730.

Ekbote, S.D. (2003). Efficacy of prochloraz 45 EC against fruit rot and dieback of chilli. *Indian Journal of Plant Protection*, 31(1): 139–140.

Ekbote, S.D., Jagadeesha, R.C. and Patil, M.S. (2002). Reaction of chilli germplasm to fruit rot disease. *Karnataka Journal of Agricultural Sciences*, 15(4): 717–718.

Gahukar, K.B., Raut, J.G. and Deshmukh, R.N. (1989). Seed infection in relation to pericarp infection of *Colletotrichum dematium* in chilli. *PKV Research Journal*, 13(1): 52–53.

Gupta, J.S., Agarwal, S.P. and Dixit, R.B. (1983). Incidence of pathogenic conidia of *Colletotrichum capsici* on the surface of chilli. *Indian Phytopathology*, 36: 357–359.

Hayden, N.J. and Maude, R.B. (1992). The role of seed borne *Aspergillus niger* in transmission of black mold of onion. *Plant Pathology*, 41: 573–581.

Hegde, G.M. and Anahosur, K.H. (2001).Effect of sodium chloride treatment on the disease development, quality and biochemical constituents of chilli fruits affected by *Colletotrichum capsici* (Sydow) Butler and Bisby. *Karnataka Journal of Agricultural Sciences*, 14(3): 681–685.

Hegde, G.M. and Kulkarni, S. (2001). Seed treatment to control damping off of chilli caused by *Colletotrichum capsici* (Sydow) Butler and Bisby. *Karnataka Journal of Agricultural Sciences*, 14(3): 829–830.

Hegde, G.M., Mesta, R.K. and Anahosur, K.H. (2002). Efficacy of triazole compounds against fruit rot pathogen (*Colletotrichum capsici*) of chilli. *Plant Pathology Newsletter*, 20: 3–4.

Hesseltine, C.W. (1986). Global significance of mycotoxins. In: *Mycotoxins and Phytotoxins*, Eds., Steyn, P.S. and Vleggaar, R., Amstredam, Netherlands: Elsevier. p. 1–18

Hingole, D.G.and Kurundkar, B.P. (2004). Fungicidal evaluations and economics in controlling anthracnose of chilli. *Journal of Soils and Crops*, 14(1): 175–180.

Huffaker, C.B. And Messenger, P.S. (1976).*Theory and practices of biological control*. London, U.K., Academic Press, Inc., pp. 788.

Jayasekhar, M., Eswaramurth, S. and Natarajan, S. (1987). Effect of certain fungicides on chili fruit rot. *Madras Agricultural Journal*, 74(10–11): 479–480.

Jeyalakshmi, C. and Seetharaman, K. (1998a). Biological control of fruit rot and dieback of chilli with plant products and antagonistic organisms. *Plant Disease Research*, 13(1): 46–48.

Jeyalakshmi, C., Durairaj, P., Seetharaman, K. and Sivaprakasam, K. (1999). Biocontrol of fruit rot and dieback of chilli using antagonistic microorganisms. *Indian Phytopathology*, 51(2): 180–183.

Jindal, K.K., Gupta, S.K. and Shyam, K.R. (1994).Studies on germination, vigour and microflora of bell pepper seeds from healthy and diseased fruits. *Indian Journal of Mycology and Plant Pathology,* 24(3): 227–228.

Joi, M.B., Deshmukh, D.P. and Khadatare, R.M. (2004). Efficacy of RIL 006/C1, a new fungicide for the control of anthracnose and fruit rot of chilli. *Annuals of Plant Protection Sciences,* 12(2): 463–464.

Karade, V.M., Ghule, S.T. and Patil, D.K. (1999). Bioefficacy of Medicinal plant extracts for seed treatment on seed infection by certain fungal pathogens. *New Botanist,* 26(1/4): 53–60.

Khare, D., Atri, D.C. and Shrivastava, C.J. (1995). A new fruit rot disease of chilli. *Indian Phytopathology,* 48(1): 111.

Khodke, S.W. and Gahukar, K.B. (1993). Fruit rot disease of chilli caused by *Alternaria alternata* Keissal. in Maharastra. *PKV Research Journal,* 17(2): 206–207.

Khodke, S.W. and Gahukar, K.B. (1995). Fruit rot of chilli caused by *Colletotrichum gloeosporioides* Penz. in Amravati District. *PKV Research Journal,* 19(1): 98–9

Kolte, S.O., Korekar, V.B. and Lokhande, N.M. (1994). Biocontrol of dieback and fruit rot of chillies (*Capsicum annuum*) caused by *Colletotrichum capsici* under field conditions. *Orissa Journal of Agricultural Research,* 7(3/4): 80–84.

Krishan K. J., Gupta, S.K. and Shyam, K.R. (1994). Studies on germination, vigour and microflora of bell pepper seeds from healthy and diseased fruits. *Indian Journal of Mycology and Plant Pathology,* 24(3): 227–228.

Kumaran, R.S., Gomathi, V. and Kannabiran, B. (2003). Fungitoxic effects of root extracts of certain plant species on *Colletotrichum capsici* causing anthracnose in *Capsicum annuum. Indian Phytopathology,* 56(1): 114–116.

Liang, L.Z. (1990). Seed borne *Fusarium* of chilli and their pathogenic significance. *Acta Phytopathologica Sinica,* 20(2): 117–121.

Ling, L. and Lin, K.R. (1944). On the occurrence of *Colletotrichum capsici* in China. *Indian Journal of Agricultural Sciences,* 14: 162–167.

Lokesha, S. and Shetty, H.S. (1990).In vitro screening of chilli (*Capsicum annuum*) seed samples for *Colletotrichum dematium* and its control. Indian *Journal of Agricultural Sciences,* 60(7): 489–490.

Malraja, *E.G.E.P.* and Narayanaswami, R. (1998). Effect of certain fungicides on the incidence of *Colletotrichum capsici* in chilli varieties K–1 and K–2. *South Indian Horticulture,* 36(4): 205–206.

Mandal, S.M.A. and Beura, S.K. (2003). Chemical control of leaf curl, anthracnose and ripe fruit rot in chilli. *Indian Journal of Plant Protection,* 31(1): 137–138.

Marchionatto, J.B. (1935).Notes on plant pathology, *Revista Facultad de Agronomia Universidad Nacional de La Plata,* 19: 407–426.

Mathew, S.K., Wahab, M.A. and Devi, S.N. (1995). Seasonal occurrence of chilli (*Capsicum annuum* L.) disease in Kerala, India. *Journal of Spices and Aromatic Crops,* 4: 86–87.

Mathur, S.B. and Sarbhoy, A.K. (1977). Physiological studies on *Alternaria alternata* from sugarbeet. *Indian Phytopathology,* 30: 384–387.

Meenugupta and Garg, R.C. (2002). Epidemiological studies on ripe fruit rot of chilli caused by *Colletotrichum capsici* (Syd.) Butler and Bisby. *Indian Journal of Hill Farming,* 15(1): 126.

Mello, A.F.S., Machado, A.C.Z. and Bedendo, I.P. (2004). Development of *Colletotrichum gloeosporioides* isolated from green pepper in different culture media, temperatures and light regimes. *Scientia Agricola*, 61(5): 542–544.

Mesta, R.K., 1996, Studies on fruit rot of chilli (*Capsicum annum* L.) caused by *Colletotrichum capsici* (Sydow.) Butler and Bisby. M. Sc. (Agri.) Thesis submitted to University of Agricultural Sciences, Dharwad.

Misra, A.P. and Mahmood, M. (1960). Effect of carbon and nitrogen on growth and sporulation of *Colletotrichum capsici* (Sydow.) Butler and Bisby. *Journal of Indian Botanical Society*, 39: 314–321.

Mridha, M.A.U. and Chowdary, M.A.H., 1990a, Efficacy of some selected fungicides against seed borne infections of chilli fruit rot fungi. *Seed Research*, 18(1): 98–99.

Muneem, K.C., Verma, S.K. and Pant, K.C. (1995). Performance of some chillies against leaf spot, powdery mildew and fruit rot in Kumaon hills. *Indian Phytopathology*, 48(2): 206.

Naik, M.K. and Rawal, R.D. (2002). Disease resistance in horticultural crops. In: *Resource Management in Plant Protection*, Vol–I Eds. Sarath Babu *et al.* Plant Protection Association of India, Hyderabad, p–303, p: 64–84.

Nandita P. and Sunita, C. (2004). Evaluation of seed health and germination of bell pepper. *Journal of Mycology and Plant Pathology*, 34(2): 368–370.

Natesh, N., Vyakaranahal, B.S., Shekhar, G.M. and Deshpande, V.K. (2005). Influence of growth regulators on growth, seed yield and quality of chilli Cv. Byadagi kaddi. *Karnataka Journal of Agricultural Sciences*, 18(1): 36–38.

Oh, B.J., Kim, K. and Youngsoon, K. (1998). A microscopic characterization of the infection of green and red pepper fruits by an isolate of *Colletotrichum gloeosporioides*. *Journal of Phytopathology*, 146: 301–303.

Padaganur, G.M. and Naik, K.S. (1991). Mycoflora of chilli seeds from fruit rot infected and healthy fruits. *Current Research*, 23(9): 183–184.

Padhi, N.N. and Rath, G.C. (1973). Sporulation of *Alternaria solani* in pure culture. *Indian Phytopathology*, 26: 495–501.

Palarpawar, M.Y. and Ghurde, V.R. (1997). Influence of different nitrogen sources on growth and sporulation of *Colletotrichum capsici* and *Colletotrichum curcumae*. *Journal of Mycology and Plant Pathology*, 27(2): 227–228.

Papavizas, G.C. and Lumsden, R.D. (1980). Biological control of soil borne propogules. *Annual Review of Phytopathology*, 18: 389–413.

Patil, C.U., Korekar, V.B. and Peshney, N.L. (1993). Effect of dieback and fruit rot on the yield of chilli. *PKV Research Journal*, 17(1): 60–63.

Potebinia, A., 1910, Beitage Dar. Micromycetam flora Miltel Russlands. *Annuals of Mycology*, Berlin, 8: 42–98.

Prabhavathy, K.G. and Reddy, S.R. (1995). Post-harvest fungal diseases of chilli (*Capsicum annuum*) from Andhra Pradesh. *Indian Phytopathology*, 48(4): 492.

Prasad, B.K., Deepa, R.S., Manoj, K. and Naresh, N. (2000). Decay of chilli fruits in India during storage. *Indian Phytopathology*, 53(1): 42–44.

Perane, R.R. and Joi, M.B. (1989). Control of fruit rot and die–back of chilli by seed treatment and spray. *Journal of Maharashtra Agricultural Universities* 14: 368.

Rahman, M.K., Islam, M.R. and Hossain, I. (2004). Effect of Bion, Amistar and vitavax on anthracnose of chilli. *Journal of Food, Agriculture and Environment*, 2(2): 210–217.

Rai, I.S. and Chohan, J.S. 1966. Studies on variation and perpetuation of *Colletotrichum capsici*(Syd.)Butler and Bisby causing fruit rot of chillies in the Punjab. *J. Res.* 2: 32–36.

Ramamurthy,V. and Samiyappan, R. (2001).Induction of defense–related genes in *Pseudomonas fluorescens* treated chilli plants in response to infection by *Colletotrichum capsici*. *Journal of Mycology and Plant Pathology*, 31(2): 146–155.

Roy, A., Bordoloi, D.K. and Paul, S.R. (1998).Reaction of chilli (*Capsicum annuum* L.) genotypes to fruit rot under field conditions. *PKV Research Journal*, 22(1): 155.

Roy, K.W., Killebrew, J.F. and Ratnayake, S. (1997). First report of *Colletotrichum capsici* on bell pepper in Mississippi. *Plant Disease*, 81(6): 693.

Saccardo, P.A. (1884). Sylloge fungorum, 3: Padua.

Salunkhe, D.K. (1985). *Storage processing and nutritional quality of fruits and vegetables*. Crawnwood Parkway, Olio, USA, pp: 91–166.

Sanford, G.B. (1926).Some factors affecting the pathogenicity of actinomycetes scabs. *Phytopathology*, 16: 525–527.

Seaver, F.J.,Chardon, C.E. and Toro, R.A. (1932). Supplement to mycology ex. Scientific survey of Puerto Rico and Virgin Islands. *New York Academic Sciences*, 8: 209.

Setty, T.A.S., Uthaiah, B.C., Rao, K.B. and Indiresh, K.M. (1988). Chemical control of seed microflora of chilli. *Plant Pathology Newsletter*, 6(1–2): 22.

Sharma, Y.K. and Thakore, B.B.L. (1999). Control of dieback of chillies with iprobenfos and its residues. *Journal of Mycology and Plant Pathology*, 29(3): 294–298.

Shivakumar,A.P. (2006).Management of post harvest diseases of chilli..M.Sc. (Agri.) thesis submitted to University of Agricultural Sciences, Dharwad, Karnataka, (India).

Singh, H.P.,Kaur, S. and Singh, J. (1993).Determination of infection in fruit rot (*Colletotrichum capsici*) of chilli (*Capsicum annuum*). *Indian Journal of Agricultural Sciences*, 63(5): 310–312.

Singh, S.N., Yadav, B.P., Sinha, S.K. and Ojha, K.L. (1997). Reaction of chilli genotypes against dieback disease caused by *Colletotrichum capsici*. *Journal of Applied Biology*, 7(1/2): 62–64.

Small, W. (1924).Annual report of the government Mycologist Report. Department of Agricultural Uganda for the year ending 31st December, 29: 1922.

Small, W. (1926). On the occurrence of species of *Colletotrichum*. *Transaction of British Mycological Society*, 11: 112–137.

Smith, R.W. and Crossan, D.F. (1958). The taxonomy, etiology and control of *Colletotrichum peperatum* (E and E) E and H and *Colletotrichum capsici* (Syd.) B. and B. *Plant Disease Reporter*. 42: 1099–1103.

Sulochana, K.K., Rajagopalan, B. and Wilson, K.I. (1992). Fungicidal control of fruit rot of chilli caused by *Colletotrichum capsici* (Syd.) Butler and Bisby. *Agricultural Research Journal of Kerala*, 30(1): 65–67.

Surekha, C., Tribhuwan, S. and Dalbir, S., 1990, Histopathology of *Colletotrichum dematium* infected chilli seeds. *Acta Botanica Indica,* 18(2): 226–230.

Suryawanshi, A.V. and Deokar, C.D. (2001a). Effect of fungicides on growth and sporulation of fungal pathogens causing fruit rot of chilli. *Madras Agricultural Journal,* 88(1/3): 181–182.

Sydow, H. (1919). Aufzahlung einiger in der provizen Kwangtung. Und kwangsi (Sub–China) gesammelter pilze. *Annuals of Mycologia,* 17: 140–143.

Thind, T.S. and Jhooty, J.S. (1985). Relative prevalence of fungal diseases of chilli fruits in Punjab. *Indian Journal of Mycology and Plant Pathology,* 15: 305–307.

Vinaya,H. (2008).Studies on seed borne aspects of anthracnose of chilli and its management..M.Sc. (Agri.) thesis submitted to University of Agricultural Sciences, Dharwad, Karnataka, (India)

Venkataravanappa, V. and Nargund, V.B. (2002). Evaluation of different fungicides against *Colletotrichum gloeosporioides* causal organism of anthracnose of Mango. Paper presented in *Annual Meeting and Symposium on Plant Disease Scenario in Southern India,* held at UAS, Agriculture College, Bangalore, December 19–21, 2002 pp. 56.

Vijayalakshmi, M. and Rao, U.V. (2003).Fungi association with chilli fruits during development. *Journal of Mycology and Plant Pathology,* 33(3): 451–452.

Weindling, R. (1932). *Trichoderma lingorum* as a parasite of other soil fungi. *Phytopathology,* 22: 837–845.

Yenjerappa, S.T., Kulkarni, S., Ravikumar, M.R. and Jawadagi, R.S. (2002). Efficacy of different fungicides and bio–fungicides against fruit rot in chilli. *Plant Pathology Newsletter,* 20: 32–33.

Plant Diseases Management in Horticultural Crops (2011) *Pages* **243–259**
Editors: **Shahid Ahamad, Ali Anwar and P.K. Sharma**
Published by: **DAYA PUBLISHING HOUSE, NEW DELHI**

Chapter 19

Cultural Practices: An Ecological and Economical Approach for Plant Disease Management

***Vaibhav K. Singh*[1], *Shailbala*[1], *Jameel Akhtar*[2] and *Bijendra Kumar*[3]**

[1]*Department of Plant Pathology, G.B.P.U.A.& T., Pantnagar – 263 145, Uttarakhand*
[2]*Department of Plant Pathology, Birsa Agricultural University, Ranchi – 834 006, Jharkhand*
[3]*College of Forestry and Hill Agriculture, Ranichauri – 249 199, G.B.P.U.A.& T, Uttarakhand*

Introduction

Cultural practices as a crop management procedure represents the oldest and most broadly applicable approach with farmers for prevention of losses in crop due to diseases and other causes (Anonymous, 1968). In spite of development of chemical methods and resistant varieties, this approach has excellent promise for future also.

In early stage of development of agriculture, the farmers had realized from observations and experiences that crop look sick when grown on the same land year after year, when land was not left fallow or when there was excess of moisture or other mismanagement. By proper adjustment of practices in the cultivation of the crop they had been avoiding these situation.

Successful use of cultural practice for disease control can be made only when complete knowledge on the nature of pathogen and its behavior in different conditions of environment, cropping system etc. are know. Although resistant varieties and fungicides are very important tool in management of diseases, their efficacy can be further improved and can be more lasting and economical by modification of cultural practices.

Cultural practices are the only feasible method of disease control in crops which give lower return per unit area or of which resistant varieties are not known. In India chickpea, pea, pigeon pea and mung bean are examples of such crops. Cultural practices often offer the opportunity to alter the environment, the condition of the host, and/or the behavior of the causal agent, to achieve economic management of disease. On this basis disease control by cultural practices is mainly preventive. These practices reduce the density and activity of inoculum. Integration of cultural practices with host resistance, fungicides and bio-control agents may be necessary to provide options for controlling economically important plant diseases (Baker, 1983).

The Concept and Applications

Katan (1996) has classified cultural practices into three categories:

- ☆ Practices, which are usually, applied for agricultural purpose not connected with crop protection, such as fertilization and irrigation. They may or may not have a positive or a negative side-effect on disease incidence.
- ☆ Practices, that are used solely for disease control, such as sanitation and flooding.
- ☆ Practices, which are used for both agricultural purpose and for disease control, such as crop rotation, grafting and composting.

The effect of cultural practices on disease has dual purpose: to develop suitable practices as control methods and to obtain information regarding their impact on disease when they are used as agricultural practices in order to avoid negative side-effects. Cultural practices may be employed, before or after planting. Deep ploughing and flooding are used before planting while irrigation and fertilization can be applied several times during the crop season for disease management.

The Procedure for Disease Management through Cultural Practices

Singh (2000) has described several procedures for disease management through cultural practices, kept under, three categories:

Inoculum Free Seeds and Planting Materials

- ☆ Management of seed borne inoculum
- ☆ Production and use of disease free propagating materials:
    - — Drying and ageing of seed
    - — Cleaning of seed
    - — Adjustment of harvesting time

Adjustment of Cultural Practices

- ☆ Crop rotation
- ☆ Monoculture
- ☆ Mixed cropping
- ☆ Intercropping
- ☆ Field fallowing
- ☆ Deep ploughing/Summer ploughing
- ☆ Mulching/Soil solarization
- ☆ Management of date of planting

- ☆ Management of optimum plant population:
  - — Plant density or stand
  - — Adjustment of depth of seeding
- ☆ Management of soil acidity and alkalinity
- ☆ Soil amendments
- ☆ Fertilizer Management
- ☆ Irrigation Management
- ☆ Heat therapy

Sanitation

- ☆ Management of crop debris
- ☆ Management of diseased plant parts:
  - — Rouging
  - — Pruning of orchard
  - — Eradication of alternate and collateral hosts
- ☆ Vector Management

Disease Free Seeds and Planting Materials

Management of Seed Borne Inoculum

A large number of fungal, bacterial and virus pathogens are transmitted through true seed or vegetative propagating materials such as seed tubers, cutting, bulbs, grafts, rhizomes etc. For effective disease control this source of primary inoculum must be taken care of. Seeds carry the pathogens as:

1. *Internally seed-borne inoculum*: Loose smuts of wheat and barley, ascochyta blight of chickpea, phomopsis blight of brinjal etc.
2. *Externally seed-borne inoculum*: Covered smuts of barley and oats, grain smut of sorghum, bunts of cereals etc.
4. *Contaminant or Admixture with Seeds*: Ear cockle of wheat (*Anguina tritici*), cysts of golden nematode of potato (*G. rostochiensis*), oospores of *Albugo candida*, sclerotia of ergot fungi etc.
5. *Through Nursery Raised Planting Stock*: Nursery stock infected with powdery mildew (*Podosphaera leucotricha*) is a major source of introduction of disease in apple orchard. Transmission of parasitic fungi (*Pythium, Phytophthora*) is common in papaya through seedlings raised in infested soil. Phytophthora root and collar rot of citrus seedlings are transmitted through infected planting stock and infested nurseries (Naqvi, 1994).

This can be successfully reduced by seed treatments with chemical/physical methods and use of healthy seeds and nursery planting materials.

Production and Use of Disease Free Propagating Materials

Following practices are followed to produce and use pathogen-free planting material:

1. *Drying and Ageing of Seed*: Some pathogens do not tolerate drying of seeds. For examples, when seed is properly dried before storage, the myceliums of downy mildew of maize (*Peronosclerospora sacchari*) present in freshly harvested grains are collapses.

2. *Cleaning of Seed*
   - ✰ Cleaning of seed is done by hot air blast that remove the dust also and by hand.
   - ✰ Dip seeds in 20 per cent common salt solution, the debris and nematode cysts or cockle float on the surface and can be skimmed off by hand.
3. *Adjustment of Harvesting Time*: Time of harvesting affects cleanliness of the seed.
   - ✰ Delayed harvesting of grain crops in temperate regions give the pathogens more time for contaminating the seed.
   - ✰ Harvesting of potato when the leaves are still green allows the late blight pathogen to contaminate tubers which carry it to the next season.

Such situation can be avoided by suitable alternation in the time of harvest of the crop.

Adjustment of Culture Practices

Crop Rotation

Crop rotation is the growing of economic plats in recurring succession and in definite sequence on the same land (Curl, 1963).

Since ancient time farmers have been aware of the fact that soil become sick and unsuitable for cultivation of a crop when the same crop is grown continuously on the same land. This can be due to several reasons such as:

1. Exhaustion of a particular essential nutrient due to excessive use of same type of crop.
2. Accumulation of organic acid and toxic substances released by the crop.
3. Easy survival of pathogen due to regular presence of susceptible host.

To overcome these problems crop-rotation is necessary in plant disease management. Reduction of survival ability or suppression of specific soil-borne pathogens by crop-rotation can be attributed to one or more of the following factors:

1. The pathogens having limited survival ability in soil restricted only to their host residue are starved out due to continuous absence of their living host or their inoculum density is reduced to innocuous level.
2. Different type of crops in the rotation has different physical, chemical and biotic effect on the soil environment. Some of these effects may be unfavorable for survival and growth of pathogen. The contribution of antagonistic micro-organisms in the biotic environment is increased by rotation.
3. In addition to direct effect on the pathogen, better growth of plant grown in rotation, due to better nutrients availability helps in disease avoidance.

Benefits

Growing crop in rotation has many benefits, such as:

1. Better use of nutrients,
2. Desirable effects on soil texture with deep rooted crops alternating with shallow rooted crops, *viz.*, a cereal with legume,
3. Water economy, in particular, conservation of water in years of fallow

4. Weed control in row crops in which in growth tillage to remove weeds could be practiced, alternated with crops sown by broadcasting, crop likely to inhibit weeds by their rapid growth and dense foliage alternated with slow growing crops having sparse foliage,
5. Suppression of soil-borne pathogens.

Reduction in Soil-borne Diseases

- ☆ Wilt of pigeonpea, chickpea, cotton, linseed
- ☆ Red rot and wilt of sugarcane
- ☆ Ergot and smut of pearl-millet
- ☆ Bunt of wheat
- ☆ Leaf smut and bunt of rice
- ☆ Molya disease of wheat and barley
- ☆ Root-knot of vegetable crops
- ☆ In wheat rotation with non susceptible crop such as oat, barley, chickpea and cluster bean reduces soil infestation of *Urocystis agrogyri* to a considerable extent; hence the flage smut can be checked by crop rotation technology (Bedi, 1957).
- ☆ Crop rotation with sugar beat and fodder maize after every 5-6 years intervals helps in wilt management on pea (Yainkov, 1989)
- ☆ The crop rotation of Pigeonpea – fallow – capsicum, Pigeonpea – fallow – tobacco, Pigeonpea – capsicum – capsicum, was found effective in minimizing Fusarium wilt of pigeonpea (Dasgupta, Chakravorti and Sengupta, 1992).
- ☆ Long rotation with graminaceous crops between potatoes are a major recommendation against bacterial wilt caused by Ralstonia solanacearum (Verma and Shekhawat, 1991).
- ☆ Take-all of wheat (caused by *Gaeumannomyces graminis*) and soybean cyst nematode (*Heterodera glycines*) are two examples of soil-borne diseases that are easily managed by short rotations of 1 and 2 years, respectively, out of susceptible crops, which may include susceptible weed hosts such as grasses in the case of take-all.

Table 19.1: Effect of Short-Term Rotation on some Pathogens

Beneficial Crop	*Pathogen Reduced or Eliminated*	*Preceding Crop*	*Country*
Rice	*Verticillium dahliae*	Cotton	USA
Pea	*Gaeumenomyces graminis*	Wheat	France
Sudan grass	*Pseudomonas solanacearum*	Tomato	-
Barley	*Meloidogyne incognita*	Cotton	USA
Groundnut	*Meloidogyne incognita*	Tomato	USA
Legume, sesame, wheat	*Pratylenchus indicus*	Rice	India

Factors Affecting Crop Rotation

The effectiveness of crop rotation in disease control will depend on the nature of the pathogen and the crop, the agricultural practices involved, soil properties and other biotic and aboitic factors.

The factors that reduce effectiveness of crop rotation in controlling soil-borne diseases include:

- ☆ Wide host range of the pathogen,
- ☆ Pathogen having effective mechanisms for survival in absence of host,
- ☆ Pathogen producing large inoculum densities as resting structure,
- ☆ Crop that are susceptible to several diseases,
- ☆ Crop which stimulate formation of resting structures,
- ☆ Frequent infestation of soil with pathogen from external sources,
- ☆ Soils those are conducive to diseases,
- ☆ Poor weed management

Monoculture

Monoculture is cultivation of a single or closely allied crop species in annual or seasonal succession, with interruption only by fallow or intermittent growing of green manure crop or application of soil amendments, not necessarily after each crop (Patil, 1981).

Monoculture may also exert selection pressure on pathogens, resulting in emergence of new pathotypes and helping in survival such as:

- ☆ *Verticillium albo-atrum* and *V. dahliae* causing wilt of cotton
- ☆ Root knot and cyst nematodes

Mixed Cropping

Mixed cropping is simultaneous cultivation on more than one crop in the same plot. Mixed cropping such as cultivation of wheat + barley, wheat+ chickpea, pigeonpea + sorghum and cotton + mothbean. Reduce the economic loss from disease. Since the same pathogen does not attack both crops in the mixture, at least one crop is saved if other is badly damaged by a disease. In addition, mixed crops reduce the incidence and spread of disease. The reduction in disease incidence in a mixed crop can be attributed to one or more to the following causes:

1. Due to reduction in number of host plants, there is sufficient spacing between them and chances of contact between foliage or roots of diseased and healthy plants are greatly reduced. Therefore, the root pathogen is unable to spread from diseased to healthy roots and spread of foliar pathogen is also reduced to a great extent.
2. The root of non-host plant may act as barrier obstructing the movement of pathogen in soil. They may release toxic substances from their roots which may suppress the growth of pathogen attacking the main crop. For examples- HCN in root exudates of sorghum is toxic to *F. udum*, the pigeonpea wilt fungus.
3. Due to reduced number of host plant in a mixed crop the susceptible area for air-borne, foliar pathogen if decreased. Therefore, there is less primary infection and less production of secondary inoculum for spread of the disease.
4. By proper selection of crop for the mixture, soil environment can also be changed to one that is not favourable for the pathogen. Control of root-rot of cotton by growing cotton with mothbean is an example.

5. The soil-borne pathogen is not uniformly distributed in field soil. Generally they are randomly present as dormant structure. Activation of this dormant structure is often dependent on contact with the host roots, the chances of which are highly reduced in a mixed crop due to spacing between plants such as:
   - ✰ Mixed cropping of pigeonpea with *Crotolaria medicaginea* was observed to reduce wilt incidence (Upadhyaya and Rai, 1981).
   - ✰ Mixed cropping either with non-host or antagonistic crop such as a tagetes, sesame, mustard etc. has also been reported to reduce root-knot and reinform nematode.

Intercropping

- ✰ Pea intercropping with mustard, linseed, wheat, chickpea and barley delays the appearance and buildup of powdery mildew.

Table 19.2: Effect of Intercropping with Sorghum on Pigeonpea Wilt

Intercropping	*Per cent Wilt Incidence in Different Years*			
	I	*II*	*III*	*IV*
P-P-P-P	86	85	64	91
SP-SP-SP-SP	55	18	21	28

P: Pigeonpea; SP: Sorghum + Pigeonpea (1:1).

Source: Natarajan *et al.*, 1985.

Field Fallowing

It is the practices of keeping the land fallow. Fallowing is of different types, such as dry fallowing, wet fallowing and flood fallowing.

1. *Dry Fallowing* for pathogen management is generally restricted to extensive farming where other methods of disinfection of soil are not economical. It is generally combined with specific crop rotation, soil-amendments and fertilizers and tillage practices. Such fallowing frequently is very useful in reducing population of parasitic nematodes.
2. *Combined Clean Fallow* and growing barley has reduced population of *M. incognita* in cotton and one fallow period between rice crops is sufficient to reduce the root lesion nematode (Pratylenchus indicus) to an acceptable level.
3. *Wet Fallowing* is usually practiced for weeks not months. The purpose of such fallowing is to make the pathogen propagules in or on the soil germinate, and become susceptible to attack of saprophyte. It may be considered for pathogens like *Sclerotium rolfsii* and *Verticillium dahliae* where sclerotia germinate quickly under the influence of alternate drying and wetting of the soil. Wet fallowing is also partly successful in reducing population of *Pythium myriotylum*, provided the population is not too high.

Deep Summer Ploughing

Ploughing during summer months exposes the resting spores of many fungi and nematodes to solar energy which are being killed due to natural heating and desiccation.

Three ploughing (15 cm deep) in summer months at the interval of 15 days decreased population of *Tylenchorhynchus vulgaris*, *Haplolaimus indicus*, *Helicotylenchus indicus* and *Pratylenchus zeae* (Haque and Prashad, 1983).

Table 19.3: Effect of Summer Ploughing on Nematode Population in Wheat (Per cent decrease over control)

Treatment	*Haplolaimus indicus*	*Helicotylenchus indicus*	*Tylenchorhynchus vulgaris*	*Pratylenchus zeae*	*Yield (kg)*
One ploughing	78.56	18.00	261.90	57.08	7.70
One ploughing	104.43	211.77	331.09	200.75	7.70
One ploughing	157.97	224.12	369.30	300.00	7.80

Mulching/Soil Solarization

Mulching or Covering Top Soil: with organic residues often helps in reducing plant diseases. Here only those materials should be considered which are not related to the crop to be protected. Mulching with rice husk (2.5 cm thick) greatly reduced the splashing of inoculum of R. solani and lowered the web blight severity in bean (Galindo *et al.*, 1983).

Soil solarization: In this management tactics, the solar energy is preserved with the help of transparent polyethylene sheet to increase soil temperature (10-15 °C above normal temperature) enough to kill the most of the soil-borne pathogens and weeds also (Khulbe, 2000, Akhtar, *et al.*, 2008).

Soil solarization with transparent polyethylene sheet mulch (25µm) for 40 days was found effective for the control of collar and root rot of strawberry caused by *Sclerotium rolfsii* (Bhardwaj *et al.*, 2004).

Fungal diseases such as damping-off, root rots, stem rots, fruit rots, wilts and blights caused by *Pythium* spp., *Phytophthora* spp., *Fusarium* spp., *S. rolfsii*, *R. solani*, *Sclerotinia sclerotoirum*, *T. basicola* and *Verticillium* spp. have been successfully managed by soil solarization

Nematode diseases such as *Ditylenchus dipsci*, *Globodera rostochiensis*, *Heterodera* spp., and *Meloidogyne* spp. have been successfully managed by soil solarization.

Bacterial canker of tomato (*Clavibacter michiganensis* sub-sp. *michiganensis*) is successfully managed by Soil solarization for 1-2 months (Tjamon *et al.*, 1992; Akhtar *et al.*, 2008).

Adjustment of Date of Sowing

The choice of sowing date in relation to crop disease has one principal aim, *viz.* to reduce to a minimum the period over which infective agent meets the susceptible stage of the host.

Favourable environment (temperature, moisture, light, soil reaction, age of the host) are essential for growth and pahtogenicity of a disease causing agent. In vector transmitted virus the peak population of the vector is reached at a particular period. Environment, if favourable for the pathogen is same as for the host, disease incidence is likely to be high. So, date of sowing should be altered in such a way that the susceptible stage of plant growth does not coincide with the environment highly favourable for the pathogen.

There are a number of examples where early or delayed sowing of the crop enables the host to escape critical period of disease incidence.

- ☆ Early sowing of wheat is reported to reduce the infection of the flage smut (Bedi, 1957).

- ☆ Late planting of potato, reduces the late blight severity.
- ☆ Okra yellow vein mosaic virus is not common in crops planted during February-March because the vector population is very low or absent.
- ☆ Delay in sowing increased the disease incidence of Karnal bunt (Singh and Prasad, 1978).
- ☆ Pea and chickpea sown in October usually suffer heavily from root rot and wilt. When these crops are sown late, the disease are not so severe or almost absent. At the same time sowing of pea in October helps in escaping the damage, caused by rust and powdery mildew in plains.

Table 19.4: Effect of Date of Sowing on the Incidence of Sclerotinia Blight of Chickpea

Date of Sowing	*Per cent Diseased Plant*
October 15	66.65
October 22	70.50
October 29	46.00
November 05	50.00
November 12	41.00
November 19	35.00
November 26	10.50
December 03	18.75

Source: Singh, 1970.

Optimization of Plant Population

Plant Density or Stand

The plant population depends upon rate of sowing and density of stands. In present day production technology much emphasis on is laid out on high plant population for getting higher yield. High yield from high plant population is possible only if optimal dose of fertilizer has been applied and there is no incidence of disease spread. If such diseases are present, the dissemination of the pathogen will be facilitated by thick plant stand.

Due to high plant density, humidity in the field is always high, temperature is low and there is lack of aeration and pathogen grows rapidly. Damping off, late blight of potato, downy mildew of grapevine is some of the diseases, which spread fast in closed spaced planting. There are also examples where dense sowing helps in disease reduction:

- ☆ Tomato leaf curl virus transmitted by *Bemisia tabaci*.
- ☆ Cucumber mosaic transmitted by *Aphis gossypii*.
- ☆ Wilt caused by *Verticillium albo-atrum* and *V. dahliae* in cotton.

This is due to reduction of effective inoculum per plant in proportion to the increase in number of plant per unit area in the densely sown field.

Adjustment of Depth of Seeding

Varying the depth of sowing enables the host to avoid the pathogen.

- ☆ Shallow sowing reduces the damping-off of seedlings.
- ☆ Deep sowing increases the incidence of barley stripe and head smut of maize.
- ☆ Deep sowing reduces the infection of chickpea blight (*Ascochyta rabiei*).

Maximum diseases occur when seeds are placed 4-5 cm deep and minimum when planting is done at 2-3 cm.

Management of Soil Acidity and Alkalinity

Most fungi, bacteria and nematodes can tolerate the pH range in which their host plants grow. This provides a method for control of the specific diseases. However, directly or indirectly soil pH affect the speed of disease cycle and thereby, the disease intensity in a larger number of cases.

- ☆ Common scab of potato (*S. scabies*) is sensitive to acidity and significantly decreased if soil pH is brought down to below 5.2 (lowering of soil pH by application of sulphur).
- ☆ Club root of crucifers (*P. brassicae*) raising the ph of field soil to 7 or above by adding lime (calcium carbonate) or gypsum provide good control of the disease.

Soil Amendments

This is one of the cheapest and effective methods of alternation of soil environment. The materials used in amendments include dry or green crop residues oil-cakes, meals, sawdust, wood shavings, compost etc. these materials are allowed to decompose in the field themselves where the pathogen are supposed to be present. The decomposition of organic matter helps in alternation of physical, chemical and biotic conditions of the soil. The altered conditions reduce the inoculum potential of the pathogen.

There are three pathways of action of the organic amendment in controlling soil borne disease-

1. Disease avoidance through modification of physico-chemical environment of soil in the favour of the host.
2. Eradication of the pathogen and control of disease through effect of the decomposition product and microbial metabolite on the pathogen and the host.
3. Eradication of pathogen through direct antagonism.

The material used for organic amendment includes dry or green crop residues oil-cakes, meals, sawdust, wood shavings, compost etc. The list of soil-borne diseases that have been controlled in glasshouse, microplots or field plots by organic amendment of soil is quite exhaustive. Some of the important ones are given in Table 19.5.

Fertilizer Management

Nutrient can affect the relationship between crop and pathogen in many ways. Manipulation of crop nutrient therefore gives the farmers a valuable tool for managing crop health. The two most immediate objectives of nutrient practice to protect crop from pathogen are-

1. Avoidance of strain, particularly nutrient strain which may directly lower the resistance of the crop to pathogen.
2. Manipulation of nutrient to the relative disadvantage of the pathogen.

Table 19.5: Suppression of Soil-borne Diseases by Organic Amendment

Crop and Disease	*Pathogen*
Fungal diseases	
Wilt of pigeonpea	*Fusarium oxysporum* f. sp. *udum*
Wilt of banana	*Fusarium oxysporum* f. sp. *cubense*
Wilt of pea	*Fusarium oxysporum* f. sp. *pisi*
Wilt of chickpea	*Fusarium oxysporum* f. sp. *ciceri*
Root rot of cotton	*Macrophomina phaseolina*
Take-all of wheat	*Ophiobolus graminis*
Soft rot of ginger	*Pythium aphanidermatum*
Black scurf of potato	*Rhizoctonia solani*
Common scab of potato	*Streptomyces scabies*
Nematodal diseases	
Cereal cyst nematode	*Heterodera avenae; Heterodera major*
Potato cyst nematode	*Heterodera rostochidensis*
Root knot of nematode vegetables	*Meloidogyne javanica; Meloidogyne javanica*
Citrus nematode	*Tylenchulus semipenetrans*

Source: Singh and Sitaramaiah, 1973.

Deficiency of calcium promotes wilt incidence of tomato. Deficiency of zinc predisposes maize to attack of downy mildew. So, proper nutrient management can therefore reduce incidence of these diseases.

Heavy dose of N_2 predisposes potato to late blight. Higher dose of nitrogenous fertilizer has also been found to increase the susceptibility of wheat crop to rusts, mildews, blight, and karma bunt (Mehrotra, 2003).

High nitrogen fertilizer is recommended for control of sugar beet root rot (*Sclerotium rolfsii*). The control is attributed to accumulation of NH_3 and change in the soil pH towards acidity by use of urea.

Table 19.6: Role of Fertilizer Application on Plant Disease Management

Fertilizer	*Diseases Reduce*	*Reference*
Phosphorus	Take-all of barley (*G. graminis*)	Mattingly and D.B., 1974
Potassium	Potato late blight and early blight, Rice sheath blight, blast and brown leaf spot and wilts of cotton, melon and tomato	Palti and Huber, 1980
Calcium	Bean root rot (*R. solani*)	Backer and Cook, 1974
Molybdenum	Late blight of potato and Ascochyta blight of beans and peas	Mudich, 1967
Manganese	Late blight of potato	Patil, 1981
Iron	Rice brown spot	Ou, 1985
Silicon	Rice blast (*P. oryzae*)	Ou, 1985

Irrigation Management

The advantage of irrigation in relation to disease can be best exemplified by the case of common scab of potato caused by *Streptomyces scabies*. Scab attack on potato tuber is prevented by maintaining soil moisture near field capacity during tuber formation. This favourable effect is due to increase in bacterial flora antagonistic to *S. scabies*.

Other example is charcoal rot fungus (*Macrophomina phaseolina*). It attacks potato when soil temperature is high and there is water stress. By irrigating the field, soil temperature is brought down, stress is removed and disease is suppressed.

On the other hands, heavy irrigation in juvenile stage of plant growth makes it susceptible to attack of *Pythium*. So frequent but low quantity of irrigation is therefore always recommended for reducing chances of damping-off.

Cereal rusts usually are more severe when crop grown in wet soil than in relatively drier soil. Vascular wilt appears aggravated after irrigation. Therefore, at the plant stage when this pathogen can attack the crop irrigation should be avoided.

Soil moisture is related to many diseases. For example, wet soil favors club root of crucifers silver scurf of potatoes and *Cercosporella* on wheat while dry soil increases severity of white mould of onion, common scab of potatoes and fusarium diseases of cereals (Colhoun, 1973). Damping–off diseases caused by *Pythium* spp. can be reduced by maintaining a dry soil surface as zoosporic fungi such as *Pythium* and *Phytophthora* depend on soil water for zoospores release and motility.

The incidence of root rot of chilli peppers caused by *P. capsici* in plots receiving alternate-row irrigation was significantly less than in plants with irrigation of every row (Biles, *et al.*, 1992).

Irrigation will be useful for management of diseases favoured by water stress such as charcoal rots caused by *M. phaseolina*. (Gaffar and Erwin, 1969).

Heat Therapy

The heat therapy is effective in case of internally seed-borne diseases, where the mycelium remains is the embryo in dormant form. Hot water treatment eliminates mycelium of *Ustilago segetum tritici*, the causal agent of loose smut, from infected embryo. Heat therapy of wheat seed by hot water also effectively checks seed infection by Alternaria alternate causing leaf blight.

Hot Water Treatment

Hot water treatment is widely used for the control of seed-borne pathogens, especially bacteria and viruses. A list of some important seed-borne diseases, claimed to have been controlled by hot water treatment, is given in Table 19.7.

Sanitation

Field sanitation is a main part of the disease management through cultural practices. The inoculum present on few plants in the field may multiply in the soil or on the plant and in due course of time may appear to cause epidemic in next season. Therefore, plant bearing such pathogen or plant debris should be removed as early as possible. This measure includes:

1. Management of crop debris
2. Management of diseased plant parts
   - ☆ Rouging,

- ✰ Pruning of orchard
- ✰ Eradication of alternate and collateral hosts

3. Management of vectors

Table 19.7: Control of Seed-borne Pathogens by Hot Water Treatment of Seed

Crop	*Disease*	*Causal Organism*	*Treatment*
Brassica spp.	Black rot	*X. campestris* pv. *campestris*	50°C for 20 or 30 min
Cucumber	Seedling blight	*Ps. syringae* pv. *lachrymans*	50°C and 75 per cent RH for 3 days
Tomato	Black speak	*Ps. syringae* pv. *tomato*	52°C for 1 hrs
Peanut	Testa nematode	*Aphlenchoides arachidis*	60°C for 5 min
Pearl millet	Downy mildew	*Sclerospora graminicola*	55°C for 10 min
Potato	Potato phyllody	*Phytoplasma*	50°C for 10 min
Rice	Udbatta	*Ephelis oryzae*	54°C for 10 min
	White tip	*Aphlenchoides besseyi*	51-53°C for 15 min
Sugarcane	Grassy shoot	*Phytoplasma*	54°C for 2 hrs
	Ratoon stunting	*Clavibacter xyli* ssp. *xyli*	50°C for 3 hrs
	Red rot	*C. falcatum*	54°C for 8 hrs
	Mosaic	*Virus (potato virus Y group)*	52°C for 20 min

Source: Chaube and Singh, 1990.

Management of Crop Debris

After harvesting of crop, most of the crop-debris is left in the field. Infected crop-debris not only serves as source of perennation of pathogen, it serves also as a substrate for multiplication of inoculum has been found to help in the control of many diseases.

The fungi causing downy mildew of maize, jowar, bajra, pea, white blister of crucifers, powdery mildew of pea and cereal are example of pathogen which survive through their sexually produced oospores or cleistothecia in crop-debris.

Paddy straw harbors many pathogens such as true sclerotia of false smut, sheath blight, stem rot and rice bunt fungus as well as spore of leaf smut of rice. Late blight of potato besides surviving in seed tubers also survives as mycelium and sporangia in crop debris. Apple scab (*V. inaequals*) fallen leaves play the major part in providing inoculum in the next spring.

Destruction of crop-debris by burning immediately after harvest reduces the amount of inoculum surviving through debris.

Burning is an effective means of eradicating pathogens and is often required by law to dispose of diseased elm trees affected by Dutch elm disease (DED) (Figure 19.1), citrus trees infected by citrus canker (Figure 19.3). Propane flaming can effectively destroy *Verticillium* microsclerotia in mint stems (Figure 19.2), and flaming potato stems prior to harvest may prevent tuber infection by the late blight pathogen (Figure 19.4). However, burning agricultural fields is controversial because the smoke creates human health and safety and environmental concerns.

Figure 19.1: Elm Trees Affected Dutch Elm Disease

Figure 19.2: Propane Flaming Destroy *Verticillium* Micro-Sclerotia in Mint Stems

Figure 19.3: Citrus Trees Infected by Citrus Canker

Figure 19.4: Flaming Potato Stems Prior to Harvest Prevent Tuber Infection by Late Blight

Management of Diseased Plant Parts

Rouging

Regular removal of diseased plants or plant parts from a population has been found effective in reducing the primary hosts and spread of many diseases. This method is one of the effective recommendations in the control of virus disease of field crop. The rouging of diseased plants not only checks the spread of disease but it reduces the amount of surviving structure also.

In production of virus-free potato tubers for seed and in control of soybean and legume virus, rouging is an important recommended practice.

For control of loose smut of wheat and for production of disease free seed, rouging is always recommended in seed plot. Rouging is feasible and economical in generally small sized fields and when the incidence of disease is not very high.

Pruning of Orchard

A number of important pathogens of fruit trees and vines are capable of perennation on dormant part on plant mostly on bud scales. This includes powdery mildew of grape, apple, downy mildew of apple and pears etc. and malformation of mango.

Pruning is always carried out after harvest and when the tree is reaching dormancy, a stage least susceptible for fresh infection.

Eradication of Alternate and Collateral Hosts

Plant pathogens usually have a wide host range and they used weeds, wild host plants or self sown host plants as means of their active survival during absence of main crop. Therefore, keeping the field free from additional host of pathogen is a major sanitary cultural practice.

Pathogen belonging to heterocious rust of cereals, rust of apple and pine has been found to survive on self-grown wild alternate hosts.

Powdery mildew of cucurbits (*Sphaerotheca fuliginae*) persists during winter on wild cucurbit plant or out of season cultivated plants. Kans grass is a collateral host of sugarcane smut (*Ustilago scitaminae*) and sugarcane downy mildew (*Peronosclerospora sacchari*). Root knot nematode infects a large number of Solanaceous weeds and from these sources the inoculum can reach the cultivated vegetables.

Table 17.8: Plant Pathogens Aided by Weeds in Survival and Spread

Crop	*Disease*	*Pathogen (S)*	*Wild or Weed Host*
Cucurbits	Powdery mildew	*Sphaerotheca fuliginae* and *Erysiphe cichoracearum*	Different cucurbitaceous weeds
Cotton	Root knot	*Meloidogyne incognita*	*Cyperus esculentus*
Maize	Downy mildew	*Sclerophthora rayssiae*	
var. *zeae*	*Digitaria sanguinalis*		
Rice	Sheath blight	*Rhizoctonia solani*	*Cynodon dactylon, Echinochloa* sp.,
	Tungro	*Virus*	Wild rice *Echinochloa* sp.,
	Bacterial leaf blight	*Xanthomonas oryzae*	*Leersia hexandra*
Wheat	Take all	*Gaumannomyces graminis*	*Agropyon repens*

Vector Management

Many diseases spread by means of their insect vectors or by nematode and other vectors. A number of viral, bacterial and some of the fungal diseases may get introduced through the visits of their respective vectors to the hosts. For this reason, eradication of such pathogens should include this measure to get the optimum result.Conclusion

A thorough familiarity with the ecology and crop husbandry is absolutely necessary for plant disease management and appropriate modification of innovations in crop husbandry practices have furnished the foundations for success of all types of crop cultivation. Now our agriculture involves several cultural practices for different conditions. Every cultural practice should be evaluated as a potential tool for plant disease management.

Among many potential promising methods, cultural practices can play an important role in achieving the goal with the objective to provide healthy seed for pathogen free soil and obtain healthy stand of the crops. Therefore, our march towards ecologically harmonious and sustainable agriculture must continue.

References

Akhtar, J. Abdul Khalid and Kumar, B. (2008). Soil solarization: A non-chemical tool for plant protection. *Green Farming*, 1(7): 50–53.

Akhtar, J., Abdulmajid Ansari, Tiu, K.R. and Chaube, H.S. (2006). Potential of soil solarization in the management of seedling diseases of vegetable nurseries. In: *Ecofriendly Management of Plant Diseases*, (Eds.) Shahid Ahmad and Udit Narayan. Daya Publishing House, New Delhi, p. 80–94.

Anonymous (1968). *Plant Disease Development and Control, Vol. 1: Principles of Plant and Animal Pest Control*, Publ. (1959), Nat. Acad. Sci., Washington, DC, p. 250.

Baker, K.F. (1983). The future of biological control of plant diseases. In: *Challenging Problem in Plant Health*, (Eds.) T. Kommedahl and P.H. Williams. American Phytopathological Society (APS) Press, St. Paul, Minnesota, p. 217.

Bedi, K.S. (1975). Fightening the flage smut. *Indian Forming*, 7: 25–27.

Bhardwaj, Umang and Raj, H. (2004). Mulching with transparent polythene and root dip in fungicides for the management of collar and root rot of strawberry. *Indian Phytopath.*, 57(1): 48–52.

Chaube, H.S. (2001). Soil solarization and management of seedling deseases of horticultural crop. Technical Bulletin Directorale of Exp. Sta. G.B.P.U.A.&T, Pantnagar, Uttarakhand, p. 142.

Chaube, H.S. and Singh, U.S. (1991). *Plant Disease Management: Principles and Prac:tices.* CRC Press, New York, p. 329.

Chaube, H.S. and Akhtar, J. (2005). Role of cultural practices in plant disease management In: *Integrated Pest Management: Principle and Applications,* Vol. 1, (Eds.) Amerika Singh, O.P. Sharma, D.K. Garg. CBS Publishers and Distributors, Bangalore, p. 70–89.

Chaube, H.S. and Pundhir, V.S. (2005). *Crop Diseases and their Management*. Prentice Hall of India Private Ltd. New Delhi, pp. 217–229.

Chaube, H.S. and Singh, U.S. (1990). *Plant Diseases Management: Principles and Practices*, CRC Press, US, p. 319.

Chaube, H.S., Akhtar, J. and Vishwakarma, S.N. (2003). Effect of soil solarization on microbiological properties, disease suppressiveness and fertility of soil. *Indian Phytopath.*, 56(3): 353.

Curl, E.A. (1963). Control of plant diseases by crop rotation. *Bot. Rev.*, 29: 413.

Dasgupta, S., Chakravorti, A. and Sengupta, P.K. (1992). Effect of crop sequence on the population of *Fusarium udum*. Bulletin Soil. *J. Mycopathological Res.*, 30: 37–42.

Galindo, J.J., Abawi, G.S., Thruston, H.D. and Galvez, G., 1983. Effect of mulching on web blight of bean in Costa Rica. *Phytopathology*, 73: 610–615.

Haque, M.N. and Prasad, D. (1983). Effect of agronomical practices and DD application on nematode population and yield of wheat and maize. *Indian J. Nematol.*, 13: 226–229.

Katan, J. (1996). Cultural practices and soil-borne disease management. In: *Management of Soilborne Diseases*, (Eds.) R.S. Utkhde and V.K. Gupta. Kalyani Publisher, India, pp. 100.

Khulbe, D. (2000). Soil solarization: Effect on soil's nutrients, microorganisms and plant growth response. *Ph.D. Thesis*, G.B.P.U.A.&T., Pantnagar, Uttarakhand.

Kumar, A., Lal, H.C., Akhtar, J. and Chabue, H.S. (2009). Concepts and practices in integrated plant disease management. In: *Plant Disease Management for Sustainable Agriculture*, (Ed.) Shahid Ahmad. Daya Publishing House, New Delhi, pp. 1–10.

Mehrotra, R.S. (2003). *Plant Pathology*, 2nd edn. Tata McGraw Hill Publication Company Ltd., New Delhi, pp. 771.

Patil, J. (1981). *Cultural Practices and Infectious Crop Diseases*. Springer-Verlag, Berlin, pp. 243.

Rao, V.K. and Krishnappa, K. (1995). Soil solarization for the control of soil-borne pathogen complexes with special reference to *Meloidogyne incognita* and *Fusarium oxysporum* f. sp. *ciceri*. *Indian Phytopath.*, 48: 300–303.

Singh, A. and Prasad, R. (1978). Date of sowing and meterological factor in relation to occurrence of karnal bunt of wheat in U.P. tarai. *Indian J. Mycol. Pl. Pathol.*, 8: 2.

Singh, R.S. and Sitaramaiah, K. (1973). Control of plant parasitic nematode with organic amendments of soil, *Exp. Sta. Res. Bull. No. 6*, G.B. pant Univ. Agric. and Tech, Pantnagar (UK).

Singh, R.S. (2000). *Plant Disease Management*. Oxford and IBH, New Delhi.

Upadhyaya, R.S. and Rai, B. (1981). Effect of mixed cropping and soil treatment on incidence of wilt diseases of pigeonpea. *Plant and Soil*, 62: 309–312.

Verma, R.K and Shekhawat, G.S. (1991). Effect of crop rotation and chemical soil treatment on bacterial wilt of potato, *Indian Phytopathology*, 44: 5.

Verma, R.K. and Shekhawat, G.S. (2001). Effect of crop rotation and chemical soil treatment on bacterial wilt of potato, *Indian Phytopath.*, 44: 5.

Yainkov, I.I. (1989). Fusarium wilt of peas. *Zashchita Rostini* (Moskova), 4: 18.

Plant Diseases Management in Horticultural Crops (2011) *Pages* **260–265**
Editors: **Shahid Ahamad, Ali Anwar and P.K. Sharma**
Published by: **DAYA PUBLISHING HOUSE, NEW DELHI**

Chapter 20

Management of Chilli Fruit Rot and Anthracnose in Kashmir Valley

Ali Anwar, Lubna Masoodi, Shahzad Ahmad and Farah Rasool

Division of Plant Pathology,
S.K.University of Agriculture Sciences and Technology of Kashmir,
Shalimar, Srinagar – 191 121, J&K

Chilli (*Capsicum annuum* L.) is an important commercial crop of India, being grown in an area of. 0.9 million hectares. Among fungal diseases, anthracnose fruit rot (*Colletotrichum capsici* (Syd.) Butler and Bisby) is an important and serious fungal disease of chilli crop causing upto75 per cent fruit infection (Arvind *et al.*, 2007). Different species of phytopathogenic fungi in the genus Colletotrichum cause fruit rot in quality and quantity of peppers and chilli fruits which are only associated with ripe or ripening fruits. Since that a more aggressive form of the disease has become established in chilli growing areas in India and abroad. This form attacks and damaging chilli fruits at any stage of its development where crops grown during the warm, wet season in the tropics and subtropics threaten the profitability of crop in areas where it becomes established.

Ramachandran *et al.* (2007) have also been recorded that anthracnose incited by *Colletotrichum* spp is one of the most damaging disease of chilli in India. The severity of the disease varies depending on cultivars grown and weather conditions prevailing in a particular region. In severe instances, the preharvest and post harvest infections together account for more than 50 per cent of the crop losses. Two species, namely *C. capsici* and *C. gloeosporioides* were known earlier to cause anthracnose in chilli. Surveys conducted recently revealed the presence of *C.acutatum* whileas *C. capsici* was the most predominant species in the major chilli growing states namely Karnataka and Andhra Pradesh. Among 92 isolates of *Colletotrichum* collected from different chilli growing areas, 53 are identified as *C.capsici*, 38 as *C. gloeosporioides* and one was found to be *C. acutatum*. The fungus is both internally and externally seed-borne. Sowing such contaminated seeds results in pre emergence and post emergence damping-off of seedlings in nursery and field. These infected seedlings form the primary sources of inoculum. The fungus survives in an active form on the stems and branches causing die-back symptoms. The wet conditions caused due to monsoon rains that occur during the June-October period help in the outbreak and spread of the disease.

Symptoms:

All crop growth stages are affected, including post harvest stages. Symptoms occur primarily on ripening fruit often where fruit is touching the soil or plant debris. On ripe fruit there are small, sunken circular depressions up to 4-6 cm in diameter (Figure 20.1). The center of the lesions becomes tan in color while the tissue beneath the lesion is lighter-colored and dotted with many dark-colored fruiting bodies of the fungus that form concentric rings in the lesion (Figure 20.1). When disease is severe, lesions may coalesce. In older lesions, black structures called acervuli may also be observed. With a hand lens, these look like small black dots; under a microscope they look like tufts of tiny black hairs. Green fruit can also be infected but symptoms would not be appeared until the fruit ripens at harvest

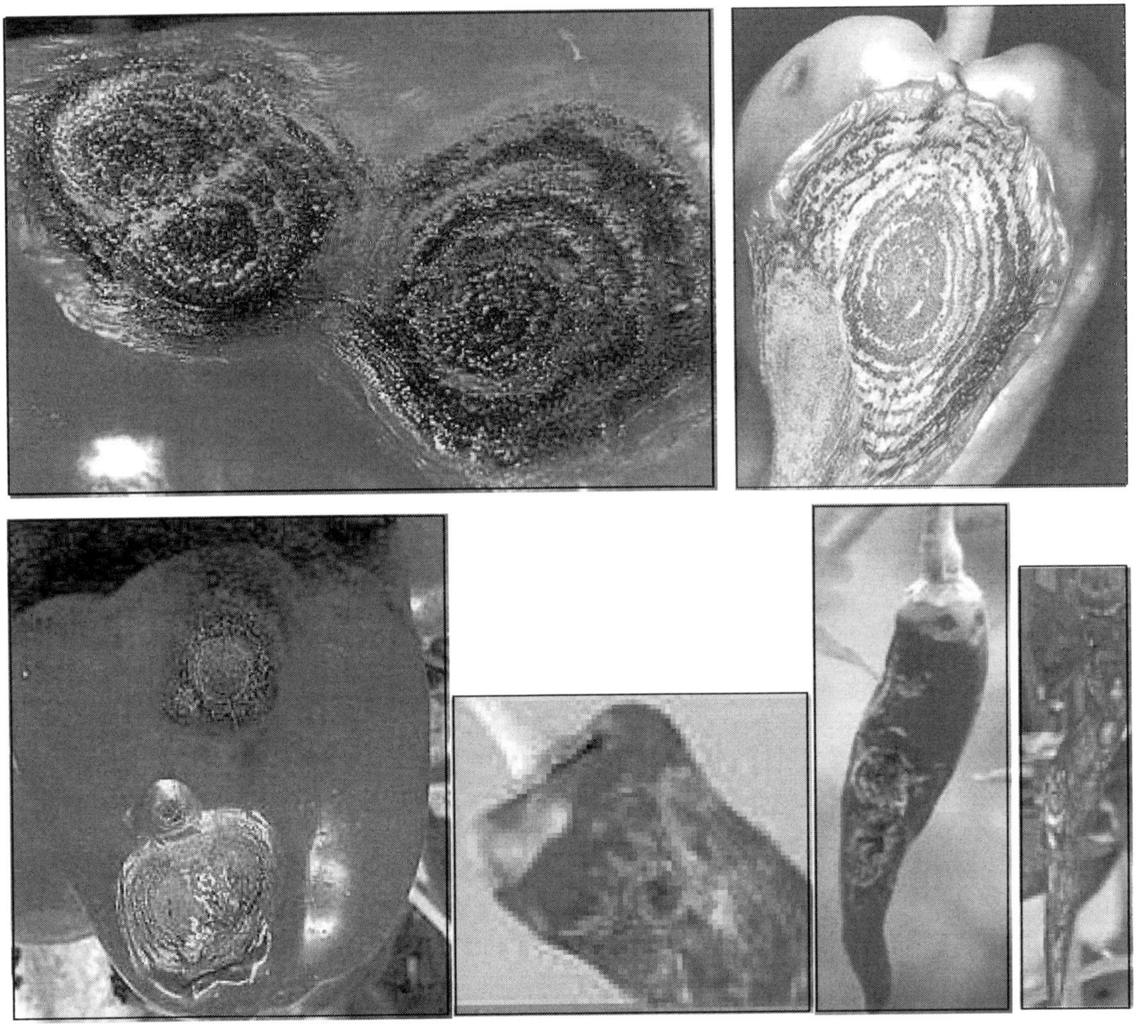

Figure 20.1: Fruit Rot with Concentric Rings Developed on Fruits

time. Such an infection is called latent. Symptoms on young fruit caused by *C. acutatum* are appeared as small, irregular shaped gray-brown spots with dark brown edges.

The pathogen forms spores quickly and profusely to spread rapidly throughout a crop, resulting in up to 100 per cent yield loss. Lesions are also appeared on stems and leaves as irregular shaped brown spots with dark brown edges. In some cases, the lesions are brown, not orange, and then black from the formation of setae and sclerotia (a dark, fungal survival structure).

Host Range

Among the *Colletotrichum* spp. that affect pepper, *C. gloeosporioides* has the widest host range among solanaceous crops and various biotypes have been reported on hosts. *C. acutatum* has caused severe fruit and foliar damage in several chilli growing areas.

Disease Cycle and Epidemiology

The fungus survives in and on seeds (Manandhar *et al.*, 1995). Anthracnose is introduced into the field on infected transplants or it can survive between seasons in plant. Spores of Colletotrichum released from a fungal fruiting body (acervulus) with numerous black, spines (setae) on crop debris or on weed hosts. New spores are produced within the infected tissue and then are dispersed to other fruit. Infection usually occurs during warm, wet weather. Temperatures around 80° F (27° C) are optimum temperature for disease development, although infection is occurred at both higher and lower temperatures. Severe losses occur during rainy weather because the spores are washed or splashed to other fruit resulting in more infections. The disease is more likely to develop on mature fruit that is present for a long period on the plant, although it can occur on both immature and mature fruit.

Variability in Pathogen

Traditionally, identification and characterization of Colletotrichum species have been based on morphological characters, such as sizeand shape of conidia and appressoria; existence of setae; the teleomorph state and cultural characters such as colony colour, growth rate and texture (Smith and Black, 1990). Selvakumar (2007) has developed 12 isolates of pathogens from plant parts showing symptoms either die back, anthracnose or fruit rots depending on expression during crop seasons. Mainly *C.capsici* was found to be associated in causing anthracnose. In addition *C. dematium*, *C. gloeosporioides*, *C. graminicola* and *C. atramentarium* have also been isolated. The colonies of pathogen are circular, smooth, and white having thick texture. The colour of the colony varies from gray, greenish to white. The growth rate of colony can also be varied from media to media depending on the composition. The size of the conidia range between 25-26 x 3.2-3.72um and the variation is not significant among isolates. The conidia are formed at the tip of conidiophores, which is unbranched. The conidium is hyaline, single celled and fusoid. Conidiophores are unbranched and aseptate.

Mohan *et al.* (2007), have also studied the distribution and diversity of 72 *Colletotrichum* spp. strains isolated from diseased fruits collected at the fruit ripening stage from the major hot pepper growing regions of Southern India (Andhra Pradesh, Karnataka, Tamil Nadu and Maharashtra provinces). Most of the isolates produced slightly fluffy type colonies, and rest either fluffy or flat colonies. The majority of isolates produced white or whitish-grey mycelia when cultured on PDA. The isolates which produced either pink or orange mycelia were *C. gloeosporioides* (*Cg*). In general, Cg isolates grew faster than *C.capsici* (Cc) ones. Conidial size is also more variable among Cg than among Cc isolates. AFLP fingerprinting showed the isolates to be highly diverse. Similarly, Ruchi *et al.* (2007) have been recorded variability in colony formation of the pathogen on different nutrient media with grey to dark brown, on pH light grey to dark grey. In particular, sequence analysis of the internal transcribed spacer (ITS) regions which lie between the 18S and 5.8S genes and the 5.8S and 28Sgenes, has proved useful in studying phylogenetic relationships of Colletotrichum species because of their comparative variability (Sreenivasaprasad *et al.*, 1994; 1996). However, there is no report concerning the use of these genes to differentiate between the *Colletotrichum* species involved in chilli anthracnose. A combined application of molecular diagnostic tools along with traditional morphological characterization is an appropriate and reliable approach for studying *Colletotrichum* species complexes (Cannon *et al.*, 2000). Than et al. (2008) differentiated isolates of chilli anthracnose from Thailand into three species: *C. acutatum, C. capsici* and *C. gloeosporioides*, based on morphological characterization, sequencing based on rDNA-ITS region and partial beta tubulin gene and pathogenicity testing.

Predisposing Factors

The pathogens are seed-borne and may also persist on alternate hosts such as other solanaceous crops (tomato, potato, and eggplant), cucumber, and many other cultivated crops and weeds. These pathogens will also readily persist in crop debris and in weeds, in some cases (*e.g. C. coccodes*) as resistant fungal structures called sclerotia. The pathogens can increase in number under continuous cultivation of pepper, tomato or potato. Secondary cycles of anthracnose development during the growing season arise from spores produced on diseased fruit or leaves. Anthracnose fungi occur primarily on leaves and twigs (Figure 20.2).

On deciduous trees these fungi over winter in infected twigs. Water-splash or wind-driven rain is required for dispersal of fungus spores or micro sclerotia on soil particles. Wounds in fruit are not required for infection but wetness is needed for spore germination and infection. The optimum temperature for fruit infection is 20–24°C with fruit surface wetness, although infection may occur from 10 to 30°C. However, the longer the period of fruit surface wetness the greater the anthracnose

Figure 20.2

severity. Fruit that are at or near the soil surface are the most likely to become infected by rain-splash or direct soil contact. Overhead irrigation will favor development of anthracnose because of increased relative humidity and increased duration of dew periods.

Management

Control of the disease is through integrated management techniques. The disease should not be introduced on infected plants. Only seeds that are pathogen-free should be planted. Transplants should be kept clean by controlling weeds and solanaceous volunteers around the transplant houses. The field should have good drainage and be free from infected plant debris. If disease was previously present, crops Anthracnose caused by *Colletotrichum* sp. on Pepper should be rotated away from solanaceous plants for at least 2 years (Roberts *et al.*, 2001). Sanitation practices in the field include control of weeds and volunteer peppers plants. Resistance is available in some varieties of chili peppers but not in bell peppers. For bell pepper production, choose cultivars that bear fruit with as hotter ripening period which may allow the fruit to escape infection by the fungus. Wounds in fruit from insects or other means should be reduced to the extent possible because wounds provide entry points for *Colletotrichum* spp. and other pathogens like bacteria that cause soft rot. For late-maturing peppers, when disease is present, apply a labeled fungicide several weeks before harvest. Check the with your county Extension agent for a suitable, labeled product. The disease can be controlled under normal weather conditions with a reasonable spray program. At the end of the season, infected plant debris from the field must be removed or deep ploughed to completely cover crop diseases (Agrios, 2005).

Select seed from anthracnose-free fruit or treat seeds with a fungicide. Hot water treatment is recommended to destroy seed-borne fungi. Soak seed at 52°C for 30minutes. Following treatment, plunge the hot seeds into cold water, dry on paper, and dust with thiram.

The most effective fungicides are chlorothalonil (1.5 per cent) and thiophanatemethyl (0.5 per cent) against the anthracnose and fruit disease. Freshly harvested seed withstands heat treatment better than one or two-year-old seed. Use healthy transplants. Sanitize flats if reusing them for transplant production. Broad-spectrum fumigants can be applied to soil in seedbeds to control the pathogens but annual re-application may be necessary if re-contamination of the soil by the fungus occurs.

Avoid potato, soybean, tomato, eggplant, and cucurbits as rotation crops. Rotate with non solanaceous crops for three years. Mulch to reduce soil splash onto fruit and lower leaves. Minimize or avoid overhead irrigation to reduce periods of wetness on pepper fruit. Harvest fruit as soon as it ripens since anthracnose develops more readily as the fruit ages. Weed regularly and avoid injuring pepper fruit. Allow infested crop debris to decompose completely by deep ploughing of crop residues before planting again. If only a few plants are affected by the disease, these can be removed from the field and disposed of. Avoid planting over lapping pepper crops near by. Apply protectant fungicides to plants starting when the first fruit are set. This will prevent or minimize the occurrence of infections. Control by fungicide application will depends upon proper dosage and good coverage of fruit. Some sources of resistance have been identified; check with your local extension agent for possible resistant cultivars.

Biological control of fruit rot and dieback of chilli with plant products tested in many laboratories and field trials showed that the crude extract from rhizome, leaves and creeping branches of sweet flag (*Acorus calamus* L.), palmorosa (*Cymbopogon martinii*)oil, *Ocimum sanctum* leaf extract, and neem (*Azadirachia indica*) oil could restrict growth of the anthracnose fungus (Jeyalakshmi and Seetharaman, 1998; Korpraditskul *et al.*, 1999).

References

Agrios, G.N. (2005). *Plant Pathology*, 5th edn. Academic Press, San Diego, p. 922.

Arvind, A., Deshpande and Rawal, Ram D. (2007). Resistant sources of chili (*Capsicum annuum* L.) anthracnose fruit rot disease [(*Colletotrichum capsici* (Syd.) Butler and Bisby] against different isolates collected from commercial chili growing areas of India. In: *First International Symposium on Chili Anthracnose*, Hoam Faculty House, Seoul National University, Seoul, Korea, pp. 46.

Cannon, P.F., Bridge, P.D. and Monte, E. (2000). Linking the past, present, and future of colletotrichum systematics. In: *Colletotrichum: Host Specificity, Pathology, and Host Pathogen Interaction*, (Eds.) D. Prusky, S. Freeman and M. Dickman. APS Press, St. Paul, Minnesota, p. 1–20.

Garg, Ruchi, Roy, B.K., Pandey, K.K. and Kumar, Virendra (2007). Effect of different media, pH, and temperature on growth and sporulation of *Colletotricum capsici*. In: *First International Symposium on Chili Anthracnose*, Hoam Faculty House, Seoul National University, Seoul, Korea, pp. 44.

Jeyalakshmi, C. and Seetharaman, K. (1998). Biological control of fruit rot and die-back of chilli with plant products and antagonistic microorganisms. *Plant Disease Research*, 13: 46–48.

Korpraditskul, V., Rattanakreetakul, C., Korpraditskul, R. and Pasabutra, T. (1999). Development of plant active substances from sweetflag to control fruit rot of mango for export. In: *Proceeding of Kasetsart University Annual Conference*, Kasetsart University, Bangkok, p. 34.

Manandhar, J.B., Hartman, G.L. and Wang, T.C. (1995). Anthracnose development on pepper fruits inoculated with *Colletotrichum gloeosporioides*. *Plant Disease*, 79.

Mohan Rao, A., Nanda, C., Prathibha, V.H. and Ramesh, S. (2007). Diversity and distribution of hot pepper *Colletotrichum* spp. isolates in Southern India. In: *First International Symposium on Chili Anthracnose*, Hoam Faculty House, Seoul National University, Seoul, Korea, pp. 41.

Ramachandran, N., Madhavi Reddy, K. and Rathnamma, K. (2007). Current status of chili anthracnose in India. In: *First International Symposium on Chili Anthracnose*, Hoam Faculty House, Seoul National University, Seoul, Korea, pp. 25.

Roberts, P.D., Pernezny, K. and Kucharek, T.A. (2001). Anthracnose caused by *Colletotrichum* sp. on pepper [Online]. *Journal of University of Florida/Institute of Food and Agricultural Sciences*. Available from http://edis.ifas. ufl.edu/PP104

Smith, B.J. and Black, L.L. (1990). Morphological, cultural, and pathogenic variation among Colletotrichum species isolated from strawberry. *Plant Disease*, 74(1): 69–76.

Selvakumar, R. (2007). Variability among *Colletotrichum capsici* causing chili anthracnose in north eastern India. In: *First International Symposium on Chili Anthracnose*, Hoam Faculty House, Seoul National University, Seoul, Korea, pp. 38.

Sreenivasaprasad, S., Mills, P. and Brown, A. (1994). Nucleotide sequence of the rDNA spacer 1 enables identification of isolates of *Colletotrichum* as *C. acutatum*. *Mycological Research*, 98: 186–188.

Sreenivasaprasad, S., Mills, P., Meehan, B.M. and Brown, A. (1996). Phylogeny and systematics of 18 *Colletotrichum* species based on ribosomal DNA spacer sequences. *Genome*, 39(3):499–512. [doi:10.1139/g96–064].

Than, P.P., Jeewon, R., Hyde, K.D., Pongsupasamit, S., Mongkolporn, O. and Taylor, P.W.J. (2008). Characterization and pathogenicity of *Colletotrichum* species associated with anthracnose disease on chilli (*Capsicum* spp.) in Thailand. *Plant Pathology*, 57(3): 562–572.

Plant Diseases Management in Horticultural Crops (2011) *Pages* **266–274**
Editors: **Shahid Ahamad, Ali Anwar and P.K. Sharma**
Published by: **DAYA PUBLISHING HOUSE, NEW DELHI**

Chapter 21

Plant Disease Resistance Genes: Concepts and Connections

Prashant Vikram[1] and Alok Singh[2]

[1]Research Scholar, IRRI, DAPO BOX 7777, Metro Manila, Philippines
[2]Reader, Department of Genetics and Plant Breeding, T.D. College, Jaunpur, India

Unlike animals, plants do not have a circulating antibody system. Contrarily, they encode a large array of genes, the R genes which autonomously maintain constant vigilance against pathogens. It is now well understood that the interaction of *R-avr* genes lead to the disease resistance reaction. This is in accordance with Flor's "Gene for Gene" hypothesis in which R genes correspond to the genes for resistance in plants against the virulence genes present in the pathogens. The term virulence genes and avirulence genes is somewhat confusing. When the gene product of pathogen confers lack of virulence, the term Avirulence genes is used and vise versa for the virulence one.

The avirulence gene (*Avr*) codes for an elicitor molecule or protein controlling the synthesis of an elicitor. On the other hand resistance gene (R) codes for a receptor molecule which recognizes the Elicitor. Thus their product act as putative receptors that specifically binds to the corresponding pathogen derived *avr* dependent ligand. Formation of this receptor-ligand complex initiates signaling cascade for the defense responses which ultimately slows the progress of pathogen. The defense responses in plants could be of various types like hypersensitive response (rapid cell death at the site of infection), an oxidative burst, cell wall strengthening or tissue reinforcement, antibody production and the induction of defense gene expression. Oxidative burst is rapid generation of active oxygen species like super oxide anion (O^{2-}), hydroxide radical ($OH^{\bullet}$), H_2O_2. It is one of the very early defense mechanisms triggered by infection. These are local responses which can in turn trigger a long lasting systemic response. This we call systemic acquired resistance. After a plant has been infected by one pathogen and recovered from infection, it can show remarkable resistance to future infections by the same or other pathogens for days. Interestingly it is somewhat similar to the immunity in animal system. Such kind of acquired resistance we call as Systemic Acquired Resistance. This acquired resistance is for broad spectrum of pathogens.

R-avr Gene Interactions

R and *avr* genes encode their respective protein which may be translocated in the plant cells through Type III secretions-Hrp and Hrc proteins (bacteria) or by some unknown mechanism (fungi). After entering in to the plant cells these proteins (virulence and avirulence) target the host R-protein and the interaction leads to defense responses, metabolism and other plant purposes. It should be noted here that the virulence protein could also be targeted towards extracellular protein. In the due course of interaction there might be three possibilities. First, the plant cell expressing a R protein (protein encoded by the R gene) not recognizing the pathogens virulence proteins. Consequently the defenses are not activated and cellular metabolism is altered. Secondly, virulence protein could be recognized by the host R protein and interaction of both of them triggering the signal transduction for activation of defense responses. It would be quite interesting, if *R-avr* gene interactions are not explained on the basis of above two possibilities. As the remedy of the case "Guard Hypothesis" has been formulated by Vander Biezen and Jones. According to them R protein in the plant interacts with another plant protein known as "Guardee". The complex then detects the pathogen virulence protein resulting in defense responses. This model was initially proposed for Arabidopsis RPM-1 gene (R gene) which requires another protein RIN4 (guardee) against *Pseudomonas syringae* type III effector molecules. Later some more cases were discovered (Table 21.1).

First Case

▲ + ■ → Plant R protein could not recognize pathogen protein

= Defenses suppressed → metabolism of host cell altered.

Second Case

▲ + ■ → Plant R protein recognizes pathogen protein

= Signal Transduction proceeds

= Defenses activated → HR, oxidative bursts etc.

Guard Hypothesis

■ + * → ▲ Modified host protein

▲ + ■* → Modified host protein recognizes the pathogen protein.

= Signal Transduction proceeds.

= Defenses are activated → HR, oxidative bursts etc.

Symbols

▲ → Pathogen protein/Ellicitors.

■ → Host R-Protein/Receptors.

* → Guard protein.

■* → Modified host protein.

Table 21.1: Examples of Guard Hypothesis

Guard	*Guardee*	*Interaction*	*Pathogen Effector*
RPM 1 (CC-NBS-LRR)	RIN 4	Phosphorylation	AvrRpm1, AvrB
RPS 2 (CC-NBS-LRR)	RIN 4	Degradation	avrRpt2
RPS 5 (CC-NBS-LRR)	PBS 1 (Kinase)	Cleavage	AvrPphB
Cf 2 (receptor domain)	Rcr 3 (Cysteine Protease)	Inhibition	Avr2
Prf (CC-NBS-LRR)	Pto (Kinase)	Conformational change	AvrPto

Conserved Domains and ClassesDomains are explained as the secondary structures of proteins responsible for their functional aspects. The R gene products are specified by these amino acids domains like GGVGKTTL, IITTR, FLLDDV (where letters F, L, D, V, I T R G, K correspond to the specific amino acids). The most interesting fact in this context is that the R gene products have several conserved domains. On the basis of these domains R genes have been classified in to several classes (Figure 21.1). They may be classified on the basis of the molecular characteristics of their products too. This criterion could divide them in seven classes as described in the Table 21.2.

Table 21.2: Different Classes of R Gene Products

Class	*Gene*	*Plant*	*Reference*
TIR-NBS-LRR (Toll-Inter-leukin homologous)	*N* *L6* *RPP5*	Tobacco Flax Arabidopsis	Witham *et al.*, 1994 Lawrence *et al.*, 1995 Parker *et al.*, 1997
CC-NBS-LRR (homologous to coiled coil domain)	*RPM1* *RPS2* *Prf*	Arabidopsis Arabidopsis Tomato	Grant *et al.*, 1995 Bent *et al.*, 1994; Mindrinos *et al.*, 1994 Salmeron *et al.*, 1996
e-LRR (Extra cellular Leucine Rich Repeats)	*HS*$^{Pro-1}$ *Cf-2* *cf-9*	Sugarbeet Tomato Tomato	Cai *et al.*, 1997; Hammond Kossacck and Jones, 1997 Dixon *et al.*, 1996 Jones *et al.*, 1994
eLRR–Kinase (e-LRR with cytoplasmic serine threonine kinase domain)	*Xa-21*	Rice	Song *et al.*, 1995
KIN (cytoplasmic serine threonine kinase) domain)	*Pto*	Tomato	Martin *et al.*, 1993
KIN-KIN (STK extracellular and ctytoplasmic both)	*Rpg1*	Maize	Brueggeman, R *et al.*, 2002
CC (Intracellular Coiled coil/ Leucine zipper domain)	*RPW 8*	Arabidopsis	Xiao *et al.* (in press)
e-LRR-PEST-RME	*Ve2*	Tomato	McDowell *et al.*, 2003
e-LRR-RME	*Ve1*	Tomato	McDowell *et al.*, 2003

e-LRR-PEST-RME: Extra cellular Leucine Rich Repeats domain, domain for protein degradation (PEST) and domain for receptor mediated endocytosis (RME); e-LRR-RME: Extra cellular Leucine Rich Repeats domain and domain for receptor mediated endocytosis (RME); TIR: Toll Inter Leukin Receptor. It has sequence homology to intracellular domain of *Drosophila*.

Note: e-LRR-PEST-RME and e-LRR-RME belong to the e-LRR group if we say categorize them. Therefore total number of classes would be seven itself.

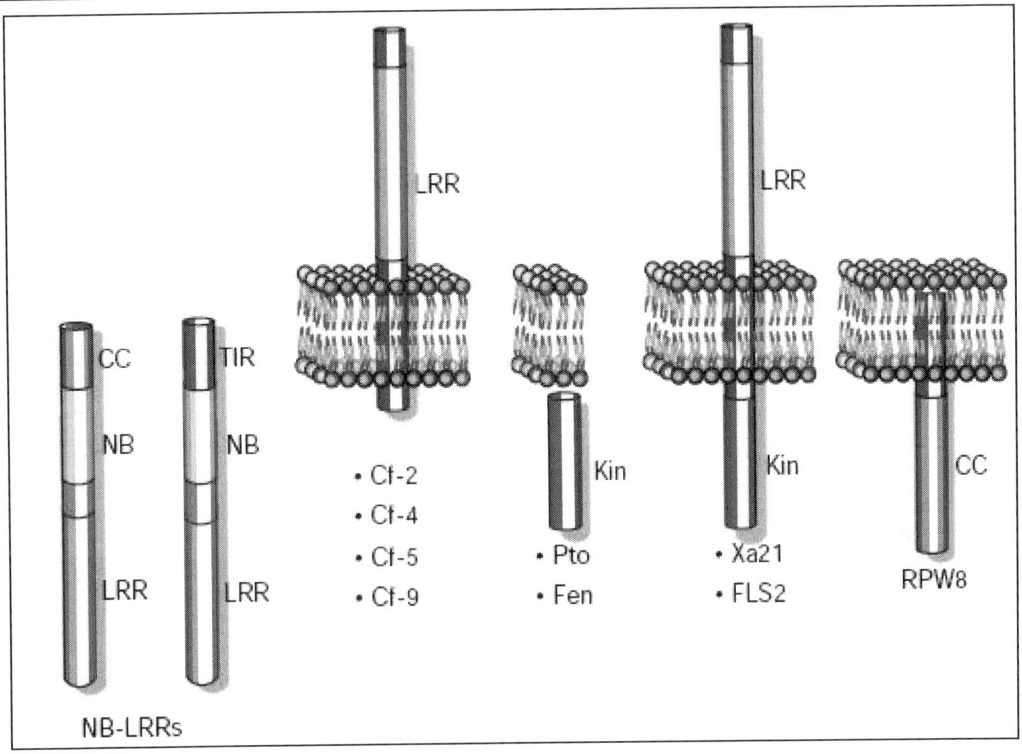

Figure 21.1: Different Classes of R Genes (Downloaded from www.intl-pag.org)

The major class having about three fourth of total R genes is NBS-LRR class. Genes of this group encode cytoplasmic receptor like proteins that contains a nucleotide binding site at N terminus and a leucine rich repeat at C terminus. These repeats are tandem repeats and have their role in recognition specificity. Examples of this class are *RPS-2, RPP-5 and RPM-1* (Arabidopsis), *Prf, 12-C* (Tomato), *N* (Tobacco), *L-6* (flax) and *Cre-3* gene of wheat. NBS-LRR class may be sub grouped in TIR-NBS-LRR and CC-NBS-LRR. Former is having homology to Toll Interleukin Receptors (TIR) and later with the Coiled Coil domain of metazoans at their N terminus. The third class of R genes encodes an extra cytoplasmic protein with Leucine rich repeats. This includes tomato *Cf-2, Cf-9* genes of tomato and a sugar beet nematode resistance gene HS^{Pro-1}. Another The fourth category encodes a transmembrane receptor with an extra-cellular LRR domain and an intra-cellular serine-threonine kinase (STK). Rice *Xa21* gene belongs to this class. Although *Xa-1* gene has NBS-LRR type of structural organization. Another class of R gene is the one which does not have either LRR or NBS region in it. It includes *Pto* gene of tomato with only an intracellular kinase domain (STK domain). This gene depends on *Prf a* NBS-LRR containing protein for its function. Maize *Rpg-1* gene represents the sixth class with two intra-cellular kinase domain. Lastly, Arabidopsis *RPW 8* protein has an intra cellular coiled coil domain anchored with the membrane.

All of R genes except maize *Hm1* gene encode protein contain domains involved in signal transduction.It should be noted that *Hm-1* gene product directs the chemical detoxification.The gene is also known for absence conserved domains. It would be quite easy to classify the R genes on basis of their translated protein product as listed in the Table 21.2.

Isolation and Cloning of the *R* Genes

More than 30 *R*-genes have been cloned since the isolation of first *R*-gene *Hm1* from maize in 1992 (Table 21.3). Most of the *R*-genes isolated so far have been isolated through either map based cloning or gene tagging approach. In the map based cloning we first find a marker near our gene of interest *i.e.* the R gene. Then we go for the co segregating markers among them and thereafter develop the overlapping clones (contings) containing flanking regions of that marker. They are then hybridized with the genomic library (using BAC/YAC) and ORFs (Open Reading Frames) are worked out. This is followed by further transformational or mutational analysis in way to go through a candidate gene. As an example Arabidopsis has about 166 NBS-LRR genes at 91 loci, indica rice has 500-700 genes at more than 142 loci, japonica has 500-700 at more than 52 loci but in soybean this number is around 1500-200 at 334 loci. Similarly, in the gene tagging approach (Transposon or *T*-DNA tagging) we need the mutants which are not a practical case with every crop. Noticeably, these techniques require some prerequisites like the saturated maps, mutants etc which restricts their use up to the limited number of crops. For most of the crops and their wild relatives such information is not available. As the solution of this case resistance genes analogs are identified using the sequence similarity

Table 21.3: Examples of Some Cloned Disease Resistance Genes in Different Host-pathogen Interactions

R Gene	*Plant*	*Pathogen*	*Avr Gene*	*Structure*	*Reference*
Hm 1	Maize	*Cochliobolus carbonum*	None	Toxin reductase	Johal and Briggs (1992)
Pto	Tomato	*Pseudomonas syringae*	*avr Pto*	Protein Kinase	Martin *et al.* (1993)
Xa21	Rice	*Xanthomonas campestris*	Unknown LRR	Protein Kinase	Song *et al.* (1995)
RPS2	Arabidopsis	*Pseudomonas syringae*	*avrRPt2*	CC-NBS-LRR	Bent *et al.* (1994) Mindrinos *et al.* (1994)
RPM1	Arabidopsis	*P. syringae pv maculicola*	*avrRPM1,avrB*	CC-NBS-LRR	Grant *et al.* (1995)
N	Tobacco	*TMV*	unknown	TIR-NBS-LRR	Witham *et al.* (1994)
L_6	Flax	*Melamspora lini*	unknown	TIR-NBS-LRR	Lawrence *et al.* (1995)
Cf-9	Tomato	*Cladosporium fulvum*	*Avr9*	LRR	Jones *et al.* (1994)
Cf-2	Tomato	*Cladosporium fulvum*	*Avr2*	LRR	Dixon *et al.* (1996)
Prf	Tomato	*P.syringae.pv.tomato*	*AvrPto*	CC-NBS-LRR	Salmeron *et al.* (1996)
I2	Tomato	*Fusarium oxysporum f sp. lycopersici*	unknown	NBS-LRR	Ori *et al.* (1997)
RB	Potato	*Phytopthora infestans*	unknown	NBS-LRR	Sanchez and Bradeen, 2006

As we discussed earlier that the certain functional domains are highly conserved even among distantly related *R* genes. *R*-proteins with NBS-LRR, extracellular-LRR and LRR-Kinase are even considered as the ubiquitous ones in plants as revealed by the genomics studies. It has been reviewed also that about 75 per cent *R* genes belong to the NBS-LRR class (*i.e.* they are having NBS domains), the most abundant one. Lots of work has been done through the PCR based techniques using degenerate *R*-gene primers working on variety of crops simultaneously. Obviously these primers are based on the conserved domains so as to make them the degenerate ones. The single degenerate primer thus can amplify *R* genes from different crops, their wild relatives and varieties as well. Using this approach a large number of Resistance genes analogs (RGA) have been identified. We call them analogs here because they amplify regions similar to the region from where the primer has been designed. There might be some sequence dissimilarity which will be the next chapter to study. This is thus an easier way to go through the candidate *R*-genes (genes suspected to be involved in the expression of trait) among crop plants.

What to Do with these Genes Now?

R genes known till date are either oligogenic or polygenic in nature. They have been deployed in different plant varieties using the back cross method (FIG) and gene pyramiding. In the back cross method we simply back cross the resistance ones to the line of interest so as to have a nearly isogenic line with desired level of resistance for a disease. This method is time consuming and requires a resistance parent (may be wild or cultivated variety) which is always not the case.

Gene pyramiding is the most popular approach for introducing the disease resistance in crops now days. It is because at a time more than one gene could be incorporated for the purpose. Now the question arises, why couldn't we have so many resistant varieties for different diseases? Problem is infact their durability. Very often *R*-genes lack durability due to single loss of function mutation in the corresponding *avr* gene in pathogen, so that the *R*-proteins could no more be able to recognize them. In the due course of evolution there is always like a "tug of war" between *R* and *avr* genes making the deployment difficult all the time. It is therefore pyramiding multiple *R*-genes into the individual plant lines is one of the best alternative. In such situation pathogen would require the accumulation of mutations in the multiple *Avr* genes and their cumulative effect would be required then to escape detection. This kind of accumulation of mutation and their cumulative effect is most unlikely.

Development of multiline is also an alternative towards *R*-gene transfer. In this method a mixture of lines each one expressing a different *R*-gene(s) is sown simultaneously in a plot, so that they may result in a multiline having broad spectrum of disease resistance. As discussed above, in this case also the *Avr* genes would have a strong selection pressure for mutations against their respective *R* genes. To avoid this possibility a susceptible line can be included in the mixture. This approach could be an alternative but not likely to become popular due to complexity in the process.

Discussion could not be summed up without transgenics at all as they are the thrust area of research now days. Resistance has been developed in some of the crops via this approach using genes conferring the pathogenesis related proteins, *R*-genes, antimicrobial peptides and transcriptional regulators (Table 21.4). *Glucanase and chitinase* genes have been transferred in several crops such as alfa-Alfa, canola, grapevine, kivi-fruit, potato, rapeseed, rice, tobacco, tomato and wheat for a number of diseases. The gene stacking approach (transferring more than one gene at a time) with Glucanase and chitinase also followed for several diseases like *Glucanase chitinase* against many fungal and bacterial diseases like Bacterial Leaf Blight and Sheath Blight.

Transgenics have been developed using the *R* genes with variable degree of success for example *RPM1*, *N*, *Xa21*, *RPP27*, *Cf-9* and *Pto*. *R* genes are like a cluster of genes that are positively regulated by transcriptional activation and negatively by RNA silencing. Another indirect approach one can follow is the development of transgenics over expressing the up-regulating elements and vice-versa. It has been reported recently in *Arabidopsis thaliana* that the over expression of transcriptional factors using transgenic approach could be an alternative for development of the disease resistant transgenic plants. However there is a long way to go with question of success too.

Table 21.4: Strategies Followed in Genetic Engineering for Disease Resistance

Strategy	*Examples*	*Remarks*
Constitutive expression		
	Pyramiding R genes	Durable resistance *e.g. RPM1, N,Xa21,RPP27, Cf-9,Pto*
	PR genes	Can increase resistance but unable to activate whole defense response. *e.g. PR1–11*
	Antimicrobial genes	Used for gene stacking approach.
Local expression		
	Transcriptional activators	Require pathogen inducible promoter and can activate the cluster of genes for defense responses.
	Toxic genes	They can stop pathogen growth but require pathogen inducible promoter.
	Ellicitor or *Avr* genes	Trigger to activate successful defense but require pathogen inducible promoter *e.g. flagellin, HrpA* and *Y, VirB1, 2* and *5* (bacterial) and *chitinase* and *glucanase* (Fungal) genes.
	Virus coat protein	PRSV coat protein gene.
RNA interference	Silencing of genes for pathogenesis	They can efficiently target the pathogens.
Gene knockouts or Mutations	Mutations of negative regulators of defense *e.g.* in *Mlo* locus.	

Conclusion

The interaction of plant disease resistance genes with pathogen proteins for virulence and avirulence reaction is still a matter of interest. Lots of facts are yet to be known. So far inducing resistance in plants is concerned we have number of options today. Also, most of the genes are specific to certain pathogens, their level of expression and durability among the plant varieties is questionable somehow. But pyramiding multiple resistance genes in the crop plants is the leading strategy. Understanding their mode of expression and heritability with the help of advanced molecular techniques our felt need today.

References

Bent, A.F., Kunkel, B.N., Dahlbeck, D., Brown, K.L., Schmidt. R., Giraudat, J., Leung, J. and Staskawicz, B.J. (1994). *RPS2* of *Arabidopsis thaliana* a leucine rich repeat class of plant disease resistance genes. *Science,* 265: 1856–1860.

Brueggeman, R. (2002). The barley stem rust-resistance gene Rpg1 is a novel disease-resistance gene with homology to receptor kinases. *Proc. Natl. Acad. Sci.,* 99: 9328–9333.

Cai, D., Kleine, M., Kifle, S., Hans Joachim, H., Sandal, N.N., Marcker, K.A., Klein lankhorst, R.M., Salentijn, E.M.J., Lange, W., Stiekema, W.J., Wyss, U., Grundler, F.M.W. and Jung, C. (1997). Positional cloning of a gene for nematode resistance in sugar beet. *Science,* 275: 832–834.

Dangl, J.L. and Jones, J.D. (2001). Plant pathogens and integrated defence responses to infection. *Nature,* 411: 826–833.

Dixon, M.S., Jones, D.A., Keddle, J.S. and Chopra, V.L. (1996). The tomato *Cf-2* disease resistance locus comprlses two functional genes endocidng leucine-rich repeat proteins. *Cell*, 84: 451–459.

Grant, M.R., Goldlard, L., Straube, E., Ashfield, T., Lewald, J., Sattler, A., Innes, R.W. and Dangi, J.L. (1995). Structure of the Arabidopsis *RPM7* gene enabling dual specificity disease resistance. *Science*, 269: 843–846.

Grover Anita and Gowthaman R. (2003). Strategies for development of fungus resistant plants. *Current Science*, 84(3).

Hammond-Kosack, K.E. and Jones, J.D.G. (1997). Plant disease resistance genes. *Annu. Rev. Plant. Physiol. Plant Mol. Biol.*, 48: 575–607.

Hankuil Yi and Eric J. Richards (2007). A cluster of disease resistance genes in *Arabidopsis* is coordinately regulated by transcriptional activation and RNA silencing. *The Plant Cell*, 19: 2929–2939

Heath, M.C. (2000). Nonhost resistance and nonspecific plant defenses. *Curr. Opin. Plant Biol.* 3: 315–319.

Johal, G.S. and Briggs, S.P. (1992). Reductase activity encoded by the *HM7* disease resistance gene in maize. *Science*, 258: 985–987.

Jones, D.A., Thomas, C.M., Hammond–Kosack, K.E., Balint, P.J. and Jones, J.D. (1994). Isolation of the tomato *Cf–9* gene for resistance to *Cladosporium fulvum* by transposon tagging. *Science*, 4: 789–93.

Lawrence, G.J., Finnegan, E.J., Ayliffe, M.A. and Ellis, J.G. (1995). The *L6* gene fort flax rust resistance is related to the Arabidopsis bacterial resistance gene *RPS2* and the tobacco viral resistance gene *N. Plant Cell*, 7: 1195–1206.

Martin, G.B., Brommon S.S.H., Chunwonge, J., Frary, A., Ganal, M.W., Spivey, R., Wu, T., Earle, E.D. and Tanksley, S.D. (1993). Map based cloning of a protein *Kinase* gene conferring disease resistance in tomato. *Science*, 262: 1432–1436.

McDowell, J.M., and Woffenden, B.J. (2003). Plant disease resistance genes: recent insights and potential applications. *Trends in Biotechnology*, 21 (4): 178–183.

Mindrinos, M., Katagiri, F., Yu, G.L. and Ausubel, F.M. (1994). The *A. thaliana* disease resistance gene RPSP encodes a protein containing a nucleotide–binding site and Leucine rich repeats. *Cell*, 78: 1089–1099.

Mundt, C.C. (2002). Use of multiline cultivars and cultivar mixtures for disease management. *Annu.Rev.Phytopathol.* 40: 381–410.

Ori, N., Eshed, Y., Paran, I., Presting, G., Aviv, D., Tankley, S., Zamir, D. and Flurr, R. (1997). The I2–C family from the wilt disease resistance locus I2 belongs to nucleotide binding Leucine–rich repeat super family of plant resistance genes. *Plant Cell*, 9: 521–532.

Parker, J.E., Coleman, M.J., Szabo, V., Frost, L.N., Schmidt, R., Vander Biezen, E.A., Moores, T., Dean, C., Daniels, M.J. and Jones, J.D.G. (1997). The Arabidopsis downy mildew resistance gene *RPP 5* shares similarly to the Toll and Interleukin–1 receptors with *N* and *L6*. *Plant Cell*, 9: 879–894.

Pink, D.A.C. (2002). Strategies using genes for non–durable resistance.Euphytica 1, 227–236.

Salmeron, J.M., Oldroyd, G.E.D., Rommens, C.M.T., Scofield, S.R., Kim, H.S., Lavelle, D.T., Dahlbeck, D. and Staskawicz, B.J. (1996). Tomato *Prf* is a member of leucine rich repeat class of Plant disease resistance genes and lies embedded with in the *Pto* Kinase gene cluster. *Cell*, 86: 123–133.

Sanchez, M.J. and Bradeen, J.M. (2006). Towards efficient isolation of R gene orthologs from multiple genotypes: Optimization of long range–PCR. *Molecular Breeding*, 17: 137–148.

Song, W.Y., Wang, G.L., Gen, L.L., Kim, H.S., Pil, Y., Holsten, T., Gardner, J., Wang, B., Zhai, W.X., Zhu, L.H., Fauquet, C. and Ronald, P. (1995). A receptor kinase–like protein encoded by the rice disease resistance gene, *Xa 21. Science*, 270: 1804–1806.

Sarah J. Gurr and Paul J. Rushton. (2005). Engineering plants with increased disease resistance: what are we going to express? Trends in Biotechnology 23(6): 275–282.

Van der Biezen, E.A. and Jones, J.D.G. (1998). Plant disease–resistance proteins and the gene–for–gene concept. Trends Biochem. Sci. 12, 454–456.

Whitham, S., Dinesh-Kumar, S.P., Choi, D., Hehl, R., Corr, C. and Baker, B. (1994). The product of the Tobacco Mosaic virus resistance gene *N:* Similarity to Toll and Interleukin–1 receptor. *Cell*, 78: 1101–1115.

Xiao, S. The Arabidopsis genes RPW 8.1 and RPW 8.2 confer induced resistance to powdery mildew diseases in tobacco. *Mol. Plant Microbe Interact.* (In press).

Plant Diseases Management in Horticultural Crops (2011) *Pages* **275–280**
Editors: **Shahid Ahamad, Ali Anwar and P.K. Sharma**
Published by: **DAYA PUBLISHING HOUSE, NEW DELHI**

Chapter 22

Root Rot: A Threat to Apples in Kashmir Valley

Farah Rasool, Shahzad Ahmad and Khurshid Ahmad

Division of Plant Pathology,
S.K.University of Agricultural Sciences and Technology of Kashmir,
Shalimar, Srinagar – 191 121, J&K

White root-rot is the disease of the roots which remain covered with white fungal mycelial growth particularly in the rainy season resulting in the death of the plants and thus attains a serious status. It is one of the most important soil borne disease of apple having a host range of about 158 plant species belonging to over 45 families (Ito and Nakamura, 1984), comprising of fruit plants, vegetables, forest trees and field crops.The diseases was first reported in Germany in 1883 (Hartig, 1883) and in 1891 it was reported from France on grapes (Viala, 1891). It was reported on apple for the first time in 1900 in Norwich and in 1913 from Canterbury (Salmon and Wormold, 1913). The fungus was first reported in India in Uttarakhand in1929 (Bose and Sindhan, 1976) and latter on in Himachal Pradesh (Agarwal, 1961). The annual losses in apple alone have been worked out to be 1.3 million (Agarwal and Sharma, 1966).The disease is endemic at 1800-2400m altitude with soil pH 6.1-6.5 (Gupta *et al.*, 1995).

Symptoms

The pathogenic fungus can be easily diagnosed by observing under and above ground parts of the plant. The earliest above ground manifestation is bronzing of the leaves, diminution in their size and a stunted tree growth resulting in the progressive decline in the vigour as a whole or certain branches. It is followed by wilting, defoliation and death of the tree (Jain, 1961 Agarwala and Sharma, 1966 Gupta 1977, Sztejnberg *et al.*, 1987). The symptoms are more pronounced and damage is greater in nursery where the pathogen produces root rot and wilt symptoms in apple seedlings.

In fruit trees, these symptoms are usually associated with heavy bloom, and fruiting next year. In the succeeding years, few leaves emerge and much of the fruits fail to reach maturity and dieing back of branches is quite evident. Infected trees often persist for 2-3 years depending upon the infection intensity in the roots. Severely infected trees may succumb within a single season (Naltran, 1927,Jain 1961, Agarwal and Sharma, 1966 Gupta, 1977.Sztejnberg *et al.*, 1987) The symptoms are more pronounced and damage is more in the nurseries. The above ground symptoms are not distinctive because of similar symptoms produced by other root maladies.

The below ground expression of the pathogen on the plant roots is the final diagnostic feature. The lateral roots turn dark brown and are covered with a greenish grey or white mycelia mat having a flocculent web of whitish strands or ribbons during monsoon season. The cortical cells are ruptured resulting in disruption of the plant system leading to the death of the tree and hence expressing disease on above ground plant parts (Nattrass (1972, Agarwala and Sharma, 1966) During rainy season white mycelia mat can be seen in the soil adhering to the roots. Nattrass (1927) observed that white flocculent web disappears after the death of the host, leaving the root surface dotted over with small rounded black sclerotia. Agarwala and Sharma (1966) reported that the fungus does not infect xylem tissues of apple roots but the bark is completely damaged, however, Gupta (1977) isolated the pathogen from the woody parts of diseased apple roots.

Causal Organism

The disease is caused by *Rosellinia necatrix* Berl.ex Prill (Anamorph: *Dermatophora necatrix* Hartig).However the perfect stage of the fungus has not been reported from India. The fungus is distinguished on the basis of pear shaped swellings at each septum of the fungal hyphae (Gupta and Sharma, 1999: Sharma and Sharma, 2004). The fungus produces scattered black; microsclerotia measuring 130x98 µm.Two types of mycelial characteristics have been described. Small black microsclerotia sometimes develop on the rotten bark, which tend to unite and form microsclerotial sheets. These sclerotia are rough, irregular, composed of compact mass of the interwoven septate hyphae with no differential cells. The rhizomorphs are not well defined. The pathogen thrives over a wide pH range, however, 6.1-6.5 pH is most favourable. Structural changes in the pear shaped swellings at the septum at different pH regimes have also been recorded (Agarwala and Sharma, 1966).

The ascostromata of *Rosellinia necatrix* are superficial, ostiolate, often as dense swarms on a common mycelial mat or subiculum, subglobose, smooth and dark. Asci are filiform, elongated, cylindrical (250-380x8-12µm), long stalked, unitunicate and having eight ascospores, the ascospores are pigmented, aseptate, often with a minute hyaline appendages. Conidiophores are produced on brown, rigid synnemata (up to 1.5mm long). The stipe (40-300µm thick) is composed of flexuous intertwined branched threads, often branched dichotomously towards the apex. Conidia measure 3-4.5x2-2.5µm which are aseptate, solitary, simple, ellipsoidal or obvoid, hyaline to pale brown and smooth. (Gupta and Sharma, 2000).

Disease Cycle and Epidemiology

The resting stage of the fungus is inevitably represented by microsclerotia which under suitable conditions give rise to fine white exploration hyphae. The fungus survives in the form of mycelium or sclerotia in the infected roots. Infection of the new roots takes place by the fungal mycelium present on the debris or by the contact of new plant roots with old dead roots. The infection by *Rosellinia necatrix* on the apple roots takes place in four phases: external proliferation in the form of mycelial thallus or strings of undifferentiated cells; formation of perenation sclerotium consisting of a melanized pseudoparenchymatous cortex and a white prosenchymatous medulla; and internal proliferation in the form of sub corticular prosenchymatous strings (Tourvielle de Labrouhe, 1982). The infection by *D. necatrix* occurs when the fungus comes in contact with the fine roots and ultimately whole root system is devoured. The pathogen ruptures the cortical and phloem cells were damaged at various points in later stages. The cavities are created by the hyphal cells of the organism. There is no distortion of xylem cells (Agarwala and Sharma, 1966). The rate of lesion development and rotting depends upon prevailing temperature, cultivar, scion/rootstock interaction, season and age of the plant. Seedlings start wilting within 6 weeks and within 10 weeks after inoculation of *R.necatrix* culture (Merwe and matthee, 1974).

The fungus contains active polyphenoloxidase, which oxidizes phenols to quinines (Sztejnberg *et al.*, 1983). Mercuri (1928) showed that *R. necatrix* also produced toxic compounds in filtrate, which necrotized vine roots. It is thermostable with more than two compounds, one inhibiting root growth at low concentration and the other bud growth at high (Abe and Kono, 1954). The fungus produces lot of pectolytic and cellulolytic enzymes which help in the degradation of host tissue in advance Gupta and Gohain, 1982 and also toxins, which help in killing the bark. Toxin production was maximum in woody parts followed by diseased roots and bark. However, no work has been done on characterization of dialyzable and stable at pH 5.5-8.5. The host reaction to filtrate was almost similar to that of fungus host range but it is difficult to categorize the toxin produced by such a polyphagous fungus to be host specific toxin Gupta and Gohain, 1982. The mycelium spreads from upper side to lower side of the roots or from infected to healthy roots through infested soil and rain water. The disease is most serious in water –logged acidic soils at pH 6.1 to 6.5. The disease severity is positively correlated with moderate temperature (15-25°C) and high moisture. Five to twenty year old trees arc most susceptible to disease. The plants infected with apple chlorotic leaf spot virus are more susceptible to white and violet root rot diseases (Arial *et al.*, 1990)

Management

Cultural Practices

The cultural practice recommended for the management of white root rot of apple from time to time include organic amendments which alter the physical, chemical and biotic conditions of the soil. Haneda *et al.* (1985) reported that the growth of *R.necatrix* was arrested by organic materials. The incorporation of neem, mahua and castor cakes prior to the inoculation of *D.necatrix* were effective in reducing the incidence of root rot Sharma (1993). Addition of neem cake in the soil one week prior to inoculation of pathogen resulted insignificant disease reduction and symptoms are also delayed (Sharma and Gupta, 1995; Sharma, 2000). Soil amendment with neem cake and deodar needles has been reported to reduce the incidence of the disease (Gupta and Jindal, 1989; Sharma and Gupta, 1996b). Hot water treatment of infected seedlings at 45°C for one hour before plantation, digging isolation trenches and removal of rotten roots followed by application of disinfectant paste. In order to save the trees in early stages of attack the diseased roots should be trimmed off and destroyed. Replanting should be done in clean soil and infested soil should be fallowed with frequent cultivation to starve out the infection hyphae. Burning and heat drying of infected roots have also been reported to increase the life of the trees (Agarwala and Sharma, 1975, Jain, 1961). Soil solarization has also been found effective in reducing the fungus inoculum in soil (Freeman *et al.*, 1986). Soil moisture plays a major role in affecting the intensity of the disease; hence, ill drained soils should be improved by following central drainage system. The acidic soils should be amended by applying lime for some years (Jain, 1961).

Chemical Control

The success in controlling the soil borne diseases lies in the fact that the required and effective levels of the fungicide reach to the point of infection in the soil. Initially, broad spectrum chemicals such as carbon bisulphide, choropicrin, calcium cynamide, formaldehyde etc. were recommended for checking root rot of apple, though these had been found effective but none could give satisfactory control at economic level of application (Cohen *et al.*, 1981; Laville and Vogel, 1984). With the advent of systemic fungicides, the old recommendations were replaced with fungicides such a benomyl, aureofungin, carbendazim etc. Artificially infected apple plants in posts recovered with the application of carbendazim (0.1 per cent) even when all the roots were destroyed (Gupta, 1977). The fungicide should be applied through deep holes (15-20 cm) made in the basin of tree at 30 cm distance from each other. Prior to the application of chemical the tree basin is brought to a soil moisture level of 30 to 40 per cent. The treatment is done 3 to 4 times every year during rainy season. Sousa (1985) observed preventive as well as curative effects of benomyl and carbendazim and reported the efficacy of role of phorate and its combination with carbendaim and Dithane M-45 in reducing the disease in nursery.

Resistant Rootstocks

Almost all the rootstocks have shown susceptibility to the pathogen, however, in the field the degree of susceptibility varies (Gupta and Verma, 1978). Recently, Sharma and Gupta (1966d) observed some degree of resistance against the pathogen in three rootstocks namely MM109,M16 and MM104, but until and unless, a rootstock with better resistance is developed, the root rot problem will persist. Modern biotechnological approaches are required to be exploited for combating this pernicious soil borne disease of apple

Table 22.1: Resistance Sources Against *D. necatrix*

Resistance	*Cydonia oblonga* *Malus baccata* (*Japan*) *M. baccata* Rohroo *M. floribunda* *Prunus persica, P, cornuta* *P. carasodes, pyrus pashia* var. Kumaoni *Rus Pippin*	Ram, 1982 Gupta dna Singh 1985
Moderately resistant	*Prunus mariana, P. myrobalan,* *Corylus avellana, Malus floribunda,* *M. toringodes, spe 262,Sel 69*	Sousa, 1985 Gupta and Singh, 1985, Agarwal and Sharma, 1966 Denardi and Berton, 1995,
Moderately susceptible	M9,MM106,MM102	Sousa, 1985

Biocontrol

Use of various antagonists in the management of disease has been explored and few universal antagonists such as *Trichoderma viride, T. harzianum* (Mercer and Krik, 1984, freeman *et al.*, 1986a; Sztejnberg *et al.*, 1987, *Enterobacter aerogenes* (Gupta and Jindal, 1989) have been found to protect the plants from *D necatrix*. These antagonists inhibited the growth of *R. necartix* and covered the pathogen culture with different degrees of sporulation. In aftificial inoculated or natural infested soils, isolate T-8 of T. harzianum reduced the colonization of avocado leaf disc, It also resulted in reduction in inoculum density of *R. necatrix* in artificial inoculated autoclaved soil and natural infested En-Zurium soil with increasing concentration of *T. harzianum* isolate t-8 (0.1, 5, 10g/kg soil) over a 6 week period. The disease WAS not detected at all with a 10 g preparation in comparison to 35 per cent colonization in control. Subsequently, the mortality of almond seedling reduced to 20 per cent from 75 per cent when 5 g preparation of isolate T-8 was added to naturally infested soil. This isolate wither competed more successfully for food substrate or behaved as a mycoparasite of the pathogen (Freeman *et al.*, 1986a) Bacterial antagonists like *Enterobacter aerogenes* and Bacillus subtilis have been found to inhibit *D. necatrix* and reduce disease up to 45 days of inoculation (Gupta and JINDAL, 1989). Sharma and Gupta (1996c) reported that repeated application of these antagonists enhanced their efficacy against the pathogen and effect is long.

References

Agarwal, R.K. (1961). Problem of root rot in Himachal Pradesh and prospects of its control with antibiotics. *Himachal Hortic.*, 2: 171–178

Anonoymous (1937). Some diseases of apples. *Mysore Agric. Cal.*, 25 p.

Cooley, J.S. (1936). *Sclerotium rolfsii* as a disease of nursery apple trees.*Phytopathology*, 26: 1081–1083.

De Sousa, A.J.T, Guillaumin, J.J., Sharples, G.P. and Whalley, AJ.S. (1995). *Rosellinia necatrix* and White root rot of fruit trees and other plants in Portugal and nearby regions. *Mycologist*, 9: 31–33

Freeman, S., Sztehnberg, A. and Chet, I. (1986). Evaluation of *Trichoderma* as a biocontrol agent for *Rosellinia necatrix*. *Plant Soil*, 94: 163–170.

Garg, R.C and Gupta, V.K. (1989). Reaction of apple rootstocks to *Phytophthora cactorum*. *Indian Phtopathol.*, 42: 580–581.

Gupta, G.K. and Agarwala, R.K. (1973). Field testing of soil fungicides for the control of seedling blight of apple. *Indian J. Mycol. Plant Pathol.*, 3: 109–111.

Gupta, V.K. and Mir, N.M. (1983). Field testing of apple rootstocks against *Phytophthora cactorum*: A new technique. *J. Tree Sci.*, 2: 81–83.

Gupta, V.K., Rana, K.S. and Mir, N.M., 1985. Variability in *Phytophthora cactorum* in India. In: *Ecology and Management of Soil Borne Plant Pathogens*, (Eds.) C.A. Parker, A.D. Rovira, A.J. Moore, P.T.W. Wong and J.F. Koilmongen. American Phytopathological Society, USA, pp. 167–171.

Gupta, V.K. and Sharma, R.C. (1999). Concepts and practices in the management of soil borne diseases. In: *Modern Approaches and Innovations in Soil Management,* (Eds.) D.J. Bagyaraj, A.M. Verma, K.K. Khanna and H.K. Kehri. Rastogi Publications, Meerut, pp. 333–334.

Gupta, V.K. and Sharma, S.K. (1999). Soil borne diseases of apple and their management. In: *Diseases of Horticultural Crops: Fruits*, (Eds.) L.R. Verma and R.C. Shama. Indus Publishing Co., New Delhi pp. 89–104.

Gupta, V.K. and Verma, K.D. (1978). Comparative susceptibility of apple rootstocks to *Dematophora necatrix*. *Indian Phytopathol.*, 31: 377–378.

Gupta, V.K. and Rana, K.S. (1982). Effect of soil factors and amendments on the viability of *Phytophthora cactorum* sporangia and oospores in apple orchard soil. *J. Tree Sci.*, 1: 64–70.

Ito, S.I. and Nakamura, N. (1984). An outbreak of white root rot and its environmental conditions in the experimental arboretum. *J. IPM For. Sci.*, 66: 262–267

Lee, S., Chong, B., Jang, H., Kim, K.H. and Choi, Y. (1935). Incidence of soil borne diseases in apple orchards in Korea. *J. Plant Pathol.*, 11: 32–138.

Mcintosh, D.L. (1975). Proceedings of the 1974 APDW Workshop on crown rot caused by *Phytophthora cactorum* in planting of apple trees aged 1 to 7 years. *Plant Dis. Rep.*, 59: 539–541.

Prasad, R.D., Rangeshwaran, R. and Kumar, P.S. (1999). Biological control of root and collar rot of sunflower. *J. Mycol. Plant Pathol.*, 29: 184–188.

Sharma, M. and Sharma, S.K. (2002). Control of white root rot of apple caused by *Dermatophora necatrix* with *Bacillus* sp. *Plant Dis. Res.*, 17: 308–312

Sharma, M. (2000). Non-chemical methods for the management of white root rot of apple. *Ph.D. Thesis*, UHF, Solan, 96p.

Sharma, R.C., Sharma, S. and Sharma, J.N. (1999). Present concepts and strategies of disease management in pome and stone fruits. *Int. J. Trop. Plant Dis.*, 17: 67–80.

Sharma, S.K. and Gupta, V.K. (1995). Management of root rot of apple through soil amendments with plant materials. *Plant Dis. Res.*, 10: 164–167.

Sharma, S.K. (1993). Studies on management of white root rot of apple. *Ph.D. Thesis*, UHF, Solan, 146 p.

Singh, R.S. and Pandey, K.R. (1996). Effect of green and mature plant residues and compost on population of *Pythium aphanidermatum* in soil. *Indian Phytopathol.*, 19: 367–372.

Sztejnberg, A., Greeman, S., Chet, I. and Katan, J. (1987). Control of *Rosellinia necatrix* in soil and in apple orchard by solarization and *Trichoderma harzianum*. *Plant Dis.*, 77: 365–369.

Viala, P. (1891)). Monograph Duepourridie (Dematophora) des Vigens. Paris.

Wazir, F.K., Meladul, K., Qureshi, J.A., Barech, A.R. and Kakar, K.M. (2000). Effect of physical and biological control measures on Colt rootstock of cherry grown on crown gall infested soil. *Sarhad J. Agric.*, 16: 49–51.

Plant Diseases Management in Horticultural Crops (2011) *Pages* ***281–289***
Editors: **Shahid Ahamad, Ali Anwar and P.K. Sharma**
Published by: **DAYA PUBLISHING HOUSE, NEW DELHI**

Chapter 23

"Yellow Vein Mosaic": A Threat to Okra Cultivation in Jammu

Ranbir Singh, Harish Kumar and V. K. Razdan

Division of Plant Pathology, Faculty of Agriculture, Main Campus, Chatha, Sher-e-Kashmir University of Agricultural Sciences and Technology, Jammu – 180 009

Okra [*Abelmoschus esculentus* (L.) Moench] originated in tropical Africa (Purseglove, 1987) is an important vegetable crop widely cultivated in different parts of the world (Ali *et al.*, 2000; Gandhi *et al.*, 2006). It is good source of vitamin, protein, carbohydrates, fats, minerals, iron and iodine and has medicinal property for people suffering from renal colic, leucorrhoea and general weakness (Hamon, 1988; Baloch *et al.*, 1990; Norman, 1992; Diaz and Ortegon, 1997; Schippers, 2002). Fruits of okra are considered useful for the control of goitre, due to presence of high iodine contents in it. In Turkey, its leaves are used in the preparation of medicines against inflammation (Mehta, 1959; Saimbhi, 1993). The okra crop is attacked by a number of fungi, bacteria, viruses, mycoplasma, nematodes and insects and among them, plant viruses are the most important factors causing severe constraint in the production and productivity of the crop (Gupta *et al.*, 2003; Kareem and Taiwo, 2007). Among the viral diseases, yellow vein mosaic virus is the most dreaded disease causing severe losses in terms of yield as well as quality (Capoor and Varma, 1950). In India, the disease was first reported by Kulkarni in 1924 from Bombay province in Maharashtra and the viral nature was established by Uppal *et al.* (1940) who named it as yellow vein mosaic virus disease (YVMV). In 1999, Arumugam and Muthukrishnan reported that the disease prevails in all plains as well as in the lower hills of India. Information regarding the status, epidemiology and management aspect of yellow vein mosaic disease of okra is scanty in Jammu and Kashmir State and needs immediate attention.

Yellow vein mosaic disease of okra is caused by yellow vein mosaic virus (YVMV) belonging to geminiviridae family. The viral nature of disease was established by Uppal and his coworkers in 1940. The typical symptoms of the disease are induced by DNA-β particles when inoculated on okra (Joyce and Ramakrishnan, 2003; Kirthi *et al.*, 2004; Kon, 2009).

Disease Incidence and Losses

Sastry and Singh (1973) estimated that early infection within 20 days after germination caused yield reduction up to 90 per cent, however, the plants infected at 35 to 50 days following germination showed losses of 83 and 49 per cent, respectively. Sharma and Arora (1993) estimated the losses between 50-90 per cent due to YVMV. Nath and Saikia (1993) reported 94.48 and 32.65 per cent loss in yield when the okra crop was infected at 35 and 63 days after sowing, respectively suggesting that early infection caused heavy yield losses as compared to late infection. Similarly, Mazumder *et al.* (1996) reported that marketable yield obtained from early sown crop was more than late sown crop and suggested that the areas where the disease was prevalent, crop should be sown early to avoid the synchronization of virulent vector and susceptible phase of the crop. Fajinmi and Fajinmi (2010) observed low virus infection in okra plots netted for 21 days after emergence as compared to control plots (90 per cent).

Arumugam and Muthukrishnan (1999) reported that YVMV caused severe damage when the infection occurred before 30 days after sowing (DAS). Similarly, Pun and Doraiswamy (2000) reported 95.7 per cent yield losses in plants inoculated one week after emergence. Ahmed and Patil (2004) conducted periodical survey to determine the incidence of YVMV on different okra cultivars and found that the crop recorded high disease incidence during summer season (58.14 per cent) as compared to rainy season (15.08 per cent). In a field experiment, least disease incidence was observed in early sown okra cultivars at closer spacing and highest on late sown at wider spacing.

Symptomatology

Capoor and Varma (1950) observed that characteristic symptoms of the disease appeared as homogenous interwoven network of yellow vein enclosing islands of green tissues within. Initially, the leaves exhibited only yellow coloured veins, but in the later stages the entire leaf turned yellow due to partial or complete destruction of chlorophyll content. Infected plants remained stunted and produced very few small leaves. According to Bhagabati and Goswami (1992), the crop remained symptomless up to 14 days after sowing and the initial symptoms were only visible at 21 days after sowing. Most of the affected plants develop thickening of veins on their lower sides bearing one or two fruits, often deformed, small, pale in colour and tough in texture are produced on infected plants (Singh, 1990). Sinha and Chakarbarti (1978) found that in cultivar Pusa Sawani, the disease appeared at intervals from 33 to 75 days after sowing resulting in considerable loss in plant growth, fruit number and seed yield per plant. However, the disease retarded the growth and development of okra plants in terms of leaf area, fruit length, fruit weight and fruit volume. (Bhagabati *et al.*, 1998; Debnath and Nath, 2002).

Transmission and Host Range

An appreciable amount of work has been done on transmission and host range of the disease by various workers (Verma *et al.*, 1966; Ahmed, 1978; Ahmed *et al.*, 1980). Capoor and Varma (1952) found the yellow vein mosaic virus (YVMV) was not sap transmissible, but under artificial condition it could be transmitted by grafting. However, under natural conditions it can only be transmitted by insect vector whitefly (*Bemisia tabaci*) in a persistent manner (Phadvibulya *et al.*, 2009), and even a single viruliferous vector could transmit the virus from diseased to healthy plants (Boonsirichai *et al.*, 2009).

The population build up of whitefly (*B. tabaci*) varied with the host plant (Lal, 1981). Hilje *et al.* (1993) made detailed studies on the ecology of whitefly (*B. tabaci*) that included various host plants for observing the vector and host relationship and found that whitefly could survive on okra, tomato, chilli, tobacco, melon, pumpkin, watermelon, cucumber, beans, cotton and sweet potato. Symptoms produced by the virus on congress grass (*Parthenium hysterophorus* L.) and *Ageratum conyzoides* (Linn.) were similar to those in okra (Singh and Singh, 1999). Although whitefly has very wide host range, few hosts belonging to the family solanaceae, cucurbitaceae and malvaceae are considered to be most preferred (Singh *et al.*, 1962).

Etiology and Epidemiology

Etiology and epidemiology of whitefly-transmitted, virus disease of okra was studied under electron microscopy of material from *Abelmoschus esculentus* infected with *Hibiscus* yellow vein mosaic virus revealed the presence of small, spherical or geminate virus like particles in the form of aggregates in the phloem sieve nuclei. Dry hot weather with little or no rainfall is conducive for the disease development of yellow vein mosaic and also for the multiplication of the vector. Temperature has been found to be related positively, whereas relative humidity was negatively correlated with disease incidence and whitefly population (Seif, 1980). Spread of yellow vein mosaic virus has shown a steep rise during the early growth stages of the crop and the extent of final infection depended on the degree of initial infection (Khan and Mukhopadhyay, 1986). Sharma *et al.* (1987) in a field study reported that the incidence of YVMV on okra increased with decreasing temperature in September compared with August which showed a significant negative correlation between temperature and virus incidence.

Singh (1990) observed that population of whitefly and virus incidence was more when the atmospheric temperature remained high and relative humidity remained low; contrary to this, low temperature and high humidity were detrimental for the whitefly population resulting in poor disease spread. A positive correlation has been reported by Nath (1994) between disease incidence, temperature, relative humidity, rainfall, number of rainy days and whitefly population on mung bean (*Vigna radiata* (L.) Wilczek).

Pun *et al.* (2005) reported positive correlation between disease incidence and whitefly population with maximum temperature, minimum temperature and sunshine hours, whereas negative influence with morning relative humidity, wind velocity and total rainfall. Ali *et al.* (2005a) observed that with rise in minimum temperature there was a corresponding increase in disease intensity and whitefly population and with rise in the relative humidity there was a corresponding decrease in whitefly population.

Management Studies

Management of yellow vein mosaic virus of okra has been attempted by several workers by controlling the vector using chemical insecticides or by growing resistant varieties.

Screening of germplasm

Use of resistant varieties is the most accepted method of disease management due to its non-hazardous effect on the environment. Chauhan *et al.* (1981) assessed forty six strains of okra against YVMV disease and found that no strain showed resistance. Maximum rate of infection was recorded in Pusa Sawani (75.80 per cent) while the lowest mean value for infection was shown by IC-9273 (17.80 per cent). Bora *et al.* (1992) screened different genotypes of okra against YVMV and found that GOH-4 and GOH-5 were highly resistant for commercial cultivation. Harender *et al.* (1993) calculated losses in fruit and seed yields in five okra varieties and observed that Punjab-7 showed least disease severity and pest incidence with increased yield.

Mohapatra *et al.* (1995) in a field experiment evaluated improved and hybrid varieties of okra against the disease and reported that Pusa Sawani was the most susceptible while HRB-9-2, DOV-91-4 and Pashupati were tolerant. Sangar *et al.* (1997) reported that the incidence of the disease was recorded higher during rainy season while Arka Anamika, Arka Abhay and Parbhani Kranti showed high, moderate and resistant reaction against the disease, respectively.

Reghupathi *et al.* (2000) evaluated the reaction of okra germplasm to YVMV and reported that KS-404, HRB 9-2, Hy-8, P-7, Parbhani Kranti, Selection-10, Selection-4, Okra No. 6, Lorm-1, VRO-3 and Punjab-7 were resistant.

Bhagat *et al.* (2001) found that the rate of infection of disease was higher in Vaishi Vadhu and Pusa Sawani as compared to Parbhani Kranti. Rehman *et al.* (2002) after screening different okra varieties/lines under field experiment found that out of the 24 varieties/lines, nine (GE, PK, CHR, PA × CHR, CHR × PK, CHR × DLPG, PK × PG, PK × AM Red, GV × DLPG) were immune, whereas three (DLPG, PG and PG × PK) were highly susceptible. After screening twelve okra germplasm for resistance to YVMV under field conditions.

Singh *et al.* (2002) in a field experiment evaluated performance of different varieties of okra to YVMV and observed that Arka Abhay, Ambica Local, Ratna Raj and Green Gold were resistant. Ali *et al.* (2005a) cultivated different okra cultivars like Pahuja, Safal, Subz Pari and Surkh Bhindi in a field trial and found that Surkh Bhindi was highly resistant, Subz Pari and Safal were moderately resistant and Pahuja was tolerant to disease. Tiendrebeogo *et al.*, 2010 reported that the overall yield losses due to YVMV were significantly higher in accessions of local cultivars (26 - 55 per cent) than commercial ones (4.4–9.6 per cent).

Chemical Management

Chemical control is the primary method for managing diseases transmitted by whiteflies. Whitefly population and disease incidence was minimized by one soil application of furadon @ 1.5 kg *a.i.* per hectare at the time of sowing followed by 4-5 foliar sprays of either dimethoate (0.05 per cent) or metasystox (0.02 per cent) or nuvacorn (0.05 per cent) at 10 days interval (Shastry and Singh, 1973; Singh, 1977). Seed treatment with agrosan followed by granular application of aldicarb and foliar spray of endosulfan at 30 days after sowing were best treatments for the control of whiteflies (Chelliah *et al.*, 1976).

Application of thimet (Phorate) @ 0.1 per cent at 15 days interval gave the best control of insect vector followed by furadon (carbofuran) @ 0.03 per cent (Krishna *et al.*, 1983). Phorate (0.75, 1.5 and 2.25 kg *a.i.* ha^{-1}), carbofuran (0.3, 0.6 and 0.75 kg *a.i.* ha^{-1}), oxydemeton-methyl (0.313 kg *a.i.* ha^{-1}), monocrotophos (0.45 kg *a.i.* ha^{-1}), endosulfan (0.438 kg *a.i.* ha^{-1}), phosalone (0.438 kg *a.i.* ha^{-1}) and malathion (0.625 kg *a.i.* ha^{-1}) were best chemicals for the control of *B. tabaci* (Nandihalli and Thontadarya, 1986). Rustamani *et al.* (1994) evaluated four insecticides namely thiodan 35 EC (endosulfan), anthio 25 EC (formothion) and curacron 35 EC (profenofos) against *B. tabaci* and found that formothion was significantly more toxic. Borah *et al.* (1996) found that dimethoate and malathion could be used in different insecticidal schedules along with cotton as trap crop for an effective and economic control of white flies and minimizing the incidence of yellow mosaic virus of green gram. Spraying with monocrotophos and cypermethrin (0.02 per cent) reduced the incidence of YVMV (Singh *et al.*, 1998). Naik *et al.* (2009) reported that combination treatment of thiomethoxam (0.0025 per cent) with novaluron (0.05 per cent) and azadirachtin (0.15 per cent) were highly effective in reducing the population of *B. tabaci* and incidence of the disease. Similarly, in a field study Salam (2009) found that seed treatment with imidacloprid (5g/kg) followed by two sprays @ 0.24 ml/l recorded least disease incidence and whitefly populations.

Plant Products

Some plant extract like neem extract have a number of properties useful for insect management (Schmutterer, 1990; Koul, 1992). Alcohol extracts have been found superior to aqueous ones and those from callistemon, datura, agave and ginger gave good degree of suppression of symptoms against theYVMV disease (Chowdhury *et al.*, 1992). Coudriel *et al.* (1985) reported that neem seed kernel extract acted as a strong repellent against *B. tabaci* in cotton.

Aqueous leaf extracts of tobacco (2 per cent), *Ipomoea cornea* (5 per cent) and a seed extract of *Azadirachta indica* and *Pongamia bragla* (5 per cent) gave a similar level of control compared to Endosulfan (0.06 per cent) and Monocrotophos (0.05 per cent). Adiroubane and Letchoumanane (1998) conducted field experiments to evaluate the efficacy of three plant extracts, sacred basil (*Ocimum sanctum*), malbar nut (*Adhutoda vesica*), Chinese chaste tree (*Vitex negudo*) against whitefly (*B. tabaci*) and found that all the treatments reduced the vector population significantly

Natural plant products, acetyl salicylic acid and insecticides were found to be effective against whitefly population build-up and YVMV disease incidence in the field. (Mariappan and Saxena, 1983). The incubation period of the virus in plants treated with leaf extracts of *Prospos chilensis* and *Bougainvillea spectabilis* increased to 19.1 days and 19.3 days respectively, compared with 10.4 days in control plants (Pun *et al.*, 1999). Singh *et al.* (1999) also reported that spraying of Asafoetida (*Ferula assafoetida*) plant extract to okra crop in rainy season showed strong insect repellent activity resulting in reduction of the disease.

Pun *et al.* (2000) studied the effect of virus inhibitory chemicals and neem products against YVMV and suggested that neem oil, neem seed kernel extract and barium chloride were most effective in reducing the virus infection by 88.3, 86.7 and 75.0 per cent, respectively.

Debnath and Nath (2002) used plant products like powder of root, seed, leaf and root extract of drumstick (*Moringa oleifera*), neem (*A. indica*) leaf powder, turmeric (*Curcuma longa*) rhizome powder and keekar (*Acacia nilotica*) bark powder, for environment friendly management of YVMV and whitefly population and found neem leaf powder and keekar bark powder as promising plant extracts in reducing infection in okra.

Referencess

Adiroubane, D. and Letachoumanane, S. (1998). Field efficacy of botanical extracts for controlling major pests of okra. *Indian Journal of Agricultural Sciences*, 68: 168-170.

Ahmed, Z. and Patil, M. S. (2004). Incidence of yellow vein mosaic virus on different okra cultivars in Karnataka. *Karnataka Journal of Agricultural Sciences*, 17: 615-616.

Ali, M., Hossain, M. Z. and Sarkern, N. C. (2000). Inheritance of Yellow vein mosaic virus (YVMV) tolerance in cultivar of okra *(Abelmoschus esculentus* L. Monech). *Euphytica*, 111: 205-209.

Ali, S., Khan, M. A., Habib, A., Rasheed, S. and Iftikhar, Y. (2005a). Correlation of environmental conditions with okra yellow vein mosaic virus and *Bemisia tabaci* population density. *International Journal of Agriculture and Biology*, 7: 142-144.

Arumugam, R. and Muthukrishnan, C. R. (1999). Screening of bhendi (okra) cultivars for resistance to yellow vein mosaic disease [India]. *Indian Journal of Horticulture*, 35: 278-280.

Baloch, A. F., Qayyum, S. M. and Baloch, M. A. (1990). Growth and yield performance of okra (*Abelmoschus esculentus* L.) cultivars. *Gomal University Journal of Research*, 10: 191. (Original not seen. Cited by Below. Patil, P. D. and Mehetre, S. S. 2007. *Sustainable Pest Management*. pp 239-248.)

Bhagabati, K. N. and Goswami, B. K. (1992). Incidence of yellow vein mosaic virus disease of okra (*Abelmoschus esculentus* L. Moench) in relation to whitefly (*Bemisia tabaci* Genn.) population under different sowing dates, *Indian Journal of Virology*, 8: 37-39.

Bhagabati, K. N., Sarma, U. C., Saikia, A. K. and Dutta, S. K. (1998). Effect of yellow vein mosaic virus infection on some morphological parameters in bhendi (*Abelmoschus esculentus* L. Moench). *Journal of Agricultural Sciences*, 11: 94-96.

Bhagat, A. P., Yadav, B. P. and Prasad, Y. (2001). Rate of dissemination of okra yellow vein mosaic virus disease in three cultivars of okra. *Indian Phytopathology*, 54: 488-489.

Boonsirichai, K., Phadvibulya, V., Adthalungrong, A., Srithongchai, W., Puripunyavanich, V. and Chookaew, S. (2009). In: Shu, Q. Y. (eds.). *Induced Plant Mutations in the Genomics Era*. pp 352-354. Food and Agriculture Organization of the United Nations, Rome.

Bora, G. C., Saikia, A. K. and Shadeque, A. (1992). Screening of okra for resistance to yellow vein mosaic virus disease. *Indian Journal of Virology*, 8: 55-57.

Borah, R. K., Nath, P. D. and Deka, N. (1996). Effect of insecticides and crop trap on the incidence of white fly (*Bemisia tabaci* Genn.) and yellow mosaic virus in green gram (*Vigna radita* L. Wilczek). *Indian Journal of Virology*, 12: 75-77.

Capoor, S. P. and Varma, P. M. (1950). Yellow vein mosaic of *Hibiscus esculentus* L. *Indian Journal of Agricultural Sciences*, 20: 217-230.

Chauhan, M.S., Duhan, J.C. and Dhankar, B.S. (1981). Infection of genetic stock of okra to yellow vein mosaic virus. *Haryana Agricultural University Journal of Research*, 11: 45-48.

Chelliah, S., Murugesan, S., Sivakumar, C. V. and Ramakrishnan, L. (1976). Combination treatments for the control of insect pests, mite, virus-vector, nematodes, fungal and viral diseases of bhendi [*Abelmoschus esculentus* (L.) Moench]. *Madras Agricultural Journal*, 63: 345-349.

Chowdhury, A. K., Biswas, B. and Saha, N. K. (1992). Inhibition of bhendi (okra) yellow vein mosaic virus (BYVMV) by different plant extracts. *Journal of Mycopathological Research*, 30: 97-102.

Coudriel, D. L., Prabhakar, N. and Meyerdirk, D. E. 1985. Sweet potato whitefly (Homoptera: Aleyrodidae): Effects of neem seed extract on oviposition and immature stages. *Environmental Entomology*, 14: 776-779.

Debnath, S. and Nath, P. S. (2002). Management of yellow vein mosaic disease of okra through insecticides, plant products and suitable varieties. *Annals of Plant Protection Sciences*, 10: 340-342.

Diaz, F. A. and Ortegon, M. A. S. (1997). Fruit characteristics and yield of new okra hybrids. *Subtropical Plant Sciences*, 49: 8-11.

Fajinmi, A. A and Fajinmi, O. B. 2010. Incidence of okra mosaic virus at different growth stages of okra plants (*Abelmoschus esculentus* (L.) Moench) under tropical condition. *Journal of General and Molecular Virology*, 2 (1): 28-31.

Gandhi, P. I., Gunasekaran, K. and S.A., T. (2006). Neem oil as a potential seed dresser for managing homopterous sucking pests of okra (*Abelmoschus esculentus* L. Moench). *Journal of Pest Science*, 79: 103-111.

Gupta, I. D., Malathi, V. G. and Mukherjee, S. K. (2003). Genetic engineering for virus resistance. *Crop Science*, 84: 341-354.

Hamon, S. (1988). Evolutionary organization of its kind *Abelmoschus* (okra). Co-adaptation and evolution of two species grown in West Africa, *A. esculentus* and *A. caillei*. Paris, Orstom, D. T. P. Works and Documents. 46: 191.

Harender, R. Bhardwaj, M. L., Sharma, I. M. and Sharma, N. K. (1993). Performance of commercial okra (*Abelmoschus esculentus*) varieties in relation to diseases and insect-pests. *Indian Journal of Agricultural Sciences*, 63: 747-748.

Hilje, L., Cubillo, D. and Segura, L. (1993). Ecological observations on the whitefly *Bemisia tabaci* (Genn.) in Costa Rica. *Manejo Integrated Depladges*, 30: 24-30.

Joyce, J. and Ramakrishnan, U. (2003). Bhendi yellow vein mosaic disease in India is caused by association of a DNA b satellite with a begomovirus. *Virology*, 305: 310-317.

Kareem, K. T. and Taiwo, M. A. (2007). Interaction of viruses in cowpea: Effect on growth and yield parameters. *Virology Journal*, 4: 15.

Khan, M. A. and Mukhopadhyay, S. (1986). Studies on the seasonal spread of yellow vein mosaic virus disease of okra. *Indian Phytopathology*, 38: 688-691.

Kirthi, N., Priyadarshini, C. G. P., Sharma, P., Maiya, S. P., Hemalatha, V., Sivaraman, P., Dhawan, P., Rishi, N. and Savithri, H. S. (2004). Genetic variability of begomoviruses associated with cotton leaf curl disease originating from India. *Archives of Virology*, 149: 2047-2057.

Kon, T., Rojas, M. R., Abdourhamane, I. K. and Gilbertson, R. L. (2009). Roles and interactions of begomoviruses and satellite DNAs associated with okra leaf curl disease in Mali, West Africa. *Journal of General Virology*, 90: 1001-1013.

Koul, O. (1992). Neem allelochemicals and insect control. In: Rizvi, R. S. J. H. and Rizvi, V. J. (eds.). *Allelopathy; Basic and Applied Aspects*. pp 389-412. Chapman and Hall Ltd., London.

Krishna, A., Neema, A. G., Nadkarni, P. G. and Bhalla, P. C. (1983). Evaluation of chemical for control of leaf curl of chillies. In: Muthukrishnan, C. R., Muthuswamy, S. and Arumugam, R. J. (eds.). *Proceedings of National Seminar on the Production Technology of Tomato and Chillies*. pp 167-169. Tamil Nadu Agricultural University, Coimbatore.

Lal, S. S. (1981). An ecological study on the whitefly, *B. tabaci* (Genn.) population on cassava (*Manihot esculenta* Cratz). *Pestology*, 5: 11-17.

Mariappan, V. and Saxena, R. C. (1983). Effect of custard apple oil and neem oil on survival of *Nephotettix virescens* (Homoptera: Cicadellidae) and rice tungro virus transmission. *Journal of Economic Entomology*, 76: 573-576.

Mazumder. N., Borthakur, U. and Choudhury, D. (1996). Incidence of yellow vein mosaic virus of Bhindi [*Abelmoschus esculentus* (L.) Moench] in relation to cultivar and vector population under different sowing dates. *Indian Journal of Virology*, 12: 137-141.

Mehta, Y. R. (1959). Vegetable growing in Uttar Pradesh. *Bureau of Agricultural Information*. pp 23. Lucknow, Uttar Pradesh.

Mohapatra, A. K., Nath, P. S. and Chowdhury, A. K. (1995). Incidence of yellow vein mosaic virus of okra [*Abelmoschus esculentus* (L.) Moench] under field conditions. *Journal of Mycopathological Research*, 33: 99-103.

Naik, V. C. B., Rao, P. A., Krishnayya, P. V. and Chalam, M. S. V. (2009). Seasonal incidence and management of *Bemisia tabaci* Genn. and *Amrasca biguttula biguttula* Ishida of brinjal. *Annals of Plant Protection Sciences*, 17: 9-13.

Nath and Saikia, A. K. (1993). Assessment of yield loss due to yellow vein mosaic of bhindi (*Abelmoschus esculentus* L. Moench.) in Assam. *Journal of Agricultural Science*, Assam, 6: 87-88.

Nath, P. D. (1994). Effect of sowing time on the incidence of yellow mosaic virus disease and whitefly population on green gram. *Annals of Agricultural Research*, 15: 174-177.

Norman, (1992). *Medical Botany*. pp 39. Rand and Clark Pub. Co. Edinburgh

Phadvibulya, V., Boonsirichai, K., Adthalungrong, A. and Srithongchai, W. (2009). In: Shu, Q.Y. (eds.). *Induced Plant Mutations in the Genomics Era*. pp 349-351. Food and Agriculture Organization of the United Nations, Rome.

Pun, K. B. and Doraiswamy, S. (2000). Influence of weather factors on the incidence of okra yellow vein mosaic virus disease. *Journal of the Agricultural Science Society of Northeast India*, 13: 91-96.

Pun, K. B., Doraiswamy, S. and Balasubramanian, G. (2005). Prediction of whitefly population and okra yellow vein mosaic virus disease incidence in okra. *Indian Journal of Virology*, 16: 19-23.

Pun, K. B., Doraiswamy, S. and Jeyarajan, R. (2000). Screening of virus inhibitory chemicals and neem products against okra yellow vein mosaic virus. *Indian Phytopathology*, 53: 95-96.

Pun, K. B., Sabitha, D., Jeyaran, R. and Doraiswamy, S. (1999). Screening of plant species for presence of antiviral proteins against okra yellow vein mosaic virus. *Indian Phytopathology*, 52: 221-223.

Purseglove, J. W. (1987). *Tropical Crops Dicotyledons*. pp 95-109. Longman Singapore Publishers (Ptv) Ltd., Singapore.

Regupathi, N., Veeragavathatham, D. and Thamburaj, S. (2000). Reaction of okra (*Abelmoschus esculentus* L. Moench) culture to bhendi yellow vein mosaic virus disease. *South Indian Horticulture*, 48: 103-104.

Rustamani, M. A., Hussain, T., Baloch, H. B., Talpur, M. A. and Mal, K. (1994). Comparative effectiveness of different insecticides in controlling whitefly and mosaic disease of chillies. *Proceedings of Pakistan Congress of Zoology*, 14: 61-64.

Saimbhi, M. S. (1993). Agro-techniques for okra. pp. 529-536. In: Chadha, K.L. and Kalloo, G. (eds.). *Advances in Horticulture: Vegetable Crops Vol. 5*. Malhotra Publication House, New Delhi, India.

Salam, S. A., Patil, M. S., Byadgi, A. S. (2009). IDM of mung bean yellow mosaic disease. *Annals of Plant Protection Sciences*, 17 (1): 157-160.

Sangar, R. B. S. (1997). Field reaction of bhindi varieties to yellow vein mosaic virus. *Indian Journal of Virology*, 13: 131-134.

Sastry, K. S. M and Singh, S. J. (1973). Restriction of yellow vein mosaic virus spread in okra through the control of vector, whitefly (*Bemisia tabaci*). *Indian Journal of Mycology and Plant Pathology*, 3: 76-80.

Schippers, R. (2002). African indigenous vegetables. An overview of the cultivated species. Natural Resources Institute/ACP-EU, Technical Centre for Agricultural and Rural Cooperation, CD ROM 214, Chatham, UK.

Schmutterer, H. (1990). Properties and potential of natural pesticides from the neem tree Azadirachta indica. *Annual Review of Entomology*, 35: 271-298.

Seif, A. A. (1980). Seasonal fluctuation of adult population of whitefly, *Bemisia tabaci* on cotton and its relationship with weather parameter. *Journal of Cotton Research and Development*, 5: 181-189.

Sharma, B. R. and Arora, S. K. (1993). Improvement of okra. pp. 343-364. In: Chadha, K.L. and Kalloo, G. (eds.). *Advances in Horticulture: Vegetable Crops Vol. 5*. Malhotra Publication House, New Delhi, India.

Sharma, B. R., Sharma, O. P. and Bansal, R. D. (1987). Influence of temperature on incidence of yellow vein moasaic virus in okra. *Vegetable Science*, 14: 65-69.

Shastry, K. S. M. and Singh, S. J. (1973). Field evaluation of insecticides for the control of whitefly (*B. tabaci* Genn.) in relation to the incidence of yellow vein mosaic of okra. *Indian Phytopathology*, 26: 129-138.

Singh, A. (1977). *Practical Plant Physiology*. pp. 66. Kalyani publishers, Ludhiana, Punjab.

Singh, A. K., Sanger, R. B. S. and Gupta, C. R. (2002). Performance of different varieties of okra to yellow vein mosaic virus under field conditions in Chhattisgarh. *Progressive Horticulture*, 34: 113-11a6.

Singh, B. R. and Singh, A. K. (1999). Viral diseases on congress grass from Uttar Pradesh. *International Journal of Tropical Plant Diseases*, 17: 165-168.

Singh, D., Verma, N. and Naqvi, Q. A. (1998). Control of yellow vein mosaic virus of okra by seed selection and insecticidal application. *Indian Journal of Virology*, 14: 121-123.

Singh, H. B., Joshi, B. S., Khanna, P. P. and Gupta, P. S. (1962). Breeding for field resistance to yellow vein mosaic in bhindi. *Indian Journal of Genetics and Plant Breeding*, 22: 137-144.

Singh, R., Mishra, R. C., Shahi, S. K. and Dikshit, A. (1999).Insect repellent activity of *Asfoetida* to prevent yellow vein mosaic virus infection in okra crop [*Abelmoschus esculentus* (L.) Moench]. *Plant Protection Bulletin*, 51: 35-37.

Singh, S. J. (1990). Etiology and epidemiology of whitefly transmitted virus disease of okra in India. *Plant Disease Research*, 5: 64-70.

Sinha, S. N. and Chakrabarti, A. K. (1978). Effect of yellow vein mosaic virus infection on okra seed production. *Seed Research*, 6: 67-70.

Tiendrebeogo, F., Traore, V. S., Lett, J. M., Barro, N., Konate, G., Traore A. S. and Traore, O. (2010). Impact of okra leaf curl disease on morphology and yield of okra. *Crop Protection*, 29 (7): 712-716.

Uppal, B. N., Varma, P. M. and Capoor, S. P. (1940). Yellow vein mosaic of bhindi. *Current Science*, 9: 227-228.

Verma P. M., Rao G. G. and Capoor S. P. (1966). Yellow mosaic of *Corchorus trilocalaris*. *Science and Culture*, 32: 466.

Plant Diseases Management in Horticultural Crops (2011) *Pages* **290–298**
Editors: **Shahid Ahamad, Ali Anwar and P.K. Sharma**
Published by: **DAYA PUBLISHING HOUSE, NEW DELHI**

Chapter 24

Nonpathogenic Disorders in Tomato and their Management

F.A. Khan[1] and Shahid Ahamad[2]

[1]Division of Post-Harvest Technology
Sher-e-Kashmir University of Agricultural Sciences and Technology of Kashmir, Shalimar Campus, Srinagar
[2]Krishi Vigyan Kendra, SKUAST-J, Rajouri – 185 131

Among the horticultural crops vegetables have an important position and are a high protective food of dietary complex of human beings. For balanced diet supplementation of vegetables along with cereals and pulses is a necessary step towards complete food. In recent past the production of vegetables have gone up due to adaptation of modern technology and fertilization formulation but still do not show any parallelism with consumption. Tomato is one of the most popular and widely grown vegetable in the world having a tremendous nutritional value and sufficient quantity of antioxidant compounds. However, there are several parasitic (infectious) and non parasitic (non infectious) disorders that distort plants and blemish fruits which results in both qualitative as well as quantitative losses. Non-parasitic or physiological disorders are the result of plant responses to abnormal environmental conditions. The disorders that result from some conditions which occurred during the pre and/or post harvest seasons that lead to undesirable changes in the fruit quality are known as physiological disorders. Thus, *Physiological disorders are problems that are not caused by insects or diseases but rather by climatic factors (temperature, rain, humidity, sun light, etc.) and management practices (training and pruning, irrigation, fertilization, harvest procedure, etc.) that change the micro-climate endured by the plant or plant part*. In some cases these environmental factors are unknown. In other cases, the causes are known, but are difficult to correct. Generally, good cultural practices that ensure consistent plant growth will reduce the occurrence of physiological disorders. Also crop varieties differ in their susceptibility to these disorders. This article focuses on non parasitic physiological disorders of tomato during pre and post-harvest period and their possible control measures.

Catface

Symptoms

Catface is a condition involving *malformation* (misshaped fruit) and scarring, particularly at blossom end of the fruit (Plate 24.1, Figure 1). It may also appear as an enlargement or perforation of the blossom scar, though the fruit shape is normal. Affected fruits are puckered with swollen protuberances and can have cavities extending deep into the flesh.

Causes

The causes of catfacing are not definitely known and there actually may be more than a single cause. It is generally agreed that any disturbance to flowers can lead to abnormally shaped fruits. Exposure of the blossoms to cold temperatures prior to anthesis has been linked to an increase in the appearance of catface. Cloudy and cool weather at blooming time may cause the blossom to stick to the developing fruit, resulting in catfacing. Cool or cold temperatures that occur about 3 weeks before bloom can increase the amount of catfacing. Extreme heat, drought, low temperature, and contact with hormone-type herbicide sprays may be causes of flower injury. In general, joistless varieties are more prone to catfacing than jointed varieties. Pruning of plants and high nitrogen may also contribute to the appearance of the disorder. Heavy pruning in indeterminate varieties has been shown to increase catfacing but this has not been shown to happen in short-stake varieties. In indeterminate varieties, catfacing is thought to be related to reduction in auxins in the plant from removing the growing points. Catfacing occurs most commonly on the large- fruited varieties. Damage from the herbicide 2,4- D may produce similar symptoms.

Management Practices

There are no effective controls for this disorder. Catfacing should decline with the arrival of warmer weather. Keep herbicides away from flowers, avoid the plants from cold temperature and plant less susceptible varieties. Try to prevent spray drift from undesirable chemicals and in the case of little leaf, prevent soils from becoming waterlogged.

Growth Cracks

Symptoms

Growth cracks appear as splitting of the outer layer or epidermis of the fruit (Plate 24.1, Figure 2). Two different forms of cracking occur in tomato fruit. Radial cracking originates from the stem end and progresses toward the blossom end. Concentric cracking occurs in a ring or rings around the stem scar. It is possible to have both types on the same fruit. Cracking occurs when the internal expansion is faster than the expansion of the epidermis and the epidermis splits. Varieties differ greatly in their susceptibility to cracking. Cracking can occur at all stages of fruit growth but usually appear towards fruit maturity at the mature green stage or in less susceptible varieties at the red ripe stage. The earlier the growth crack develops, the larger it is likely to be once the fruit is harvested. These are the possible entrance points for decay-causing organisms. Cracking is more of a problem in vine-ripe operations than in a mature-green operation.

Causes

Growth cracks result from extremely rapid fruit growth brought on by periods of abundant rain and high temperatures, especially when such weather conditions follow drought. Fruit exposed to direct sun are most susceptible. Fruit cracking is most common on the large-fruited varieties, such as 'Beefsteak. Wide fluctuation in air temperature can also increase cracking.

Management practices

Selecting crack resistant cultivars such as 'Jetstar' and providing a uniform supply of moisture to the plants (through irrigation management and the use of plastic mulch) is the best defense against growth cracking. Cracking may also be reduced by maintaining good foliage cover, since exposed fruit are more susceptible.

Internal White Tissue

Symptoms

Internal White tissue, yellow eye, yellow shoulder, and green shoulder appear to represent a range of symptom severity for a single problem, yellow shoulder disorder (YSD). Fruit affected by this disorder usually show no outward symptoms. The YSD is characterized by discolored yellow or green sectors under the fruit peel (Plate 24.1, Figure 3). The disorder involves abnormal fruit development and is not a delay in fruit ripening. The discoloration (ranging from a few millimeters to the top 1/3 of the fruit) is caused by a failure of green chloroplast in tissue affected by YSD to develop into red chloroplast. This modification is accompanied by a more random cell orientation and smaller cells relative to mature green fruit. These changes begin early in fruit development and cannot be reversed by delaying harvest. Understanding that YSD involves altered fruit development rather than delayed fruit ripening is important to management strategies: the problem must be treated before it is seen in the field.

YSD affects both the appearance and nutritional value of tomato. Fruits may not show external discoloration but can still have the disorder. Standards for grading processed vegetables set very strict limitations on the amount of discolored tissue allowed in peeled tomato products. However, grading practices rarely detect internal disorders before peeling. Thus premium prices may be paid for tomato that do not yield an acceptable amount of high grade fruit. Processing efficiency is also reduced because the tomato peel adheres to the discolored flesh after peeling; extra labor is then required to sort and divert tomatoes into products that have a lower return. Ultimately, both the growers and processors would be negatively affected by YSD.

Causes

The causes of YSD are not yet understood but the disorder appears to be related to potassium availability in the soil. Weather (temperature, light intensity and rain) and genetics (variety) have also some implication on the occurrence of the disorder.

Management practices

Adequate potassium fertility early on in fruit development is important in controlling the appearance of the disorder. Also, selecting varieties that have reduced susceptibility to the disorder is also advised. Proper water management, fertility, disease control and variety selection are key factors in reducing losses due to these physiological disorders of tomato fruits. By maintaining crop health prior to and during fruit development, the highest quality fruit can be produced. High quality fruit can be assured of commanding premium prices in the market place and are always in demand.

Blossom End Rot (BER)

Symptoms

Blossom-end rot is a physiological disorder involving a calcium deficiency that is either due to poor uptake or translocation into the fruit. The disorder can appear on fruit at any stage of development, but it is most common when fruit are one-third to one-half grown. The initial symptom appears as small water soaked spot at the blossom end of the fruit that increases in size and becomes dry and dark-brown. Discolored tissue shrinks until the affected area is flat or concave and tomato flesh is conspicuously rotted (Plate 24.1, Figure 4). It may also occur internally with no visible symptoms on outside of fruit.

Causes

Low soil Ca, high N rates, using ammonical sources of N, high concentrations of soluble K and Mg in the soil, high salinity, low humidity, inadequate soil moisture, excess soil moisture, damage to root system by nematodes, disease, mechanical means or heavy pruning may increase BER. Occurrence increases dramatically when calcium levels in the soil system drop below 0.08 per cent. Eventually, secondary organisms invade the affected tissue and cause the fruit to rot. In greenhouse production not cycling the irrigation system at night can increase BER, since night is an important time of Ca uptake.

The amount of calcium salt available to the plant decreases rapidly in the presence of excessive salts such as potassium, magnesium, ammonium, and sodium. Extreme fluctuation in moisture can also reduce the availability of calcium salts needed by the plant. Heavy applications of nitrogen fertilizers and abundant rain cause rapid and luxuriant plant growth and predispose the fruit to blossom-end rot, especially during periods of dry, hot weather.

Blossom-end rot is caused by a lack of calcium in the developing fruit. Wide fluctuations in soil moisture levels impair calcium uptake by the root system. Excessive nitrogen fertilization and root pruning may also contribute to blossom-end rot.

Management Practices

Blossom-end rot can be minimized by maintaining a uniform supply of moisture through irrigation and soil mulches, incorporating fertilizers high in super phosphates and low in nitrogen prior to planting, and avoiding root pruning by not cultivating within 1 foot of the plants.

Gray Wall (Blotchy Ripening)

Symptoms

Blotchy ripening is a physiological disorder characterized by the randomized development of green or green-yellowish areas on the surface of red tomato fruits. Areas showing blotchy ripening have less organic acids, SSC, and starch.

Internally gray wall is characterized by dark necrotic areas usually in the vascular tissue of the outer walls (Plate 24.1, Figure 5). The necrosis is sometimes present in the cross-walls and very infrequently in the center pith area of the fruit. Outward symptoms show up as grayish appearance caused by partial collapse of the wall tissue; hence the term gray wall. It typically develops on green fruit prior to harvest but can develop later. Fruits affected are typically not marketable due to blotchy appearance as fruit ripens.

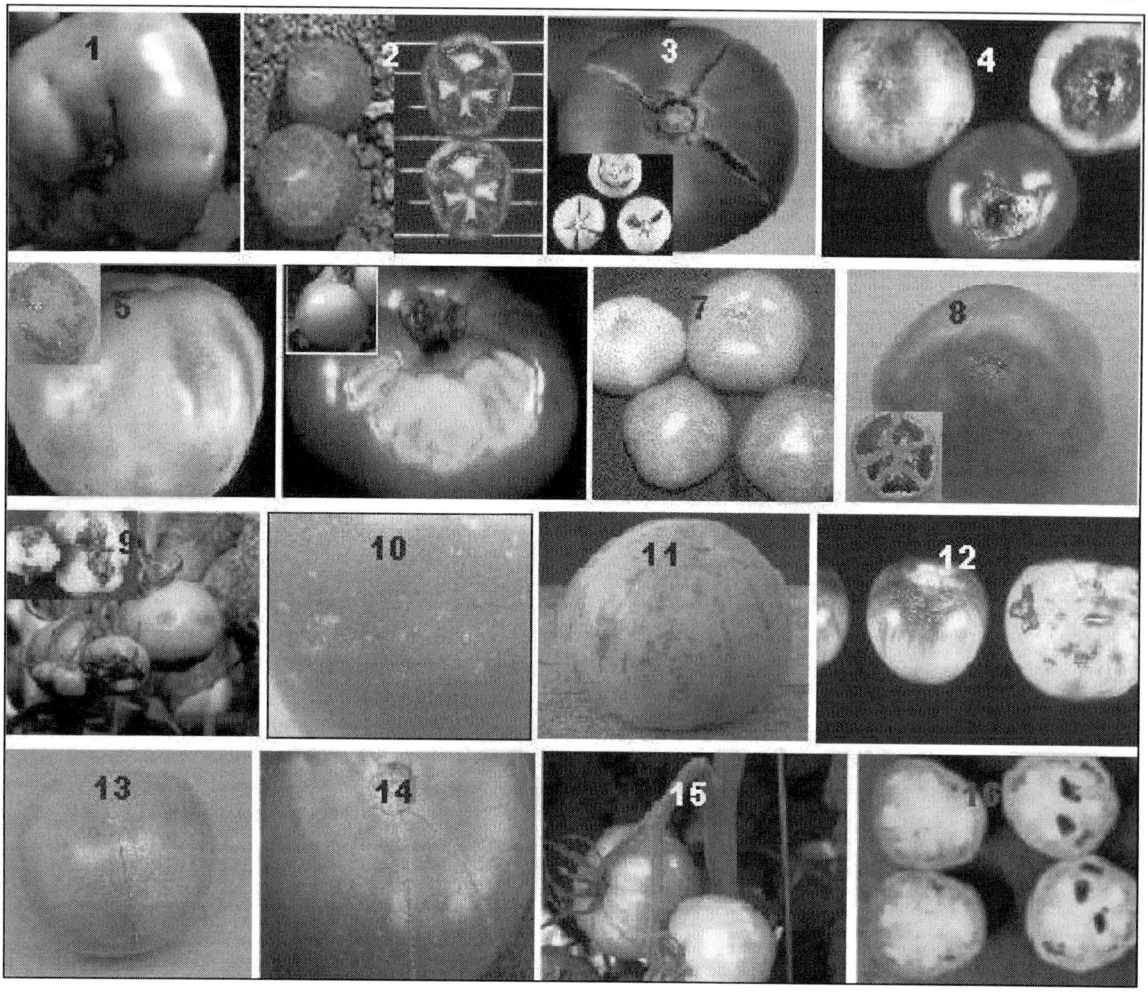

Plate 24.1: Photograph Showing Symptoms of Different Physiological Disorders of Tomato Fruits

(1) Catface; (2) Growth cracks; (3) Internal white tissue; (4) Blossom end rot-BER; (5) Grey wall or blotchy ripening; (6) Sunscald; (7) Chilling injury; (8: Puffiness; (9) Herbicide injury; (10) Pox and fleck; (11) Zebra strip; (12) Rain check; (13) Zippering; (14) Gold speck or russeting; (15) Penducle kinking and (16) Heat stress.

Causes

Cause is not completely understood. Apparently, the development of this disorder is related to the availability of potassium and inorganic nitrogen in the soil system.

There are variety differences in susceptibility. Gray wall is more of a problem during cool and short days. High N may increase the problem and adequate K may reduce the problem.

Sunscald

Symptoms

Sunscald is most commonly seen on green fruit. It is associated with excessive exposure to the sunlight and the resultant elevated tissue temperature during fruit development, disrupting lycopene synthesis and resulting in the appearance of yellow areas in the affected tissues that remain during the ripening process.

It can be broken down into 2 types, sub lethal and lethal. Sub lethal sunscald can be described as a yellow, hard area usually on the shoulder of the fruit (Plate 24.1, Figure 6). This occurs when tissue temperature rises above about 86° F. The high tissue temperature will not allow the red pigment to develop nor the flesh to soften but allows the yellow pigments to develop. With lethal sunscald, the tissue turns white and dies. Many times the dead tissue will turn black from fungi that are feeding on the dead tissue. Lethal sunscald occurs when tissue temperatures rise above 104° F.

Causes

Damage usually occurs when fruits are suddenly exposed to sunlight. This most frequently occurs after a harvest or a storm when leaves are moved around and fruit exposed. Over pruning can also increase sunscald problems especially with fruit in the upper part of the plant.

Management Practices

Good spray programs to ensure good foliage cover can reduce the problem. Growers at times may use a sun screen material such as snow or surround to help reduce sunscald. Plants grown in wire cages provide good foliage protection. Also, control foliar diseases of tomato which defoliate the plants and expose the fruit to direct sunlight.

Internal White Tissue

Symptoms:

Visual symptoms of chilling injury include failure to ripen and develop full color and flavor, irregular (blotchy) color development, premature softening, surface pitting, browning of seeds, and increased decay (especially Black mold caused by *Alternaria* spp.) (Plate 24.1, Figure 7).

Causes

Tomato fruit are chilling sensitive at temperatures below 10°C (50°F) if held for longer than 2 weeks or at 5°C (41°F) for longer than 6-8 days. The recommended storage temperature varies with the maturity stage. Chilling injury is cumulative and may be initiated in the field prior to harvest.

Management Practices

Mature-green fruit will ripen normally at 13 to 21 °C (55 to 70 °F). On the other hand, ripe tomato fruits can be stored at 10 °C (50 °F), without visible symptoms of chilling injury, although flavor and aroma is negatively affected.

Puffiness

Symptoms

The outer wall of the fruit is normal, but the tomato is hollow inside. One of the seed cavities is usually empty (Plate 24.1, Figure 8). When this problem is slight, it may be impossible to detect puffiness until fruit are cut. Severe puffy fruit will appear to be flat-sided or angular in nature. When fruit are cut, open cavities are observed between the seed gel area and the outer wall. Fruits are also very light in relation to size.

Causes

This problem is caused by any factor that affects fruit set including inadequate pollination, fertilization, or seed development. Extreme high or low temperatures, excessive nitrogen fertilization, and heavy rains may interfere with normal pollination, resulting in puffy fruit. Puffiness occurs most frequently on early fruit.

Management Practices

No effective controls, however, use of "hot set" varieties can reduce the problem but even these have limitations when night temperatures remain above about 75° F. Puffiness should decline later in the summer.

Herbicide Injury

Symptoms

Contaminated plants show one or more of the following symptoms depending on the degree of exposure and age of plant at exposure. Older leaves are excessively pointed, down-curved, or rolled with prominent light-colored veins; young leaves do not fully expand and are narrow and elongated with parallel veins; stems are split, distorted, or brittle; and fruits are catfaced or irregularly shaped (Plate 24.1, Figure 9). Plants exposed to small amounts of phenoxy herbicides will outgrow the symptoms without seriously reducing yield or fruit quality. Harvest might be delayed, however. Plants do not recover from severe.

Causes

Tomatoes are very sensitive to injury from herbicide chemicals. The most common injury symptoms observed are caused by phenoxy herbicides such as 2,4-D (2,4-dichlorophenoxyacetic acid) and dicamba (substituted benzoic acids). These chemicals are growth regulators, hormone-type and weed control chemicals. Tomato plants usually come in contact with the chemical through spray drift or the use of a sprayer that once contained the herbicide.

Pox and Fleck

Symptoms

In most cases when a fruit is affected by disorders are found together but are considered separate problems. Pox is described as small cuticular disruptions found at random on the fruit surface (Plate 24.1, Figure 10). The number can vary from a few to many. Fleck, also known as Gold Fleck, develops as small irregular shaped green spots at random on the surface of immature fruit which become golden in color as fruit ripens. Number of spots can vary from few to many.

Causes

Fruits severely affected with pox and fleck are not marketable. Both conditions seem to be genetic in nature, but are difficult to breed out of a variety since the disorders only show up under certain environmental conditions. There seems to be some differences of opinion as to the conditions inducing this problem. There are differences between varieties as to susceptibility to pox and fleck.

Zebra Stripe

Symptoms

Zebra stripe can be characterized as a series of dark green spots arranged in a line from the stem end to the bloom end (Plate 24.1, Figure 11). At times it seems the spots coalesce together and form elongated markings. Many times the dark green areas will disappear when fruit ripens.

Rain Check

Symptoms

Rain check can be described as tiny cracks that develop on the shoulder of the fruit (Plate 24.1, Figure 12). These cracks can vary from just a few to almost complete coverage of the shoulder. The cracks feel rough to the touch and affected areas can take on a leathery appearance and not develop proper color as fruit ripens. Green fruits are most susceptible, followed by breakers and are not affected at all. Damage occurs most often on exposed fruit after a rain.

Causes:

Exact cause is not known, but appears to be related to exposure of the fruit to water. The problem is more severe when heavy rains occur after a long dry period. There are differences among varieties to susceptibility to rain check. Also, varieties with good leaf coverage usually have fewer rain checks.

Zippering

Symptoms

Zippering is described as a fruit having thin scars that extend partially or fully from the stem scar area to the blossom end (Plate 24.1, Figure 13). The longitudinal scar has small transverse scars along it. At times there may be open holes in the locules in addition to the zipper scar.

Usually an anther that is attached to the newly forming fruit causes the zipper scar. Some people feel that a zipper is formed when the "blooms" stick to the fruit and do not shed properly but this may not be a cause. The only control is to select varieties that are not prone to zippering.

Goldspot/Goldspeck/Russeting

Symptoms

Fruit skins are rough and not shiny. Areas around the stem remain orange (Plate 24.1, Figure 14).

Causes

Possible causes of this disorder are excess fruit calcium and/or high Ca/K ratios, high P, high humidity and high average temperatures.

Management Practices

To prevent russeting control excess Ca uptake, allow top growth or suckers to develop, minimal deleafing, lower humidity, reduce watering, avoid large fluctuations between day and night temperatures and protect fruit from sun. Maintain minimal EC of 3.0, avoid large changes in EC and avoid high fruit growth rates late in fruit maturation.

Peduce Kinking

Truss bent sharply which resulted in yield losses up to 17 per cent. Distal fruit more affected. Fruit more affected by late (>10 mm diameter) kinking, so may grow out of early kinking problems (Plate 24.1, Figure 15). Possible causes of this disorder are low light, cultivar and too high temperatures during the vegetative phase (may result in vcertical or stick trusses).

Heat Stress

Symptoms

The disorder may be identified by seedless fruit. Small fruit may stay on the plant but not enlarge (Plate 24.1, Figure 16). Blossoms fall off or turn brown. Fruit may be misshapen as in cold stress. Flowers are numerous but small.

Conclusion

Tomato fruit can be affected by several non parasitic disorders that distort and blemish fruits. Non parasitic or physiological disorders are the result of plant responses to abnormal environmental conditions. In some cases these environmental factors are unknown while in other, the causes are known. Once symptoms are visible physiological disorders are difficult to correct, however, unlike the "parasitic" diseases caused by fungal, bacterial, or viral organisms, these disorders are not infectious or a threat to the entire field. Therefore, one should focus on preventive measures. A number of steps may be taken to minimize the occurrence of physiological disorders which include use of resistant cultivars, good cultural practices and post-harvest handling. Consult Extension publications and specialists for information on varietal susceptibility to physiological disorders.

References

Anonymous (2002). *Tomato Disorders: A Guide to the Identification of Common Problems.* http://aggie-horticulture.tamu.edu/tomatoproblemsolver/index.html.

Anonymous (2003). *Tomato Disorders.* http://nhb.gov.in/bulletin_files/vegetable/tomato/tom003.pdf.

Gleason, M.L. and Edmunds, B.A. (1914). *Tomato Disease and Disorders.* http://www.extension.iastate.edu/Publications/PM1266.pdf.

Iowa State University (1997). Physiological disorders of tomato: *Horticulture and Home Pest News*, pp. 114. http://www.ipm.iastate.edu/ipm/hortnews/1997/7-18-1997/tomdis.html

Olson, S.M. (2008). *Physiological, Nutritional and other Disorders of Tomato Fruit.* http://edis.ifas.ufl.edu/HS200.

Zitter, Thomas A. and Reiners, Steve (2004). *Common Tomato Fruit Disorders.* http://vegetablemdonline.ppath.cornell.edu/NewsArticles/Tom_ComDis.htm.

Plant Diseases Management in Horticultural Crops (2011) *Pages* **299–311**
Editors: **Shahid Ahamad, Ali Anwar and P.K. Sharma**
Published by: **DAYA PUBLISHING HOUSE, NEW DELHI**

Chapter 25

Nonpathogenic Diseases of Potato Tubers

***F.A. Khan*[1] *and Shahid Ahamad*[2]**

[1]*Division of Post Harvest Technology*
Sher-e-Kashmir University of Agricultural Sciences and Technology of Kashmir,
Shalimar Campus, Srinagar
[2]*Krishi Vigyan Kendra, SKUAST-J, Rajouri – 185 131*

Good disease management is critical to the successful production of the potato (*Solanum tuberosum* L.). The potato plant is susceptible to at least 75 diseases and nonparasitic disorders, many of which consistently cause yield losses in potato production areas. The management of many production problems depends on the identification of diseased or disorders. The objective of this chapter is to aid in the diagnosis of those noninfectious diseases and defects that most often result in production problems. Various external and internal physiological disorders that affect growth and production of potato tubers are discussed. Symptoms of these disorders, as they appear on the tubers are depicted with help of photographs.

External Disorders

Deformations

Tuber growth rates often fluctuate in response to widely varying growing conditions causing malformations. There are several types of tuber deformations or malformations such as such as bottlenecks, dumbbells, pointed ends, and knobs (Plate 25.1, Figures 1a–c). Tubers with these defects are also referred to as "being rough." They are due to problems associated with apical buds and longitudinal growth. Harvest quality or grade is affected and tubers are discarded as culls.

These deformations are primarily due to high temperature or water stress in the field that cause constricted growth in the bud, middle, or stem end portion of the tuber, depending upon the extent of the stress and the stage of growth at which it occurs. The severity of the deformation increases with higher temperatures and longer high-temperature periods. Basically, high temperatures, above 80 °F, decrease cell division and lower the supply of carbohydrates available to the tuber. Other factors that exacerbate temperature-induced deformation are excessive nitrogen application before a high temperature period, uneven nutrient or moisture supply, hail and frost. Potato varieties vary considerably in their sensitivity to high-temperature field stress. The rule of thumb is that longer-tuber varieties are more susceptible than rounder-tuber varieties. For example, long whites such as Kennebec are more susceptible than round whites as Atlantic. Russet Burbank is one of the most sensitive of all varieties.

During a period of stress, longitudinal growth slows or may even stop. When favorable conditions return, tuber growth resumes, "stop and go" growth. Therefore, appearance is due to irregular longitudinal growth, often because of a constriction. Dumbbells have a constriction in the middle. Growth disruption in dumbbell and elongated (hot dogs) tubers occurred at mid-bulking. Kidney-shaped tubers tend to have a slighter mid-section constriction and tend to curve, giving the kidney-shape appearance. Elongated tubers show little lateral growth; they also tend to be enlarged and curved. In case of pointy stem-end or bottleneck growth disruption occurred during early bulking and the constriction is at the stem end. Often dumbbells and pointy bud-ends are associated with jelly end rot or glassy end.

Knobbiness also sometimes referred to a secondary growth is due to stimulated growth of lateral buds in one or more eyes. Unlike pointy end and dumbbells, there are no internal defects or rots associated with the formation of knobs. Knobby potatoes are considered culls, lowering marketable yield.

Management Practices

Management approaches for preventing malformed tubers include promoting uniform growth by establishing uniform stands, avoiding large fluctuations in nitrogen availability, maintaining available soil water content above 70 per cent, and general avoidance of cultural practices, such as late cultivation, that may alter tuber growth patterns.

Tuber Cracking

Potato tuber cracking is a physiological disorder of the potato tuber in which the tuber splits while growing (Plate 25.1, Figure 2). Potato tuber cracking can take the form of growth cracks, elephant or alligator hide, skin checking, or any other symptom showing a cracked appearance of the skin. Symptoms can be either superficial or affect a major portion of the tuber. Growth cracks generally start at the bud or apical end of the potato and can extend lengthwise. Growth cracks can vary in severity from appearing as a surface abrasion to a split through the tuber. The severity depends on the stage of growth the initial cracking occurred. Even though cracking does not usually predispose the tuber to rotting, growth cracks can negatively impact potato tuber quality. Growth cracks make fresh-market tubers unattractive. Severe growth cracks can even impact the quality of chip processing potatoes. Growth cracks appear as large, open fissure in the tuber skin and flesh and are caused by irregularities in tuber growth, especially in response to widely fluctuating water supplies. Other factors, including virus infection and herbicide injury, can also cause tuber growth cracks. Elephant or alligator hide develops during the growing season and appears as shallow, corky cracks on the tuber skin. The primary cause of this condition is unknown, but contributing factors may include high temperature, high soil organic matte excessive soil moisture and fertilization.

Longitudinal fissures develop when the core tissue inside the tuber grows faster than the outer tissues, the periderm. In other words, the internal pressure of the tuber is greater than the tensile strength of the skin or surface tissue. The resulting fissure or crack can extend the entire length of the tuber and may be shallow or a half inch deep. Growth cracks undergo wound-healing and show a characteristic suberized appearance. When healed, they seldom become infected with a pathogen. Growth cracking can also be associated with plant infection by a few relatively uncommon viruses such as yellow dwarf virus and spindle tuber viroid. More importantly, growth cracking can be caused by exposure to members of two new herbicide families, *imidazolinones* such as *Pursuit* and *sulfonylureas* such as *Ally* and *Accent*.

Management Practices

Three cultural practices to reduce the incidence of growth cracks are uniform plant spacing, adequate soil moisture with consistent irrigation scheduling and uniform fertilization especially avoiding excessive and late applications of nitrogen. Maintaining of adequate soil moisture is especially important during the bulking stage when the plants are large and tubers are rapidly expanding. The only recourse when the incidence of severe growth cracks is high is to select out tubers with severe growth cracks prior to packing and/or shipping.

Chaining

Tuber chaining refers to the initiation followed by limited growth of tubers at the nodes of a stolon after the apical tuber at the end of the stolon was initiated and has begun growing. The disorder's name refers to the chain appearance of the series of little tubers along the stolon. Tuber chaining affects harvest yields. The small tubers formed compete with the primary tuber for nutrients and thereby lowers the weight of the marketable tuber. There is no effect on fresh market quality of the primary tuber except for reduced tuber size. For processing, the texture and starch content may not be desirable. Tuber chaining may also interfere with harvest separation of the primary tuber from stolon. The cause of tuber chaining is high soil temperature. Soil temperatures around the daughter tubers that are above 75 °F, especially early to mid bulking or tuber growth promote tuber chaining. Once this disorder has begun, the return of cooling soil temperature around tubers will not overcome the disorder. In other words, the chaining and nutrient competition will continue for the rest of the season, thereby yields are lowered. On a physiological level, once apical dominance is lost by the primary tuber, it does not return. Low soil moisture does not cause this disorder but is usually associated with high soil temperature.

The development or initiation of these tubers is due to a break down of apical dominance of the primary tuber. This is a hormonal phenomenon associated with auxin (IAA) concentrations basipetal from the primary tuber. The sizing of the primary tuber is inhibited due to the lessened carbohydrate (sucrose) taken in because of competition with the secondary tubers. The primary tuber, in severe cases, can have a glassy or soft texture and have a low specific gravity making it undesirable for processing. A related phenomenon is sprouting from buds on the nodes of the stolon. The sprout does not affect quality but will compete for nutrients much like tuber chaining, and is also related to loss of apical dominance.

Management Practices

Since air temperature cannot be controlled, the best practices to avoid over-heating of the soil are to plant deep, to hill and to maintain a good trapezoidal structure to the row. Seed pieces should be at

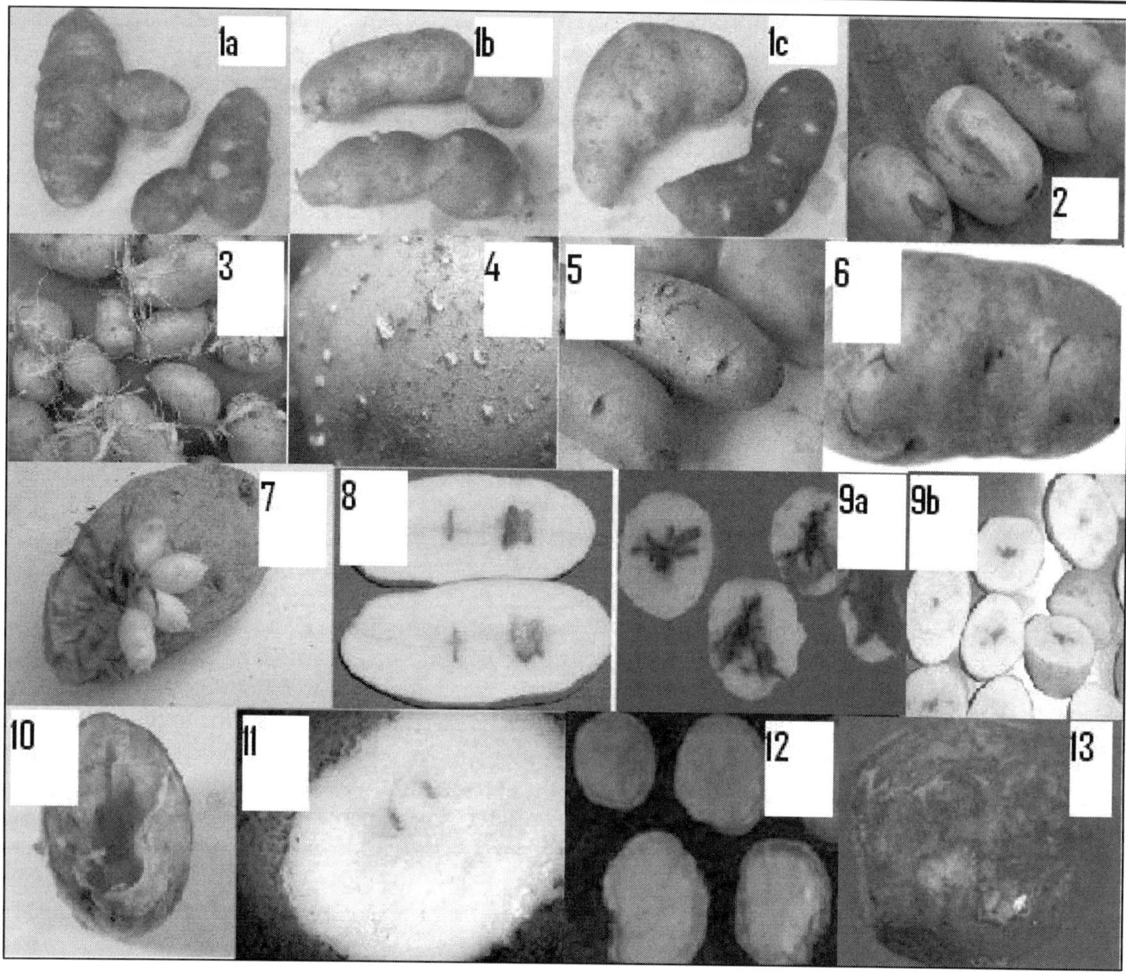

Plate 24.1: Photograph Showing Symptoms of Nonpathogenic Diseases of Potato Tubers

(1) Deformations; (2) Tuber Cracking; (3) Hair Sprouting; (4) Swollen Lenticels; (5) Greening; (6) Thumbnail Cracks; (7) Little Tuber Disorder; (8: Brown Center or Hollow Heart; (9) Blackheart; (10) Heat Necrosis; (11) Vascular or Stem-end Discoloration; (12) Chilling; (13) Freezing Injury.

least six inches below the surface; eight inches would not hurt. In severe hot weather, cooling the ground with irrigation may be necessary.

Hair Sprouting

Premature sprouting of tubers late in the season or early in storage affects quality of the harvested tubers (Plate 25.1, Figure 3). Besides appearance, the sprouts soften the tuber by taking up its nutrients and lowering the starch content. The taste of the tuber is uneven with a sweeter taste near the sprout. This makes the tuber unacceptable for processing since frying will turn this area dark. The cause of hair sprouting is high soil temperature. The conditions for this disorder are similar to that for tuber chaining except that exposure to high soil temperature is toward the end of the growing season, late bulking or plateau stages. Once sprouted, the sprouts will grow under good conditions. Hair sprouting is also a symptom of infection with *Colletotrichum atramentarium*, related to black dot, and associated with *aster yellow* carried by leafhoppers and *psyllid yellow* carried by potato psyllids.

Sprouts appear from the eyes of the primary tuber. The hair sprout is very thin, hair-like, and is too weak to emerge or grow much. A related to hair sprouting is heat sprouting which may remain underground, or emerge and become green and leafy. Tuber sprouting is a hormonal phenomenon involving the lack of development of tuber dormancy, related to abscisic acid (ABA).

Management Practices

Practices to minimize this disorder are the same as for tuber chaining. Since air temperature cannot be controlled, the best practices to avoid over-heating of the soil are to plant deep, to hill and to maintain a good row structure. In severe hot weather, cooling the ground with irrigation may be necessary.

Swollen Lenticels

Lenticels are pores in the skin of tubers (botanically, they are stomata's). They are involved in gaseous exchange of oxygen and carbon dioxide during respiration and photosynthesis. Besides giving an unmarketable appearance to the tuber, the major problem is that an entrance to pathogenic organisms, bacterial soft rot, pink rot and leak is created. This disorder is caused by exposure of the tuber to very wet conditions in the field or in storage. The swelling seems to be related to oxygen deprivation by the watery film covering the lenticels.

Swollen or enlarged lenticels (water spot, water scab) develop when tissue below the lenticels swell and burst through the protective covering of the lenticels (Plate 25.1, Figure 4). This forms a corky mass around the lenticels. The disorder is somewhat reversible if the wet period is short.

Management Practices

Avoid over-watering. Avoid harvesting low, swampy spots in the field. Pick fields with good drainage. Avoid condensation in storage. Keep storage well ventilated.

Greening

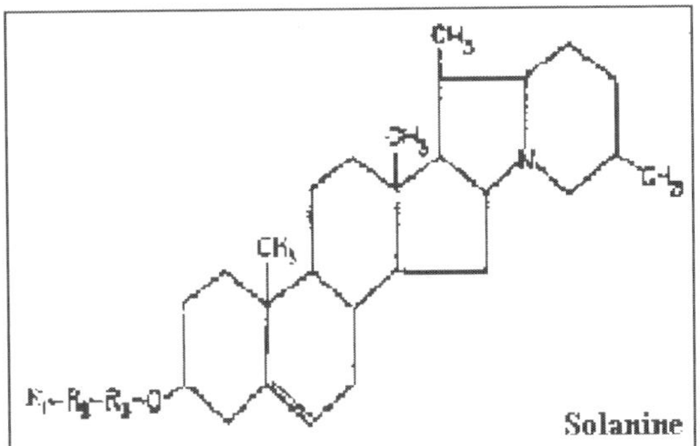

Solanine

Light-induced formation of a green surface color, resulting from chlorophyll accumulation, is known as greening (Plate 25.1, Figure 5). There are two facets to the question of green potatoes. One is the market appearance of potatoes and the other is health concerns dealing with eating a green potato. Marketing appearance problems are associated directly with greenness which is due to *chlorophyll biosynthesis. Health concerns are due to a parallel biosynthesis of a glycoalkaloid called solanine.* Greening of 5 per cent of a lot of tubers is considered as 'damaging' and the lot will be graded down. Green pigment chlorophyll is completely safe and is found in all plants, lettuce, spinach etc. When the potato greens, *solanine* increases to potentially dangerous levels. Increased solanine levels are responsible for the bitter taste in potatoes after being cooked. Solanine biosynthesis occurs parallel but independent of chlorophyll biosynthesis; each can occur without the other. Unlike chlorophyll, light is not needed for solanine formation but is substantially promoted by it. The formation of solanine in potato is localized to the skin, usually no deeper than an eight of an inch (3 mm). Tubers growing at or close to the soil surface may become green from direct exposure to sunlight or from light penetrating through cracks in the soil surface. This is usually an intense green color on a limited part of the tuber surface. Greening also occurs from extended exposure to low light levels in storage or on store shelves. This situation usually produces a lighter and more diffuse coloration of the entire tuber.

The potato plant also has the interesting ability to produce its own protective chemicals which can make it lethal to insects, animals and fungi which attack it. These protective chemicals (glycoalkaloids) are at high levels in the leaves, stems and sprouts of the potato plant and are normally at very low levels in potato tubers. However on exposure to light the potato tuber will produce elevated levels of these protective glycoalkaloids, with the highest levels being in the sprouts as they emerge from the tuber. Potatoes will also produce high levels of glycoalkaloids (such as solanine) in response to bruising, cutting and other forms of physical damage, as well as to rotting caused by fungi or bacteria. In these instances high levels of glycoalkaloids are present in the potato. However in non-damaged potatoes, greening is a warning sign.

Many factors play a role in greening, of potato tubers in the field. Exposure to light in the field occurs when potato tubers protrude from the ground. The factors affecting greening of potatoes include more than light exposure. Greening is affected by variety, maturity and age, temperature, intensity and quality of light, and duration of light. Potatoes also develop more greening under light exposure, when temperatures are higher, *e.g.*, 68 °F versus 41 °F. Retail packaging can also contribute to increased greening. Consumers want to be able to view produce prior to purchase. Packaging materials have changed over time from burlap and other opaque materials to transparent bags which allow exposure to light during retail storage and display. White skinned varieties often green more readily than the red or russet varieties. The latter can green also but it may be masked and not as easily detected. Immature potatoes and those recently harvested green more readily due to lack of a thick outer skin.

Management Practices

Several cultural practices can be used to minimize greening, including proper seed planting depth and hilling, and rolling during vine kill to close soil cracks. In storage, potatoes should not receive prolonged exposure to light.

Thumbnail Cracks

The name of defect "thumbnail cracks," or "air checks." derived from the arched shape of the crack in the tuber skin, which resembles a thumbnail (Plate 25.1, Figure 6). Thumbnail cracks in fresh-market potatoes are a quality problem because they significantly reduce the appeal of potatoes to consumers.

Published information identifies two major causes. One reported cause is rough handling of tubers. Thumbnail cracks can develop from a slight injury that just breaks the tuber skin but does not damage the underlying flesh. The force of impact resulting in a thumbnail crack is much less than that causing damage typically referred to as shatter bruise. After the damage occurs, tubers subjected to low humidity, such as in retail store displays, will form the characteristic thumbnail cracks. Minor tuber damage occurring during harvest usually heals rapidly in storage under high relative humidity (95 per cent or more), lessening the chance of thumbnail cracks becoming visible later. Therefore, damage that occurs during packing, shipping, and handling by retailers is probably more important than harvest damage in the development of thumbnail cracks. Thumbnail cracks may also occur in the absence of impact injury when warm, highly hydrated (crisp) tubers are moved into colder air. As the tuber cools, the flesh cannot hold as much moisture, and the tuber skin cracks to relieve internal pressure. The small break in the tuber skin then develops into a thumbnail crack when the tuber is placed in dry air. Thumbnail cracks that occur under these conditions are often referred to as air checks.

Thumbnail cracking is shallow and random. Appearance is that of semi-circular breaks, half-moons, and is associated with exposure of very turgid (hydrated) tubers to very dry conditions. Surface splitting or air cracking is a slight separation of the skin. This is associated with exposure of very turgid tubers to sudden cold temperatures. Both these types of cracking are associated with harvest but should not be confused with shatter bruise which has a distinct appearance and is caused by impacts. Thumbnail and air cracks heal slowly and are subject to infection and dehydration in storage.

Management Practices

Harvest tubers when pulp temperatures are 50° to 60°F and tuber hydration level is midway between dehydrated (limp) and hydrated (crisp). Harvesting tubers too cool and subjecting them to rapid dehydration increases tuber susceptibility to thumbnail cracking. Handle tubers as gently as possible by making certain conveyors on harvesters, pilers, and all packing equipment are operated at full capacity. Avoid excessive drop heights on all equipment. in storage at 50° to 55°F with 95 per cent or more relative humidity for 2 to 3 weeks to promote rapid wound healing. High humidity plays a crucial role in promoting wound healing during the first few weeks of storage after harvest. Relative humidity below 95 per cent delays wound healing and may promote the development of thumbnail cracks.

Little Tuber Disorder

When many tubers form off the seed piece before sprout emergence, the disorder is called little tuber disorder (Plate 25.1, Figure 7). Small tubers develop at the same time as the seed piece sprouts. This disorder will affect plant growth and reduce yields. Little tuber disorder develops in storage due to aging of seed tubers. Upon planting little tubers are initiated directly at the eyes of the seed piece or on very short stolons while the seed piece is sprouting. This can also occur in storage prior to planting. Physiologically, it is induced by a break down of apical dominance of the eyes on the tuber and of sprouts on the stolons. The mechanism is similar to that for tuber chaining.

This disorder is due to physiological aging of seed tubers as a response to high late-season storage temperature. It occurs when seed piece temperature is greater than 68°F and is planted in soil that is less than 50°F. Little tuber disorder can also occur when sprouted seed tubers or pieces are placed in cold storage and then planted. Elevated concentrations of the gases carbon dioxide (CO_2) and ethylene (C_2H_4), a gaseous hormone, may induce little tuber disorder as well as be involved in tuber chaining and heat sprouting. All of these disorders are due to changes in hormone levels.

Management Practices

Since the disorder is caused by seed tuber aging in warm storages, the key practice is storing seed potatoes at less than 40°F, but above temperatures causing chilling or freezing damage. Avoid long storage of seed tubers. Avoid physiologically aging seed such as rough handling (bruising), poor ventilation or elevated temperatures. Do not plant in cold (<50°F), dry (<60 per cent FC) soil.

Internal Disorders

Brown Center or Hollow Heart

Brown center and hollow heart are two facets of the same disorder but can occur independently of each other. Brown center is characterized by a region of cell death in the pith of the tuber which results in brown tissue (Plate 25.1, Figure 8). Hollow heart is characterized by a star shape hollow in the center of the tuber. Brown center frequently precedes the development of hollow heart. Brown center and hollow heart can cause serious losses in crop quality and economic return to the grower. The disorders make cut fresh-market tubers unattractive and can reduce repeat sales. Severe hollow heart negatively impacts the quality of chip processing potatoes. However, both disorders are reported as not harmful and do not effect the taste or nutrition of the tuber.

This physiological disorder is caused by rapid tissue growth after a cool temperature and moisture stress resulting in an irregular brown cavity in the center of the potato. Hollow heart and brown center is a more serious problem during cool, wet weather. Certain varieties are more prone to this disorder. Hollow heart tubers are edible. An association of stem-end hollow heart with potassium deficiency has also been reported. Brown center and hollow heart arise at a higher incidence when growing conditions abruptly change during the season. This can arise when the potato plants recover too quickly after a period of environmental or nutritional stress. When the tubers begin to grow rapidly, the tuber pith can die and/or pull apart leaving a void in the center.

There is no outer way to detect it. Tubers are cut to detect this disorder. It is characterized as a small one-eighth to one inch diameter, brown, circular or elliptic, opaque area with a diffuse border along the longitudinal tuber axis. In round to oval tubers, its usually at the tubers center; with long or oblong tubers, there may be two brown areas, one at each end. Brown areas are distinct but have a smooth, gradual change to unaffected tissue. Depending on the speed of growth resumption after stress, brown center may or may not develop into hollow heart. Hollow heart appears as a lens- or star-shaped, irregular cavity in the center of round tubers such as Atlantic or at either or both stem and bud ends of long tubers such as Russet Norkotah. The internal walls are white to tan. The cavity is larger with larger tubers and is mostly seen in very large tubers. No rot is associated with the disorder.

Management Practices

To reduce the incidence of brown center and hollow heart, uniform soil moisture should be maintained during the season especially soon after tuber initiation. Ensure the crop in nutrient balance. Apply potassium fertilizers sufficiently. Nitrogen applications should be split into several applications before row closure to achieve steady plant and tuber growth. Every effort to improve even plant spacing by reducing seed skips at planting should also be made.

Blackheart

The blackheart of potato is a nonparasitic disease commonly found in storage godowns. This is due to high storage temperature and low oxygen supply. Rare in early-crop potatoes due to typical marketing practices; In conditions of restricted airflow and high respiration, tubers held above 15°C (rapidly above 20°C) develop an internal brown discoloration which eventually becomes deep black. Insufficient oxygen reaches the interior of the tuber under these conditions.

Blackheart can develop around harvest, in storage and in transit as inside trucks. It is caused by an oxygen deficit at the center of the tuber. Oxygen deprivation results in asphyxiation, loss of respiration, and death of cells. Any pre-harvest, storage and transit condition that prevents oxygen from reaching the tuber center will result in blackheart. These conditions are commonly poor ventilation, water-logging, long exposure to high field temperatures (>90°F) before harvest, and prolonged storage at low temperatures (<35°F). Tubers used as seed can have lower vigor and stand since tuber starch reduced in bulk and may not support emergence.

The center of affected tubers is initially pink then turns dark brown to blue black with an irregular pattern, and the border of the discolored area is usually very distinct (Plate 25.1, Figures 9a–b). Darkened areas of the tuber are usually fairly firm, in contrast to those of tubers affected by Pythium leak, which are spongy, however, when temperature is greater than 65°F it may turn soft and inky. Affected tissues do not smell, and shrinking of the tissue may result in the formation of a cavity in the center of the tuber. Blackheart develops when tubers are held in a low–oxygen environment or when gas diffusion through the tubers is slowed because of extremely cold (32° F) or warm (96°–104° F) temperatures. This condition can also develop in the field when soils are flooded or in poorly aerated storages. Because seed–piece size is effectively reduced by the death of affected tissues, plant stand and vigor are likely to be reduced

Management Practices

Good ventilation in storage prevents blackheart. Avoid poorly drained ground. Avoid flooded areas. Avoid closed bins, deep piles and poor ventilation. Avoid poorly aerated trucks and storage. Avoid temperature extremes, inhibit diffusion of oxygen through tuber.

Heat Necrosis

Internal heat necrosis (IHN) is a physiological tuber disorder that causes an unacceptable browning of the tuber tissue and can increase economic loss to the growers. The disorder is caused by elevated soil temperatures during the latter stages of growth and development of the tuber. Acid soil and low soil calcium also have a role in developing this disorder. If the vines and leaves are still actively growing and green during this period of elevated temperatures, water and nutrients are translocated from the tuber to supply the plant. The vascular system of the tuber is stressed and cannot sustain the evapotranspirational demands of the plant. Under these conditions, it is reported that the vascular ring deteriorates and becomes necrotic. The three suspected leading causes of IHN in tubers are temperature, soil moisture, and plant nutrition.

Necrotic areas are mostly found in and around the vascular ring usually coalescing and radiating to the center (pith). The necrosis appears as light tan, dark yellow to reddish brown flecks or specks; these may be dark brown or black in the most severe cases (Plate 25.1, Figure 10). They resemble the necrosis seen with chilling injury. Flecks usually cluster near the center toward the bud end and can appear similar to blackheart. The flecks are firm; there's no breakdown or rot. Vascular tissue is usually not affected but, in some cases, flecks may be confined to the vascular tissue at the bud end. The symptoms are more prevalent at the bud (apical) end of the tuber than at the stem end. The exterior of the potato tuber does not show visible signs of IHN. This disorder does not affect the nutritional value of the tuber, but the economic impact can be significant due to off-grade quality.

Management Practices

Prevention of internal heat necrosis requires maintaining adequate soil moisture, especially during hot periods, applying adequate calcium in the tuber-forming zone in the soil (particularly in sandy or low calcium soils), and managing fertilization, irrigation, and other cultural practices to promote uniform vine and tuber growth. Add lime to acid soils. Don't store tubers in ground long after vine death. Schedule irrigation for seasonal growth.

Vascular or Stem-end Discoloration

This physiological disorder is typified by a shallow, brown discoloration in the vascular system (phloem and xylem) near the stolon attachment (Plate 25.1, Figure 11). Intensity can vary with season and cultivar but typically does not extend greater than 0.5 to 1.0 inch into the tuber. The disorder may or may not be visible at harvest but may develop during storage. It can easily be confused with net necrosis and laboratory analysis for potato leaf roll virus (PLRV). This non-pathogenic disorder is associated with low soil moisture and rapid, 1-3 days, vine death due to acid, frost, or mechanical kill. High temperature stress at vine kill increases the severity of the disorder. Tubers near full maturity are most susceptible. The disorder may be seen in the field before harvest and develops during the first two months of storage

Disorder is characterized by a light tan to reddish brown speckling or a dark brown streaking in the vascular tissue. The speckling or streaking usually extends to about a half inch from the attachment point, but in severe cases may extend in the vascular system the length of the tuber. This disorder is readily mistaken for vascular discoloration due to Verticillium wilt or leaf roll net necrosis (pathogenic causes).

Management Practices

Practices for reducing vascular discoloration are difficult to elucidate because causal factors have not been clearly identified. However, cultural practices should include avoiding vine killing when the plants are subject to moisture or high temperature stress, or if the vines have not begun to senesce. Irrigation before vine kill may often reduce the potential for development of disorder. Irrigate prior to vine desiccation. Desiccate vines gradually, 1-2 weeks for complete kill.

Jelly End or Glassy End

The physiological disorder commonly referred to as sugar ends is also known as translucent ends, glassy ends, or jelly ends. It is a serious concern for processing potatoes and mostly affects varieties with long tuber type such as Russet Burbank. This disorder is most associated with water deficit in plants due to low soil temperature, high air temperature, dry winds, and too much top growth with respect to tuber growth. It may occur early in the season around tuber initiation or late in the season near harvest. It may be associated with too rapid vine desiccation. The stem end tends to be pointy and is also associated with a dumbbell tuber shape. Jelly end describes the flaccidity, wiggliness, of the stem end. Upon drying, the stem end becomes leathery and shrivels; the skin becoming wrinkled. The tissue at the stem end will have a glassy or opaque appearance hence the term glassy or translucent end. The affected area is sharply bordered from healthy tissue and usually extends less than two inches into the tuber. Affected tubers usually are pointed on the stem end, although bud ends also have been observed with translucent end. Tuber flesh from the affected area is translucent or glassy, devoid of starch, and usually high in reducing sugars. Tubers usually break down in storage or, if processed, produce dark fry products due to elevated sugar levels.

Inadequate soil moisture during early bulking coupled with high temperature seems to cause this disorder. Jelly end appears due to a lack of starch at the stem end resulting in a low specific gravity. Reducing sugars like glucose are high. The disorder is most prevalent in pointy and dumbbell-shaped tubers and in long tuber shaped varieties as Russet Burbank. Two general causal theories have been proposed. Starch breakdown or removal from the stem end when growth renews after stress. And, starch produced in the leaves is either not broken down and transported to tubers or, in tubers, the starch breakdown product is not reassembled into starch for some reason.

Management Practices

Irrigate to maintain uniform growth. Monitor soil moisture to schedule irrigation. Allow gradual vine desiccation. Irrigate somewhat after vine kill. Don't plant after sugar beets because of low crop residue and compaction. Planting after corn or small grains is preferred with shallow incorporation of residue for good soil aeration and water infiltration. Avoid highly salinated (sodium) fields especially with high levels of residual nitrogen. Avoid over nitrogen fertilization. Avoid tillage practices that compact soil. Last cultivation should be before emergence and first irrigation should be after emergence. Some cultivars ('Lemhi', 'Nooksack') are less susceptible.

Heat and Internal Sprouting

Hair or spindle sprouting may occur in the field or in storage. It is associated with hot and dry conditions during late tuber growth (plateau stage). This disorder is also associated in seed from plants affected by psyllid yellows (caused by toxin injected by potato psyllid) and aster yellows (caused by mycoplasma injected by aster leafhopper). The presence of heat sprouts was found to be in inverse ratio to the relative length of rest period. The greatest prevalence of heat sprouting was associated with a short rest period and the smallest prevalence with a long rest period. Heat sprouting is most prevalent in late-maturing plants.

Internal sprouting is a disorder that occurs in storage. As the tubers break dormancy, sprouts typically will grow away from the tuber, but because of an external growth restriction, the sprouts grow into the tuber rather than outward. This can occur from a sprout from an eye on the underside of the tuber growing up and penetrating it or from deep eyes when the sprout grows sideways into the tuber. The tip of the internal sprout is commonly necrotic, brown. When it occurs, it usually is from an eye with multiple sprouts (rosette). This disorder occurs primarily with physiologically aged tubers and occurs when storage temperatures are above 55°F. The two major causes associated with internal sprouting are application of below effective levels of CIPC, a sprout inhibitor, and high pile pressure late in the storage season.

A small thin sprout grows prior to normal dormancy break. Premature sprouting will cause a breakdown of starch to reducing sugars below the sprouted eye. Hot conditions late in the season will inhibit dormancy of mature tubers and push them to sprout. Experiments were reported that short exposure (few hours) to 100°F is able to induce heat sprouting of mature tubers.

Management Practices

To prevent heat and internal sprouting, plant deep and avoid high setting varieties Adequate turgidity of the stored tubers should be maintained, dirt and debris should be eliminated from the stored potato pile, and the proper rate and timing of sprout inhibitor applications should be used.

Chilling

Chilling injury to tubers can occur in the field, in storage or in transit due to exposure to low, non-freezing temperatures. Storage at temperatures near 0°C (32°F) for a few weeks may result in a mahogany discoloration of internal tissue in some varieties. Much longer periods of storage are generally required for chilling injury. Surface injury appears as diffuse brown to black patches. Patches may be slightly shiny and are prone to molds. Affected internal tissue appears smokey-grayish and diffuse similar to early stages of leak (Plate 25.1, Figure 12). A net necrosis, brown specks, in the vascular ring at the stem end is common ("mahogany browning"). It is similar to leaf roll net necrosis. But, unlike the latter which is scattered in tuber (star burst effect) chilling net necrosis is confined in and around the vascular ring. (Phloem tissue is the most sensitive to chilling.) In severe cases, the specks enlarge and are blackened. Brown streaks into tuber may appear. Blackening of the phloem closely resembles virus leafroll net necrosis.

Only a few hours of exposure to temperatures of 32 to 35°F is needed for chilling injury. The disorder is absent when temperatures are above 38°F. Immature tubers, harvested too soon after vine desiccation, are mostly affected. Chilling impairs and delays wound healing. Poor sprouting occurs with affected seed tubers. Chilling lowers internal quality and reduces storage life. Affected tissue turns sweet due to breakdown of starch storage organelles (vacuoles) and subsequent breakdown of starch to glucose by cellular enzymes. Therefore, affected tubers are not suitable for processing. Affected tubers should not be used for seed because they sprout poorly and there is potential of seed decay in the ground.

Management Practices

Avoid exposure to temperatures below 38°F. There are varietal differences in tolerance to level of coolness and its duration. Harvest before soil temperature cools to near freezing. Store seed and table potatoes at 38-40°F. Ventilate to dry out chilled tubers and avoid possible rot. Keep storage insulation dry. Don't store against an outside wall.

Freezing Injury

Freeze injury to tubers can occur in the field, in storage or in transit due to exposure to below freezing temperatures, <28°F. In storage of seed and table stock tubers, freezing is usually associated with poor ventilation and poor insulation. Potatoes that are expos freezing (below 28°F) temperatures can show multiple symptoms. Freezing damage is difficult to diagnose while the tubers are frozen because they show no obvious symptoms than the surface being hard. Prior to thawing, affected tuber area is hard. As the tubers begin to thaw, the first obvious sign of frost damage is free moisture (weeping effect) on the outside of the tuber (Plate 25.1, Figure 13). The next phase is cellular breakdown causing the tissue to turn brown, gray, or black. Typically, a distinct line is visible between healthy and frozen tissue. Frozen tissues tend to disintegrate and will eventually dry out. the underlying tissue will be bluish-grayish and the margin will be sharply defined, not diffuse. Also, unlike chilling, there will not be a net necrosis.

Ice crystals form in cells upon freezing. These puncture the cell and organelle membranes. Upon thawing, the cells loose structural integrity and spill out their contents. Bacterial soft rot commonly will attack thawed affected tissue in storage. Don't use for seed due to poor sprouting and high susceptibility to soft rot.

Control Measures

To minimize frost or freezing damage, proper hilling procedures should be used to minimize tuber exposure to cold air temperatures. If possible, potatoes should be harvested before potential frost exposure. In addition, potatoes should be transported in favorable conditions and storage temperatures maintained above 37°F.

References

Anonymous. *Potato Education Guide*, University of Nebraska, Lincoln. http://www.panhandle.unl.edu/potato/html/physiological_disorders.htm.

Bohl, W.L. and Thornton, M.K. (2006). *Thumbnail Cracks of Potato Tubers*. University of Idaho Extension. CIS 1129 Issued in furtherance of cooperative extension work in agriculture and home economics, Acts of May 8 and June 30, 1914.

Craig, Y.G., Mccord, P.H., Haynes, K.G. and Rikki, S.S.B. (2008). Internal heat necrosis of potato: A review. *American Journal of Potato Research*, 85: 69–76.

Folsom, D. (1947). Performance of greeninmg of potato tubers. *American Journal of Potato Research*, 24: 10.

Hiller, L.K. and Thornton, R.E. (1993). Management of physiological disorders. In: *Potato Health Management*, (Ed.) R.C. Rowe. APS Press, St. Paul, pp. 87–94.

Hochmuth, G.J., Hutchinson, C.M., Maynard, D.N., Stall, W.M., Kucharek, T.A., Webb, S.E., Taylor, T.G., Smith, S.A. and Simonne, E.H. (2001). Potato production in Florida. In: *Vegetable Production Guide for Florida*, (Eds.) D.N. Maynard and S.M. Olson. Vance Publishing (http://www.gov.pe.ca/af/agweb/index.php3?number=1001565).

Hooker, W.J. (1981). *Compendium of Potato Diseases*. American Phytopathological Society, St. Paul, pp. 125.

Hutchinson, C.M. (2003). Horticultural Sciences Department, Florida Cooperative Extension Service, Institute of Food and Agricultural Sciences, University of Florida. (http://edis.ifas.ufl.edu.).

McCann, I.R. and Stark, J.C. (1989). Irrigation and nitrogen management effects on potato brown center and hollow heart. *HortScience*, 24(6): 950–952.

Patil, B.C.., Salunkhe, O.K. and Singh, B. Metabolism of solanine and chlorophyll in potato tubers as affected by light and specific chemicals. *Journal of Food Science*, 36: 474–476.

Sieczka, J.B. and Thornton, R.E. (1993). Commercial potato production in North America. *Potato Association of America Handbook*. Revision of American Potato Journal Supplement volume 57 and USDA Handbook 267 by the Extension Section of the Potato Association of America.

Stark, J.C., Olsen, N., Kleinkopf, G.E. and Love, S.L.. *Tuber Quality*. http://www.ag.uidaho.edu/potato/production/files/tuber%20qua%20l%20i%20t%20y%20–%20web.pdf.

Wierseme, S.G. (1985). *Technical Information Bulletin*. Physiological development of potatoes seed tubers. Training and communication department, CIP, Lima, Peru.

Plant Diseases Management in Horticultural Crops (2011) *Pages 312–344*
Editors: **Shahid Ahamad, Ali Anwar and P.K. Sharma**
Published by: **DAYA PUBLISHING HOUSE, NEW DELHI**

Chapter 26

Yellow Vein Mosaic Disease of Okra and its Management Strategies

***C.P. Khare*[1], *D.R. Bhandarker*[1], *J.N. Srivastava*[2], *M.P. Thakur*[1] and *V.S. Thrimurthi*[1]**

[1]*Division of Plant pathology, Indira Gandhi Agricultural University, Raipur, C.G.*
[2]*Division of Plant Pathology,*
Sher-e-Kashmir University of Agriculture Sciences and Technology, Jammu

Among the vegetables, okra (*Abelmoschus esculentus* L.) Moench) is the sixth important vegetable crop cultivated throughout the country. In world, area and production under okra is 8.33 million ha. and 54.28 million tones, respectively, while the productivity is 6.5 tones/ha. (FAOSTST, 2007). Okra occupies 4.09 million ha. area in India with production of 41.93 million tones having productivity 10.25 tones/ha (Anonymous, 2009). Okra is considered to be originated from Africa or Asia. However, Zeven and Zhukovasky (1955) believed that it had been originated from India. Their view appears to be correct from the Sanskrit words 'Tindisha' and 'Gandhmula' to designate *bhendi*.

Okra grows well in areas where day temperature ranges between 25-27°C. It is susceptible to drought and low night temperature; grow well in light type of soil with fairly high organic matter content. Okra fruits have medicinal value and are used for fevers, catarrhal attacks and irritable stales of genitourinary organs, chronic dysentery and bland mucilage generally in the form of soup (Basu and Ghosh, 1943, Pal *et al.*, 1952 and Nadkarni, 1972).

Okra is vulnerable to many fungal and viral diseases. It is spontaneously infected by different fungal diseases like powdery mildew (*Erysiphe cichoracearum*) and (*Leveillula taurica*), leaf blight (*Rhizoctonia solani*), sooty leaf spot (*Cercospora abelmoschi*), dieback (*Colletotrichum dematium*), root rot and collar rot (*Macrophomina phaseolina*), damping off (*Pythium* spp.), and wilt, bacterial disease such as blight (*Xanthomonas* spp.) and viral diseases like yellow vein mosaic (Capoor and Verma, 1950), enation mosaic, enation leaf curl (Singh and Dutta, 1986), okra leaf curl and okra mosaic (Lana, 1976), yellow mottle in okra (Torre *et al.*, 2003) and fruit distortion mosaic disease of okra (Reddy *et al.*, 2003).

Amongst the various constraints in cultivation of okra, viral diseases, particularly yellow vein mosaic is major one. The disease reported before 85 years by Kulkarni (1924) in Maharashtra had appeared in epidemic form for first time in Maharashtra in 1950 (Capoor and Verma, 1950). Subsequently, epidemic outbreak of the disease had been reported from different states where okra is cultivated (Jha and Mishra, 1955, Tripathi and Joshi, 1967, Chelliah *et al.*, 1975 and Saikia and Bhagbati, 1994). Presently, the disease is occurring in all okra growing states of India some times in endemic and mostly in epidemic form threatening the cultivation of okra.

Historical Background

Perusal of the ponderous literature has revealed that the disease in the past had been called by different names from different places. Kulkarni (1924) first time designated it as yellow mosaic of okra. It was named as yellow vein mosaic of okra by Uppal *et al.* (1940). Capoor and Verma (1950) established viral nature of the pathogen causing the disease.

Bertus (1942) from Srilanka named the disease as mosaic of okra. Fernando and Udurawana (1942) from Srilanka reported the disease as yellow vein banding of okra. In Trinidad, it was called as mosaic disease of *Hibiscus esculents* by Owen (1946). The disease also referred as vein clearing of okra by Anonymous (1961).

Besides yellow vein mosaic virus, the okra has been reported to be infected by okra mosaic Tymovirus (Lana and Bozarth, 1975) and enation leaf curl virus (Singh, 1990). The initial symptoms induced by these two viruses found to be similar therefore there is a possibility of committing the mistake in diagnosing.

Symptomatology

The symptoms of the disease described by different workers have been found to be almost similar with minor differences. The difference in symptoms varies to the type of cultivars infected, age of the cultivar at the time of infection, climate, strain of the virus and inoculum load etc. The common symptoms of the disease described on leaves and fruits were as follows.

The first symptom of the disease was vein clearing initiated from small veins and extended to the large ones followed by chlorosis. In severe cases veinal chlorosis was followed by yellowing of veins and leaves (Kulkarni, 1924; Uppal *et al.*, 1940; Fernando and Uduravana, 1942; Owen, 1946; Capoor and Verma, 1950; Nariani and Seth, 1958; Raychaudhari and Nariani, 1977). Sometimes, the yellow network of veins was followed by thickening of veins and veinlets (Nariani and Seth, 1958; Raychaudhari and Nariani, 1977 and Singh *et al.*, 1977). Sometimes yellow vein banding followed by interveinal clearing and development of minute enations on the aerial side of leaves was also evident (Fernando and Uduravana, 1942). In few cases complete yellowing of leaves was also reported (Singh and Shrivastava, 1974).

The fruits arise from diseased plants were malformed and bleached (Kulkarni, 1924; Fernando and Uduravana, 1942 and Raychaudhari and Nariani, 1977). The diseased plants remained stunted and beared few fruits which were hard and beared few fruits, smaller in size (Mandhar and Singh, 1972 and Singh *et al.*, 1977). The disease besides leaves and fruits has been reported to infect the flowers also. The diseased plants beared less and small flowers with enations on petals. The flowers showed delay in opening (Sonwane, 2004).

The characteristics symptoms of the disease are clearing of the veins and veinlets which may thickened under severe condition. The chlorosis extends to the interveinal area. Whole leaf turned yellow under severe infection. The fruits remain dwarf, malformed and yellowish green in color that may reduce the marketability of the produce (Gupta and Thind, 2006).

Pathogen

Harrison *et al.* (1991) identified the gemini virus particles by immunosorbent electron microscopy in extracts of okra yellow vein mosaic virus infected leaves in India. They further studied that whiteflies (*Bemisia tabaci* Genn.) transmit this virus, by its pattern of reaction with the panel of monoclonal antibodies. Okra yellow vein mosaic virus.

The virus causing yellow vein mosaic in okra is a Gemini virus. It belongs to genus Begomovirus and family Geminiviridae. It is *ss*DNA spherical virus with 20 nm diameter (Mortelli, 1992). The virus is transmitted by whitefly (*Bemisia tabaci* Genn.) in persistent manner by both nymph and adults (Verma, 1952; Muniyappa, 1980 and Mukhopadhyay, *et al.*, 1994). Verma (1952) reported females were slightly better vectors than that of male. The minimum incubation period for the virus was 7 hours.

The causal agent of yellow vein mosaic virus disease is a geminivirus of 18nm x 30nm in size Gupta and Thind (2006) and Handa and Dutta (1993) and it has a close relationship with Indian cassava mosaic geminivirus (ICMV). Bhendi yellow vein mosaic virus is more closely related to Indian Cassava mosaic virus than African Cassava mosaic virus.

The monopartite begomo viruses cause the yellow vein mosaic disease of okra which comes under the family Geminiviridae and a small satellite. DNA-ß induces typical symptom of yellow vein mosaic disease.

Pun *et al.* (1999) detected OYVM virus in infected okra plants by direct antigen coating ELISA (DAC-ELISA) using polyclonal antibodies (PAbs) raised against African cassava mosaic virus (ACMV) and Indian cassava mosaic virus (ICMV). The reaction of antigen extract from infected okra leaves was stronger with PAbs to ACMV. Using Pabs and DAC ELISA, OYMV could also be detected in 5 weed plant species (*Acalyphae indica, Croton bonplandianus, Hibiscus rosasinensis* and *Parthenium hysterophorus*). The titre of the virus in all the 5 species was invariably lower than in okra. DAC-ELISA on nitrocellulose membrane could be successfully employed to detect OYVMV in infected okra plant and single viruliferous white flies, *Bemisia tabaci*.

According to Fauquet *et al.* (2005), the taxonomic position of yellow vein mosaic virus as follows.

- ✰ Biota
- ✰ Viruses
- ✰ DNA viruses
- ✰ *ss*DNA viruses
- ✰ Family- Geminiviridae
- ✰ Genus–Begomo virus
- ✰ Virus–Okra yellow vein mosaic virus

Vector

Whitefly

Insects are small flies; approximately 1/8 of an inch long is completely white and can fly, when shaken. Insects pierce the outer layer of plant and extract the liquid contained inside and exuded sticky, sugar concentrate of plant juice that can promote mould growth.

Martin *et al.* (2000) reported that adult white flies measure from 1-3 mm in length, are four-winged and fully mobile with a feeding rostrum and seven segmented antennae.

Bemisia tabaci the vector of virus is a pest of glass house crops in many parts of the world, especially tomato, capsicum, poinsettia, Hibiscus, Gerbera and Gloxinia. The white fly has a very wide host range within a number of families (Compositae, Covolvulaceae, Cruciferae, Cucurbitaceae, Euphorbiaceae, Leguminosae, Malvacea, Solanaceae etc.) (www.wikipedia. com).

Martin *et al.* (2000) reported that the majority of white flies colonize only dicotyledonous angiosperms and a smaller but significant, number feed on monocots, particularly grasses and palms.

Transmission of Whitefly

Unlike other viruses, yellow vein mosaic in okra has been reported to be unique in having only one vector whitefly as well as one species of it *viz.*, *Bemisia tabaci* (Verma, 1952; Verma, 1955a; Raychaudhari and Nariani, 1977 and Muniyappa, 1980).

The both nymph and adults had ability to transmit the virus in persistent manner (Verma, 1952; Muniyappa, 1980 and Mukhopadhyay *et al.*, 1994). Though, both male and females have been found involved in transmission, the females were more efficient than males (Verma, 1952).

The virus acquisition, retension and inoculation by vector have been found mostly similar with few exceptions. Acquisition access period was reported as 4-6 hours (Raychaudhari and Nariani, 1977), 12-24 hours (Verma, 1952). Preliminary period of four hours seemed to improve the efficiency of whiteflies as a vector (Verma, 1952).

Inoculation feeding period of 30 minutes was sufficient to transmit the virus but a five minute probe was not sufficient (Verma, 1955a and Verma, 19955b).

Mukhopadhyay *et al.* (1994) studied the transmission of the virus by whitefly. They reported that minimum inoculation access period of 30 minutes; optimum inoculation access period of 8 hrs; retension of the virus in the vector for 7 to 10 days and number of viruliferous insects for optimum infection as 5.

Occurrence and Distribution

It is mainly confined to the Indian sub-continent, subsequently it also reported from Srilanka (Park, 1930 and Fernando and Udurawana, 1942), Trinidad (Owen, 1946), Bangladesh (Ali *et al.*, 2000) and Pakistan (Rehman and Ahmad, 2002).

In India, it was first reported from Maharashtra East while, Bombay Presidency (Kulkarni, 1924), subsequently, it was reported from Bihar (Jha and Mishra, 1955); Uttar Pradesh (Tripathi and Joshi, 1967); Gujarat (Tsering and Patel, 1990); Assam (Nath and Saikia, 1993); Madhya Pradesh (Sanger, 1997); West Bengal (Mukhopadhyay *et al.*, 1994) and Nagaland (Singh *et al.*, 1994b).

The disease outbreak has been found to be in endemic and epidemic forms in all the okra growing states in India. The incidence has been found to vary from location to location and season to season.

The incidence of yellow vein mosaic of okra has been reported to the tune of 50 to 100 per cent (Park, 1930; Capoor and Verma, 1950; Jha and Mishra, 1955; Tripathi and Joshi, 1967; Chelliah *et al.*, 1975 and Saikia and Bhagbati, 1994).

The incidence was found to vary from season to season. Sanger (1997) reported more incidences in rainy season (73 per cent) than in summer season (25 per cent).

Bhagbati and Goswami (1992) reported that the disease incidence reached its maximum (100 per cent) in May and June sown crop while, it was observed minimum (17 per cent) in the crop sown in the month of October. It was gradually decreased from May to October.

Epidemiology

Natural Hosts Range of Vector Whitefly and Yellow Vein Mosaic Virus

Both virus and vector have been reported to have very wide host range that includes crops, ornamentals and weeds. It is important for both virus and vector as they help the virus to perpetuate and vector to harbour and multiply. Further, these hosts act as reservoirs of the virus and become source of primary inoculum. Thus, these natural hosts are important in determining the epidemic outbreak of yellow vein mosaic of okra.

The crops *viz.*, *Hibiscus moschatus*, *H. subdarrifa*, *Abelmoschus esculentus*, *H. canabinus*, *Hibiscus vitifolius*, *Hibiscus manihot*, *A. critinus*, *A. tuberculantus*, *A. ficulneus*, *Hibiscus pandurformis* and *Malvastrus tricuspidatum* have been reported as the natural host of both yellow vein mosaic virus and whitefly (Capoor and Verma, 1950; Vasudeva and Samraj, 1948; Nariani and Seth, 1958; Gunthilgaraj *et al.*, 1977; Atiri, 1983; Khan and Mukhopadhyay, 1986; Jayrajan *et al.*, 1988; Cheema, 1991; Jambhale and Nerkar, 1992; Mathew and Muniyappa, 1992; Dahal *et al.*, 1993 and Mukhopadhyay *et al.*, 1994).

Ornamental plants such as hollyhock and *Croton Sparciflorus* have been reported as the natural host of both yellow vein mosaic virus and whitefly (Gunthilagaraj *et al.*, 1977; Atiri, 1983 and Mukhopadhyay *et al.*, 1994).

The weeds *viz.*, *Ageratum conzoides* Linn., *Vemonia cinera*; wild bhendi, *H. tetraphyllus*, *Althaea rosea*, *Sida* spp. and *Parthenium hysterophorus* have been reported as the natural hosts of both yellow vein mosaic virus and whitefly (Capoor and Verma, 1950; Vasudeva, 1957; Gunthilgaraj *et al.*, 1977; Atiri, 1983; Jayrajan *et al.*, 1988; Cheema, 1991; Jambhale and Nerkar, 1992; Mukhopadhyay *et al.*, 1994; Pun *et al.*, 1999a and Singh and Singh, 1999).

The crops *viz.*, *Capsicum annum*, *Nicotiana rustica*, *Sesamum indicum*, tomato, beans, cotton, *Glycine max*, *Cajanus cajan*, *Vigna sinensis*, pumpkin, *Phaseolus vulgaris*, *Vigna mungo*, *Vigna radiata*, *Lablab purpureus*, *Nicotiana tabacum* and *Carica papaya* etc. were found to be the hosts of only whitefly (Shrivastava *et al.*, 1985; Verma, 1989; Venugopalrao and Reddy, 1989; Mohanty and Basu, 1990; Tresing and Patel, 1990; Harrison *et al.*, 1991; Burbon *et al.*, 1992; Dahal *et al.*, 1993, Nandihalli *et al.*, 1993; Mukhopadhyay *et al.*, 1994; Aslam and Gebara, 1995; Hirano *et al.*, 1995; Sanchez *et al.*, Razvi *et al.*, 1999; Jayshree *et al.*, 1999; Biji *et al.*, 2000; Foda *et al.*, 2000; Naik *et al.*, 2003 and Rathi, 2001).

The flower plants *viz.*, *Xanthicum straumarium*, *Zinia elegans* and *Cosmos sulphurus* (Shrivastava *et al.*, 1977, Shrivastava *et al.*, 1985; Mukhopadhyay *et al.*, 1994; Razvi *et al.*, 1999, Biji *et al.*, 2000, Naik *et al.*, 2003 and Rathi, 2001), and the weeds such as *Digera alternifolius*, *Euphorbia jeniculata*, *Aubergines cassara*, *Amaranthus viridis*, *Amaranthus hypochondriacous*, *Amaranthus tricolor*, *Alternenthra philasiraides*, *Alysicarpus monifer*, *Bidens bitenata*, *Borreria stricta*, *Cassia toniflora*, *Cleome viscosa*, *Commelina bengalensis*, *Crotaloria striata*, *Eclipta alba*, *Emillia sonchifolia*, *Euphorbia birth*, *Galinsoga parviflora*, *Lagascea mollis*,

Leueus asperma, *Parthenium hysteraphorus*, *Phyllanlnus fraternous* and *Achyanthus asperma* harboured only whiteflies. (Mukhopadhyay *et al.*, 1999; Aslam and Gebara, 1995; Hirano *et al.*, 1995; Sanchez *et al.*, 1997; Razvi *et al.*, 1999, Jayshree, 1999; Bi ji *et al.*, 2000; Foda *et al.*, 2000; Naik *et al.*, 2003 and Rathi, 2001).

According to Gupta and Thind (2006) this pathogen had a wide host range which includes *Croton sparsiflora*, *Malvastrum tricuspidalum*, *Abelmoschus manihot*, *Althaea rosea*, *Hibiscus tetraphyllus* and *Ageratum* species.

Whiteflies maintained continuous cycle through its wild hosts. Though a single whitefly was able to transmit the virus, the minimum number of flies required to produce 100 per cent infection was about 10. The whitefly which fed on diseased plants for 12-24 hours remained viruliferous throughout their infestation (Verma, 1952 and Markose and Peter, 1990).

The multiple regression analysis increase of one insect brought about 18.5 per cent of disease (Chelliah *et al.*, 1975).

Shastry *et al.* (1978) reported that yellow vein mosaic and its whitefly vector were positively correlated with temperature, rainfall and relative humidity, seasonal fluctuation in the incidence and severity of disease corresponded closely to the fluctuation in seasonal population of vector.

Singh, (1980) reported that the incidence of yellow vein of okra was positively correlated to whitefly vector. Hot weather was much congenial for disease development and spread with multiplication of vector.

Singh, (1990) reported that highest vector densities were consistently recorded during the hot season from March-June months. Considerably, the lowest population level was recorded from July onwards till February and almost nil adult whiteflies observed during October month. The whitefly populations and disease incidence was nil during January, February and December months, respectively.

Bhagabati and Goswami (1992) observed white fly populations were highest in the crops sown in May and incidence of okra yellow vein mosaic bigeminivirus was highest (100 per cent) in crop sown in May and June. Disease incidence and populations of *Bemisia tabaci* were least in October sown crops. They further observed that the lowest disease incidence occurred on okra sown at the beginning of October (16.7 per cent) and the highest on crops sown in May and June (100 per cent), with incidence in the February and March crops of 36.5 and 54.2 per cent, respectively.

Nath *et al.* (1992) reported the lesser whitefly population and disease incidence in the crop sown during February 10 to March 10 and as sowing date advances the disease incidence and whitefly population increased. There was positive correlation between whitefly and disease incidence, temperature, relative humidity and rainfall.

Borad *et al.* (1993) reported that *Bemisia tabaci* and yellow vein mosaic disease was low in August to October. The vector population reached maximum size during the first week of October; symptoms appeared 1 week after infection with *Bemisia tabaci.* The disease increased progressively with the corresponding increase in vector populations.

A strong positive correlation was obtained between per cent age of disease incidence and whitefly (*Bemisia tabaci*) population (Nath and Saikia, 1995).

Muzumdar *et al.* (1996) reported that the incidence of yellow vein mosaic virus disease of okra and its vector *Bemisia tabaci* in three cultivars Pusa Sawani, Parbhani Kranti and M-31 for consecutive two years (1992 and 1993) revealed lesser disease incidence and whitefly population in the crop sown during February 25 to March 20 in comparison to sowing dates delayed from April 15 to July 25. Simple correlation studies revealed positive significant association between disease incidence and whitefly population.

Disease Infection

Chelliah *et al.* (1975) reported that the heavy losses occurred (about 88 per cent) during infection by this virus in 30 days old crop.

Pun and Doraiswamy (1999) reported that the per cent infection by okra yellow vein mosaic virus decreased with the age of inoculation. One week age inoculum showed 100 per cent infection upto 3 week age plant and subsequently decreased to 83, 66, 51 and 31 per cent when infected at 4, 5, 6 and 7 week, respectively. Lesser the age of inoculation, greater was the damage incurred in terms of plant growth and yield.

Bhagat *et al.* (2001) reported that the rate of infection was higher in Vaishali Vadhu and Pusa Sawani as compared to Parbhani Kranti. It was almost five times in Pusa Sawani in 40 days old plants during 1993. Maximum rate of disease development was recorded between 35 to 45 days after sowing irrespective of varieties in both the years.

Sonwane, (2004) reported that okra crop was found equally vulnerable to the infection of yellow vein mosaic virus at all the growth stages. The infection of okra by yellow vein mosaic virus was more dependent on whitefly activities than the age of the crop.

Disease Progress

The quantum of initial infection and rate of spread found to be season dependent phenomenon. The gradient study on the spread showed stiff rise during early growth stages of the crop. In the interrelationship studies, it was found that the extent of final infection in the field was related with the extent of initial infection. The rate of spread per day on the other hand was independent on the quantum of initial infection (Khan and Mukhopadhyay, 1985a).

Mukhopadhyay *et al.* (1994) reported that yellow vein mosaic disease appeared within 13-14 days, when sown in February but it takes 26-37 days in case of December sown crop. The rate of spread was maximum in February-March sown crops and minimum in April-May sown crops.

The spread of yellow vein mosaic virus disease was slow in February sown crops and symptom was first appeared at 4 weeks after sowing. The disease incidence increased slowly reaching 43.7 per cent at 12 weeks after sowing. In May sown crops symptoms were first appeared 2 weeks after sowing and from 5 weeks onward incidence increased rapidly reaching 100 per cent 9 weeks. Disease symptoms in the September sown crop first appeared 3 weeks after sowing and 77.5 per cent incidence was recorded 12 weeks after sowing (Saikia and Bhagbati, 1994).

Sonwane (2004) reported that the disease progress curve for yellow vein mosaic disease of okra was initiated before 30 days after sowing in *kharif* and 70 days after sowing in summer and it lasted throughout the season. The disease progressed steadily in *kharif*. However, the progress trend of the disease in summer was not exactly as it was in *kharif*.

Distribution of Infected Plants

Saikia and Bhagbati (1994) investigated the spread of yellow vein mosaic disease of okra. The field investigation found that the distribution of yellow vein mosaic infected plants were highly random at the initial appearance of the disease in field. The extent of disease spread was related with the amount of initial infection and the age of the plant at which infection takes place. The results suggested that the disease was first introduced into the field from outside and there after the spread was within the okra field.

Sonwane *et al.* (2004) reported that in both the seasons *kharif* and *rabi*, the distribution of infected okra plants was random upto 60 days after sowing. Random distribution of infected okra plants by the virus in early growth period indicated that the disease was simple interest disease.

Effect of Climate on Yellow Mosaic Disease

Climatic factor have no direct effect on the development of yellow vein mosaic of okra but they affect the activities of the whiteflies and thus have indirect bearing on the development of disease.

Chelliah *et al.* (1975) reported that simple correlation studies showed negative association between disease incidence and relative humidity in all the three growth periods *i.e.* 30, 45 and 60 days old crop growth and positive association between maximum, minimum temperature and the disease incidence in 45 and 60 days old crop. Multiple regression analysis led to the conclusion that in 30 days old crop, increase of whitefly population by one number brought about 18.5 per cent of the disease, while 1 per cent decrease in relative humidity increased the disease incidence by 1.2 per cent. In 45 days old crop, minimum temperature alone exerted positive influence on disease incidence during 45-60 days of crop age, increase in minimum temperature by 1°C resulted in 6.3 per cent increase of the disease.

Singh *et al.* (1977) studied the effect of dry weather on development of yellow vein mosaic. Temperature between 30-32°C during March-June had been reported to be conducive for development of yellow vein mosaic disease while less incidence (10-30 per cent) during rainy and winter season and a rapid variation in severity from 50-60 per cent during summer has been reported.

Incidence of okra yellow vein mosaic disease and its vector whitefly activities were positively correlated with temperature, rainfall and relative humidity (Shastry *et al.*, 1978).

The development of the yellow vein mosaic of okra has been reported to have positive correlation with rainfall and relative humidity. The dry and hot weather with little or no rainfall was formed very much congenial condition for disease development and spread (Singh, 1986 and Singh, 1990).

Sharma *et al.* (1987) studied the effect of temperature on the incidence of yellow vein mosaic virus on 6 varieties of okra was assed over a period of 6 years. Incidence increased with decrease in temperature in September compared with August. A significant negative correlation coefficient between temperature and virus incidence was observed. This was particularly evident in the resistant varieties which were free of virus in August but showed virus symptoms in September. It was concluded that resistance to Bhendi yellow vein mosaic virus was influenced by temperature and could therefore be under the control of polygenic system.

The influence of different climatic factors *viz.*, temperature (maximum and minimum), relative humidity (am and pm), bright sunshine hours and wind velocity (kmph) has been reported to be favourable for development of okra yellow vein mosaic virus and its incidence (Singh and Singh, 1989 and Singh *et al.*, 1994b).

Singh, (1990) observed that dry and hot weather with little or no rainfall was very much congenial for disease development and spread.

Seasonal fluctuation in yellow vein mosaic disease incidence, symptoms expression and severity of disease corresponded closely to the fluctuation in seasonal abundance of the whitefly vector has been reported (Chelliah *et al.*, 1975; Sastry *et al.*, 1978; Singh, 1980; Sharma *et al.*, 1987 and Singh 1990).

The climatic factors *viz.*, relative humidity and temperature have been reported to be important for development of yellow vein mosaic of okra (Kumar and Kadian, 1992). They reported that maximum temperature (32±2°C) and minimum temperature (25±2°C) and relative humidity (40 per cent) were congenial for incidence of the yellow vein mosaic disease.

Rate of spread of yellow vein mosaic disease of okra were maximum in February/March sown crops and minimum in April/May sown crops depending upon the vector activity or in other words meteorological conditions. Maximum and minimum relative humidity and rainfall were negatively correlated and maximum temperature was positively correlated to the vector population and disease incidence (Mukhopadhyay *et al.*, 1994). They also reported early sowing of okra crop in February favoured the first appearance of visible symptoms within 13-14 days, when temperature was slightly elevated whereas, the delay in symptom expression upto 20 to 27 days in December sown crop was observed when temperature was lower.

The studies of Muzumdar *et al.* (1996) revealed the positive significant association between disease incidence and whitefly population, temperature, relative humidity (pm), rainfall and number of rainy days.

Pun *et al.* (2000a) reported that a highly positive correlation existed between sunshine hour and disease incidence while, morning relative humidity and wind velocity had a highly significant negative association with disease. Non-significant association of maximum temperature and total rainfall with disease incidence was positive and negative, respectively. Simple correlation studies however showed that maximum and minimum temperature also exerted a highly significant positive effect on the disease incidence.

Kumawat *et al.* (2000) reported that the infestation of whiteflies in okra started in the fourth week of July and reached peak in the fourth week of September. Relation of whitefly with abiotic factors *viz.*, minimum and maximum temperature, relative humidity and rainfall were also assessed was significantly correlated with maximum temperature.

Kalita *et al.* (2005) reported that whitefly population per plant was positively and significantly correlated with both single and combined infection of okra yellow vein mosaic virus and okra leaf curl virus. Among weather variables, maximum and minimum temperature, wind speed and total rainfall were negatively correlated, whereas morning relative humidity and sunshine hours were positively correlated with disease incidence. Evening relative humidity had negative and positive correlation with disease incidence under normal rainfall condition in 2001 and low rainfall condition in 2002, respectively.

Disease Development and Vector Population

Mohanty and Basu (1990) observed that the ability of *Bemisia tabaci* to transmit okra yellow vein mosaic Gemini virus was not significantly affected when the insects were reared on different host plant. The performance of the vector was, however, most consistent when it was reared on aubergines (egg plant).

Nath *et al.* (1992) reported that the incidence caused by OYVM bigeminivirus was lowest when population of vector *Bemisia tabaci*, were low. They recorded the positive association of vector population with disease incidence, relative humidity, temperature and rainfall.

Borad *et al.* (1993) revealed that the symptoms of okra yellow vein mosaic virus appeared in one week after infestation with *Bemisia tabaci*. The disease per cent age increased progressively with the corresponding increase in vector population. Both adult of *B. tabaci* and incidence of YVMV were observed at 16 and 20 days after seed sowing. The relationship between adult white fly and incidence of YVMV is highly significant.

Shivpuri *et al.* (2004) concluded that the white fly is a serious pest of okra grown part of India. It is polyphagous pest causing crop losses in three ways (i) as a phloem feeder (ii) honey dew secreted by white flies serves as medium for growth and development of fungi (iii) as an efficient vector of Gemini viruses.

Shivpuri *et al.* (2004) further suggested that the adhoc economic threshold level (ETL) of 5 white flies per leaf was fix for spray of insecticides. The numbers of nymph as well as adult were counted on six leaves selected at random from upper, middle and lower portion of the plant for computing population of white flies.

Ali *et al.* (2005) observed the correlation of environmental conditions with okra yellow vein mosaic virus disease severity and *Bemisia tabaci* population on commercially grown varieties of okra.

Yield Losses

The yellow vein mosaic disease of okra has been reported to cause enormous losses both qualitative and quantitative.

Quantitative Losses

The yellow vein mosaic disease of okra has been reported to cause enormous losses in yield of okra from different regions of the country. Late incidence of the disease reduced the yield of fruits by 25 per cent while early infections secured 100 per cent loss (Capoor and Verma, 1952).

Newton and Pieris (1953) from Ceylon reported the yield of okra have been fallen by 50 per cent due to the infection of yellow vein mosaic disease.

The yellow vein mosaic disease caused losses from 25 to 85 per cent depending upon the time of infection of the okra plants (Anonymous, 1961).

The yellow vein mosaic disease reduced yield of okra depending upon stage at which infection occurred, the yield losses were 93.80 per cent, when plants infected at 35 days after germination and it decreased to 83.63 and 49.36 per cent, when crop was infected at 50 and 65 days after germination (Shastry and Singh, 1974a). Same result was reported by Chelliah and Murugasen (1976a) that, when infection of this disease occurred on 30 days old okra crop resulted in 88 per cent loss in yield.

Singh *et al.* (1977) studied the effect of growth stages at the time of infection on losses. They found that plants infected within 35 days after sowing exhibited about 93 per cent loss in yield. They also found loss in yield by 83.63 and 49.36 per cent, when the plants were infected at 50 and 65 days after germination. Sinha and Charkrabarti (1978) found 86 per cent losses when crop was infected 33 days after sowing and decreased to 32 per cent when crop was infected at 75 days after sowing. Their studies further revealed the fact that virus had no any effect on seed germination.

Jambhale and Nerkar (1986) reported that yellow vein mosaic virus reduced the yield of okra by 10 to 93 per cent.

Kumar and Kadian (1992) reported the yield reduction in okra due to yellow vein mosaic disease was to the tune of 51.56 to 96.82 per cent.

Nath *et al.* (1992) concluded that the fruit yield of okra was negatively correlated with disease incidence.

Harender *et al.* (1993), observed incidence of pest and disease (*Cercospora* sp., *Erysiphae chicharaceorum, Heliothis armigera* and okra yellow vein mosaic virus) and calculated losses in fruit and seed yield together with insect and pests in 5 varieties of okra. The results reveal that the Punjab showed least disease severity and pest incidence with lowest reduction in marketable fruit and seed yield, 22.1 and 22.5 per cent respectively under north Indian condition.

Nath and Saikia (1993) reported that the yellow vein mosaic virus causes loss of okra fruits as much as 94.42 per cent and minimum 32.65 per cent.

Gupta and Thind (2006) reported that the yield losses due to the disease were 49.3 to 93.8 per cent coupled with reduction in number of fruit and seed yield per plant.

Pun and Doraiswamy (1999) confirmed that lesser the age of plant at the time of whitefly inoculation greater will be the damage incurred in terms of yield. Yield losses being 95.7 per cent when plants inoculated at 1 week age and 17.4 per cent when plants inoculated at 7 week age.

Bhagat (2000) reported that the plants infected at 30 days after sowing reduced more plant height, fruits number per plant, fruit length and weight than the plants infected at 60 days after sowing.

Qualitative Losses

The disease not only reduced the yield but also deteriorated the quality of fruits, which in turn affected the market value of the fruits.

Singh and Singh (2000) observed that the disease causes severe loss in yield as well as quality of the marketable fruits was adversely affected.

The fruits of the disease infected plants have been found smaller in size than fruits from healthy plants. They were bleached and distorted (Nariani and Seth, 1958).

Effect of Disease on Plant Growth Character

The growth of okra plant was retarded and few leaves and fruits were produced when infection occurred within 35 days of germination reported by Sastry and Singh, (1974).

Sinha and Chakrabarti, (1978) reported that plant growth, fruit number and seed yields/plant were adversely affected by yellow vein mosaic virus in okra plant.

Singh *et al.* (2003) observed the performance of 5 okra hybrids and reported DVR 2 had the longest fruits and highest number of fruits, plant height, stem diameter, dry weight of shoots and fresh and dry weight of roots per plant. Whereas, DVR-1 the highest fruit weight per plant and total fruit yield/ha. They also observed that HIHB-0090, DVR-1 and DVR-2 were free from nematode and yellow vein mosaic virus infection.

Effect of Disease on Morphological Parameters of Okra

Bhagabati *et al.* (1998) reported that leaf area, fruit length, fruit weight and fruit volume were drastically reduced by virus infection. Moisture content of both diseased leaves and fruits was higher than that of healthy okra plant at all growth stages.

Ndunguru and Rajabu, (2004) reported significant ($P < 0.05$) variation in yield components between virus infected and healthy plants. Diseased plant height was reduced 19.5 per cent, number of fruits 34 per cent and petiole length 32.1 per cent as compared to healthy ones.

Management of Yellow Vein Mosaic Virus Disease of Okra

By Using Chemical/Insecticides

Different insecticides application *viz.*, sprays, soil applicants, seed dressers have been assessed for controlling the whitefly vector aimed for prevention of yellow vein mosaic disease of okra. The role played by the insecticides for the management of yellow vein mosaic of okra is described as follows.

Nene, (1973) conducted an experiment to find out the insecticides that could kill in the shortest possible time the whitefly, *Bemisia tabaci* Genn. a vector of viruses. Ambithion, baythion, malathion and parathion among non-systemic insecticides and monocrotophos among systemic insecticides gave a quicker kill of whitefly adults but none of the insecticides could kill whitefly adults in less than 20 minutes. Non phytotoxic mineral oil at 2 per cent quickly killed the adults by immobilizing them in less than 3 minutes after they settled on the oil sprayed leaf surface.

The study revealed that protecting the crop in early stages by application of insecticides was very important because if the crop was not protected in early stages *i.e.* immediately after germination the disease incidence would be more resulting in comparatively low yields even if the crop was sprayed regularly after 20 days of germination (Shastry and Singh, 1973). Similar results were reported by Satyanarayana *et al.* (1992). They reported that not only the chemical was important but also the number of applications was equally important in the management of disease.

Chakrabarti and Mukhopadhyay, (1977) evaluated 6 insecticides *viz.*, methyl demeton (Metasystox 0.02 per cent), phosphomidon, endosulfan, carbofuran and fenitrothion against yellow vein mosaic disease on Cv. Pusa Sawani. The insecticides significantly reduced the number of diseased plants and whiteflies, irrespective of dates of sowing. The total yield of treated plants was also significantly higher than that of the control.

Insecticides like oxyelemeton methyl (Metosystox 0.02 per cent), dimethoate (Rogar 0.05 per cent), monocrotophos (Nuvacron 0.05 per cent) were proved effective in reducing whitefly population as well as incidence of yellow vein mosaic disease of okra (Singh *et al.*, 1977).

Singh and Singh, (1989) observed that the insecticide phosphomidon (0.02 per cent) found effective as compared to methyl demeton (0.025 per cent) and furatox (15kg/ha) in reducing disease incidence and whitefly population in all the three dates of sowing *i.e.* in March, April and May.

Neither systemic nor contact insecticides delayed or reduced the incidence of okra yellow vein mosaic virus in field was reported by Dahal *et al.* (1992). They also reported that the disease appeared 3 weeks after sowing and incidence reached to comparable levels in both treated and untreated plots after 45 to 60 days differed significantly between various dates of sowing.

Borah and Nath, (1995) reported that spray of dimethoate 0.03 per cent at 15 and 30 days after germination reduced whitefly population. Spraying of diamethoate (0.03 per cent) treatment had significantly lowest incidence of yellow vein mosaic virus followed by dimethoate (0.03 per cent) at 15 days after germination + spraying of malathion (0.05 per cent) at 25 to 30 days after germination.

Singh *et al.* (1998) studied the effect of two insecticides monocrotophos 36 per cent SL (0.02 per cent) and cypermethrin 25 per cent EC (0.02 per cent) and seed selection in 3 successive experimental years (1992-94). They reported that the disease incidence decreased upto 0.44 per cent in 1994 compared with 1.77 per cent in 1992. The Aleurodidae population was also greatly reduced by sprays of cypermethrin 26 per cent EC and monocrotophos 36 per cent SL.

Keshwal and Khatri, (1999) reported that spraying insecticide along with growth regulator like Tricontanol controlled the yellow vein mosaic of okra and also reported that hybrid okra varieties were more susceptible to the virus than the local cultivars.

Pun *et al.* (2000b) conducted the experiments to determine the effect of virus inhibitory chemicals *i.e.* acetyl salicylic acid @ 200 and 500 ppm, salicylic acid @ 500 ppm, benzoic acid @ 500 ppm and barium chloride @ 1000 ppm and neem products (neem oil @ 3 per cent aqueous solution) and neem seed kernel extract @ 5 per cent aqueous solution on the control of yellow vein mosaic virus infection of okra Cv. Pusa Sawani. The treatments were applied as sprays. All the chemicals and neem products significantly reduced okra yellow vein mosaic infection. Among the chemicals barium chloride was most effective (76.76 per cent reduction in infection in comparison with the control) followed by acetyl salicylic acid @ 200 ppm and salicylic acid @ 500 ppm, which gave reductions of 75.0 and 73.3 per cent, respectively. Neem oil and neem seed kernel extract reduced virus infection by 88.3 and 86.7 per cent, respectively.

Debnath *et al.* (2002) reported that four sprayings of metasystox (demeton s-methyl) 15, 30, 45 and 60 days after sowing recorded the highest yield advantage with comparatively lowest disease incidence.

Soil application of aldicarb granules (1.00 kg a.i./kg/ha) a week after sowing and spraying with endosulfan (0.02 per cent) at 30 and 50 days after sowing were effective in reducing incidence of okra yellow vein mosaic (Chelliah and Murugasen, 1976b). Similar results were reported by Uttasami *et al.* (1977).

Basha and Balsubramanyam (1982) observed that the application of aldicarb (0.04 per cent or 0.5 kg a.i./ha) on 15th day of sowing followed by spraying with 0.03 per cent endosulfan twice on both 15th day and 45th day of sowing or spraying monocrotophos (0.04 per cent) or endosulfan (0.05 per cent) four times at fortnightly intervals commencing from 15 days after sowing were found to be effective in reducing the yellow vein mosaic disease of okra.

Khan and Mukhopadhyay, (1985b) studied that soil application of furatox-10G followed by four foliar sprays of methyl demeton @ 0.03 per cent at an interval of 15 days from the date of sowing reduced okra yellow vein mosaic disease incidence to 59.66 and yield upto 59.46 q/ha whereas, average disease incidence in control was recorded 81.22 per cent; average number of whitefly population as 231 and total yield reduced to 23.80q/ha.

Spray schedule of 4-6 sprays soon after emergence of seedlings with systemic insecticides like Rogar (1.5 ml/lit.), dimecron or nuvan (1 ml/3 lit) and or two applications of thimet or dysyston granules to soil has been recommended to prevent spread of the disease (Markose and Peter, 1990).

Atiri *et al.* (1991) reported that the treatments with synthetic pyrethroids, lambacyhalathrin at 15 g a.i./ha and aqueous neem solution @ 467 lit/ha significantly reduced incidence and severity of yellow vein mosaic disease. Treatment with cypermethrin + diamethoate mixture (3:25) at 280g a.i./ha apparently had effect on disease incidence and severity.

Bhagat *et al.* (1997) reported the soil application of carbofuran @ 1 kg a.i./ha at sowing and 20 days after sowing followed by one spray of oxydemeton methyl @ 250 ml a.i./ha 45 days after sowing significantly reduced the vector population, disease incidence and increased the yield of okra.

Application of systemic granular insecticides for the control of yellow vein mosaic through the vector control has been successful. The disease could be effectively minimized by the furrow application of aldicarb and carbofuran granules @ 1.0 kg a.i./ha at the time of sowing. These treatments have accounted for 13.8 per cent and 76.2 per cent reduction in the disease incidence, respectively over control. Phorate, fensulphothion and mephespholan have also reduced the incidence by more than 50 per cent over the control (Palaniswamy *et al.*, 1973).

Shastry and Singh, (1974b) conducted an experiment to restrict the spread of YVMV by controlling its vector population, indicated that four sprays each of parathion (0.02 per cent), oxydemeton methyl (0.02 per cent) and Dimethoate (0.05 per cent) at 10 days intervals starting just after the germination of the seeds or application of Phorate10 G (15 kg/ha) at the time of sowing. The treated seeds not only reduced the vector (Whitefly- *Bemisia tabaci*) but also restricted the spread of the virus to a greater extent. The application of insecticides during the seedling stage resulted in higher incidence of YVMV.

Singh *et al.* (1977) reported that the soil application of carbofuran granules at the time of seed sowing was useful for the control of yellow vein mosaic virus and in increasing the yield.

Soil application of methyl phosphorodithioate (furtex 10G) 15kg/ha or phosphomidon (0.02 per cent) or methyl demeton (0.025 per cent) checked the yellow vein mosaic significantly. Lowest incidence was recorded in phosphomidon (0.02 per cent) protected plot followed by methyl demeton (0.025 per cent) and furatex (15 kg/ha) was reported by Singh and Singh, (1989).

Application of carbofuran granules to the soil at the time of sowing followed by 2 sprays of monocrotophos at fortnightly intervals to control the whitefly vector (*Bemisia tabaci*) was effective. Early spraying was important to prevent seedling infection during the first 20 days after sowing (Murthy and Reddy, 1992).

Handa *et al.* (1993) reported that one or two applications of carbofuran 3G was slightly better than two applications of phorate 10G at the time of sowings and within 20 days later reduced the disease incidence. Similar results for soil application of carbofuran were reported by Bhagat *et al.* (1997).

Efficacy of actamiprid at four doses against whitefly on okra was evaluated by Sonalkar and Sonalkar, (1999). They reported that actamiprid @ 20g a.i./ha reduced the whitefly population significanty.

According to Kumar and Moorthy (2000), seed treatment with imidacloprid (Gaucho 600 FS) at 5, 9 and 12 ml/kg seed failed to prevent the transmission of yellow vein mosaic virus by *Bemisia tabaci*.

Mote *et al.* (1994) reported that imidacloprid 15 g/kg seed treatment was found promising in controlling whiteflies and per cent yellow vein mosaic virus of okra.

Bhagat *et al.* (2001) observed that the susceptible stage of the crop from 35 to 50 day after sowing must be supplemented with systemic insecticide to reduce white fly population and thereby reducing disease and security to obtain good harvest.

Bhargava *et al.* (2001) used two formulations of imidacloprid 600 FS and 70 WP as seed dresser in okra @ 5 and 9 ml/kg seeds and 5, 7.5 and 10 g/kg seeds, respectively. Both formulations were found effective in controlling whitefly (*Bemisia tabaci*) and gave higher yield.

Shivpuri *et al.* (2004) tested imidacloprid as seed dresser with monocrotophos and endosulfan as spray for the control of yellow vein mosaic of okra during two *kharif* seasons, transmitted by whitefly (*Bemisia tabaci*) They found that the treatment of imidacloprid 600 FS @ 5ml/kg seed was most promising when tried as seed dresser along with one spray of monochrotophos followed by three sprays of endosulfan.

Ali *et al.* (2005) reported that imidacloprid significantly reduced the whitefly population and neem extract and biological control agent were also found to be effective against white fly compared to distilled water and untreated control.

The vector population and disease incidence can be reduced by application of 4 to 5 foliar spray of Dimethate (0.05 per cent) or Metasystox (0.02 per cent) or Monocrotophos (0.05 per cent) followed by 1 or 2 spray of mineral oil (2 per cent) (www.krishiworld.com).

Biological Management

Spraying of Plant Extracts

The antiviral principles in non host plant species were effective against different plant viruses have been reported by several workers (Okuyma *et al.*, 1978).

Chaudhary *et al.* (1992) made extracts of callistemon, datura, agave and ginger in water and in alcohols and used in controlling the yellow vein mosaic virus of okra. Alcohol extracts were superior to aqueous ones in preventing infection by yellow vein mosaic and extract of callistemon, datura, agave and ginger (*Zingiber officinale*) gave a good field performance. A lower rate of disease dissemination was recorded in treated plants than in the controls sprayed with water only. Mortality of the vectors (*Bemisia tabaci*) was 20-80 per cent when they were confined for 30 minutes in a cage with plants treated with the extracts.

Pun *et al.* (1999b) screened different plant extracts against yellow vein mosaic of okra and reported that spraying with leaf extracts of *Prosapsis chilensis* and *Bougainvillea spectabilis* before whitefly inoculation were highly effective in reducing yellow vein mosaic virus infection. They reduced the incidence to the extent of 83 and 81 per cent, respectively over control. The extracts not only reduced the disease incidence but delayed it by 8 days. They also evaluated 10 plant extracts for their efficacy against yellow vein mosaic virus. Among them, *Bougainvillea spectabilis* showed maximum inhibition (98.3 per cent) of the virus transmission followed by *Boerhavia diffusa* (91.7 per cent) over control, whereas *Vitex nigundo* showed minimum reduction of 56.7 per cent over control.

In rainy season spraying of asafoetida plant extract to an okra crop was tested for the control of the whitefly vector. The asafetida formulations @ 1-3 per cent concentration *in vitro* and in field trials in Allahabad, Uttar Pradesh, India showed strong insect repellent activity against *A. biguttula biguttula* and reduced yellow vein mosaic virus infection levels in okra (Singh *et al.*, 1999).

The effect of virus inhibitory chemicals (acetyl salicylic acid 200 and 500 ppm, benzoic acid @ 500 ppm, barium chloride @ 1000) and neem products (neem oil @ 3 per cent aqueous solution and neem seed kernel extract @ 5 per cent aqueous solution) on the incidence of okra yellow vein mosaic disease was studied by Pun *et al.*, 2000b). They reported that all the chemicals and neem products significantly reduced okra yellow vein mosaic virus infection. But the neem products were found to be more effective than the virus inhibitory chemicals in reducing the infection. Neem oil and Neem seed kernel extract reduced virus infection by 83.3 and 88.7 per cent, respectively.

To determine an environmentally safe management of okra yellow vein mosaic disease experiment was carried by Srabani Debnath *et al.* (2002). They used tolerant cultivars, effective scheduling of efficient insecticides and plant product based vector control measures and reported that F_1 hybrids showed highest yield, lower degrees of virus infection followed by open pollinated variety.

Cultural Management

Intercropping

Khan and Mukhopadhyay, (1985a) reported that mixed cultivation of pumpkin with okra enhanced both total incidence of yellow vein mosaic virus disease and its spread in okra crop. So, pumpkin should not be grown along with okra.

The intercropping of okra with cowpea and mungbean also significantly reduced the whitefly population and disease incidence and increased the yield (Singh and Singh, 1989).

Mukhopadhyay *et al.* (1994) reported that the yellow vein disease incidence was reduced by growing trap crops of cucumber and brinjal.

Different Dates of Sowing

Sowing of okra during the last week of June and application of 0.02 per cent methyl demeton may give good harvest in comparison with early or late sowing of the crop with same pesticide schedule (Chakrabarti and Mukhopadhyay, 1977).

Singh and Singh (1989) reported that the disease incidence was low (0 to 6.5 per cent) in March sown than in April sown (2.3 to 18 per cent) and May sown (3 to 27 per cent) crop, respectively.

Early May or late August sowing was highly desirable for avoidance of yellow vein mosaic and to minimize the loss in yields was suggested by Markose and Peter, (1990).

Dahal *et al.* (1992) reported that the rates of yellow vein mosaic disease incidence were similar among the treated plots but differed significantly between various dates of sowing. Onset and spatial development of the disease varied with the time of sowing. Incidence was lower in May sowing than in that of June or August. The disease severely reduced both pod and seed yield, respectively.

Whitefly population were highest in May sown crop and incidence of yellow vein mosaic disease was highest (100 per cent) in May and June sown crops. Disease incidence and whitefly population were lowest in October sown crop (Goswami and Bhagbati, 1992).

Gupta (1992) found that okra crop sown in February- March remained free of disease upto 90 days while in May and June crops were unaffected for 60 days and April sown crop were unaffected for 30 to 60 days after sowing.

Lesser disease incidence and whitefly population were recorded in the crop sown during February 10 to March 10 and as the sowing date advances the disease incidence and whitefly population increases while the fruit yield decreases (Nath *et al.*, 1992).

Borad *et al.* (1993) reported that population of *Bemisia tabaci* in okra fields and disease incidence of yellow vein mosaic was low in August-October in 1988-1989. In both the years, the population reached at the maximum during the first week of October. The disease incidence was 41 and 90 per cent for 26 February and 8 April sowing, respectively. Plants sown on 24 April showed only 5 per cent infection.

Singh *at al.* (1994a) reported that seeds of okra Cv. Pusa Sawani were sown at weekly intervals during June and July in 1989 and 1990 in Uttar Pradesh. Early sowing of crop exhibited a higher per cent age of yellow vein mosaic virus infection and lower yield than late sowing in both years.

The influence of 15 different sowing dates from February to March on okra yellow vein mosaic disease on okra Cv. Pusa Sawani varied from 75 to 91 per cent in plots sown between early April and end of June. Infection in plots sown during February to the end of March was progressively less. The lowest yield of okra was obtained from the plots sown in May and June. A strong positive correlation was obtained between per cent of disease incidence and whitefly (*Bemisia tabaci*) population (r=0.085) whereas, a strong negative correlation was obtained from disease incidence and fruit yield (r=-0.84) reported by Nath and Saikia (1995).

Muzumdar *et al.* (1996) reported the lowest disease incidence and whitefly population were revealed in crop sown between February 25 and March 25 compared with sowing dates of April 15 to

July 25. The total and marketable yield was maximum in early sown crops rather than crop sown after 15 April and number of unmarketable okras increased with delayed sowing.

Gulshanlal *et al.* (2001) studied on the performance of three cultivars *viz.*, Parbhani Kranti, Pusa Sawani and Punjab-7 under 3 sowing dates (16 June, 29 June and 12 July) were conducted during *Kharif* season of 1996-97. Parbhani Kranti obtained the highest green pods yield (85.5 q/ha) followed by Pusa Sawani (80.4q/ha) and Punjab-7 (72.5 q/ha). Punjab-7 exhibited the lowest yellow vein mosaic virus infection (0.3 per cent), while Pusa Sawani showed the highest infection (41.4 per cent). Sowing of okra on 16 June produced the highest pod yield (92.1 q/ha); the green pod yield decreased by 9.95 and 33.37 per cent for 29 June and 12 July sowing dates, respectively. Parbhani Kranti sown on 16 June produced the highest green pod yield.

Sonwane (2004) reported that dates of sowing in *Kharif* had no marked influence on the infection rate. However, in summer it was evident due to late occurrence of the disease. The progress of the yellow vein mosaic virus of okra in early sown crop in both *Kharif* and summer was slower than the late sown okra. The early sown okra had more whitefly population in *Kharif* and less population in summer than okra sown on later dates. The early sown okra though differed in respect of whitefly activities recorded significantly less incidence of yellow vein mosaic and significantly more yield.

Screening for Resistance

Management of Yellow Vein Mosaic Disease of Okra through Resistance

Development of virus resistant varieties has been considered the most effective, economical and reliable means for controlling viral diseases. The first step in any virus resistance programme was to identify germplasm possessing immunity or resistance to the viruses. In okra, use of resistant and tolerant varieties has also been reported to be the best way to overcome from yellow vein mosaic of disease.

Nariani and Seth (1958) reported that no cultivars of okra were immune to the yellow vein mosaic virus however; a few resistant cultivars to yellow vein mosaic virus have been developed (Arumugai and Chelliah, 1975; Gunthilgaraj *et al.*, 1977; Singh and Thakur, 1979; Atiri, 1983 and Jambhale and Nerkar, 1992).

Thampan and Vijayan (1964) reported few fines of okra were resistant to okra yellow vein mosaic virus infection.

Singh and Thakur (1979) crossed wild species A. *manihot* and A. *manihot* sub sp. *manihot* with susceptible A. *esculentus* the hybrid produced was considered as resistant.

Sharma and Sharma (1984a) identified A. *manihot* sub sp. *manihot* as highly resistant to yellow vein mosaic virus. The species was found to carry latent infection.

Jambhale and Nerkar (1986) reported Cv. Parbhani Kranti to possess field resistance. The tolerance of Cv. Parbhani Kranti was also confirmed by Sharma and Sharma (1984a); Dhankar *et al.* (1996); Pulliah *et al.* (1998); Chandra Deo *et al.* (2000); Raghupathi *et al.* (2000) and Bhagat *et al.* (2001). Its resistance was again reported by Singh (2000); Senthilkumar (2005) and Satheeshkumar *et al.* (2005).

Sharma and Sharma (1984b) evolved Punjab Padmini by crossing resistant variety EC-31380 (Asunton kolo) from Ghana with Pusa Sawani and Pusa Reshma which was claimed to be resistant to yellow vein mosaic disease also in (1993) Sharma *et al.*, confirmed its resistance.

The okra varieties P-7, PB-57 and Sel-10 were considered as resistant to yellow vein mosaic bigeminivirus in Indian resistance breeding programme (Gill *et al.*, 1991).

Bora *et al.* (1992) reported Arka Anamika and other five genotypes were free from yellow vein mosaic disease and yielded high similarly, its resistance was confirmed by Sanger (1997); Sanigrahi and Chaudhary (1998); Suresh Kumar (2000); Indurani *et al.* (2003) and Senthilkumar (2005). Shrivastava *et al.* (1995) reported Arka Anamika as moderately resistant.

Harender Raj *et al.* (1993) reported that, Punjab-7 possess resistance to yellow vein mosaic virus of okra. Contrast finding was reported by Dhanju and Cheema (1995). They rated Punjab-7 as susceptible.

Srivastava *et al.* (1995) reported 'Varsha Uphar' was free of the yellow vein mosaic virus. Its resistance was also confirmed by Dhankar (1996) and Sateesh Kumar *et al.* (2005).

Arka Abhay reported as resistant to yellow vein mosaic virus of okra (Sanger, 1997). Its resistance was also confirmed by Sanigrahi and Chaudhary (1998).

Adhunik, a resistant hybrid to yellow vein mosaic virus of okra, the most suitable hybrid for cultivation in Uttar Pradesh of India during *Kharif* was reported by Shepherd and Winsten (1999).

Ragupathi *et al.* (2000) reported that disease was absent in the highly resistant cultivars BO 1 and HRB 55 and resistant cultivars KS 404, HRB 9-2, HY 8, PY, Parbhani Kranti, Sel 10 and Sel 4 however, BO 2 was susceptible.

Srabani Debnath and Nath (2003) reported that AM-4-5 showed less infection (4.44 per cent) of yellow vein mosaic in year 1999 and tolerance to yellow vein mosaic virus in 2000 (0 per cent) with lowest mean yield (185 g/ha).

Indurani *et al.* (2003) reported Arka Anamika and PA4 x Varsha Uphar had no symptoms of yellow vein mosaic and yellow vein mosaic per cent age was minimum in PA 4 and VOHD-1.

Azad bhendi-1, a new okra cultivar developed from Pusa Sawani and Parbhani Kranti exhibited higher resistance to yellow vein mosaic virus, exhibited higher yield (100-125 q/ha) and earlier fruiting (40-42 days) (Yadav *et al.*, 2004).

Parbhani Kranti and Arka Anamika were resistant to the yellow vein mosaic disease, while Sel-4 and Sel-10 were resistant to whitefly and Hy-8 was moderately resistant (Senthilkumar, 2005).

Sateesh Kumar *et al.* (2005) reported that Parbhani Kranti, Varsha Uphar given higher yield and were resistant to yellow vein mosaic virus disease. A mutant EMS-8 apart from higher yield also showed resistant to the yellow vein mosaic virus of okra.

Five heterotic okra hybrids possessing resistance to yellow vein mosaic virus were further advanced for F_2, F_3, F_4, F_5 and F_6 generations. The progenies belonging to Sel-4 x Sel-10, Sel-4 x P-7 and P-7 x Sel-10 were free from yellow vein mosaic disease. The progenies belonging to family EMS-8 x BO-1 and PS x P-7 were highly susceptible. The progenies belonging to Sel-4 x Sel-10 were superior for yield, fruit quality and disease resistance (Fugro and Rajput, 1999).

Bhagat (2000) reported that the yield and growth parameters were less affected in the resistant cultivar Parbhani Kranti while in Vaishali Vadhu (susceptible) and Pusa Sawani (highly susceptible) cultivars they were more due to incidence of the yellow vein mosaic disease. Also Bhagat *et al.* (2001) reported rate of dissemination of yellow vein mosaic virus in okra cultivations as reported in above result.

Ali *et al.* (2001) crossed okra variety IPSA Okra-1 tolerant to yellow vein mosaic virus with 3 susceptible genotypes. A grafting test revealed that the tolerance in IPSA okra-1 was genetic and not due to escape. The F_1 hybrids were tolerant to yellow vein mosaic virus. From the segregation pattern for disease reaction in F_2 and backcross generations of 3 crosses, it could be hypothesized that, the tolerance of yellow vein mosaic virus in IPSA orka-1 was quantitative.

The F_1 hybrid of the resistant lines was resistant and that of susceptible parents were susceptible. The studies indicated that, resistance to disease was monogenic and dominant. Maximum number of fruits and yield per plant was recorded by hybrid 410 x 407 followed by 409 x 421 and 409 x 408 involving resistant x resistant and resistant x susceptible crosses, respectively (Rattan *et al.*, 2000).

Singh (2000) studied suitability of okra varieties for cultivation in Uttar Pradesh. He found that variety Parbhani Kranti gave higher yield, fruit weight in addition it was also resistant to yellow vein mosaic virus in all the years of study.

Ten hybrids of okra exhibited desirable and significant heterosis over the susceptible check (PI-4960702) for yellow vein mosaic virus. Hybrids involving at least one immune to disease to parent (Arka Abhay or Arka Anamika) were all found to be immune except Arka Abhay x Pusa Sawani as moderately resistant was reported by Suresh Kumar (2000).

Genetics of yellow vein mosaic virus resistance in okra for different crosses was studied by Vashisht *et al.* (2001) and they reported that additive gene effect was more significant than dominant gene effect. Scaling tests revealed that the presence of epitasis except for virus grading in cross-1 (Cross between resistant cultivars).

The variety of okra Pusa Makhmali is highly susceptible, Punjab No. 13 and Pusa Sawani is susceptible, Punjab Padmini and Parbhani Kranti is tolerant and Arka Anamika is resistant to yellow vein mosaic virus of okra (www.ficciagroindia.com).

Screening of Okra Lines for Resistance Against Yellow Vein Mosaic Disease of Okra

Varietal resistance in okra to yellow vein mosaic virus has been reported by several researchers.

Sharma and Sharma (1981) found that new lines evolved in addition to resistance were also superior in horticultural traits compared with the commercially cultivated variety 'Pusa Sawani'.

Verma and Mukharjee (1955) tested 43 genotypes against okra yellow vein mosaic virus. Some of these showed resistance to the virus. Three types of symptoms *viz.* yellow vein, yellow lamina and green mosaic were observed on the genotypes of 37 varieties tested for the resistance, none were found to be resistant. However, 3 lines *viz.* F3, L1, Buk and AE-7 were found less susceptible recording less than 10 per cent infection. Similar results were reported by Gunthilgaraj *et al.* (1977).

Ninety four lines of germplasm were tested for source of resistance to whitefly in okra. Of them, three lines *viz.* IC-1542, Sel-1-1 and Sel-2-2 exhibited resistance to yellow vein mosaic disease by *B. tabaci*. *A. manihot*, a wild species from Ghana was almost immune to okra yellow vein mosaic disease of okra (Sandhu *et al.*, 1974).

Screening of 46 cultivars for yellow vein mosaic disease was carried out by Chauhan *et al.* (1981). They reported that, cultivar IC-1342 showed the highest yield and number of fruits/plant of the 46 cultivars. The highest incidence of okra yellow vein mosaic virus noticed in Pusa long green-1 and Parkins long green while, the lowest was in IC-9273. None of the cultivars was found resistant.

Khan and Mukhopadhyay (1986) screened five varieties under field conditions, S1-1 showed lowest incidence of infection (24-36 per cent) and gave the highest yield (40.36 q ha^{-1}).

Dhankar *et al.* (1989) reported that difference in disease incidence and yield potential were observed on screening 97 genotypes against okra yellow vein mosaic virus under natural condition in the field.

The cultivars IC-9273, Bavaria, 3(1) and IC-23592 were resistant. The highest yielding genotype was selection-2 with a moderate disease reaction.

Screening of 24 okra varieties for susceptibility of yellow vein mosaic virus under field condition in Uttar Pradesh was carried out by Singh and Gupta (1991) and reported that none of the varieties was highly resistant, 3 were resistant, 8 moderately susceptible, 3 susceptible and 10 as highly susceptible.

Arora *et al.* (1992) observed differences among the 157 advanced germplasm lines and 7 cultivars/ hybrids evaluated in the field for reaction to okra yellow vein mosaic virus over 2 years. Incidence of yellow vein mosaic virus was highest (100 per cent) in Pusa Makhmali compared with only 0.64 per cent in the resistant line EMS-8.

Fourteen okra varieties screened against okra yellow vein mosaic virus. None was immune to disease. Okra Vishal was resistant, M-31 and S-1-1 were moderately resistant and Pusa Sawani was susceptible (Nath and Saikia1992).

Mohapatra *et al.* (1995) reported that Pusa Sawani was the most susceptible and recorded 100 per cent infection while, varieties like HRB-9-2, DOV-91-4 and Pashupati showed tolerance at least under field condition.

Screening of 51 okra hybrids and their 20 parents against okra yellow vein mosaic virus, only one parent of Parbhani Kranti and 11 hybrids were highly resistant to yellow vein mosaic virus, P-7 was moderately resistant while, rest of the parents and hybrids were susceptible or highly susceptible to okra yellow vein mosaic virus (Dhankar *et al.*, 1996).

Poopathi *et al.* (1996) evaluated thirteen *Abelmoschus esculentus* against okra yellow vein mosaic virus. AROH-2 and Parbhani Kranti were tolerant and asymptomatic; Hybrid-8 showed the highest yield.

Nath *et al.* (1999) reported that Pusa Sawani was highly susceptible to the disease when sown in pre *Kharif* (76 per cent) and of *Kharif* (80 per cent) seasons. They recorded Parbhani Kranti, Arka Abhay, Arka Anamika, VRO-3 and Okra No.6 as for cultivation in the new alluvial zone of West Bengal.

Screening of 18 okra genotypes for resistance against okra yellow vein mosaic virus was carried out by Pathak *et al.* (1999). They developed by crossing six female lines (IC-9275, Arka Abhay, Parbhani Kranti, 7 D-2, P-7 and Pusa Makhmali) with three testers *viz.* Punjab Padmini, EC-16511 and HB-55. Out of 72 cross progenies, 59 were resistant or moderately resistant. All male testers and female lines were found to be resistant to moderately resistant except one female line *i.e.* Pusa Makhmali.

Batra *et al.* (2000) screened open pollinated okra varieties against yellow vein mosaic virus of okra and reported that Okra No.6, LORM-1, VRO-3 and P-7 were found free from disease reaction whereas, VR0-4 showed mild reaction. Okra hybrid DVR-1 and DVR-2 were free from disease, Okra No.6 among open pollinated and AROH-8 among hybrids given the highest yield.

Philip *et al.* (2000) selected fifty okra progenies from superior plants of the F_3 generation of cross between *A. esculentus* var. Kiran with *A. monihot*. The F_1 parents seeds were subjected to gamma irradiation and tested in experiment and reported that most of F_4M_4 progenies exhibited lower mean values for average fruit weight compared to the cultivated parent. In addition, the incidence of yellow vein mosaic disease was less in the F_4M_4 generation.

Twelve okra cultivars including the highly susceptible Pusa Sawani and MDU-1 were evaluated in Tamil Nadu, India by Raghupati *et al.* (2000) and they reported that the disease was absent in the highly resistant cultivars BO-1 and HRB-55. Resistant cultivars were KS-404, HRB-9-2, Hy-8, P-7, Parbhani Kranti, Sel-10 and Sel-4 with the disease incidence of 0.51, 0.82, 1.26, 1.68, 2.87, 3.83 and 8.69 per cent, respectively. BO-2 was susceptible, MDU-1 and Pusa Sawani recorded a disease incidence of 90.83 and 91.53 per cent, respectively.

Singh and Singh (2000) screened improved genotypes of okra which were treated with gamma rays (15, 30, 45 and 60 KR) and EMS (0.25, 0.50, 0.75 and 1.00 per cent) in M2, M3 and M4 against yellow vein mosaic virus, respectively. Among all doses of mutagens, 45 and 60 KR gamma rays and 0.75 and 1.0 per cent EMS imported high resistance to resistant plants in both M_2 and M_3 generations. Hence, incidence of yellow vein mosaic virus showed dose dependent relationship and increase in doses of mutagens decreased the disease infection.

Different lines were screened against yellow vein mosaic virus disease of okra under field condition in different regions of India. Among these lines, the F3, Bulk, L, AE-7, IC-1542, Selection-1-1, Sel-22, IC-9273, Bavania, 3(1), IC-23592, EMS-8, Vishal, Parbhani Kranti, Arka Abhay, Arka Anamika, VRO-3, Okra No.6, LORM-1, P-7, DVR-1, DVR-2, KS-404, HRB-9-2, Hy-8, Sel-10, Sel-4, Bharat hari, V-606-8, Ambika local, VRO-5, NOL-101 exhibited resistance to yellow vein mosaic disease (Raghupati *et al.*, 2000; Singh *et al.*, 2002 and Vijaya, 2004).

Abdul Rehman and Waquar Ahmad (2002) reported that the varieties or lines GE, PK, CHR, PA x CHR, CHR x PK, CHR x DLPG, PK x PG, PK X AM Red and GV x GLPG were immune to yellow vein mosaic virus of okra.

In screening programme, DVR-2 recorded the longest pod and pod weight and pod yield/ha. The hybrid HIHB-0090, DVR-1 and DVR-2 were free from yellow vein mosaic virus infection (Singh *et al.*, 2003).

Neerja *et al.* (2004) screened okra hybrid against yellow vein mosaic disease and reported that hybrid NOH-1S had the highest yield (74.8 q/ha) with 11.9 per cent yellow vein mosaic disease incidence. However, NOH-1S, JNDOH-1, AROH-47 and HYOH-1 were at par with each other for yield and yellow vein mosaic virus disease incidence.

In screening programme of okra hybrids, AROH-10 showed the highest yield followed by DVR-3 and AROH-9, KOH, DVR-3 and DVR-4 and required lower number of days for pod picking and also disease free at 30 days after sowing. Highest infection was in HYOH01, HYOH-2 and PRVOH-01 (Singh and Jain, 2004).

References

Ali, M., Hossain, M.Z. and Sarker, N.C. (2000). Inheritance of Yellow Vein Mosaic Virus (YVMV) tolerance in a cultivar of okra (*Abelmoschus esculentus* (L.) Moench). *Euphytica*, 111(3): 205–209.

Ali, S., Khan, M.A., Habib, A., Rasheed, S. and Iftikhar, Y. (2005). Correlation of environmental conditions with okra yellow vein mosaic virus and *Bemisia tabaci* population density. *International Journal of Agriculture and Biology*. 7(1): 142–144.

Anonymous (1961). The vein clearing disease of bhendi. *Ind. Hort*. 5(4): 11.

Anonymous (2007). *FAO State Database*, 2007.

Anonymous (2009). *Indian Horticulture Database* 2008, 2009.

Arora, S.K., Dhanju, K.C. and Sharma, B.R. (1992). Resistance in okra (*Abelmoschus esculentus* (L). Moench) genotypes to yellow vein mosaic virus. *Plant Dis. Res.* 7(2): 221–225.

Arumugai, S. and Chelliah, S. (1975). *Abelmoschus manihot*–A resistant source of Bhendi yellow vein mosaic virus. *Madras Agric. J.* 63(5): 310–312.

Asha Shivpuri, S., Bhargava, K.K., Chhipa, H.P. and Ghasolia,R.P. (2004). Management of yellow vein mosaic virus of okra. *J. Mycol. Pl. Patho.* 34(2): 353–355.

Aslam, M. and Gebara, F. (1995).Host plant preference of vegetables by cotton whitefly, *Bemisia tabaci* (Genn.). *Pakistan J. Zoology*. 27(3): 269–272.

Atiri, G.I. (1983). Identification of resistance of okra mosaic virus in locally grown okra varieties. *Annals of Applied Biology*. 102(Suppl.): 132–133.

Atiri, G.I., Ivbijaro, M.K. and Uladele, A.O. (1991). Effect of natural and synthetic chemicals on the incidence and severity of okra mosaic virus in okra. *Tropical Agric.* 68(2): 178–180.

Basha, A.A. and Balasubramanyam, M. (1982). Control of yellow vein mosaic disease of bhendi through chemical control of the vector *Bemisia tabaci.* In All India Symposium on Vector and Vector-borne Diseases, Trivendrum, Feb. 26–28, pp: 54.

Batra, V.K., Singh, Jitendra and Singh, J. (2000). Screening of okra varieties to yellow vein mosaic virus under field conditions. *Vegetable Science*. 27(2): 192–193.

Berry, S.K., Kalra, C.L., Sehgal, R.C., Kulkarni, S.G., Sukhvir Kaur, Arora, S.K. and Sharma, B.R. (1988). Quality characteristics of seeds of five okra cultivars. *J. Food Science and Technology*. 25: 303–305.

Bertus, L.S. (1942). Plant Pathology. Adm. Rep. Dir. Agric. Ceylon. 1941, pp: 5.

Bhagabati, K.N. and Goswami, B.K. (1992). Incidence of yellow vein mosaic disease of okra (*Abelmoschus esculentus* (L.) Moench) in relation to whitefly (*B. tabaci* Genn.) population under different sowing dates. *Ind. J. Virol.* 8(1)37–39.

Bhagat, A.P. (2000). Effect of bhendi yellow vein mosaic virus (BYVMV) on growth and yield of bhendi. *J. Mycol. PI. Path.* 30(1): 110–111.

Bhagat, A.P., Yadav, B.P. and Prasad, Y. (1997). Management of bhendi yellow vein mosaic virus disease by insecticides. *J. Mycol. Pl. Path.* 27(2): 215–216.

Bhagat, A.P., Yadav, B.P. and Prasad, Y. (2001). Rate of dissemination of okra yellow vein mosaic virus disease in three cultivars of okra. *Indian Phytopath.* 54(4): 488–489.

Bhargava, K.K. and Bhatnagar A. (2001). Bio–efficacy of imidacloprid as a seed dresser gainst sucking pests of okra. *Pest Management and Econ. Zoology*. 9(1): 31–34.

Bi, Ji., Ballmer, G.R., Toscano, N.C., Madore, M.A., Dugger, P. (ed) and Richter, D. (2000). Effect of nitrogen fertility on cotton whitefly interactions. Proceedings Beltwide cotton conferences, San Antonio, USA, 4–8 Jan., 2000: 2(2000): 1135–1142.

Bora, G.C., Saikia, A.K. and Shadique, A. (1992). Screening of okra genotypes for resistance to yellow vein mosaic virus disease. *Indian J. Virol.* 8(1): 55–57.

Borad, V.K., Puri, S.N., Brown, J.K. and Butler, G.D. Jr. (1993). Relationship of *Bemisia tabaci* population density and yellow vein mosaic disease incidence in okra. *Pest Management and Econ. Zoology*. 1(1): 119.

Borah, R.K. and Nath, P.O. (1995). Evaluation of an insecticide schedule on the incidence of whitefly (*Bemisia tabaci* (Genn.)) and yellow vein mosaic in okra. *Indian J. Virology*. 11(2): 65–67.

Burbon, C., Fishpool, L.D.C., Fouqueet, C., Forgette, O. and Thouvench, J.C. (1992). Host associated biotypes within West African Population of the whitefly *Bemisia tabaci* (Genn.), (Hom–Aleurodidae). *J. of Applied Entomology*. 113(4): 416–423.

Capoor, S.P. and Verma, P.M. (1949). Bhendi mosaic and its control in Poona. *Indian Farming*. 2: 14–16.

Capoor, S.P. and Verma, P.M. (1950). Bhendi mosaic and its control in Poona. Indian Farming, 2: 14–16.

Capoor, S.P. and Verma, P.M. (1950). Yellow vein mosaic of *H. esculentus* (L.). *Indian J. Agril. Sci*. 20: 217–230.

Capoor. S.P. and Verma, P.M. (1952). Yellow vein mosaic of *Hibiscus esculentus* (L.). *Indian J. Agric. Res*. 56: 31–51.

Chadha, K.L. (2001). Handbook of Horticulture, New Delhi, ICAR Publication.

Chakrabarti, R. and Mukhopadhyay, S. (1977). Effect of some pesticides on yellow vein mosaic disease of bhendi. Pesticides, 11 (8): 19–20.

Chandra, Deo., Singh, K.P., Panda, P.K. and Deo, C. (2000). Screening of okra parental lines and their F 1 S for resistance against yellow vein mosaic virus. *Vegetable Sci*. 27(1): 78–79.

Chaudhary, A.K. Biswas, B. and Saha, N.K. (1992). Inhibition of bhendi (okra) yellow vein mosaic virus (BYVMU) by different plant extracts. *J. Mycopathological Res*.30 (2): 97–102.

Chauhan, M.S., Duhan, J.C. and Dhankar, B.S. (1981).Infection of genetic stock of okra to yellow vein mosaic virus. *Haryana Agric. Univ. Research Journal*. 11(1): 45–48.

Cheema, S.S. (1991). *Ageratum conzoides* (L.), a new host for yellow vein mosaic virus of bhendi in Punjab. *Plant Dis. Res*. 6(1): 104.

Cheillah, S, and Murugesan, S. (1976a). Estimation of loss due to yellow vein mosaic disease in bhendi. *Annamalai, Univ. Agric. Res. Ann*. 6: 169–170.

Chelliah, S. and Murugesan, S. (1976b). Seasonal incidence of bhendi yellow vein mosaic. *Annamalai Univ. Agric. Res. Ann*. 6: 167168.

Chelliah, S., Murugesan, S. and Murugesan, M. (1975). Influence of weather factors on the incidence of bhendi. *Madras Agric. J*. 62(7): 412–419.

Chevalier, J.N. (1940) Africa, its people and their culture history, New York, pp: 456.

Costa, A.S. (1969). Whiteflies as virus vectors. In: Viruses, Vectors and Vegetation, Ed. K. Maromorosch. Inter Science Publications, New York, pp: 95–119.

Dahal, G., Nenpane, F.P. and Baral, D.R. (1992). Effect of planting and insecticides on the incidence and spread of yellow vein mosaic of okra in Nepal. *International J. of Tropical Plant Diseases*. 10(1): 1 09–124.

Dahal, G., Thapa, R.B. and Dangol, D.R. (1993). Epidemics of Gemini virus disease, monitoring and partial characterization of their suspected whitefly vector *Bemisia tabaci* in Nepal. *J. Institute Agric. and Animal Sci*. 14: 55–68.

De Candolle, A.P. (1883). Originedes Plantes cultivees. Paris. pp: 150–157.

Debnath, S. and Nath, P.S. (2002). Management of yellow vein mosaic disease of okra through insecticides, plant products and suitable varieties. *Annals of Pl. Prot. Sci.* 10(2): 340–342.

Debnath, S. and Nath, P.S. (2003). Performance of okra varieties in relation to yield and tolerance to yellow vein mosaic virus. *Annals of Plant Protection Sciences.* 11(2): 400–401.

DeCandolle, A.P. (1883). Originedes Plantes cultivees, Paris. pp: 150–157.

Dhankar, B.S. (1996). Varsha Uphar : A virus resistant okra. IVIS News Letter, 2): 17.

Dhankar, B.S., Chauhan, M.S. and Nandkishor (1989). Reaction of different genotypes of okra yellow mosaic virus. *Indian J. Virol.* 5(2) pp: 94–98.

Dhankhar, S.K., Dhankhar, B.S. and Saharan, B.S. (1996). Screening of okra genotypes for resistance to yellow vein mosaic disease. *Annals Biol.* 12(1): 90–92.

Dhunja, K.C. and Cheema, S.S. (1995). Occurrence of enation leaf curl disease of okra in Punjab. *Plant Dis. Res.* 10(2): 157–159.

Fauquet, C.M., Mayo, M.A., Moniloff, J., Desselberger, U. and Ball L.A. (2005). Virus Taxonomy VIII Report of ICTV. Elsevier/Academic Press London.

Femi Lana, A. (1976). Mosaic virus and leaf curl disease of okra, Nigeria. *Pest Agric. News Sum.* 22(4): 474–478.

Fernando, M. and Udurawana, S.B. (1942). The nature of the mosaic disease of Banbakka (*Hibiscus esculentus* L.). Trop. Agriculturalist. XCVIII (1): 16–24.

Foda, M.E., Dugger, K. (ed) and Richter, D. (ed.) (2000). Population dynamics, host preference and seasonal distribution patterns of whitefly (*Bemisia tabaci* (Genn.)) in middle Egypt. 2000, Proceedings Beltwide Cotton Conferences, San Antonio, USA, 4–8 Jan., 2000: Vol. 2: 1380–1382.

Fugro, P.A. and Rajput, J.C. (1999). Breeding okra for yellow vein mosaic virus resistance. *J. Mycol. Pl. Pathol.* 29(1): 25–28.

Gill, H S., Kataria, A.S., Prakash, J. and Pierik, R.L.M. (1991). Resistance breeding under coordinated programme. International Seminar on new frontiers in Horticulture organized by Indo–American Proceedings of the Inter Horticulture– New Technologies and Application, 12.

Givord, L. and Hirth, L. (1978). Identification, purification and some properties of mosaic virus of okra (*H. esculents*). *Ann. Appl. Biol.* 74: 359–370.

Goswami, B.K. and Bhagbati, K.N. (1992). Natural incidence of yellow vein mosaic virus disease of bhendi in relation to different date of sowing. *J. Assam Sci. Society.* 34(2): 19–24.

Gowdar, S.B., RameshBabu, H.N. and Reddy, N.A. (2007). Efficacy of insecticides on okra yellow vein mosaic virus and whitefly vector, *Bemisia tabaci* (Guenn.). *Ann. Pl. Protec. Sci.* 15 (1): 116–119.

Grubben, G.S.H. (1977). Tropical vegetables and their genetic resources FAO, Rome.

Gulshan Lal, Singh, O.K., Jan, S.K. and Lal, G. (2001). Response of okra (*Abelmoschus esculentus* (L.) Moench) cultivars to varying sowing dates under Tarai foot hills of Himalayas. *Advances in Horticulture and Forestry.* 8: 129–137.

Gunathilagaraj, K., Padmanabhan and Kumarswami, J. (1977). Susceptibility of bhendi varieties to yellow vein mosaic virus. *Madras Agric. J.* 64(7): 486–487.

Gupta, K.K. (1992). Influence of sowing dates on the incidence of *Cercospora* leaf blight of bhendi in Sikkim. *Indian Phytopath.* 45 (2): 273–275.

Gupta, S.K. and Thind, T.S. (2006). Disease problem in vegetable production. Scientific Publisher, Jodhpur (India). 576 p.

Handa, A. and Gupta, D.M. (2006). Characterization of yellow vein mosaic virus of bhendi (*Abelmoschus esculentus* (L.) International Journal of Tropical Plant Diseases.

Handa, A., Gupta, M.D. and Handa, A. (1993). Management of bhendi yellow vein mosaic virus. *Indian Phytopath.* 46 (2): 123–130.

Harender Raj, Bhardwa, M.L. and Sharma, I.M. (1993). Performance of commercial okra (*Hibiscus esculents*) varieties in relation to disease and insect pests. *Indian J. Agri. Sci.* 63(11): 747–748.

Harrison, B.D., Muniyappa, V., Swanson, M.M., Roberts, I.M. and Robinson, O.J. (1991). Recognition and differentiation of seven whitefly transmitted Gemini viruses from India and their relationships to African cassava mosaic and Thailand mungbean yellow mosaic viruses. *Annual Appl. Biol.* 118(2): 299–308.

Hirano, K., Budiyanto, E., Swastika, N. and Fuji, K. (1995). Population dynamics of the whitefly, *Bemisia tabaci* (Genn.) in Java, Indonesia with special ref. to Spatio–temporal changes in the quantity of food resources. *Ecological Res.* 10(1)75–85.

Indurani, C., Veera Ragavathatham and Auxillia, D. (2003). Studies on the development of F_1 hybrids in okra (*Abelmoschus esculentus* (L.) Moench) with high yield and resistance to yellow vein mosaic virus. *South Indian Horticulture.* 51(1/6): 219–226.

Jambhale, N.D. and Nerkar, Y.S. (1986). Performance of yellow vein mosaic resistant 'Parbhani Kranti' variety of okra. Indian J. Virol., 56 (9): 677–680.

Jambhale, N.D. and Nerkar, Y.S. (1992). Screening of okra cultivars, related species and interspecific hybrid derivatives for resistance to powdery mildew. *J. Mah. Agric. Uni.* 17(1): 53–54.

Jayarajan, R., Doraiswamy, S., Sivaprokasam, K., Venkata, Rao, A. and Ramakrishnan, L. 1988. Incidence of whitefly transmitted viruses in Tamil Nadu. *Madras Agrie. J.* 75(56): 212–213.

Jayshree, K., Pun, K.B. and Doraiswamy, S. (1999). Virus vector relationship of yellow vein mosaic virus and whitefly (*Bemisia tabaci*) in pumpkin. *Indian Phytopath.* 52(1): 10–13.

Jha, A. and Mishra, J.N. (1955). Yellow vein mosaic of bhendi in Bihar. In Proc. Bihar Acad. of Agric. Sci., 4: 129–130.

Kalita, M.K. and Dhawan, P. (2006). Management of yellow vein mosaic and leaf curl diseases of okra by adjusting date of sowing and row to row spacing. Indian Journal of Agricultural Sciences. 76(12): 762–765.

Kalita, M.K., Dhwan, P. and Rishi, N. (2005). Influence of weather on whitefly population and okra leaf curl Begomovirus diseases. *J. Mycol. Pl. Pathol.* 35(1): 114–116.

Keshwal, R.L. and Khatri, A.K. (1999). Whitefly transmission of virus diseases in crops. *Bhartiya Krishi Anusandhan Patrika.* 14(1–2): 81–83.

Khan, M.A and Mukhopadhyay, S. (1975). Studies on effect of some alternative cultural method on incidence of yellow vein mosaic virus disease of okra. *Ind. J. Virol.* 1(1): 69–72.

Khan, M.A and Mukhopadhyay, S. (1985)a. Studies on the seasonal spread of yellow vein mosaic virus disease of okra. *Indian Phytopath.* 38(4): 688–691.

Khan, M.A and Mukhopadhyay, S. (1985)b. Effect of different pesticide combinations on the incidence of yellow vein mosaic virus disease of okra and its whitefly vector, (*Bemisia tabaci* (Genn.)) *Indian J. Virology.* 1(2): 147–151.

Khan, M.A and Mukhopadhyay, S. (1986). Screening of okra varieties tolerant to yellow vein mosaic virus. *Res. and Dev. Reptr.* 3(1): 86–87.

Kulkarni, G.S. (1924). Mosaic of other related diseases of crops in the Bombay Presidency. *Poona Agric. College Magazine.* XVI: 6–12.

Kumar, N. and Kadian, O.P. (1992). Studies on the effect of yellow vein mosaic virus infection on growth and yield components of okra cv. Pusa Sawani. *Ind. Phytopath.* Suppl. 43 and 44: LXXXIII.

Kumar, N.K.K. and Moorthy, D.N.K. (2000). Transmission of yellow vein mosaic Gemini virus to imidacloprid treated okra by the whitefly, *Bemisia tabaci* (Genn.), *Insect Environment.* 6(1): 46–47.

Kumawat R.L., Pareek, B.L. and Meena, B.L. (2000). Seasonal incidence of jassid and whitefly on okra and their correlation with abiotic factors. *Annals of Biology.* 16(2): 167–169.

Lana, A.O. and Bozarth, R.F. 1975. Studies on a virus induced mosaic disease of *Abelmoschus esculentus* in Nigeria. *Phytopathology.* 65: 77–86.

Madden L.V., Raymond Lovie, J.J. Abt. and Knoke, J.K. (1982). Evaluation of test for andomness of infected plants. *Phytopathology.* 72(2): 195–198.

Mandahar, C.L. and Singh, J.S. (1972). Destruction of chlorophylls *a* and *b* in virus infected leaves. *Sci. Cultivation.* 37: 485487.

Mandal, S.K., Sah, S.B. and Gupta, S.C. (2007). Management of insect pest on okra with biopesticides and chemicals. *Ann. Pl. Protec. Sci.* 15 (1): 87–91.

Markose, B.L. and Peter, K.V. (1990). Review of Research on vegetables and tuber crop okra. Pub. Unit. Directorate of Extension. *Kerala Agri. Univ. Mannuthy.* 82–87.

Martelli (1992). Classification and nomenclature of plant viruses: state of the art. *Plant disease.* 76 (5): 436–442.

Martin, F.W. (1982). A second edible okra species and its hybrids with common okra. *Annals of Botany.* 50: 277–283.

Masters, M.T. (1875). (In) Hooker, J.D. (ed) Flora of British India. Ashford Kent., 1: 320–348.

Mathew, A.V. and Muniyappa, V. (1992). Purification and characterization of Indian Cassava mosaic virus. *Journal of Phytopathology.* 135(4): 299–308.

Mayee, C.D. and Datar, V.V. (1986). Phytopathometry. Technical Bulletin–I, M.AU., Parbhani– 431402 (India).

Mohanty, A.K. and Basu, A.N. (1990). Disease transmitting ability of *Bemisia tabaci* Genn. Reared on different host plants during different periods. *Indian J. of Virology.* Vol. 6(1–2): 108–109.

Mohapatra, A.K., Nath, P.S. and Chowdhary, A.K. (1995). Incidence of yellow vein mosaic virus of okra (*Abelmoschus esculentus* (L.) Monech) under field condition. *Mycopathol. Res.* 33(2): 99–103.

Mote, U.N., Datkhile, R.V. and Pawar, S.A. (1994). Imidacloprid as seed dresser against sucking pest of okra. *Pestology*. 18(3): 5–9.

Mukhopadhyay, S., Nath, P.S., Das,,Babu, D. and Verma, A.K. (1994). Comparative epidemiology and cultural control of some whitefly transmitted viruses in West Bengal. In: *Virology in the Tropics*, (Eds.) Narayan Rishi, K.L. Ahuja and B.P. Singh. Malhotra Publishing House, New Delhi, pp. 285–304

Muniyappa, V. (1980). Whiteflies in vectors of plant pathogen (K.F. Harris and K. Maramorosch eds.). Academic Press, New York. pp: 39–85.

Murdock, G.P. (1959). Africa, its people and their culture history, New York, pp: 456.

Murthy, K.V. and Reddy, V.S. (1992). Chemical control of yellow vein mosaic of bhendi. *Ind. J. Pl. Prot*. 20(2): 198–201.

Muzumdar, N., Borthakur, U. and Chaudhary, D. (1996). Incidence of yellow vein mosaic virus of Bhendi in relation to cultivar and vector population under different sowing dates. *Indian J. Vi rol*. 12(2): 137–141.

Nadkarni, K. M. (1972). *Indian Materia Medica*. Nadkarni and Co., Bombay.

Naik, G.R., Muniyappa, V. and John, Colvin (2003). Host preference and natural occurrence of *Bemisia tabaci* (Genn.) on weed hosts–A vector of tomato leaf curl Geminivirus disease. *Indian J. Agric. Res*. 37(4): 253–258.

Nandihalli, B.S., Patil, B.V. and Lingappa, S. (1993). Population dynamics of cotton whitefly, *Bemisia tabaci* (Genn.). *Karnataka J. Agric. Sci*. 6(1): 25–29.

Nariani, T.K. and Seth, M.L. (1958). Reaction of *Abeloschus* and *Hibiscus* species of yellow vein mosaic virus. *Indian Phytopath*. 11: 137–140.

Nath, P. and Saikia, A.K. (1992). Relative resistance of okra cultivars of yellow vein mosaic virus. *New Agriculturist*. 3(2): 199–202.

Nath, P. and Saikia, A.K. (1993). Assessment of yield loss due to yellow vein mosaic of bhendi in Assam. *J. Agril. Sci. Society of N.E. India*. 6: 87–88.

Nath, P. and Saikia, A.K. (1995). Influence of sowing time on yellow vein mosaic virus of okra. *Ind. J. Mycol. Pl. Path*. 25(3): 277–279.

Nath, P.O., Gupta, M.K. and Bora, P. (1992). Influence of sowing time on the incidence of yellow vein mosaic and whitefly population on okra. *Indian Journal of Virology*. 8(1): 45–48.

Nath, P.S., Papia Bhattacharya, Malik, S.C., Pandit, M.K. and Bhattacharya, P. (1999). Studies on the incidence of bhendi yellow vein mosaic virus on some selected popular verities under field conditions. *Vegetable Science*. 26(2): 157–159.

Ndunguru, J. and Rajabu, A.C. (2004). Effect of okra mosaic virus diseases on the above ground morphological yield components of okra in Tanzania. *Scientia Horticulturae*. 99(3/4): 225–235.

Neerajd, G., Vijaya, M. and Chiranjeevi, C. (2004). Screening of okra hybrids against pest and diseases. *Indian J. Plant Prot*. 32(1): 129–131.

Nene, Y.L. (1973). Control of *Bemisia tabaci* a vector of several plant viruses. *Ind. J. Agric. Sci*. 43(5): 433–436.

Newton, W. and Pieris, J.W.L. (1953). Virus disease of plant in Ceylon. *F.A.O. Pl. Prot. Bull.* 2(2): 17–21.

Nirmal, D.O. (1983). Epidemiology and management of bud necrosis of groundnut. Ph.D. (Agri.) Thesis submitted to M.A.U., Parbhani.

Nirmal, D.O., Sirsat, V.B. and Pawar, N.D. (1993). Epidemiology and management of spotted wilt of tomato caused by tomato spotted wilt virus. Final Report of Adhoc Scheme. Dept. of Plant Pathology, M.A.U., Parbhani.

Okuyama, S., Takemi, K and Saka, H. (1978). Inhibitor of plant virus infection. Some properties of virus inhibitor in *Yucca recurvifolia. Science Report of Faculty of Agriculture, Ibraki University, Japan.* 26: 49–56.

Owen, H. (1946). Mosaic disease of malvaceae in Trinidad, B.W.1. Trop. Agric. Trin. Vol. XXIII (9): 157–162.

Pal, B.P., Singh, H.B. and Swarup, V. (1952). Taxonomic relationships and breeding possibilities of species *Abelmoschus* related to okra. *Botanical Gazette.* 113: 455–464.

Palaniswamy, P., Thirumurthy, S. and Subramanian, T.R. (1973). Effect of systemic granular insecticides on the incidence of yellow mosaic disease of okra (*Abelmoschus esculentus* L.). *South Indian Horticulture.* 21: 104–106.

Pandey, U. B. (1998). Export of vegetables. Souvenir: Silver Jubilee National Symposium on "Emerging Scenario in Vegetable Research and Development," P.D.V.R., Varanasi, 12–14, December, (1998), pp: 131–138.

Panse, V.G. and Sukhatme, P.V. (1978). Statistical Methods for Agricultural Workers. Indian Council of Agricultural Research, New Delhi.

Park, M. (1930). Report of the Mycological Division. Ceylon Dept. Agric. Tech. Report., pp: 1–6.

Pathak R., Shymal, M.M. and Singh, A.K. 1999. Screening for yellow vein mosaic virus resistant genotypes in okra (*Abelmoschus esculentus* (L.) Moech). *Prog. Hort.* 31(3–4): 166–170.

Peter G. Markham· Ian D. Bedford, Sijun Liu, Marion S. Pinner (2006) The transmission of geminiviruses by Bemisia tabaci. *Pesticide Science.* 42 (2): 123–128.

Phittp, A.M.C., Manfur, P. and Rajgopalan, B. (2000). Variability in F_4 generation of irradiated interspecific hybrids in okra (*Abelmoschus esculentus* (L.) Moench). *J. Tropical Agriculture.* 38(1–2): 87–89.

Poopathi, G., Manivannan, M. and Ramaswamy, N. (1996). A note on the comparative performance of bhendi cultivars for yield and yellow vein mosaic disease. *Prog. Hort.* 28(3–4): 143146.

Pullaiah, N., Reddy, T.B., Moses, G.J., Reddy, B.M. and Reddy, D.R. (1998). Inheritance of resistance to yellow vein mosaic virus in okra (*Abelmoschus esculentus* (L.) Moench). Indian Journal of Genetics and Plant Breeding. 58(3): 349–352.

Pun, K.B. and Doraiswamy, S. (2000)a. Influence of weather factors on the incidence of okra yellow vein mosaic virus disease. *J. Agric. Sci. Society of North East India.* 13(1): 91–96.

Pun, K.B. and Doraiswamy, S. and Jeyarajan, R. (2005). Management of Okra yellow vein mosaic virus disease and its whitefly vector. Indian Journal of Virology. 16(1/2): 32–35.

Pun, K.B. and Doraswamy, S. (1999). Effect of age of okra plant on susceptibility to okra yellow vein mosaic virus. *Indian J. Virol.* 15(1): 57 –58.

Pun, K.B., Doraiswamy, S. and Jeyarajan, R. (1999)a. Immunological detection of okra yellow vein mosaic virus. *Indian J. Virol.* 16(2): 93–96.

Pun, K.B., Doraiswamy, S. and Jeyarajan, R. (1999)b. Screening of plant species for the presence of antiviral principles against okra yellow vein mosaic virus. *Indian Phytopath.* 52(3): 221–223.

Pun, K.B., Doraiswamy, S. and Jeyarajan, R. (2000)b. Screening of virus inhibitory chemicals and neem products against okra yellow vein mosaic virus disease. *Indian Phytopath.* 53(1): 95–96.

Pun, K.B., Sabitha Doraiswamy and Doraiswamy, S. (2000). Influence of weather factors on the incidence of okra yellow vein mosaic virus disease. J. Agric. Sci. Society of North East India, 13(1): 91–96.

Ragupathi, N., Veergavathatham, D. and Thamburaj, S. (2000).Reaction of okra cultivars on bhendi yellow vein mosaic virus (BYMV) disease. *South Ind. Hort.* 48(16): 103–104.

Rathi, Y.P.S. (2001). Whitefly borne plant virus disease and their management. Plant Protection, National Conference: Plant Protection New Horizons in the Millennium Feb. 23–25, pp: 151–162.

Rattan, R.S. and Bindal, A. (2000). Development of okra hybrids resistant to yellow vein mosaic virus. *Vegetable Science.* 27(2): 121–125.

Raychaudhuri, S.P. and Nariani, T.K. (1977). Viruses and mycoplasma diseases of plants in India. Oxford and IBH Publishing Company, New Delhi, pp: 160–167.

Razvi, S.A, Azam, K.M. and Al–Raeesi, A.A. (1999). Monitoring of sweet potato whitefly, *Bemisia tabaci* (Genn.) with yellow sticky traps. *Sultan Qaboos University J. Scientific Res. Agric. Sci.* 4(1): 11–16.

Reddy, K.M., Salil Jalai and Samul, O.K. (2003). Fruit distortion mosaic disease of okra in India. *Plant Disease.* 87(11): 122.

Rehman, A. and Ahmad, W. (2002). Screening of okra vein mosaic Begomo virus under field conditions. *Pakistan J. Phytopath.* 14(1): 84–87.

Saikia, A.K. and Bhagabati, K.N. (1994). Spread of yellow vein mosaic virus disease of okra. In: Virology in the Tropics. Eds : Narayan Rishi, K.L. Ahuja and B.P. Singh (1994) Malhotra Publishing House, New Delhi, pp: 303–308.

Samarajeewa, P.K. and Rathnayaka, R.M.U.S.K. (2004). Disease resistance and genetic variation of wild relatives of okra (*Abelmoschus esculentus* L.). Annals of the Sri Lanka Department of Agriculture. 6: 167–176.

Sanchez, A., Gerand Poyey, F. and Esparza, D. (1997). Bionomics of the tobacco whitefly, *Bemisia tabaci* and potential for population increase on five host plant species. Revista–dela–facultad–de–Agronomia, Universidaddel–zulia. 14(2): 193–206.

Sandhu, G.S., Sharman, B.R., Balkar Singh and Bhalla, J.S. (1974). Sources of resistance to jassid and whitefly in okra germplasm. *Crop Improvement.* 1: 77–81.

Sangar, RB.S. (1997). Field reaction of bhendi varieties to yellow vein mosaic virus. *Ind. J. Virol.* 13(2): 131–134.

Sannigrahi, A.K. and Choudhary, K. (1998). Evaluation of okra cultivars for yield and resistance to yellow vein mosaic virus in Assam. *Environment and Ecology.* 16(1): 238–239.

Satheosh Kumar, P., Saravana, K., Sabesan, T. and Ganesan, J. (2005). Performance of okra cultivars in relation to yellow vein mosaic virus and yield (*Abelmoschus esculentus* (L.) Moench) of okra. National Seminar on ETPPSR Annamalai University, March 7–8, 2005, pp: 157.

Satyanarayana Murthy, K.V.V. and Reddy Rajaram (1992). Chemical control of yellow vein mosaic disease of bhendi. *Indian J. Pl. Prot.* 22: 198–201.

Senthil Kumar, N., Shanmugapriya, P. and Ganesan, J. (2005). Diverse sources of resistance to yellow vein mosaic virus disease (BYVMV) in okra (*Abelmoschus esculentus* (L.) Moench). National Seminar on ETPPSR Annamalai University, March 7–8, 2005, pp: 159.

Sharma, B.R and Sharma, O.P. (1984)a. Breeding for resistance to yellow vein mosaic virus in okra. *Ind. J. Agric. Sci.* 50(1): 917 –920.

Sharma, B.R and Sharma, O.P. (1984)b. Field evaluation of okra germplasm against yellow vein mosaic virus. *Punjab Hort. J.* 24: 114, 131–133.

Sharma, B.R. and Arora, S.K. (1993). Improvement of okra crop. Advances in Horticulture (Edts, Chadha, K.L. and Kalloo, G. Malhotra Publishing House, New Delhi 5 : 343–364.

Sharma, B.R., Sharma, O.P. and Bansal, R.D. (1987). Influence of temperature on incidence of yellow vein mosaic virus in okra. *Vegetable Sci.* 14(1): 65–69.

Shastry, K.M.S. and Singh, S.J. (1973). Field evaluation of insecticides for the control of whitefly in relation to the incidence of yellow vein mosaic of okra. *Indian Phytopath.* 26: 129–138.

Shastry, K.M.S. and Singh, S.J. (1974)a. Effect of yellow vein mosaic virus infection on growth and yield of okra crop. *Indian Phytopath.* 27(3): 294–297.

Shastry, K.M.S. and Singh, S.J. (1974)b. Restriction of yellow vein mosaic virus spread in okra through the control of vector, *Bemisia tabaci. Indian J. Mycol. Pl. Pathol.* 3: 76–80.

Shastry, K.M.S., Shastry, K.S. and Singh, S.J. (1978). Yellow vein mosaic virus disease of Bhendi. *Ind. Hort.* 22(1): 14–15.

Shepherd, H. and Winston, S.L. (1999). Performance evaluation of hybrids of okra (*Abelmoschus esculentus* (L.) Moench) for growth and yield during *kharif* season in Allahabad region. *Bioved.* 10 (1–2): 105–107.

Shivpuri, A., Bhargava, K.K., Chhipa, H.P., and Ghasolia, R.P. (2004). Management of Yellow Vein Mosaic Virus of Okra. *Journal of Mycology and Plant Pathology.* 34(2): 353–355.

Srivastava, B.N., Shrivastava, K.M., Singh, B.P. and Dwadesh Shreni, C. (1977). Zinnia yellow net disease transmission, host range and agent vector relationship. *Plant Disease Reporter.* 61 (7): 550–554.

Shrivastava, K.M., Aslam, M. and Rao, B.L.S. (1985). A whitefly transmitted yellow vein mosaic disease of *Cosmos sulphurens* Cav. *Current Science.* 54(21): 1126–1128.

Shrivastava, P.K., Srivastava, K.J., Sharma, H.K. and Gupta, R.P. (1995). Evaluation of different varieties of okra against yellow vein mosaic virus (YVMV). *Newsletter National Hort. Res. and Development Foundation.* 15(4): 8–10.

Singh, A.K. and Singh, K. P. (2000). Screening for disease incidence of yellow vein mosaic virus (YVMV) in okra (*Abelmoschus esculentus* (L.) Moench) with gamma rays and EMS. *Vegetable Science.* 27(1): 72–75.

Singh, A.K., Sanger, R.B.S. and Gupta, C.R. (2002). Performance of different verities of okra to yellow vein mosaic virus under field conditions in Chattisgarh. *Progressive Horticulture.* 34(1): 113–116.

Singh, A.P. (2000). To study the production efficiency of okra verities under western Uttar Pradesh condition. *Bhartiya Krishi Anusandhan Patrika*. 15(1–2): 34–38.

Singh, B.R. and Gupta, S.P. (1991). Reaction of okra lines to yellow vein mosaic. *Indian J. Virol*. 7(2): 188–189.

Singh, B.R. and Singh, A.K. (1999). Viral diseases on congress grass from Uttar Pradesh. *International J. Tropical Pl. Dis*. 17(12): 165–168.

Singh, B.R., Singh, M. and Singh, M. (1989). Control of yellow vein mosaic of okra by checking its vector through adjusting dates of sowing, insecticidal application and crop barrier. *Indian. J. Virol*. 5(1–2): 61–66.

Singh, D., Varma, N. and Naqvi, Q.A. (1998). Control of yellow vein mosaic of okra by seed selection and insecticidal application. *Indian J. Viral*. 14(2): 121–123.

Singh, D.K. and Jain, S.K. (2004). Screening of okra hybrids during rainy season under Tarai condition of Uttarakhand. *Scientific Horticulture*. 9: 111–116.

Singh, D.K., Singh, S.K. and Jain, S.K. (2003). Evaluation of okra hybrids for growth, yield and yellow vein mosaic virus. *Scientific Horticulture*. 8: 129–133.

Singh, D.K., Tewari, R.P. and Lal, G. (1994)a. Effect of sowing time on virus incidence and seed yield of okra. *Annals Agric. Res*. 15(3): 374–375.

Singh, D.K., Singh, S.K. and Jain, S.K. (2003). Evaluation of okra hybrids for growth, yield and yellow vein mosaic virus. Scientific Horticulture. 8: 129–133.

Singh, M. and Thakur, M.R. (1979). Nature of resistance of yellow vein mosaic in *Abelmoschus manihot*. *Current Sci*. 48(4): 164–165.

Singh, R. and Srivastava, R.P. (1974). Physiological changes in bhendi (*Abelmoschus esculentus*) fruit affected by yellow vein mosaic virus. *Current Sci*. 43: 89–91.

Singh, R., Mishra, R.C., Shahi, S.K., Dikshit, A., Singh, R. and Dikshit, A. (1999). Insect repellent activity of asafetida to prevent yellow vein mosaic virus infection in okra crop (*Abelmoschus esculentus* L.). *Plant Protection Bulletin*, Faridabad. 51(3–4): 35–37.

Singh, S.J. (1980). Studies on epidemiology of yellow vein mosaic virus of okra. *Indian J. Mycol. Pl. Pathol*. 10: 35.

Singh, S.J. (1986). Enation leaf curl of okra a new virus disease. *Indian J. Virol*. 2: 114–117.

Singh, S.J. (1990). Etiology and epidemiology of whitefly transmitted virus diseases of okra in India. *Plant Dis. Res*. 5(1): 64–70.

Singh, S.J. (1994). Epidemiology of enation leaf curl disease of okra. In: virology in the tropics: Eds. Narayan Rishi, KL. Ahuja and B.P. Singh, Malhotra Publishing House, New Delhi. pp: 357–361.

Singh, S.J., Shastry, K.S. and Sastry, K.M.S. (1977). Yellow vein mosaic virus disease of bhendi. *Indian Hort*. 22(1): 14–15.

Singh, S.J., Shastry, K.S. and Sastry, K.M.S. (1994)b. Study of influence of moderate temperature on whitefly population. *Indian J. Virol*. 2: 114–117.

Sinha, S.N. and Chakrabarthi, A.K. (1978). Effect of yellow vein mosaic virus infection on okra seed production. *Seed Res*. 6: 67–70.

Sipell, D.W., Bindra, O.S. and Khalifa, H. (1982). Preliminary study of relationship of leaf lobing and hair density in cotton with whitefly (*Bemisia tabaci*) population and proposal for further investigation. Gezira Agric. Res. Stn., Agric. Res. Crop Wad Medani, Sudan, Working Paper, 9: 6.

Snarma, B.H., Arora, S.K., Dhanju, K.C. and Ghai, T.R (1993). Performance of okra cultivars in relation to yellow vein mosaic virus and yield. *Indian J. Virol.* 9(2): 139–142.

Sonalkar, V. (1999). *Bemisia tabaci* (Genn.) control with acetamiprid on okra In Vidarbha region of Maharashtra. 7(1): 87–89.

Sonwane G.T. (2004). Epidemiology and management of yellow vein mosaic of okra. M.Sc. Thesis submitted to M.A.U., Parbhani. pp: 54–61.

Srabani, D. and Nath, P.S. (2003). Performance of okra varieties in relation to yield and tolerance to YVMV. Annals of Plant Protection Sciences. 11(2): 400–401.

Suresh Kumar, (2000). Genetic studies in diverse bhendi (*Abelmoschus esculentus* (L). Moench) genotypes. M. Sc. (Agri.) Thesis, submitted to University of Agricultural Sciences, Dharwad.

Thakur, M. R., Arora, S.K. and Kabir, J. (2003). Vegetable Crop, Vol. 3, 3rd Edition. pp: 209–239.

Thakur, M.R. and Arora, S.K. (1989). Punjab – A virus resistant variety of okra. *Prog. Farming.* 24: 13–14.

Thampan and Vijayan (1964). "Pusa Sawani" a resistant variety to yellow vein mosaic virus disease of bhendi. *New Agriculturist.* 3(2): 199–202.

Torre– Almarz and Monsalvao– Reyes, A.C. (2003). First report of geminivirus including yellow mottle in okra (*Abelmoschus esculentus*) in Mexico. *Plant Dis.* 87: 202.

Tripathi, S.N. and Joshi, G.S. (1967). Studies on incidence of mosaic virus in Pusa Sawani variety of bhendi. *Allahabad Farmer.* 416: 277–299.

Tsering, K. and Patel, B.N. (1990). Simultaneous transmission of tobacco leaf curl virus and yellow vein mosaic virus of *Abelmoschus esculentus* (L.) Moench by *Bemisia tabaci* (Genn.). *Tobacco Res.* 16(2): 127–128.

Uppal, B.N., Verma, P.M. and Capoor, S.P. (1940). Yellow vein mosaic of bhendi. *Current Sci.* 9: 227–228.

Uttasami, S., Chellaiah, S. and Balasubramanian, M. (1977). Insecticidal control of pests and yellow vein mosaic of bhendi. *Science and Cultivars.* 43: 510–512.

Van Borssum Waalkes, T. (1966). Blumea, 14: 1–251.

Vander Plank, J.E. (1963). Plant disease, Epidemics and Control. Academic Press, New York. pp: 349.

Vashishta, V.K., Sharma, B.R. and Dhillon, G.S. (2001). Genetics of resistance to yellow vein mosaic virus in okra. *Crop Improvement.* 28(2): 218–225.

Vasudeva, R.S. (1957). Report of the Division of Mycology and Plant Pathology. Sci. Rep. agric. Res. Inst. New Delhi, (1954). 55: 87–101.

Vasudeva, R.S. and Samraj, J. (1948). Leaf curl disease of tomato. *Phytopathology.* 38: 364–369.

Vavilov, N.I. (1951). The origin, variation, immunity and breeding of curtivalea plants. *Chron Bot.* 13(1–6).

Venugopalrao, N. and Reddy, A.S. (1989). Seasonal influence of development and duration of whitefly (*Bemisia tabaci*) in upland cotton (*Gossypium hirsutum*). *Ind. J. Agric. Sci.* 59(6): 383–385.

Verma, P.M. (1952). Studies on the relationship of bhendi yellow vein mosaic virus and its vector, (*Bemisia tabaci* Genn.). *Indian J. Agric. Sci.* 22: 75–91.

Verma, P.M. (1955)a. Persistence of yellow vein mosaic virus of *Abelmoschus esculentus* (L.) Moench in its vector *Bemisia tabaci* (Genn.) *Indian J. Agric. Sci.* 25: 293–302.

Verma, P.M. (1955)b. Ability of whitefly to carry more than one virus simultaneously. *Current Sci.* 24: 317–318.

Verma, P.M. and Mukherjee, S.K. (1955). Studies on varietal classification of virus resistance in ladies finger (*Abelmoschus esculentus*). Proc. 42nd Ind. Sci. Congr. (Baroda). pp: 371–372.

Verma, V. K., Basu, D., Nath, P.S., Das, S., Gathak, S.S. and Mukhopadhyay, S. (1989). Some ecological considerations of whitefly and whitefly transmitted virus diseases of vegetable in West Bengal. *Indian J. Virology.* 5(1–2): 79–87.

Vijaya, M. (2004). Screening of okra entries to yellow vein mosaic virus (YVMV) disease under field conditions. *The Orissa J. Horticulture.* 32(1): 75–77.

Yadav, J.R., Srivastava, J.P., Singh, B. and Kumar, R. (2004). Azad Bhendi–1 (Azad Ganga), a disease resistant variety of Bhendi (*Abelmoschus esculentus* (L.) Moench). *Plant Archives.* 4(1): 205–207.

Zeven, A.C. and Zhukovasky, P.M. (1955). Dictionary of cultivated plants and their centre of diversity. Centre of Agricultural Publishing and Documentation, Wageningen, Netherlands. pp: 219.

Plant Diseases Management in Horticultural Crops (2011) *Pages* **345–357**
Editors: **Shahid Ahamad, Ali Anwar and P.K. Sharma**
Published by: **DAYA PUBLISHING HOUSE, NEW DELHI**

Chapter 27

Bubble Diseases of White Button Mushroom and their Management

Mandvi Singh[1], Mohd. Akram[1] and A.K. Singh[2]

[1]Department of Plant Pathology, [2]Directorate of Extension,
C.S. Azad University of Agriculture and Technology, Kanpur – 208 002

Introduction

The occurrence of fungi associated with mushrooms was probably observed as early as people became interested in their cultivation. The fungi are ubiquitous and have relatively simple requirements for their growth. The required materials are readily available in nature which provides carbon, nitrogen and minerals to the fungi. Water and oxygen availability is not a problem in most of the environments, and the fungi are known for their ability to produce asexual spores abundantly. Some fungi are also able to use the mycelium and sporophores of the mushroom for their own needs, often detrimental to the mushroom by causing various diseases (Chang and Miles, 2005).

White button mushroom, *Agaricus bisporus* (Lange) Imbach cultivation in India is being done either in controlled environment cropping rooms on pasteurized compost (Figure 27.1) or under natural climatic conditions using unpasteurized (long method) or pasteurized compost. The unpasteurized compost harbours several parasitic and antagonistic fungi. Similarly, in absence of pasteurization facility, casing soil is treated with formaldehyde and in some cases not treated at all. The casing mixtures used in our country include, Farm Yard Manure (FYM), soil and spent compost (1-3 years old) in absence of standard peat moss. Untreated or unpasteurized casing invites several soil-borne pathogens of *A. bisporus.*

Among different biotic and abiotic factors adversely affecting the yield of mushrooms, fungal disorders are most important and cause maximum loss in Indian conditions. The predominant fungal flora isolated from the compost and casing samples recently from India were *Trichoderna viride,*

Figure 27.1: White Button Mushroom Crop

Verticillium fungicola, Populospora byssina, Geotrichum sp., *Penicillium* spp., *Aspergillus* spp., *Trichoderma* spp., *Chaetomium* spp. and *Pythium* sp. (Sharma, 1991, 92; Sharma and Vijay, 1991, 96). It was further observed that the steam pasteurized casing and compost samples carried less fungal flora than those treated with formaldehyde. Besides these, some workers have also reported occurrence of some other weed fungi of mushroom beds and losses caused by them (Tewari, 1990; Tiwari and Singh, 1991; Vijay and Gupta, 1992) but information on pathogens like *Mycogone perniciosa, V. fungicola* var. *fungicola, Hypomyces rosellus, Trichoderma* sp. and *Fusarium monliiformae,* is very scanty. In this chapter we are going to discuss the two important bubble diseases of mushroom *i.e.* wet bubble and dry bubble.

Economic Importance

Wet Bubble

Serious crop losses due to wet bubble have been reported from earlier days of *Agaricus* cultivation in Pennsylvania, USA. In 1930, annual losses were estimated at 15 per cent of the mushrooms; in 1974 the loss was 5.2 million pounds. Wet bubble was the most common disease in 1957 in UK. *Mycogone perniciosa* caused more losses than *Verticilium fungicola.* However, in 1974 statistics for Pennsylvania showed a greater loss from *Verticillium* sp. Suggestions regarding the reason for this change involve resistance of *V. fungicola* to the fungicide (Benomyl), which was being used widely. Other contributing factors include a lowering of hygiene standards with greater reliance on Benomyl as well as changes in cultivation procedures, especially the introduction of picking lines, which involved more movement of equipment (mainly trays) into and out of the mushroom house, increasing the chances for introduction of pathogens (Gandy, 1972). Glaser and Gapinski (1976) reported yield loss upto 99.2 per cent on inoculation.

Dry Bubble

From India the heavy incidence of dry bubble disease was reported from mushroom farm located at Chail and Taradevi (Seth *et al.*, 1973). The pathogen has been invariably isolated from the compost and casing samples collected from mushroom farms in Haryana, HP and Punjab. However, Thapa and Jandaik (1985) have recorded 25- 50 per cent disease incidence at Solan and Kasauli, and upto 15 per cent at Shimla and Chail during 1981. Wong and Preece (1987) detected the pathogen in 47 of 485 samples of baled peat and in one of 126 samples of bagged limestone arriving at an isolated mushroom farm in the UK. Spadafora (1989) have also reported the occurrence of pathogen in commercial mushroom farms in Pennsylvania, U.S.A.

Sharma and Vijay (1996) reported that it was the most common fungi isolated from 55 compost and casing samples which were collected during 1995. They reported that inoculum @1 g/kg compost at spawning caused 17.02 per cent yield loss and inoculation of compost at different doses in different times caused yield loss from 2.1 to 38 per cent. When different loads of inoculums of pathogen were inoculated in compost, the per cent reduction in the number and weight of fruit bodies varied from 2.26- 47.2 and 2.19-38.01, respectively.

Wet Bubble Disease (*Mycogone perniciosa*)

Earliest report of this disease was made by Smith (1924) who named the special masses of false tissues as 'Sclerodermoid masses' and the term still used for the deformity caused by this parasitic fungus (Figure 27.2). Nair (1975) reported that it was among the most economically important diseases. From India for the first time it was reported from few mushroom farms of Village Pinglen (Telisil Palwana) in Jammu and Kashmir State (Kaul *et al.*, 1978). Fourty per cent incidence was recorded in the year 1994-95 from Pune by Sharma and Vijay (1996).

Figure 27.2: Symptoms of Wet Bubble Disease

In this disease the vegetative mycelium of *A. bisporus* is not infected *by M. perniciosa.* The infection takes place only after formation of the rhizophores, which are the forerunners of the fruiting body. In "wet bubble" the growth of *M. perniciosa* on the fruiting body is thick, has a velvety texture, and looking white. The sporophore (fruiting body) when examined internally shows wetness and has an offensive stench with the tissues becoming invaded by bacteria. Drops of amber-colored exudates are present on these mushrooms. It has been reported by Gandy (1972) that the expression of disease symptoms is related to the stage of fruit body development and the time of infection. When infection is early, the expression of disease symptoms is severe. This may involve failure of differentiation of stipes, pilei, or lamellae (gills). What does form is a spherical mass of false tissue, which has been called sclerodermoid tissue, and this mass may be equal to or greater than 5 cm when sectioned, three zones are revealed in the sporophore beneath the hyphal layer of *M. perniciosa.* The zone closest to the sporophore surface contains brown dead tissue. Beneath this is a zone of extensive enzyme secretion, and then a zone in which the mycelium of *M. perniciosa* is present between the hyphae of the mushroom (Smith, 1972). Other workers demonstrated a significant difference in the hyphae of *M. perniciosa* growing on the surface (circa 5µm in diameter) and those growing internally (circa 25 µm in diameter).

When infection occurs on already developed sporophores, *M. perniciosa* grows over sectors of the gills and the pilei. The gills undergo hypertrophy and they eventually become covered by the mycelium of the wet bubble organism. In view of the symptoms displayed in this disease, it is not surprising that *M. perniciosa* has been found to produce enzymes that degrade the cell wall polymers of *A. bisporus* as reported by (Gandy, 1972).

Symptoms

Smith (1924) suggested that the chief factor governing the time of appearance of symptoms was the quantity of inoculum. Flectcher and Gannet (1968) reported that when infection took place before the differentiation of the stipe and pileus the sclerodermoid form resulted, whereas, infection after differentiation resulted in the production of thickened stipe with deformation of the pileus. The minimum time between inoculation and symptom production was eleven days. Geijin (1976) described that symptoms of the disease appear as a regular and often large mass of tissue known as 'Sclerodermoid mass'. This appears if the undifferentiated mushroom is infected. Bech and Kovacs (1977), Van, (1982) and Figueiredo and Mucei (1985) have also described the similar symptoms. According to Tu and Liao (1989) symptoms of the disease vary with the stage of sporophore development.

The Pathogen (*Mycogone perniciosa*)

Smith (1934), Atkins (1971) and Hsu and Han (1981) described the morphological features of the pathogen. They reported the mycelium as white to amber brown in colour and hyphae 3-4µ broad and septate conidiophores sub-verticillate to verticillately branched 145-450g x 3-4µ in size (Figure 27.3). Conidium 1-2 celled and size ranging from 12-33µ x 3-4µ and hyaline. Chlamydosphores, two-celled, upper cell 15-23µ x 14-20 µ and lower cell 12-18µ x 8-15µ in size. Sharma (1995) reported conidiophores measuring 200 x 3-5 µ and conidia measuring 5-10 x 4-5µ. The upper cell of chlamydospore measuring 15-30 x 10-20 µ and lower cell 5-10 x 4-5 µ in size.

Chaze and Sarazin (1935) produced evidence which indicated that vegetative mushroom mycelium was immune to infection by *M. perniciosa* and became susceptible only after formation of the rhizomorphs which produced the sporophore primordia. They also found that the growth of mushroom mycelium increased after contact with *M. perniciosa* while that of the latter stopped. Labrousse (1936) found no parasitic relationship between vegetative mycelium of mushroom and *M. perniciosa.* Han *et al.* (1974) and Gray and Morgan-Jones (1981) reported that in paired cultures with mushroom the latter ultimately overgrew the colony of *M. perniciosa.*

Effect of Temperature and pH on the Mycelial Growth of Pathogen

Lambert (1930) found 24°C to be the optimum temperature for the mycelial growth of the pathogen. Dough and Hung (1971) reported that the pathogen showed tolerance to a wide pH range in the acid side, and able to grew at pH 4.4. However, the growth becomes weaker or rather restricted at pH 8.4.

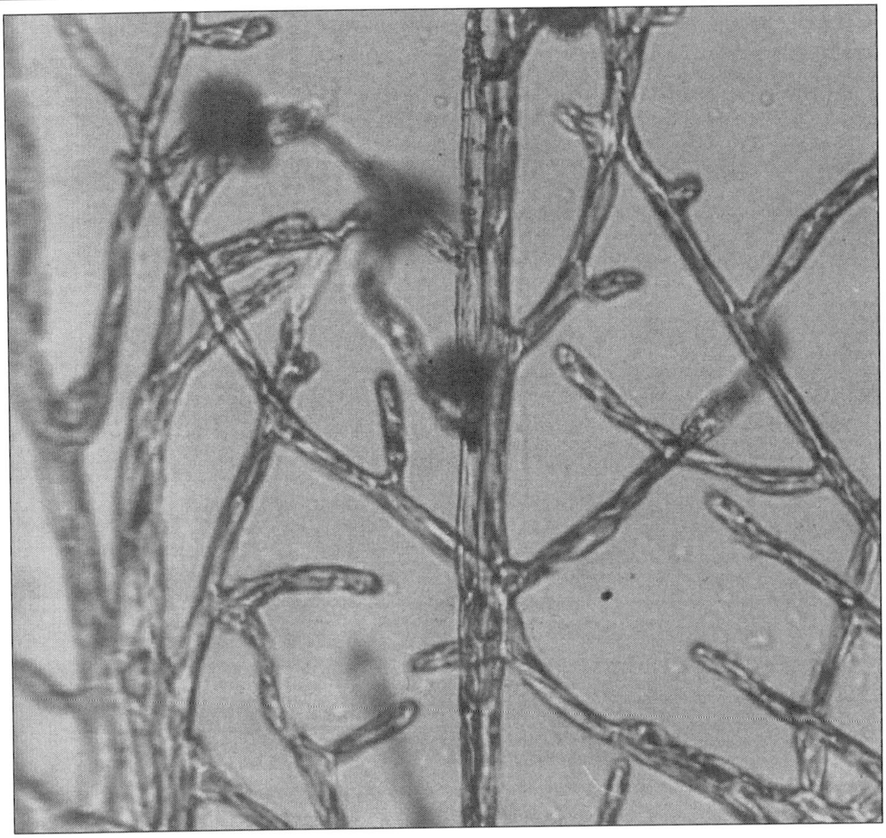

Figure 27.3: Morphological Feature of *M. perniciosa*

Hsu and Han (1981) stated that the optimum temperature for mycelial growth was 25°C. No mycelial growth was observed at the temperature below 12°C and above 32°C. The optimum pH was 6.2. In the same year Zaayen and Rutjens stated that the thermal death point for the mycelium of the pathogen was 48°C for 30 minutes. The mycelium was unable to grow at 4°C. A minimum of growth started slowly at 8°C and reached its optimum at 24°C and was considerably reduced at 26° C.

Chemical Management

Nair and Baker (1977) reported that total control of the disease can be achieved by incorporation of Benlate, Bavistin or possibly Tecto and good hygiene. Gandy and Spencer (1978) compared the two formulations of Chlorothalonil for their activity against the pathogen and its host (*A. bisporus*). It was found that the flowable formulation was more toxic than the wettable powder to both pathogen and host but more marketable mushrooms were obtained from diseased plots treated with the flowable formulation at each dose rate (1.0, 0.5 and 0.1 g a.i./m^2).

Stoller (1981) gave an account of the preparation and use of sodium hydroxymethane sulphate from formaldehyde and sodium bisulphate, forming an odourless and cheap compound, which gave almost complete control of the pathogen. Zaayen (1983) reported that prochloraz at 3 g/m^2 applied once as a spray 9 days after casing gave excellent control. Tecto flowable fungicide (450 g a.i. TBZ/dm^3 in a water dispersible suspension) satisfactorily controlled the pathogen.

Dry Bubble Disease (*Verticillium fungicola*)

Dry bubble disease" caused by *V. fungicola,* the mycelium of *A. bisporus* is killed as has been made evident by a number of studies. Many investigations of the interactions between the mushroom mycelium and *V. fungicola* have been conducted. With *V. fungicola* there was a severe necrosis of the mushroom mycelium, which was viewed as similar to the necrosis of dry bubble occurring on fruiting bodies. However, as reported by Gandy (1972) it was learned experimentally that exudates from rhizomorphs and volatile substances from *A. bisporus* had no effect on germination of the conidia of *V. fungicola.* A main generalization from comparative studies of *M. perniciosa* and *V. fungicola* interactions with *A. bisporus* was that *V. fungicola* was responsible for more mushroom necrosis than *M. perniciosa.*

The growth of *V. fungicola* on sporophores is thin, has a felty texture and is grayish white. The sporophore is dry at first, but this may change if it undergoes bacterial rot. Internal examination of the sporophore reveals a gray-brown zone of variable thickness just below the surface. In some cases, there is complete discoloration of the internal tissues. Cavities lined with spore-bearing conidiophores of *V. fungicola* are present in the sporophores. This necrotrophic mycoparasite is responsible for

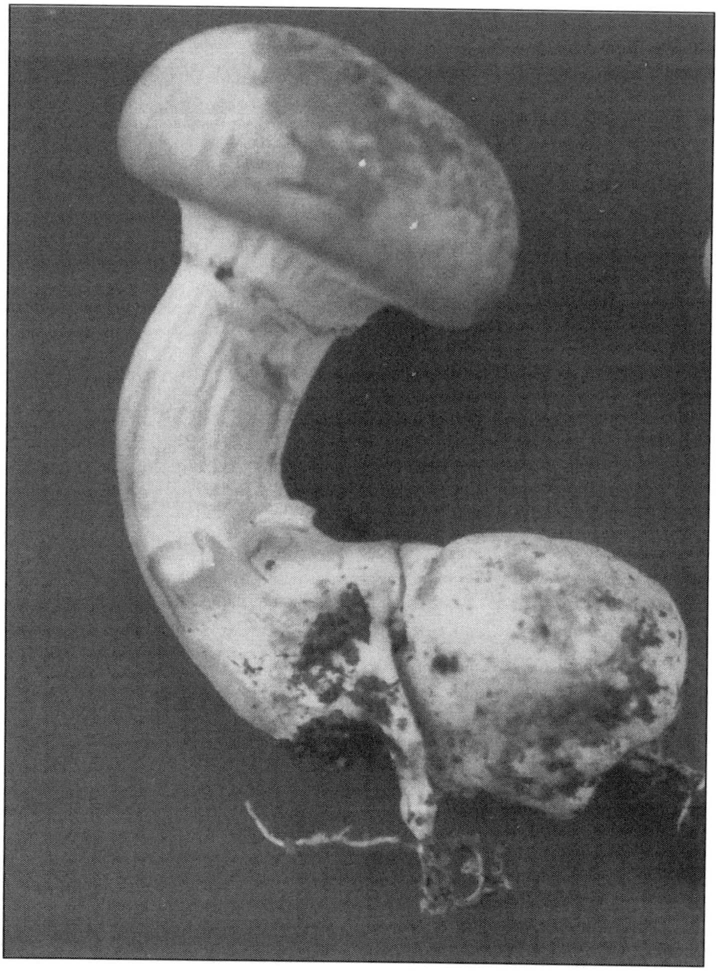

Figure 27.4: Symptoms of Dry Bubble Disease

developmental deformities, of the fruiting body, which may take the form of bulbous stipes on which vestigial pilei are located, or there may be localized lesions on the stipe, which then cause the fruiting body to bend and tissues to peel. When well-developed sporophores are infected, some superficial lesions, pinpoint in size and pale brown in color, may appear within a few days. As is the case with wet bubble caused by *M. perniciosa*, early infection caused by *V. fungicola* causes the deformity known as sclerodermoid, but in the case of *Verticillium* these sclerodermoid masses are less than 1 cm in diameter.

Symptoms

The symptoms produced by the pathogen have been described by Geijin (1976); Seth (1977) and Chen *et al.* (1981). Tu and Liao (1987) reported that the pathogen produces distortion symptoms and is characterized by the dull spots covered with fluffy mould penetrating into the cap tissues. The average time from infection to production of symptoms is 14 days.

North and Wuest (1993) described the symptoms of pathogen as brown lesions, shattered stipes and undeveloped (mummified) sporocarps. The degree of disease and symptoms observed are dependent on inoculums level and the point of development of the mushroom crop at the time of infection. Dragt *et al.* (1996) studied the infected mushroom by light microscopy and found that the hyphae growing on and in the hyphae of *A. bisporus* and appresoria of pathogen were observed.

The Pathogen (*Verticillium fungicola*)

Sharma (1995) and Dhar (1998) described the morphological features of the fungus. The fungus produces hyaline conidia, 3.5-15.9 x 1.5-5 µ on verticillately branched conidiophores (220 x 80 x 1.5 x 5.0 µ). Chaze and Sarazin (1936) reported that mushroom mycelium showed antagonism *in vitro* to the pathogen (Figure 27.5). It was confirmed by Olivier and Guillaumes (1976). In contrary to this Gray and Morgan-Jones (1981) found that the pathogen grew over the mushroom mycelium and caused severe necrosis. Vijay and Gupta (1991) reported that the pathogen was highly toxic and showing 81 per cent inhibition of *A. bisporus* mycelium.

Effect of pH and Temperature on Mycelial Growth of the Pathogen

Bech and Kovacs (1981) reported that the spores of pathogen are able to withstand higher temperature. In the dry state pathogen is capable of surviving 125°C for 10 min. In aqueous suspension spores can withstand at 44°C for 10 min. and 38°C for one hour.

The optimum temperature for the growth of pathogen is 25°C. The pathogen can not tolerate higher temperature and dies when exposed to dry heat and wet heat of 55°C for 4 min and 1 min (Tu and Liao, 1987). Bech *et al.* (1989) have also reported the effect of temperature on the growth and spore formation of pathogen.

Management

Gandy and Spencer (1976), Zaayen (1977) and Fletcher (1977) reported that chlorothalonil (as Deconil 2787, 75 per cent a.i., WP) applied as a drench, and reduced the incidence of disease. While, Geijin (1977) stated that disease can be controlled by spraying with carbendazim, benomyl or thiophanate-methyl at 100, 150 and 200 g/100m^2, respectively, in 100–500 litres water immediately after casing. Formaldehyde (at 2 litres in 100 lit. water/100 m^2) applied immediately after casing was effective. If disease reappears, formaldehyde is replaced by a systemic fungicide such as Benlate, Bavistin or Topsin-M throughout one cultivation cycle. If pathogen becomes resistant to these fungicides

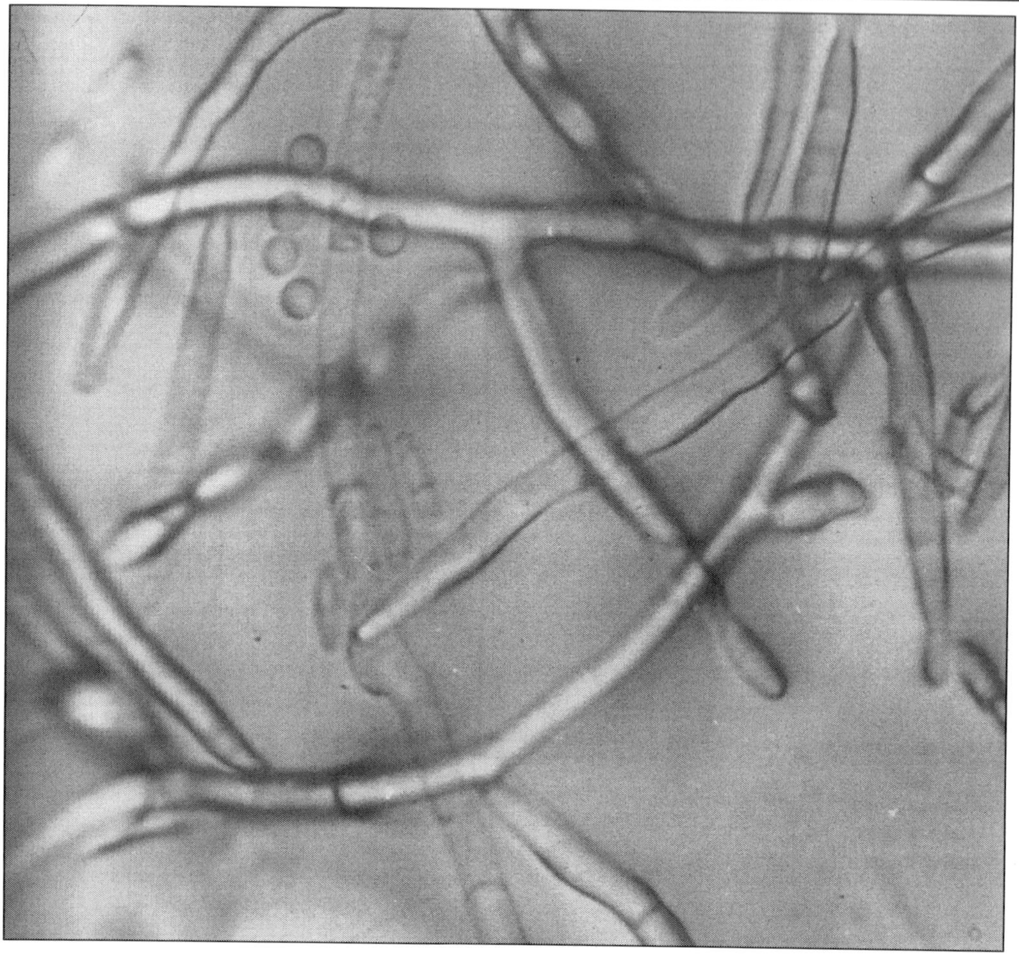

Figure 27.5: Morphological Features of *V. fungicola*

as well, Deconil (at 3 g/m^2) can be used immediately after casing and again 14 days later or Curamil (at 0.5 ml/m^2) can be applied after casing and thereafter at weekly intervals (Zaayen and Geijin, 1979).

A good control of disease on mushroom was achieved by spraying with prochloraz/manganese at 60 g/100 m^2 within 7 days of casing and subsequently at two weeks interval. Zaayen (1983) and Russel (1984) in their studies reported that Prochloraz at 3 g/m^2, applied once as a spray 9 days after casing, gave excellent control of pathogen. According to Eicker (1984), Tecto fowable fungicide (450 g a.i. TBZ dm^3 in a water dispersable suspension) satisfactorily controlled the pathogen. In *in vitro* studies it was found that it was non-toxic to mushroom mycelium. In the year 1987 he reported that prochloraz-Mn complex (50 per cent a.i.) gave satisfactory control of pathogen. For each crop two applications of the fungicide (1.5 g a.i./m^2) were made nine days after casing and a similar application at the end of first break. Gea *et al.* (1997) also reported efficacy of sporgon against pathogen. Sharma *et al.* (1997) reported that drenching of mycelial discs with, formalin (0.5-2.0 per cent concentration) inhibits the radial growth of the pathogen from 45.8-100 per cent. When cultures of the pathogen were exposed to vapours of formaldehyde (1.0-4.0 per cent) for 6-24 hours, radial growth was significantly inhibited irrespective of the concentration and exposure periods.

General Guidelines for the Management of Bubble Diseases

In order to take the most effective measures for the management of diseases of mushrooms, it is necessary to understand the size of the initial population or the initial inoculums density, the rate at which the disease develops and the time duration in which infection takes place. For the prevention and control of diseases, it is, therefore, necessary to take the following measures:

Sanitation and Hygiene

Hygiene covers all the measures which are necessary to give pests and diseases as little chance as possible of developing and spreading. Thus, hygiene and sanitation go hand in hand at all stages of growing mushrooms. Farm hygiene the best defense a mushroom grower has against mushroom pests and diseases. It should be the main defense, even more so now, as a general feeling against the use of 'chemicals' on food crops increases (Thakur, 2006).

Mushroom pathogens gain entry to a mushroom farm in a wide variety of ways like through insect-pests, raw material without being apparently visible to the naked eye. Based on the critical observation daring all the stages of mushroom production, the following steps have become a routine practice for successful cultivation of mushrooms.The location of mushroom unit should be in such a place, where effluent of chemical industries does not pollute the water and also the air is free from toxic fumes or gases.

- Floor for the preparation of compost should be cemented/tiled and covered by a roof.
- Substrates used for compost preparation should be fresh, protected from rain and mixed in exact proportions.
- Excess watering should be avoided during composting and ready to use compost should have 68-70 per cent moisture and pH 7-7.2.
- Pasteurization and conditioning of the compost should be for optimum duration at right temperatures and over/under pasteurization may not produce quality compost and invite many disease problems.
- Restrict free access of persons working in composting yard to spawning and other cleaner areas without changing the dress and foot-dip. Similarly, all machinery including tractors and fork-lift should not be moved without thorough hosing to the cleaner areas. After filling, all equipments and machinery should be thoroughly cleaned.
- Spawn should be fresh and free from all the contaminants.
- All equipments used for spawning, and the floor and walls of spawning area must be washed and disinfected.
- If fresh air is used to cool down spawn-running compost, it should preferably be filtered before it enters the room. Filters should exclude all particles of 2 micrometre and above.
- Casing mixture should be properly pasteurized (60-65°C for 5-6 h).
- Casing mixture should be stored in a clean and disinfected place. All containers, equipment and machinery used for casing should be thoroughly washed and disinfected. The main points to watch include previously cleaned machinery and tools, wearing of clean overalls and boots, clean hands or use of clean gloves, limited access to the working area, washed

floors, keeping dust to a minimum and trying not to have dusty and dirty jobs going on at the same time elsewhere on the farm. Floors must not be swept while dry.

- ☆ The pickers should use clean overalls and gloves. Picking should be started from new or cleaner crops towards older crops.
- ☆ Waste from picking, trash, stems, unusable mushrooms should be carefully collected and disposed off daily.
- ☆ Avoid surface condensation of water on developing mushrooms.
- ☆ Add chlorine to the spraying water to give a concentration of 150 ppm at every watering to manage bacterial diseases.
- ☆ Remove the heavily infected bags from the cropping rooms or treat the patches by spot application of 2 per cent formalin or 0.05 per cent bavistin.
- ☆ Maintain optimum environmental conditions in the cropping rooms to avoid abiotic disorders.
- ☆ Control insect-pests well in time to avoid the spread of pathogen by them.
- ☆ At the end of crop, cooking out at 70°C for 12 h is very essential to eliminate the propagation of any pathogen in the growing rooms in successive crops.

Use of Chemicals

It will be better to avoid the use of chemicals, if it is necessary use as mentioned below:

- ☆ Benomyl (benlate 50 WP) = Mix 240g/100m^2 with casing or add in water at 240g/200 litres/100m^2 during first watering.

  Or

- ☆ Carbendazim (bavistin) = same as for benoyml.

  Or

- ☆ Chlorothalonil (Bravo or Repulse) = Apply as spray 1 week after casing and repeat not less than 2 weeks later at 200 g in 100-200 litres/100m^2.

Conclusion

It is obvious from the information summarized above that mushrooms are adversely affected by a large number of diseases caused by biotic and abiotic agents/factor. Diseases are more in seasonal and satellite farms where unpasteurized compost is used and cultivation is done under most unhygienic conditions. Unless and until strict hygiene measures are followed and mushroom cultivation is undertaken under improved conditions, fungal diseases and competitor moulds will continue to pose a serious threat for successful cultivation and getting quality mushroom, only under special unavoidable situations, the fungicides are to be used but to a minimum level so that the residual levels do not limit their export which has a vast potential at present.

References

Atkins, P. and Atkins, F.C. (1971). Major diseases of the cultivated white Mushroom. *Mush. Grow. Assoc.*, pp. 1–20.

Beach, K. and Kovacs, G. (1977). Mycogone: A serious disease pathogen of mushroom. *Horticultural Aabstracts*, 47: 10573.

Beach, K. and Kovacs, G. (1981). Differences in germination ability and reduction to external *conditions in Mycogone perniciosa and V. Fungicola. Mush. Sci.,* 11: 381–392.

Beach, K., Jacobsen, B.D. and Kovacs, G. (1989). Investigations on the influence of temperature on growth and spore formation of *Mycogone perniciosa* and *Verticillium fungicola* two pathogenic fungi of the cultivated mushroom. *Mushroom Science, 739–751.*

Chaze, J. and Sarazin, A. (1935). Le parasitisme du champignon de couche par la mole est un phenomene reversible. Comptes rendus hebdomadaire des seances de I' Acadamie des Sciences, Paris, 200(21): 1781–1783.

Chaze, J. and Sarazin, A. (1936). Nouvelles donees biololgiques et experimentales sur la mole, maladie due champignon de couche. *Annals des Sciences Naturelles,* 10, 84 pp.

Chen, C., Cai, F. and Chang, Z. (1981). Two *Verticillium* species parasitic to *Agaricus* (Lange) Sing. *Acta Phylopathologica Sinica,* 11(4): 43–47.

Dhar, B.L. (1998). Important mushroom diseases and their management. In: *Pathological Problems of Economic Crop and their Management,* (Ed.) S.M. Paul Khurana. Scientific Publishers, Jodhpur, pp. 535–555.

Dough, T.C. and Hung, C.H. (1971). Studies on the bubble disease of cultivated mushroom. In: *Proc. 3rd Symp. Mush. Res. Imp. Prod.,* pp. 213–232.

Dragt, J.W., Gells, F.P., De-Bruijn, C. and Van Griensven, L.J.L.D. (1996). Intracellular infection of the cultivated mushroom *Agaricus bisporus* by the mycoparasite *Verticillium fungicola. Mycol. Res.,* 100: 1082–1086.

Eicker, A. (1984). A report on the use of thiabendazon for the control of fungal pathogens of cultivated mushrooms. *South African Journal of Botany,* 33: 179–183.

Figueiredo, M.B. and Mucci, E.S.F. (1985). Diseases and pests of edible mushroom (*Agaricus campestris* L.). *Biologico,* 51(4): 93–111.

Fletcher, J.T. (1977). Mushrooms, moulds and management. *Mushroom Journal,* 56*:* 252–256.

Fletcher, J.T. and Ganny, G.W. (1968). Experiments on the biology and control of *Mycogone perniciosa. Mushroom Science,* 7: 221–237.

Fletcher, J.T. and Hims H.J. (1981). Dry bubble disease control. *Mush. J.,* 100: 138–139.

Gams, W. and Hoozemans, A.C.M. ()1970. *Cladobotryum konidienformen* von *Hypomyces* Arten. *Persoonia,* 6: 95–110.

Gandy, D. and Spencer, D.M. (1976). The use of chlorothalonil for the control of benzimidazole tolerant strains of *Verticillium fungicola. Scientia Horticulturae,* 5: 13–21.

Gandy, D.G., Spencer, D.M. and Bisset, P. (1980). Control of dry bubble *Verticillium funigicola* by fungicides. Investigations into biological control of *V. fungicola.* False truffle disease of the edible mushroom. Report, Institutes, Departments of Agriculture. Pp 145–148.

Gandy, D.G. (1972). Observations on the development of *Verticillium malthousei* in mushroom crops and the role of cultural practices in its control. *Mushroom Science,* 7: 171–181.

Gandy, D.G. and Spencer, D.M. (1978). Fungicides for the control of *Mycogone perniciosa* (Magn.), the cause of wet bubble on the cultivated mushroom. *Scientia Horticulturae,* 8: 307–313.

Gandy, D.G., Duncan, C.W. and Edevards, R.L. (1954). Control of mushroom diseases. *Mushroom Sci.*, 2: 167–175.

Gea, F.J., Tello, J.C. and Honrubia, M. (1997). *In vitro* sensitivity of *Verticillium fungicola* to selected fungicides. *Mycopathologia*, 136(3): 133–137.

Geijin, J. Van de. (1976). Fungal diseases in practice. *Dieleman–Van Zaayen*, 7: 939– 943.

Geijin, J. Van. (1977). Practical control of *Verticillium* and *Mycogone. Champignon*, 186: 7–9.

Glaser, T. and Gapinski, M. (1976). Mycoses of mushrooms in the Poznan district. *Rocz-AR-Poznauiu*, 85: 59–66.

Gray, D.J. and Morgan-Jones (1981). Host-parasite relationships of *Agaricus brunnescens* and a number of mycoparasitic hypomycetes. *Mycopathologia*, 75(1): 55–59.

Han, Y.S., Kim, D.S., Jun, B.S. and Shin, K.C. (1974). Some factors affecting growth of *Mycogone perniciosa* Magn. causing wet bubble in cultivated mushroom. *Korean Journal of Mycology*, 2(1): 1–6.

Han, Y.S., Shin, K.C. and Kim, D.S. (1974). Some biological studies on *Mycogone perniciosa* Management. Causing wet bubble in cultivated mushroom, *A. bisporus* (Lange) Sing. *Korean Journal of Mycology*, 2(1): 7–14.

Hsu, Hoang-Kao and Han, You-Hsin (1981). Physiological and ecological properties and chemical properties and chemical control of *Mycogone perniciosa*. *Mushroom Science*, 11: 403–425.

Kaul, T.N., Kachroo, J.L. and Ahmed, N. (1978). Diseases and competitors of mushroom farms in Kashmir Valley. *Indian Mush. Sci.*, 1: 193–203.

Labrousse, F. (1936). Les maladies Verticilliennes du champignon de couche. (Premiere note). *Revue de Pathologic Vegetale et d'entomoligie agricole de France*, 23(2): 162–172.

Lambert, E.B. (1930). Studies on the relation of temperature to the growth, parasitism, thermal death points and control of *Mycogone perniciosa*. *Phytopathology*, 20: 75–83.

Lelley, J. (1980). Study on the possibility of controlling dry bubble (*Verticillium fungicola*) by treating the casing layer with Orbivet. *Champignon*, 225: 13–20.

Nair, N.G. (1975). Observations on three important diseases of the cultivated mushroom in New South Wales. *Plant Disease Survey*, p. 30–32.

Nair, N.G. and Baker, H.J. (1977). *Fusarium moniloforme*. J. Sheld var. intermedium Neish and Leggett and *Neurospora* sp. Benzimidazole fungicides control wet bubble disease in mushroom. *Agricultural Gazette of New South Wales*, 88(3): 24–25.

North, L.H. and Wuest, P.J. (1993). The infection process and symptom expression of *Verticillium* disease of *Agaricus bisporus*. *Can. J. Plant Pathol.*, 15: 74–80.

Olivier, J.M. and Guillames, J. (1976). Effect antagonists exerce *in vitro* par le mycelium de Psalliota bispora Lange vis-a-vis de differentes especes fongiques et bacteriennes. *Ann. Phytopathol.*, 8: 213–231.

Russel, P. (1984). Sporgon on mushrooms. *Mushroom Journal*, 141: 199–300.

Samuels, G.J. and Johnston, P.R. (1980). Benomyl and the verticillium diseases of cultivated mushrooms. *New Zealand Journal of Agricultural Research*, 23: 155–157.

Seth, P.K., Kumar, S. and Shandilya. (1973). Combating dry bubble of mushrooms. *Indian Hort.*, 18(2): 17–18.

Seth, P.K. (1977). Pathogens and competitors of *Agaricus bisporus* and their control. *Indian Journal of Mushrooms*, 3: 1, 31–40.

Sharma, S.R. (1991). Mycoflora of casing soils. In: *National Symposium on Mushroom*, Thiruvananthapurum, 22–24 January. *Indian Mushroom Proceedings*. Thiruvananthapurum, Kerala Agriculture University, pp. 56–57.

Sharma, S.R. (1991). Mycoflora of casing soils. In: *Indian Mushrooms*, (Ed.) M.C. Nair. KAV, Vellanikosa, pp. 56–58.

Sharma, S.R. (1992). Compost and casing mycoflora from mushroom farms of northern India. *Mushroom Res.*, 1: 119–121.

Sharma, S.R. (1995). Management of mushroom disease. In: *Advances in Horticulture, Vol. 13: Mushroom*, (Eds.) K.L. Chadha and S.R. Sharma. Malhotra Publishing House, New Delhi, pp. 195–238.

Sharma, S.R. and Vijay, B. (1991). Microflora from compost: Casing samples collected from different growing areas. *Annual Report*, NCMRT, Solan, p. 35–37.

Sharma, S.R. and Vijay, B. (1996). Prevalence and interaction of competitor and parasitic moulds in *Agaricus bisporus. Mushroom Res.*, 5: 13–18.

Sharma, V.P., Sharma, S. R. and Jandaik, C.L. (1997.)Efficacy of formaldehyde against some common competitors and mycoparasites of mushroom. *Mushroom Research*, 6(2): 83–88.

Smith, F.E.V. (1924). Three diseases of cultivated mushroom. *Trans. Br. Mycol. Soc.*, 10: 81–97.

Spadafora, V.J., Wuest, P.J. and Rinker, D.L. (1989). Development of *Verticillium* disease in Pennsylvania mushroom crops. *Mushroom Science*, 12: 753–763.

Tewari, A.K. (1990). Chemical control of undesirable fungi encountered during cultivation of *Agaricus bisporus* (Lange) Singh. *Thesis, Master of Science in Agriculture (Plant Pathology)*, G.B.P.U.A.&T., Pantnagar, 82 p.

Tewari, A.K. and Singh, R.P. (1991). Studies on undesirable fungi encountered from the beds of *Agaricus bisporus* (Lange) Singh. *Adv. Mushroom Sci.*, p. 21.

Thapa, C.D. and Jandaik, G.L. (1984). Studies on dry bubble disease of cultivated mushroom (*Agaricus bisporous*) in India. Incidence, symptoms, morphology and pathogenicity. *Indian Journel of Mushroom*, 10: 50–60.

Tu, C.C. and Liao, Y.M. (1987). Major diseases of cultivated mushroom and their control in Taiwan. *Mushroom Science*, 12(11): 615–626.

Tu, C.C. and Liao, Y.M. (1989). Major diseases of cultivated mushroom and their control in Taiwan. *Mushroom Science*, H: 627–636.

Zaayen A.V 1983. Fungal diseases. *Mushroom Journal*, pp. 155–156.

Zaayen, A. and Rutjens, A.J. (1981). Thermal death points for two *Agaricus* species and for the spores of some major pathogens. *Mush. Sci.*, 11(11): 393–402.

Zaayen, A. Van and Geijin, J. Van. (1979). New possibilities for control of dry bubble disease, caused by *Verticillium fungicola*, in mushroom cultivation. *Champignon*, p. 12–13.

Plant Diseases Management in Horticultural Crops (2011) *Pages 358–371*
Editors: **Shahid Ahamad, Ali Anwar and P.K. Sharma**
Published by: **DAYA PUBLISHING HOUSE, NEW DELHI**

Chapter 28

Cloning, Sequencing and Transformation of Coat Protein Gene of Papaya Ringspot Virus

A.S. Byadgi*[1]*, M. Kengnal*[1]*, S. Kunklikar*[1]*,
***V.B. Nargund*[1] *and Jameel Akhtar*[2]**

[1]*Department of Plant pathology, University of Agricultural Sciences, Dharwad – 580 005, Karnataka*
[2]*Birsa Agricultural University, Ranchi, C.G.*

Introduction

Papaya (*Carica papaya* L.) is an important fruit crop grown in many countries across the globe. It is largely used in food processing and pharmaceutical companies. In India it is grown in almost all the states but commercially cultivated in Maharashtra and Karnataka. Papaya ringspot disease caused by *Papaya Ring Spot Virus* (PRSV) is rapidly spreading in India causing up to 100 per cent loss to commercial growers and papaya based industries. PRSV is an important pathogen of papaya and cucurbits. The virus is classified into two biotypes that have virions, which cannot be distinguished in serological tests, but differ in their ability to infect papaya (Purcifull *et al.*, 1984). PRSV-P naturally infects papaya and is a major limiting factor in papaya production worldwide (Purcifull *et al.*, 1984).

The virion is flexuous rod measuring 760-800 nm x 12 nm. It consists of positive sense single stranded RNA with 9000 to 10,326 nucleotides in length excluding the poly 'A' tail encapsidated by coat protein of molecular weight 30 – 36 kD. It has a large open reading frame that starts at nucleotide position 86-88 and ends at the positions 10118 to 10120 encoding a polyprotein of 3344 amino acids. The genome is monocistronic expressing a large polyprotein, which is subsequently processed into individual functional proteins. CP sequences have been determined for several isolates of PRSV-P from different parts of the world, together with the smaller numbers of PRSV-W isolates (Bateson *et al.*, 1994; Wang *et al.*, 1994; Jain *et al.*, 1998; Silva-Rosalis *et al.*, 2000). Nucleotide and amino acids sequence divergence of up to 14 per cent and 10 per cent respectively, has been reported between these isolates. Although initial data from the USA and Australia (Bateson *et al.*, 1994) suggested that there was little variation in PRSV within these countries, more recent, albeit limited, data from Mexico (Silva-Rosales *et al.*, 2000), Brazil (Gene Bank) and India (Jain *et al.*, 1998) have suggested there may be greater sequence variation within countries.

Among the pathogen derived resistance coat protein mediated resistance was the first of its kind, exploited practically and found successful in tobacco. Two models have been proposed to explain how the presence of transgene derived coat protein inhibits the early stages of virus disassembly (Register *et al.*, 1989). First, the presence of transgene-derived coat protein in the cytoplasm of a cell may disturb the disassembly-assembly equilibrium of incoming virus particles in favour of assembly and thus effectively inhibit uncoating. Second, there may be some kind of receptor or uncoating site within cells that is responsible for initiating the disassembly of virus particles. Transgene-derived coat protein is then envisaged to block this site prior to infection and preventing virions from attaching to it.

Coat protein interaction would have the potential to suppress translation of the viral RNA-dependent RNA polymerase (RdRp) that is encoded in the 5′ open reading frame (ORF). However, it is possible that CPMP inhibits cell-to-cell movement of PVX, for which CP is an essential cofactor (Chapman *et al.*, 1992). According to Baulcombe (1996), inhibition of virion disassembly, inhibition of systemic virus movement and interaction of the CP required for virus infection cycle are the modes of expressing resistance derived from viral pathogen. Since, these have been proved experimentally using CP of Tobacco Mosaic Virus (TMV) in tobacco against TMV. The mechanism of CP mediated resistance proved its efficiency in field trails of papaya carrying CP gene of PRSV against PRSV incidence in Jamaica. Where, the transgenics expressed acceptable commercial performance and prevention of uncoating of infected virion was presumed (Teenant *et al.*, 2005).

Cloning of PRSV cp Gene

The goal of developing of any transgenic plant is based on the isolation and cloning of the gene of our interest and evaluation of the same for further transfer in to plant system, which would do desired expression in the field performance.

Cloned CP gene of PRSV was introduced into a binary cosmid vector pGA482GG having T-DNA region, flanked by NPTII and GUS genes for confirmation of transgenic, later transformed into a disarmed *Agrobacterium tumefaciens* strain C58 (Fitch *et al.*, 1993). The CP gene of PRSV-YK a severe mosaic strain originated from Taiwan was cloned along with a β-GUS behind a CaMV promoter and NOS terminator was constructed in a Ti-vector with NPT II selection marker. The Construct was transferred to *Agrobacterium tumefaciens* LBA4404 by triparental mating method (Bau *et al.*, 2003). Brazilian isolate of PRSV CP gene was cloned by Lima *et al.* (2002). They isolated total RNA from PRSV infected papaya plant and Reverse Transcriptase (RT) was performed using antisense primer (5′-AGCTAACCATGGGCGAGTATTCA GTTGCGC-3′). A total of 5μl of RT was used as the template for PCR using 100 ng of each sense (5′-ATCATTCCATG GGCGTGTTCCATGAATCAA-3′) and antisense (5′-agcta accatgggcgagtattcagttgcgc-3′) primers. The PCR product was separted on 1 per cent agarose gel electrophoresis. The product, CP gene obtained was cloned using the pGEM-T vector System I (Promega, Medison,WI) and transferred into *Esherichia coli* strain XLI, which formed blue and white colonies. Where, white confirmed of transformation.

Kunkalikar (2003) cloned CP gene of PRSV- Hydrabad strain in to pTZ57R vector of 2.88kb size with T overhang. The vector was transformed into *E. colli* strain DH5-α, multiplied and finally confirmed the transformation by white and blue colony development of which, white confirms the transformants due to presence of IPTG + X-gal in the medium. Cloning and engineering of PRSV VE and LA strains CP genes was designed by Fermin *et al.* (2004). RT-PCR was performed with collected samples using homologous primers. The PCR products were digested with *Noc I* and cloned into the cloning vector pGMM a derivative plasmid of pBlue Script II SK (+), with the leader sequence of CMV to allow for expression *in planta*, the cloned CP gene was subcloned into a plant expression vector cassette pUC 18cp express, there by putting the transgene under the control of the 35S double enhancer promoter of CaMV. After digestion with *Hind III*, the expression cassette was subcloned into the plant transformation vector pGA482G without GUS gene.

Magdalita *et al.* (2004) cloned CP gene of PRSV Philippine isolate at Kasetsart University in Thailand using pCP-LBP plasmid vector. The RNA isolated from PRSV infected leaves was converted to DNA and analyzed by PCR using MB11 and MB12 primers specific to PRSV-CP gene. The DNA was digested with *Bam H I* and *Xba I* to detect the presence of the insert, which helps in ligation with the plasmid vector. Later they sequenced the gene and studied the homology with other poty viruses.

Total RNA from PRSV TH was isolated and cDNA was synthesized by Reverse Transcriptase and subjected to PCR using 100ng of primer 5′-GCTGGGCCCATATGTGTCTG-3′ with the help of *Not I* primer adaptor. The cp gene product obtained was cloned in vector pCR® 2.1 (Invitrogen, Carlsbad, CA) (Souza Jr and Gonsalves, 2005).

Kengnal (2009) cloned CP gene of PRSV-UASD strain at University of Agricultural Sciences, Dharwad using pTZ57R/T a multiple restriction site cloning vector (2.88 kb size) with T overhang. Total RNA from PRSV infested squash leaves was extracted using Rneasy Plant mini kit and cDNA was synthesized using Omniscript RT kit (50) from Qiagen through RT-PCR. PRSV-UASD-*CP* was amplified using specific primers and a *CP* gene specific to ~ 850 bp band was confirmed in electrophoresis. The clones were confirmed by PCR amplification and restriction analysis using *Xba* I and *Bam* HI enzymes. The sequencing was done by universal M 13 primers. The confirmed clones were used for transformation in to papaya.

Sequencing and Comparison of Nucleotide Sequences of PRSV Coat Protein Gene

Development of pathogen derived resistance against plant viruses is gaining much popularity. Coat protein mediated, movement protein mediated, replication mediated and gene silencing are some of the mechanisms being found potential in this regard.

Coat protein mediated has been proved successful in deriving resistance against tobacco mosaic virus in tobacco (Powell Abel *et al.*, 1986). The success of this technique depends on relatedness of transgene with challenge virus strain (Bateson *et al.*, 1994; Tennant *et al.*, 1994; Clark *et al.*, 1995). It is essential to compare the coat protein nucleotide and amino acids sequences which provide evidence for variability in strains and could become a deciding factor for development of cp mediated transgenic plants.

Nucleotide sequence comparison of 3′ terminal regions of severe, mild and non-papaya infecting strains of PRSV was given by Wang and Yeh (1992). They compared sequences of PRSV HA 5-1 with other published sequences and found 2 strains sharing 99.4 per cent similarity in the 3′ terminal 2235 residues. They differed in 10 residues at NIb gene, resulting in changes of 5 amino acids and 2 residues in coat protein gene. Untranslated 3′ regions were identical but HA contained 2 extra nucleotides (AG) at 3′ extreme. Comparison with non papaya infecting PRSV-W strain shared 97.9 per cent similarity in 3′ terminal 2235 residues. PRSV-W and HA differed in 40 nucleotides in the coding region, resulting in 4 amino acid changes in the NIb gene and 6 in CP gene, which differed in 7 nucleotides in the 3′ untranslated region.

The origin of Australian PRSV type P and its variability was determined by Bateson *et al.* (1994). They sequenced six Australian and three Asian isolates and compared with previously reported four sequences. They observed 12 per cent variability at amino acid level among Asian isolates and between Asian and Australian isolates. No significant variability was noticed among Australian isolates rather they were closely related to each other. They noticed that between isolates coat protein core region was highly conserved and N terminus was most variable, suggesting its involvement in host specificity. C-terminus was highly conserved between isolates. They concluded the origin of PRSV type P was from type W rather an introduction.

The most common PRSV-YK strain of Taiwan was studied by Wang *et al.* (1994). They constructed cDNA library of genomic RNA of the virus. A clone of 2.96kb corresponding to the 3′ region of YK RNA had 2960 nucleotide residues and part of NIb gene, cp gene and 3′ non coding region. The sequence was compared with 3′ regions of Hawaiian severe strain PRSV-HA, mild strain PRSV-HA 5-1, non infecting PRSV-W (Florida strain) and Australian PRSV-W type. The cp region showed 89.6-90.3 per cent nucleotide similarities and 93.2-94.8 per cent amino acids similarities with others. 3′ non coding region showed 91.4 and 96.4 per cent similarities.

PRSV isolates type P and type W were characterized from India (Pune) by Jain *et al.* (1998). The comparative sequence analysis showed that 3′ UTRs in isolates P and W were 209 nucleotides in length excluding poly (A) tail, and shared 96 per cent similarity. The *CP* genes of were also similar with 87 per cent nucleotide and 93 per cent amino acids similarity. The variability in *CP* genes was confined to amino terminus of the gene. The DAG triplet associated with aphid transmissibility was present in *CP* of W isolate and was replaced by DAD in P isolate. Three Mexican isolates of PRSV were cloned and sequenced for their coat proteins and compared with other eleven isolates form different geographical locations by Silva Rosales *et al.* (2000). Mexican isolates shared higher similarity with Australian and American isolates than Asian isolates. Bateson *et al.* (2002) studied evolution and molecular epidemiology of PRSV-P. They sequenced coat protein genes of both PRSV P and W type isolates from Vietnam, Thailand, India and Philippines and compared with sequences of 28 isolates already published. They could find that mutation together with long distance movement had contributed for variation in isolates and evolution of type P from W.

Roberto *et al.* (2002) studied the homology of twelve Brazilian isolates of PRSV. The *CP* gene from these isolates were compared, they shared an average homology of 97.3 per cent at the nucleotide level among Brazilian isolates. When compared to 27 isolates from outside the Brazil in a homology tree, the Brazilian isolates were clustered with Australian, Hawaiian, Central and North American isolates, with an average degree of homology of 90.7 per cent among them.

Marilla *et al.* (2003) compared *CP* nucleotide sequences of three mild strains with three severe strains of PRSV-W type. Mild strains shared 98 to 100 per cent identity. When compared with severe strains the identity ranged form 93 to 95 per cent, except in case of PRSV-W-2R, which resulted from reversion of the mild strain PRSV-W-2. The *CP* sequence of the reverting strain showed 100 per cent identity with the sequence of its parental strain. An insertion of six nucleotides in the core region of the *CP* gene, which reflected the addition of two amino acids (Asn and Asp) in the deducted sequence of the protein was found in all mild strains.

Hema and Prasad (2004) compared the *CP* gene of south Indian strain (INP-UAS) PRSV with other reported sequences. They found N termini were variable and distinctiveness of INP-UAS was linked to its geographical location. Phylogenetic analysis also showed that the INP-UAS strain coat protein gene was relatively divergent from those of other PRSV-P isolates as it formed a separate and distinct group. This strain had deletion of 24 nucleotides that corresponded to eight amino acids in the N-terminal region of the *CP*.

Bag *et al.* (2007) studied the sequence diversity in the coat proteins of 28 papaya ringspot virus isolates originating from different locations in India. They found heterogeneity in *CP* gene length (275-289 amino acids) among the isolates from central, eastern, northern, southern and western India (up to 23 per cent). Among all the isolates, KA4, INU-01 and AP2 from southern India were unique. Maximum heterogeneity was observed in southern isolates (up to 23 per cent), followed by central (up to 11 per cent), eastern and northern (up to 10 per cent) and western (up to 7 per cent) isolates. Lack of relationship between variability and geographical origin of the isolates was also found.

Kenganal (2009) also studied the sequence diversity in the coat protein gene of 19 papaya ringspot virus isolates obtained from 19 different geographical locations. The sequence homology obtained in BLAST revealed highest homology of 94 per cent with *CP* gene sequences of PRSV isolates H (Hydrabad) and B (Bangalooru) followed by 93 per cent with Karnataka (Dharwad) strain

In vitro Regeneration in Papaya

Papaya being a dioecious plant is not so easy to regenerate *in vitro* due to its latex content in much of the plant parts and also the differential response of different parts of the plant earmarked for development of efficient protocols of regeneration, which later can be employed for genetic engineering purpose too.

Different auxins tested for callus induction revealed that efficiency of NAA was excellent followed by 2,4-D and IAA. Addition of 1mg/l NAA was found sufficient for good callus growth from 8 days old *in vitro* grown papaya seedling stem (Arora and Singh, 1978). Rapid clonal propagation was envisaged (Litz and Conover, 1978) using apices of matured field grown plants on MS medium containing 50µM Kinetin and 10µM NAA for callus induction. After two months, transferred to a proliferation medium containing 2µM BAP and 1µM NAA, which caused a 7 fold increase in total number of plants within 3 weeks. Rooting was induced by subculturing plantlets on MS + 0.5-15µM IBA. Finally successful transfer of papayas to soil was accomplished after hardening.

Shoot tips and lateral buds from female plants were tried for propagation of papaya cv. "Coorg Honey Dew". Explants were established on MS medium containing 0.5-10µM NAA + 25-50µM Kinetin, multiplication of shots on MS + 2µM BAP and rooting on MS + 10µM IBA (Rajeevan and Pandey, 1983). Lateral buds of papaya inoculated *in vitro* on MS + 4µM zeatin and 8µM 2ip induced better shoot production. Rooting was observed with 20µM IBA, but plantlets so produced remained stunted (Rajeevan and Pandey, 1986). Rapid clonal multiplication was attempted by Drew (1988) using axillary buds on Drew and Smith (DS) medium. Initially on liquid medium, followed by solid medium containing 1µM BAP + 1µM NAA. After 2-3 subcultures growth of axillary buds occurred on DS medium containing 1µM 6-BAP + 0.25µM NAA, enhanced growth was found on DS medium containing 2µM BAP + 0.5µM NAA. Roots initiated on medium containing 10µM IBA and were hardened in a high humidity (90 per cent) cabinet under glasshouse. Similar attempt was done by Winnaar (1988), he used shoot apices of nursery and orchard trees of papaya cv. "Sunrise Solo" for establishment on Murashige and Tucker basal medium with half strength inorganic salts + 0.5mg/l 6-BAP and

0.2mg/l NAA. Established explants were multiplied on same basal medium containing 0.5 mg/l BAP and 0.1mg/l NAA, which rooted on medium containing 4.0 mg/l IBA.

Immature zygotic embryos from open pollinated and selfed *Carica papaya*, collected at 90 to 114 days post-anthesis produced somatic embryos on apical domes and radical meristems after three weeks incubation on half strength MS medium + 5mg/l coconut water + 0.1 to 25 mg/l 2,4-D + 400 mg/l glutamine and 6 per cent sucrose. Zygotic embryos became embryonic and yielded somatic embryos after five months, which later germinated on MS medium containing 5 mg/l kinetin and grew large enough, shoots were rooted on vermiculite and grown in greenhouse (Fitch and Manshardt, 1990).

Mondal *et al.* (1990) tried *in vitro* propagation of papaya with shoot buds. They inoculated shoot buds from saplings and fruit bearing trees for establishment on MS medium containing 1mg/l GA3 and 2mg/l kinetin. Calli was developed on MS medium containing 1mg/l NAA and 3 mg/l kinetin followed by multiplication on MS + 0.1 mg/l NAA and BAP 0.5 mg/l and finally rooting on MS + 2 mg/l IBA. Reuveni *et al.* (1990) propagated the dioecious papaya clones *in vitro*. Establishment and proliferation was achieved on MS medium supplemented with 0.5 mg/l BAP and 0.1 mg/l NAA, addition of adenine sulphate 160 mg/l improved multiplication and shoot growth. Elongation was obtained on MS supplemented with 1.0 mg/l kinetin and 0.05 mg/l NAA and finally rooting was achieved on half strength of MS supplemented with 1.0 mg/l IBA. High rooting per cent age and high quality adventitious root systems were achieved by exposing shoots to a medium containing 10µM IBA for 2 days before transfer to a hormone-free medium containing 31 µM riboflavin and incubation for 2 days under dark before transferring to light (Drew *et al.*, 1993).

Petiole explants were found successful for establishment of callus and regeneration of plantlets through organogenesis (Hossain *et al.*, 1993). Seeds of three to four months after anthesis were collected; immature internal integuments were cultured on three successive media M1 (2,4-D 9µM), M2 (2,4-D 0.009µM + Kin 1µM) and M3 (M2 + ABA 0.5 µM), which yielded a friable calli, capable of producing embryos of maternal origin (Monmarson *et al.*, 1995). Jordan and Velozo (1996) noted the establishment of embryos after incubation of axillary buds initially on WPM medium containing 2,4-D + BAP/zeatin for six days followed by incubation on growth regulator free medium with continuous shaking at 50 rpm for three months. The cell suspension yielded a large number of somatic embryos directly at same growth conditions upon second subculture. High frequency maturation of embryonic calli was observed by Bernard *et al.* (1998) in liquid phase medium. This yielded significantly higher frequency of somatic embryogenesis and regeneration on MS medium supplemented with 3 per cent sucrose and 0.5µM ABA.

Mature zygotic embryos of *Carica papaya* L. of "Sunrise Solo" were inoculated on MS medium supplemented with 2 mg/l 2,4-D and incubated under dark at 25°C. Direct somatic embryogenesis was observed from hypocotyledonary epidemic cells after 25 days and indirect somatic embryogenesis was observed from embryonic superficial cells of pre embryonic complexes located on peripheral and internal cell layers of cultures after 49 days (Fernando *et al.*, 2001). The auxillary buds from ten-month-old plants were found best suited for establishment *in vitro* and they broke out early with minimum contamination on MS medium supplemented with 1.0 mg/l BAP (Dineshbabu *et al.*, 2002).

Agnihotri *et al.* (2004) used young inflorescence tips of female and male plants as explants on different media for different stages of regeneration on MS basal *viz.*, B1 supplemented with 0.1mg/l BAP + 0.25 mg/l IAA + 0.25 mg/l CCC + 40mg/l AdS, and 50mg/l L-glutamine gave contamination free establishment (shoot proliferation) of cultures up to 100 per cent by avoiding intervening callusing.

The developed shoots were nurtured on B2 medium supplemented with BAP 0.01mg/l + IAA 0.5 mg/l + AdS 15 mg/l. Nurtured shoots having sufficiently multiplied were subjected to pulse treatment with 10mg/l IBA (24hrs exposure) followed by rooting via three steps (step1: Bm3+ 10mg/Liba + 2 per cent sucrose for 24hr liquid culture, step2: 0.1mg/d-biotin + 0.1mg/l CCC + 0.25 mg/lIAA + 15 mg/l AdS + 2 per cent sucrose for 7 days semisolid medium and step3: BM3 + 1mg/l thiamine HCl + 0.5mg/l pyridoxine-HCl + 0.5mg/lnicotinic acid +10mg/l L-arginine-HCl +50mg/l L-glutamine + 0.25mg/lIAA. Finally the rooted shoots were transferred to sollrite for hardening.

Strong apical dominance and internal microbial contamination was a major impediment in meristem tip culture of papaya and none of the growth regulators yielded any favorable *in vitro* bud establishment. Hence, heading back of fruiting plants at the height of 75 cm was most effective to enhance the lateral bud sprouting followed by pre excision spraying with decontaminants for minimizing the contamination (Bajpai, 2004). Rimberia *et al.* (2005) derived female plants via embryo induction through anther culture on MS medium supplemented with 0.1mg/l BA and 0.1 mg/l NAA, pretreatment of embryos with water for a day was found efficient in embryo induction.

Development of Transgenic in Papaya Resistant to PRSV

Absence of host plant resistance, specificity in cross protection, and inability of mild strains to keep away the severe strain for longer period has diverted the attention towards developing new ways of resistance against the PRSV. Since, the mild strains developed severe symptoms, especially during winter period, when employed for cross protection. Though the introduction of resistant gene from wild papaya in to commercial cultivars has also been attempted but, no success was achieved (Manshardt and Wenslaff, 1989(b)).

However the advent of genetic engineering in 1990's thrown light on this problem. The concept of parasite derived resistance (PDR) proposed by Sanford and Johnson (1985) suggests that expressing genetic materials of a pathogen in a host would disrupt the essential pathogenic processes and result in resistance against the pathogen.

The first demonstration of this involved coat protein (CP) mediated resistance reported by Powell Abel *et al.* (1986). Here the CP gene of tobacco mosaic virus was transferred to tobacco, and symptom development was delayed in transgenic plants, when challenged with TMV. Since then, there have been numerous reports demonstrating that transgenic plants with viral CP genes are resistant to virus infection (Beachy *et al.*, 1990; Fitchen and Beachy, 1993). The level of resistance ranged from delay in the symptom development to immunity and the spectrum of resistance ranged from narrow to broad. The CP-mediated protection has proved innovative and adoptable method for control of plant viruses.

Transformation of PRSV-CP Gene in Papaya

The advent of genetic engineering during 1990s shed new light on successful expression of viral coat protein gene in its host, which upon integration into the host conferred resistance to the virus invaded. The practical utility of this mechanism in tobacco (Powel-Abel *et al.*, 1986) motivated to work out for the possibility of similar resistance in papaya against PRSV. *Agrobacterium* mediated transformation was achieved successfully using pTiB6S3: pM0N200/pTiB6S3 with high efficiency. Transformants were identified by screening resistance against 300µg/ml of kanamycin. Putative transformants were confirmed on the basis of nopaline production (Pang and Sanford, 1988). It paved a standardized procedure and protocol for further attempts.

The first transgenic papaya plants were regenerated from somatic embryos bombarded with DNA-coated microprojectiles (Fitch *et al.*, 1990). Immature zygotic embryos and hypocotyls section were used as explants with tungsten particles carrying chimeric NPT II and GUS genes for bombardment. Both the explants were cultured on MS medium containing 10 mg/l 2,4-D, 400 mg/l glutamine and 6 per cent sucrose. Ten putative transgenic isolates produced somatic embryos and five regenerated leafy shoots after nine months, out of which three showed expression of both NPT II and GUS activity.

Immature zygotic embryos treated with 2-4, D were bombarded with plasmid construct that contains NPT II and GUS genes flanking a PRV CP gene expression cassette. The bombarded tissue yielded a total of 20 kanamycin resistant embryo clusters over a period of two-year culture period. Only five lines regenerated into plantlets, three of which produced abnormal shoots with broom shaped leaves that resembled due to virus or herbicide induced effects. The other two lines K 19-1 and S 3-2 appeared normal (Fitch *et al.*, 1992). The integration of CP gene was confirmed by expression of NPT II, GUS genes, PCR and Northern analysis. The same gene cassette was attempted (Fitch *et al.*, 1993) for transformation by *Agrobacterium* mediation, using pGA482GG binary cosmid vector and disarmed C 58 strain of *A. tumifaciens.* The putative transformants were obtained on 150µg/ml kanamycin selective medium and were confirmed similar to that of earlier tests.

A system for the production of transgenic papaya (*Carica papaya* L.) plants using zygotic embryos and embryogenic callus as target cells for particle bombardment was described by Cabrera *et al.* (1995). Phosphinothricin (PPT) and kanamycin (NPTII) resistant genes were used as selectable markers, and GUS as a reporter gene. Selection with 100 mg/l kanamycin and 4 mg/l phosphinothricin yielded a total of over 90 resistant embryogenic colonies, where embryogenic callus was target tissue. The developed plants expressed the transgenes in the entire assay.

Regeneration of transgenic papaya (*Carica papaya* L.) has been hampered by the low rates of transformation by conventional *Agrobacterium* infection or microprojectile bombardment. Hence, transformation based on wounding of cultured embryogenic immature zygotic embryos collected 75-90 days after pollination with carborundum (600mesh) in liquid phase, was devised (Cheng *et al.*, 1996). The embryogenic tissue was vertexed with carborundum powder in sterile water for one minute before treating with *A. tumifaciens* consisting CP gene of PRSV Taiwan strain. The developed somatic embryos were transferred to the medium containing NAA, BAP and Kanamycin and subsequently regenerated into normal plants.

In order to establish a model system for introduction of foreign genes into papaya (*Carica papaya* L.) by *Agrobacterium* mediated transformation, petioles from multi shoots were used as explant. Bacterial neomycin phosphotransferase II (NPTII) gene and glucuronidase (GUS) gene were used as a selection marker and a reporter, respectively. Cross sections of papaya petioles grown *in vitro* were co-cultured with *Agrobacterium* for 2 days and putative transformed callus were identified on selection medium containing kanamycin and carbenicillin and consequently regenerated into plants via somatic embryogenesis and finally confirmed the transgenics by PCR (Yang *et al.*, 1996).

Mahon *et al.* (1996) developed an efficient transformation and regeneration system for secondary somatic embryos of an Australian papaya cultivar. Highest transient expression levels were obtained for zygotic embryos using a helium pressure of 700 kPa with the target tissue at a distance of 13.5 cm from filter unit containing the carrier particles. Somatic embryos had highest transient expression levels when bombarded using 500 kPa pressure at a distance of 7.5 to 10 cm from the filter unit containing the particles. Pre treatment of both zygotic and somatic embryos before bombardment with osmoticum improved transformation efficiency and transient expression. Untranslatable CP gene of PRSV HA 5-1 was transformed into pre synchronously developed somatic embryos (using spreading technique) by particle bombardment. The developed plants were resistant to homologous PRSV isolates from Hawaii, Australia, Taiwan, Mexico, Jamaica, Bahamas and Brazil (Gonsalves *et al.*, 1998). A reproducible and effective biolistic method for transforming papaya with PRSV CP gene was developed by Cai Wenqi *et al.* (1999). They used somatic embryo tissue aroused from excised immature zygotic embryos.

The CP gene of PRSV YK, a severe mosaic strain was transferred in to immature zygotic embryos of papaya cv. Tainung No.2 through *Agrobacterium tumefaciens* using liquid-phase wounding with carborundum. The transformed cells were germinated on the germination medium for 2 to 4 weeks, which later multiplied and rooted on MS medium supplemented with different growth regulators (Bau *et al.*, 2003). The coat protein gene of PRSV EV from Venezuela and PRSV LA from Lagunillas were transformed into papaya cv. Thailand Red, using *A. tumefaciens* strain LB 4404. The regenerated plants showed effective protection against homologous and heterologous isolates of PRSV (Fermin *et al.*, 2004). Magdalita *et al.* (2004) did an efficient transformation of PRSV CP gene using microprojectile bombardment (ballistic gene gun) into papaya somatic embryos. Using transient expression, a pressure of 1200 kPa and a distance of 12.5 cm were established as best conditions for bombardment.

Confirmation of Transformation/Transgenic Plants

In *Agrobacterium* mediated transformation of PRSV CP gene in to papaya, putative transformants were obtained on kanamycin selective medium after 2-4 weeks. Primers specific to PRSV CP (upstream primers MO928, 5'TACCGGTCTGAATGAGGAAGC3' and down stream primers MO1008, 5'GTGCATGTCTCTGTTGACAT3') amplified the 0.82kb DNA corresponding to the size of the CP gene. The western blotting analysis from the extracts of calli and shoots of putative transformants using anti-PRSV serum as the primary antibody, detected the CP gene product which confirms the integration of CP gene into papaya (Cheng *et al.*, 1996). In a similar attempt, transformants of PRSV CP gene from petiole explants were regenerated on kanamycin 150mg/l selective medium. They confirmed the integration of CP gene by amplification of 674 nt bases of GUS gene and same were reacted strongly with GUS-specific probe when analyzed by Southern blotting, both the expressions were absent in non-transformants (Yang *et al.*, 1996).

In a six-month study under glasshouse conditions, repeated mechanical inoculation of the virus was carried out on papaya carrying PRSV *CP* gene. The plants expressed a consistent resistance over entire six-month period. In field conditions severe epiphytotic was developed to check the resistance. All the plants carrying *CP* gene remained symptomless, but most of them died at the end of one-year succumbing to fungal root infection reflecting a reduced ability to remove soil moisture from deeper layer of soil. Two years after the first manual inoculation none of the control plant remained alive, where as the *CP*-R0 clone 55-1 performed exceptionally well in the field, showing virtual immunity to PRSV infection throughout the course of experiment. Occasional minor leaf mottling was reported on this clone, but mottling was transient, and there was no ELISA evidence that it was caused by PRSV. At the end of the experiment, 12 of the 20 *CP* engineered plants were still alive and without symptomless since eight plants died due to root rot (Lius *et al.*, 1997).

Papaya lines carrying replicase gene were obtained on kanamycin selection medium. The regenerated putative plants yielded a total of 1,602-bp length of replicase (RP) gene of PRSV under PCR amplification. Apart from which, the Southern blot analysis also proved positive confirming the integration of RP gene into host. These plants, carrying RP gene were cross-inoculated with PRSV on every 10 days interval. All the transgenic papayas showed diverse levels of resistance to PRSV ranging from mild symptoms to symptomless normal growth upto two years, displaying normal leaves (Chen *et al.*, 2001). Putative trasnformants of papaya having PRSV YK *CP* gene were obtained on kanamycin 100 µg/ml selection medium followed by PCR confirmation using forward primer MO926 (5'-TCTAAAAATGAAGCTGTGGA-3') and reverse primer MO1008 (5't-GTGCATGTCTCTGTTGACAT-3'). The PCR product confirmed the presence of 0.84 kbp CP gene of PRSV YK. The northern blot analysis too dictated the presence of transgene transcripts in the regenerated plants. Under glass house study of transgenic plants, where they were challenge inoculated with the corresponding severe strain, a variety of classified response was noticed, such as susceptible, moderately resistant, resistant and immune reactions with varied delay in symptom expression and severity (Bau *et al.*, 2003).

The papaya transformants containing the CP gene of Venezuelan PRSV isolate were developed and evaluated by Fermin *et al.* (2004). The R0 plants obtained on kanamycin selection medium showed no evidence of NPTII protein production by ELISA. However the CP gene was amplified by PCR and detected by Southern analysis. The R1 CP-transgenic papaya obtained from R0 crosses were tested for NPTII and inoculated with three isolates of PRSV. All 146 test plants were negative for NPTII gene by ELISA and amplification by PCR. In further analysis of challenge inoculation, 53 per cent of the plants derived in R1 showed resistance to PRSV EV, LA or HA strains. When these plants were challenged for second time with PRSV EV, half of the plants showed resistance. The similar reaction was noticed when third time challenged.

To avoid the real and perceived problems associated with chemical based selection for plant transformation, a visual selectable marker to produce transformed papaya by particle bombardment was designed by (Zhu *et al.*, 2004). They used Green Fluorescent Protein (GFP) gene as a selection marker, which upon integration into papaya callus, produced GFP protein and appeared bright and distinct under UV light visualization. The protein not only helped to identify the transformants but also increased the efficiency of transformation by 15-24 folds.

Confirmation and Evaluation of Transgenic Plants

In an exclusive experiment on field performance of transgenic papaya lines carrying coat protein gene of PRSV Taiwan (Bau *et al.*, 2004), test plants were subjected to constant attack by viruliferous aphids for a period ranging from 16 to 18 months, a normal papaya production cycle in Taiwan. A second trial was conducted under extremely high inoculum pressure. The field response of transgenics expressed high level of resistance to PRSV infection, compared with limited effectiveness of control strategies, such as cross protection and genetic PRSV tolerance identified in papaya germplasm. About 20 to 30 per cent of transgenic plants developed mild symptoms caused by PRSV and confirmed by ELISA. These plants grew normally and produced marketable fruit. In some cases plants with few chlorotic spots on upper leaves recovered and became symptomless under summer conditions. Subsequent analysis by ELISA detected no PRSV in the recovered plants.

Different reactions were observed between transgenic lines carrying either CPT or CPNT gene under high disease pressure; earlier ones exhibited mild symptoms to symptomless appearance and were moderately resistant to PRSV, while lines carrying CPNT were as susceptible as nontransformed controls. Some CPNT lines showed mild reactions and were considered weakly resistant. These lines could be regarded as tolerant to PRSV infection, provided the continued production of acceptable fruits during the field tests (Tennant *et al.*, 2005).

References

Agnihotri, S., Singh, S. K., Jain, M., Sharma, M., Sharma, A. K. and Chaturvedi, H. C. (2004). *In vitro* cloning of male and female *Carica papaya* through tips of shoots and inflorescences. *Indian Journal of Biotechnology*, 3: 235–240.

Arora, I. K. and Singh, R. N. (1978). Growth hormones and in vitro callus formation of papaya. *Scientia Horticulturae*, 8: 357–361.

Bag, S., Surekha Agrwal and Jain, R.K. (2007). Sequence diversity in the coat proteins of papaya ringspot virus isolates originating from different locations in India. *Indian Phytopathology*, 60: 244–260.

Bajpai Anju, Ramesh Chandra and Gorakh Sing. (2004). Optimizing various pre–treatment for *in vitro* bud establishment in papaya micropropagation. *Progressive Horticulture*, 36: 309–312.

Bau, H–Jiunn, Cheng Ying–Huey, Yu Tsong–Ann, Yang Sherng and Yeh Shyi–Dong. (2003). Broad spectrum resistance to different geographic strains of Papaya ringspot virus in Coat Protein gene Transgenic Papaya. *Phtopathology*, 93: 112–120.

Bau, H–Jiunn, Cheng Ying–Huey, Yu Tsong–Ann, Yang Jiu–Sherng Liou Pan–Chi, Hasiao Chi–Hsiung, Lin Chien–Yih and Yeh Shyi–Dong. (2004).Field evaluation of transgenic papaya lines carrying the coat protein gene of papaya ringspot virus in Taiwan. *Plant disease*, 88: 594–598.

Bateson, M, F., Henderson, J., Chaleeprom, W., Gibbs, A. J. and Dale, J. L. (1994). Papaya ringspot potyvirus: isolate variability and the origin of PRSV type P in Australia. *Journal of General Virology*, 75: 3547 – 3553.

Bateson, M. F., Rosemarie, E. L., Peter Revill, Worawan, C., Coung, V. Ha, Gibbs, A. J. and Dale, J. L. (2002).On the evolution and molecular epidemiology of the poty virus Papaya

Beachy, R. N. S. Loesch–Fries and Tumer, N. E. (1990). Coat protein–mediated resistance against virus infection. *Annual Review of Phytopathology*, 28: 452 – 474.

BAULCOMBE, C. D. (1996). Mechanisms of pathogen derived resistance to viruses in transgenic plants. *The Plant Cell*, 8: 1833–1844.

Bernard, C., Smith, M. A. L. and Yadava, U. L. (1998). *Journal of Horticultural Science and Biotechnology*, 73: 307–311.

Cabrera P., J. L., Vegas Garcia, A. and Herrera Estrella, L. (1995). Herbicide resistant transgenic papaya plants by as efficient particle bombardment transformation method. *Plant Cell Reports*, 15: 1–7.

Cai Wenqi, Gonsalves Carol, Tennant Paula, Fermin Gustavo, Souza Manoel, Jr., Sarindu Nonglak, Jan Fuii–Jyh, Zhu Hai–Ying and Gonsalves Dennis. (1999). A protocol for efficient transformation and regeneration of *Carica papaya* L. *In vitro Cell Development Biology–Plant*, 35: 61–69.

Chapman, S., Hills, G.J., Watts, J. and Baulcombe, D. C. (1992). Mutational analyses of the coat protein gene in potato virus X: Effects of virion morphology and viral pathogenicity. *Virology*, 191: 223–230.

Chen, G., Ye, C. M., Huang, J. C., Yu, M. and Li, B. J. (2001). Cloning of the Papaya ringspot virus replicate gene and regeneration of the PRSV–resistant papaya through the introduction of the PRSV replicate gene. *Plant Cell Reports*, 20: 272–277.

Cheng, H. Y., Yang, J.S. and Yeh, D. S. (1996). Efficient transformation of papaya by coat protein gene of papaya ringspot virus mediated by *Agrobacterium* following liquid –phase wounding of embryogenic tissue with carborundum. *Plant Cell Reports*, 16: 127–132.

Clark, W. G., Fitchen, J., Nejidat, A., Deom, C. M. and Beachy, R. N. (1995). Studies on coat protein mediated resistance to tobacco mosaic virus. II. Challenge by a mutant with altered vision surface does not overcome resistance conferred by TMV coat protein. *Journal of General Virology*, 34: 475–483.

Dineshbabu, K., Sathiamoorthy, N., Chezhiyan, N., and Kapil Deo N Singh. (2002). *In vitro* establishment of gynodioecious papaya variety CO–7 influenced by age of mother plants. *The Orissa Journal of Agriculture*, 30: 5–7.

Drew,R. A. (1988). Rapid clonal propagtion of papaya *in vitro* from mature filed grown trees., *Hort Science*, 23: 9 –11.

Drew, R. A., Mc Comb, J. A. and Considine, A. J. (1993). Rhizogenesis and root growth of *Carica papaya* L. *in vitro* in relation to auxine sensitive phases and use of riboflavin. *Plant Cell, Tissue and Organ Culture*, 33: 1–7.

Fermin Gustavo, Inglessis Valentina, Garboza Cesar, Rangel Sairo and Dagert Manuel, (2004).Engineered Resistance Against papaya ringspot virus in Venezuelan transgenic papayas. *Plant Disease*, 88: 516–521.

Fernando, A. J., Melo Murilo, Marli, K.M., Soares and Beatriz Appezzato–Da–Gloria (2001). Anatomy of somatic embryogenesis in *Carica papaya* L. *Brazilin Archives of Biology and Technology*, 44: 247–255.

Fitch, M. M. M. and Manshardt, R. M. (1990).Somatic embryogenesis and plant regeneration from immature zygotic embryos of papaya (*Carica papaya* L.). *Plant Cell Reports*, 9: 320–324.

Fitch, M. M. M., Richard, M. M., Gonsalves, D., Jerry, L. Slightom and John, C. Sanford. (1990).Stable transformation of papaya via microprojectile bombardment. *Plant Cell Reports*, 9: 189–194.

Fitch, M. M. M., Richard, M. M, Gonsalves, D., Jerry, L. Slightom and John, C. Sanford. (1992). Virus resistant papaya plants derived from tissue bombarded with the coat protein gene of papaya ringspot virus. *Biotechnology*, 10: 1466–1472.

Fitch, M. M. M., Richard, M. M, Gonsalves, D. and Jerry, L. Slightom. (1993). Transgenic papaya plants from *Agrobacterium* –mediated transformation of somatic embryos. *Plant Cell Reports*, 12: 245–249.

Fitchen, J. H. and Beachy, R. N. (1993). Genetically engineered protection against viruses in transgenic plants. *Annual Review of Microbiology*, 47: 739–763.

Gonsalves, C., Cai, W., Tennant P. and Gonsalves D. (1998).Effective development of papaya ringspot virus resistant papaya with untranslatable coat protein gene using a modified projectile transformation method. *Acta Horticulture*, 461: 311–314.

Hema, M. V. and Prasad, D. T. (2004).Comparison of the coat protein of a south Indian strain of PRSV with other strains from different geographical locations. *Journal of Plant Pathology*, 86: 35–42.

Hossain, M., Rahman, S. M., Islam, R. and Joarder, I. O. (1993). High efficiency plant regeneration from petiole explants of *Carica papaya* L. through organogenesis. *Plant Cell Reports*, 13: 99–102.

Jain, R. K., Pappu, H. R., Pappu, S. S., Verma, A. and Ram, R. D. (1998).Molecular characterization of papaya ring spot virus isolates from India. *Annals of Applied Biology*, 132: 413–425.

Jordan, M. and Velozo, J. (1996). Improvement of somatic embryogenesis in highland papaya cells suspensions. *Plant Cell Tissue and Organ Culture*, 44: 189–194.

Kengnal, M. (2009). Transformation studies on papaya ring spot virus (PRSV) coat protein gene in Papaya. *Ph.D. Thesis*, University of Agricultural Sciences, Dharwad, p. 210.

Kunkalikar, S. (2003).Molecular characterization, cloning of coat protein gene, epidemiology and management of Papaya ringspot virus, *Ph.D. Thesis*, University of Agricultural Sciences, Dharwad, p. 109.

Lima, A. C., Roberto, Souza Jr, Monoel, T., Pio–Ribeiro Gilvan and Lima, A. Albersio. (2002). Sequence of the coat protein gene from Brazilian isolates of Papaya ringspot virus. *Fitopatol. Bras*, 27: 263–267.

Litz, R. E. and Conover, R. A. (1978). *In vitro* propagation of papaya. *Hort Science*, 13: 241–242.

Lius, S., Manshardt, M. R., Fitch, M. M. M., Slightom, J. L., Sanford, C. J. and Gonsalves D. (1997).Pathogen–derived resistance provides papaya with effective protection against papaya ringspot virus. *Molecular Breeding*, 3: 161–168.

Magdalita, M. Piblito, Valencia, D. Loloita, Ocampo, T. I. D. Anna, Tabay, T. Reynaldo, and Villegas, N. Violeta. (2004). Towards development of PRSV resistant papaya by genetic engineering: *Proceedings of the 4th International Crop Science Congress, Brisbane, Australia*, www. cropscience. org. au

Mahon, R. E., Bateson, M. F., Chamberlain, D. A., Higins, C. M., Drew, R.A. and Dale, L.J. (1996).Transformation of an Australian variety of *Carica papaya* using Microprojectile Bombardment. *Australian Journal of Plant Physiology*, 23: 679–685.

Manshardt, R. M. and Wenslaff, T. F. (1989(b). Interspecific hybridization of papaya with other *Carica* species. *Journal of American Society of Horticultural Science*, 114: 689 – 694.

Marilla, G. S., Della Vecchia, Luis, E. A. C. and Jorge, A. M. R. (2003). Nucleotide sequence comparison of the Capsid protein gene of severe and protective mild strains of papaya ringspot virus. *Fitopatolgica Brasil*, 28: 678–681.

Mondal Mousumi, Gupta Sukumar and Mukharjee Baran Barid(1990).*In vitro* propagation of shoot buds of *Carica papaya* L. (Caricaceae) var. Honey Dew., *Plant Cell Reports*, 81: 609 – 612

Monmarson, S., Nicole Michaux–Ferriere and Teisson, C. (1995). Production of high frequency embryogenic calli from integuments of immature seeds of *Carica papaya* L. *Journal of Horticultural Sciences*, 70: 57–64.

Pang, S. Z. and Sanford, J.C. (1988). Agrobacterium–mediated gene transfer in papaya. *Journal of American Society of Horticultural Science*, 113: 287–291.

Powell–Abel, P., Nelson, R. S., De, B., Hoffman, N., Rogers, S. G. Fraley, R.T. and Beachy, R. N. (1986). Delay of disease development in transgenic plants that express the tobacco mosaic virus coat protein gene. *Science*, 232: 738 – 743.

Prucifull, D. E., Simone, G. W., Baker, C. A. and Hiebert, E. (1984). Papaya ringspot virus. *CMI/AAB Description of Plant viruses*, No. 292, p. 8.

Rajeevan, M.S. and Pandey, R. M. (1983). Propagation of papaya through tissue culture. *Acta Horticulture*, p 131–139.

Rajeevan, M.S. and Pandey, R. M. (1986). Lateral bud culture of papaya (*Carica papaya* L.) for clonal propagation. *Plant Cell, Tissue and Organ Culture*, 6: 181–188.

Register, J.C., Powell, P. A., Nelson, R. S. and Beachy, R. N. (1989). Genetically engineered cross protection against TMV interferes with initial infection and long distance spread of he virus. In: *Molecular Biology of Plant Pathogen Interactions* (Staskawicz, B., Ahlquist, P., and Yoder, O. eds.) New York, Liss. pp. 269–281.

Reuveni, Q., Ahlesinger, D. R. and Lavi, U. (1990). *In vitro* clonal propagation of dioecious *Carica papaya* L. *Plant Cell, Tissue and Organ Culture*, 20: 41–46.

Rimberia, K.F., Sunagawa. Haruki, Urasaki Naoya, Ishimine Yukio And Adaniya, Shinichi. (2005). Embryo induction via anther culture in papaya sex analysis of the derived plantlets. *Scientia Horticulture*, 103: 199–208.

Roberto, C. A. L., Manoel, T. Souza J. R., Gilvan Pio–Ribeiro and J. Albersio, A. L. (2002). Sequences of the coat protein gene from Brazilian isolates of Papaya ringspot virus. *Fitopatologia Brasileria*, 27: 263–267.

Sanford, J. and Johnston, S. (1985). The concept of pathogen derived resistance: Deriving resistance genes from the parasite's own genome. *Journal of Theoretical Biology*, 113: 395–405.

Silva Rosales, L., Becerra Leora, N., Ruiz Castra, S., Teliz Oritz, D. and Carrazana, J. C. (2000). Coat protein sequence comparison of three Mexican isolates of Papaya ringspot virus with other geographical isolates reveal a close relationship to American and Australian isolates. *Archives of Virology*, 145: 835–843.

Souza Jr. M.T. and Gonsalves D. (2005). Sequence similarity between the viral cp gene and the transgenic in transgenic papayas. *Pseq: Agropec. Bras., Brasilia*, 40: 479–486.

Tennant, P. F., Gonsalves, C., Ling, K., Fitch, M., Manshardt, R. and Gonsalves, D. (1994). Differential protection against papaya ringspot virus isolates in coat protein gene transgenic papaya and classically cross–protected papaya. *Phytopathology*, 84: 1359–1366.

Tennant, P. F., Ahamad, M. H. and Gonsalves, D. (2005). Field Resistance of coat protein transgenic papaya ringspot virus in Jamaica. *Plant Disease*, 89: 841–847.

Wang, C. H. and Yeh, S. D. (1992). Nucleotide sequence comparison of the 3'teriminal regions of severe, mild and non–papaya infecting strains of papaya ringspot virus. *Archives of Virology*, 127: 345–354.

Wang, C.H., Bau, H. J. and Yeh, S. D. (1994). Comparison of the nuclear inclusion b protein and coat protein genes of five Papaya ringspot virus strains distinct in geographical origin and pathogenicity. *Phytopathology*, 84: 1205–1210.

Winnaar, De Wilna. (1988). Clonal propagation of papaya *in vitro*. *Plant Cell, Tissue and Organ Culture*, 12: 305–310.

Yang, J.S., Yu, T. A. Cheng, Y. H. and Yeh, S. D. (1996). Transgenic papaya plants from *Agrobacterium*–mediated transformation of petioles of *in vitro* propagated multi shoots. *Plant Cell Reports*, 15: 459–464.

Zhu, Y. J., Agbayani, R. and Moore, P. H. (2004). Green fluorescent protein as a visual selection marker for papaya (*Carica papaya* L.) transformation. *Plant Cell Reports*, 22: 660–667.

Plant Diseases Management in Horticultural Crops (2011) *Pages 372–384*
Editors: **Shahid Ahamad, Ali Anwar and P.K. Sharma**
Published by: **DAYA PUBLISHING HOUSE, NEW DELHI**

Chapter 29

Characterization and Management of Papaya Ring Spot Virus

A.S. Byadgi*[1]*, M. Kengnal*[1]*, S. Kunklikar*[1]*,
***V.B. Nargund*[1] *and Jameel Akthar*[2]**

[1]*Department of Plant Pathology,*
University of Agricultural Sciences, Dharwad – 580 005, Karnataka
[2]*Birsa Agricultual University, Ranchi, C.G.*

Papaya ringspot disease, a deadly disease of papaya has become cosmopolitan in all papaya growing regions. No part of any papaya orchard has left without this disease across the globe. It has threatened entire papaya industry to the tune of major impact on national economies of many papaya growing nations. The disease was first time reported as viral disease on papaya in the island of Ohau in Hawaii (Parris, 1938), which was sap transmissible and later named as " Wailu" disease (Lindner *et al.*, 1945) and shown to be of viral nature by Jensen (1949) from Kailu, island of Ohau in Hawaii.

The global production of papaya is 69,37,094 MT produced in 51 countries worldwide in a total area of 3,71,345 hectares, with an average yield of 18,6810 kg/ha. India stands fourth in the production contributing 12 per cent of global production followed by Brazil, Mexico, and Nigeria Among the states in India, Karnataka ranks fifth with a production of 221,700 tones grown in an area of 3600 ha with an average yield of 167500 kg/ha during 2004-05 (Anonymous, 2006). Current national production is 23.17 lakh tones (Anonymous, 2009).

The crop is being infected by 12 viral, 8 bacterial, 33 fungal, 6 nematodal and 2 phytoplasma diseases (Nishijima, 1999) of which, ringspot disease caused by Papaya Rings Spot Virus (PRSV) has gained global importance in all the papaya growing countries. In India ringspot disease prevails in all the states wherever papaya is cultivated and in Karnataka its severity is more in northern part and recently protruding in southern part as well.

The work carried out on economic importance, distribution, symptomatology, properties of virus and management through host plant resistance, cross protection and coat protein mediated resistance has been reviewed in the following pages

Economic Importance and Distribution

PRSV is becoming increasingly severe in almost all the parts of the world. Farmers and industries depending on papaya are incurring substantial financial loss. It is severe in all the states of India wherever papaya is cultivated. Very severe incidence of ringspot disease has been recorded in different state of Northern India (Khurana and Bhargava, 1970, Surekha *et al.*, 1977, Yemewar and Mali, 1980, Barbosa and Paguio, 1982 and Lokhande *et al.*, 1992).

In Karnataka, a severe incidence of the disease ranging from 50 to 100 per cent had been recorded from northern districts and no disease in southern parts of the state (Byadgi *et al.*, 1995, Shaikh, 1996 and Hegde, 1998). But the recent surveys undertaken in the state indicated wide spread distribution of the disease in entire Karnataka except few districts (Kunkalikar, 2003). The disease has been recorded in 24 districts but three districts *viz.*, Udupi, Hassan and Kodagu were free from the incidence. The disease incidence was more than 75 per cent in Belgaum, Bijapur, Bidar and Gulbarga districts. It ranged from more than 50 to 75 per cent in Dharwad, Bangalore (Urban) and Bagalkot districts. The incidence in the range of more than 25 to 50 per cent was observed in Haveri, Bellary and Bangalore (Rural) districts. Most of the districts *viz.*, Davangere, Mandya, Shimoga, Chamarajnagar, Tumkur, Mysore, Chickmagalur, Uttara Kannada, Kolar, Chitradurga, Raichur, Gadag, Koppal and Dakshina Kannada showed disease incidence in the range from zero to 25 per cent. The disease was not observed in Udupi, Hassan and Kodagu districts. The disease incidence is more up to 100 per cent in Gulbarga, Bidar and Bijapur where papaya is cultivated on a commercial scale.

Kengnal (2009) also recorded Highest incidence of 100 per cent in Bagalkot, Bangalooru (Urban), Bellari, Bidar, Bijapur, Chitradurga, Dharwad, Gadag, Gulbarga and Haveri Districts. Lowest incidence of 10 per cent was recorded from various places of Bangalooru (Rural), Chikkamangalur, Dawangeri, Karwar, Kolhar, Mandya, Mangalore, Mysore, Raichur, shimoga and Tumkur districts. However, the highest average of 92 per cent was recorded from Bidar district. This might be because of the papaya cultivation on large scale in the district during 2002-2005 for both papain as well as for consumption purpose have served better opportunity for multiplication, survival and spread of the virus. Secondarily the warm humid climatic conditions of the district could have favored the vector build up and their movement. Hence, highest incidence was recorded here. The next highest average was found in Gulburga (86.25 per cent) followed by Bijapur (81.29 per cent).

The disease was concentrated mostly in urban areas of districts where papaya is grown only in kitchen gardens. It may be because of cosmopolitan nature of population in urban areas who bring infected papaya material like seedlings and fruits from different regions of the country for propagation and consumption which acts as primary source of inoculum and spread in such areas. The disease was severe in commercial orchards also where extreme symptoms like collapse of infected plants were observed. This may be attributed mainly to cultivation of single variety like Red Lady and planting in close spacing, which facilitate easy movement of aphids from plant to plant during their transitional flights.

The absence of disease in remote places may be due to absence of viruliferous aphids and infected plants as primary source of inoculum in the area. In newly planted big orchards, disease appears mostly on plants in border rows indicating they might have infected by viruliferous aphids from neighboring areas which tried to feed on these plants during transitional flights and in the process inoculated the virus. The per cent incidence of disease is usually low in such orchards. Older orchards will show high incidence (up to 100 per cent) indicating virus must have rapidly spread internally in these orchards once there was primary infection on few plants. The spread of the disease seems to be internal once the primary infection starts from external source. The low disease intensity in southern districts indicates that disease is slowly spreading to these districts from north. It may also be due to slow movement of vectors, weather factors and also type of varieties grown (Kunkalikar, 2003 and Kengnal, 2009).

Papaya ringspot virus is transmitted by aphid vectors. Usually no aphids colonize on papaya but they try to feed on papaya during their transitory flights and in the process virus is transmitted during proboscis made even by single aphid. There is direct correlation between aphid incidence and disease spread. The aphid incidence in turn depends on temperature *i.e.*, when temperature rises above 34- 35°C, no incidence of aphids is observed in the field.

Zhang *et al.* (1995) studied temporal dynamic of papaya ring spot disease. Epidemics of PRSV were monitored in fields. By analyzing the data of seasonal dynamics of disease development, it was shown that the greatest increase in disease rate appeared in early September. During the epidemic season, alate population of aphids occurred in 3 peaks: early June, mid August and mid October. Ten to twenty days after each peak, there was a peak in disease incidence development. The seasonal dynamics of disease development and the growth and decline of aphid population were correlated to wind velocity, temperature and R. H. by comparing the fitness of 3 epidemic models (Logistic, Gompertz and Weibull) to the data of disease incidence. It was found that the disease development was best described by the Gompertz model. With the equation of disease progress and analysis of every parameter and calculating the number of days to reach incidence of 5, 50, 90 and 99 per cent, it was found that the disease epidemic pattern of PRSV belongs to the type of low virus source, low rate of infection and long epidemic period.

The loss in yield of papaya depends on the stage of the crop infected. As usual early infection in seedling stage results in very severe symptoms leading to complete sterility and hence accounts for total loss. The loss varies from 20 to 100 per cent corresponding to late to earliest infection. Papaya plants and fruits infected by PRSV produce not only deformed fruits but also yield lesser quantity of papain from such affected fruits. The virus infection results in a significant decrease in the dry weight (25 per cent), total chlorophyll (76 per cent), total sugars (43 per cent), reducing sugars (33 per cent) and non-reducing sugars (51 per cent) as compared to healthy leaves (Khurana and Bhargava, 1970(a)).

It is well known that virus infection affects pollen fertility in plants. The PRSV infection also leads to decrease in total number of pollens with increased abnormality and sterility per cent age in diseased plant as compared to healthy plants. The flowers from healthy plant will be bigger and normal as compared to diseased plant. Pollen germination per cent age decreases drastically in infected plants accounting for lower yield in diseased plants.

Papaya Rings Spot Virus (PRSV) has been reported from every continent, wherever papaya is grown. Of the total 12 viruses reported on papaya, PRSV is the most destructive and has been the major constraint and limiting factor for its cultivation in all tropical and subtropical regions. In India the first report of PRSV was made by Capoor and Verma (1948) as papaya mosaic, subsequently it was reported in Bihar by Mishra and Jha (1955) as mosaic, from Madhya Pradesh by Garga (1963), from Uttar Pradesh by Khurana and Bhargava (1970(b)). Later Surekha *et al.* (1977) recorded incidence of PRSV in Udaipur of Rajasthan. From Marathwada region of Maharastra, Yemewar and Mali (1980) reported severe incidence of PRSV, from Punjab by Cheema and Reddy (1985) as papaya mosaic virus, from Andhra Pradesh by Susan John (1985) as papaya mosaic. Later Ramakrishna Rao (1988) from Maharashtra stated that the virus causing mosaic, leaf distortion, shoe stringing and rings on fruits of papaya was PRSV – P, a member of potyvirus group. In Karnataka Byagdi *et al*, (1995) recorded a severe incidence of PRSV from northern parts of the state, later Shaikh (1996) and Hegde (1998) recorded the incidence of PRSV from Dharwad and Belgum districts of Karnataka, respectively. Singh *et al.* (2003) recorded a virus on papaya in Eastern Uttar Pradesh causing severe mosaic and distortion of leaves followed by ringspots on fruits and confirmed the virus serologically as PRSV.

The assessment of losses due to incidence of PRSV was recorded oftenly by many workers who came across with the virus. Khurana and Bhargava (1970b) during their survey observed 75 – 100 per cent incidence of PRSV in and around Ghorakhpur district of Uttar Pradesh. Barbosa and Paguio (1982) recorded 70 per cent yield losses due to PRSV. Wan and Conover (1983) surveyed 30 papaya plantations in Southern Florida and observed incidence of the virus up to 100 per cent. Wahab (1991) reported occurrence of PRSV in 26 ha out of 74 ha which was surveyed in Johore, Malaysia. Lokhande *et al.* (1992) surveyed different cultivars of papaya and recorded the incidence of PRSV ranging from 75 – 100 per cent and noted its severity during the rainy season. Different parts of Dharwad and Belgum districts of North Karnataka were surveyed by Shaikh (1996). He recorded the PRSV incidence up to 100 per cent. Hegde (1998) also recorded 100 per cent incidence in different districts of Karnataka. Dahal *et al.* (1997) conducted a three-year (1992, 1993 and 1994) rigorous survey for incidence of PRSV on papaya in Terai and inner Terai districts of Nepal, and found an average incidence ranging from 75 to 100 per cent. Singh *et al.* (2003) surveyed in Eastern Uttar Pradesh and recorded the incidence of PRSV in every district ranging from 48 to 100 per cent.

Symptomatology

PRSV infects all the stages of papaya plants. Leaves exhibit variety of peculiar symptoms depending upon the age of the plant and environmental conditions. The initial symptoms begin with light discolouration of leaves turning towards pale yellow. Mild mosaic, mosaic, mottling, cholorotic spots, chlorotic rings, vein clearing, leaf curling, blisters, leaf distortion, shoe string formation in leaves, pale oily greasy streaks on stem and ringspots are prominently seen on leaves, fruits, stem and hence the virus named as ringspot virus.

Lindner *et al.* (1945) in his early observation found mosaic pattern on leaves and yellow rings on fruits. Later different symptoms like puckering of leaf tissues between veins and veinlets of young leaves and mosaic pattern on expanded leaves (Holmes *et al.*, 1948), swelling of leaf tissues between veins resulting in upward curling of leaves followed by chlorotic mottling, blistering of leaf tissue (Jensen, 1949) were recorded. The other symptoms like chlorotic ringspots, mosaic pattern, shoestring on leaves, blistering, and distortion of leaves in severe case, lobed and semi apocarpus conditions of fruits and overcrowding of leaves were also noticed (Bhandari, 1952). Ringspots on fruits hollow stem and malformed fruits were seen prominent in severely infected orchards (Khurana and Bhargava, 1970(b); Yemewar and Mali, (1980). In the early infested and abandoned orchards, bare plants without any fruit set with no flowering, stunted growth, beheaded appearance, tapered canopy and death of the plant were recorded (Lokhande *et al.*, 1992; Thomas and Dodman, 1993; Hussain and Verma 1994; Shaikh, 1996; Dahal *et al.*, 1997; Kunkaliker, 2003, Ranebennur, 2005). Apart from these, Singh *et al.* (2003) recorded downward turning of leaf margins, elongation and distortion of leaves with scorched appearance.

Properties

Papaya Ringspot Virus (PRSV) belongs to superkingdom Viridae, kingdom ssRNA family *Potyviridae* genus *Potyvirus* and species Papaya Ringspot Virus. The virion is a flexious rod measuring 760 – 800nm x 12nm, consists of positive sense single stranded RNA, with often pinwheel inclusion bodies in the cytoplasm of the infected host tissue. (De La Rosa and Lastra, 1983; Prucifull *et al.*, 1984; Wu *et al.*, 1983; Yeh and Gonsalves, 1985; Fauquet *et al.*, 2005). Prucifull *et al.* (1984) also recorded that PRSV is existing in two major biotypes: biotype P and biotype W. Biotype P is thought to arisen by mutation from biotype W (Bateson *et al.*, 1994). It infects both papaya and cucurbits, where as type W infects only cucurbits but not papaya.

Virion is nonenveloped, flexious rod consisting of positive sense single stranded RNA of 40S with an approximate length of 10 Kb (De La Rosa and Lastra, 1983; Yeh and Gonsalves, 1985). Molecular weight of the genome is 330 KD (Prucifull *et al.*, 1984). Genome is surrounded by coat protein of 32 – 36 KD. Yeh *et al.* (1992) sequenced complete RNA genome of PRSV, found 10326 nt in length excluding poly A tail containing a large open reading frame that starts at 86 to 88 nt position with a highly conserved sequence of AAAUAAAANANCUCAACACAACAVA at the 5′ end of the RNA. They also confirmed 80 per cent genomic similarity between the five isolates they studied. Two cleavage sites of polyprotein were determined by amino acid sequencing of N termini of helper component and cylindrical inclusion proteins. The genetic organization is similar to that of other poty viruses except the first protein processed from the N terminus of the polyprotein. There are about 21 isolates of PRSV have been reported from all over the world being PRSV–P, Su–mm, Su–sm, Su–smm, T-echen, T–wang, HA, HB, F–340, Ecuador, BDAU, BUNAU, DAYAU, WPAU, P–FL, PHAW, SRI, TAIW, THAI, VIET and P–IND. The cryptogram of PRSV is R/1, X/5.5, E/E, S/Ap (Fauquet *et al.*, 2005). PRSV becomes inactive at 54–56°C, which is its thermal inactivation point with a dilution end point of 10^{-3} and longevity *in vitro* of 3h (Conover, 1962). The buoyant density of the virus in CsCl is 1.32 g cm^{-3}. (Fauquet *et al.*, 2005)

Management

Through Host Plant Resistance

In *Carica* genera, none of the species is having complete resistance against PRSV. Although attempts are being made to identifying new sources of resistance, since from the time of devastating effects of PRSV were known. Capoor and Verma (1961) reported that, among species of *Carica*, *C. cauliflora* was immune to PRSV. Conover and Litz (1981) noticed that the plants selected as most tolerant among the Columbian types were sib mated for three generations, which resulted in tolerance from 4 to 55 per cent. Screening of large number of local and foreign papaya cultivars and lines for resistance to PRSV was practiced in India, China, Taiwan and Philippines, but out of them only F1 – TT – 5 cultivar was found to be tolerant to PRSV (Wang, 1982). Zee (1985) reported that a line 356 – 3 selected from Florida accession was the most tolerant. The tolerance was readily transferred to line 356–3 x solo papaya hybrids in a quantitative manner.

Conover *et al.* (1988) observed that Florida strain (356–3) and *C. cauliflora* were the only sources of usable resistance or tolerance to PRSV in the species *Carica papaya*. However, the genetic analysis of tolerance to PRSV disease in Line 356 – 3 has indicated multiple QTLs affecting various components of resistance. Introgression of resistant genes from wild species in to commercial cultivars has been attempted (Khupse *et al.*, 1980; Moore and Litz, 1984; Manshardt and Wenslaff, 1989(a), 1989(b): Chen *et al.*, 1991).

Sondur (1994) reported that resistance to PRSV in papaya was controlled by multiple factors and was quantitatively inherited. Shaikh (1996) screened 12 papaya cultivars against PRSV and observed that only Red Lady cultivar was tolerant to infection yielding good number and size of fruits.

Cross Protection Technique

Cross protection is the phenomenon whereby plants that are systemically infected with a mild strain of a virus are protected against the effects of infection by a more virulent related strain (Yeh *et al.*, 1988). This practice has long been known and used to control many viruses. The component in cross protection programs is the availability of a mild strain, which effectively protects against the severe virus. The practice was discovered by McKinney (1929) first time for Tobacco Mosaic Virus (TMV).

In papaya, several successful attempts have been made for effective control of PRSV using chemicals, botanicals, drugs, oils and nutrients, but none of them could control the virus successfully. Unfortunately *Carica papaya* don't have any host derived resistance against the PRSV. Though some species posses the resistance but are incompatible in interspecific hybridization with cultivated papaya (Cook and Zettler, 1970). Hence the utilization of this novel approach became more suitable in papaya against PRSV and many attempts were begun.

Two mild strains of PRSV were isolated by Lin (1980) from PRSV infected papaya by local lesion isolation on *Chenopodium amaranticolar*. However, these strains were found to be neither stable nor mild in later field trails. Rezende *et al.* (1981) detected a mild strain of PRSV designated as "Aperrecida D` oeste" from Brazil. The isolate was found to have good cross protection ability. Papaya plants preimmunized with this isolate were cross inoculated and then exposed to field conditions for four months. It did not show any increase in severe symptoms, but later it did not perform well. Yeh and Gonsalves (1984) did an attempt to obtain mild strains of PRSV from naturally occurring isolates on papaya trees with mildest symptoms in a heavily infected papaya orchard on the island of Hawaii. They were evaluated by inoculating on papaya and *Cucumis metuliferus* and by ELISA also. Out of 116 isolates collected including an isolate Su-mm from Taiwan, 94 isolates caused severe symptoms at 20 days after inoculation, 10 isolates caused mild mottling at 30 days after inoculation, but all showed severe mosaic or leaf distortion at 50 days after inoculation. Twelve isolates did not show any symptom on test plants and results of ELISA indicated no infection. Thus, no ideal mild strain was isolated from field collection. Isolate Su-mm remained mild in winter, when others were severe. However, it caused severe symptoms in summer. Isolation of mild strains of PRSV by single lesion isolates on *Chenopodium quinoa* was also unsuccessful (Gonsalves, 1989).

Kalyankar (2001) and Badag (2002) studied four mild strains *viz.*, PRSV-M1, PRSV-M2, PRSV-M3 and PRSV-M4 by inoculating on *Chenopodium amranticlolar* and noticed chlorotic local lesions by PRSV-M1 and chlorotic lesions centered with different coloured spots by rest of the three strains. They also concluded that on papaya, cross protection can be seen only if the challenge inoculation was done after 10 days of mild strain inoculation. But, they did not find any consistent results on field performance. Thus, isolation of mild strains of PRSV from natural population was found as a difficult task.

The efforts on natural selection of mild strains in PRSV failed in many instances. The endeavor was shifted to artificial induction of mild strains by mutation. The very first successful attempt of developing a mild strain of PRSV by chemical induced mutation was done by Yeh and Gonsalves (1984). They used a severe strain PRV – HA isolated from Hawaii. Crude sap from PRV – HA infected papaya was treated with nitrous acid (pH 6.0) at 20°C for 30 minutes and inoculated on *Chenopodium quinoa* for evaluation. Single lesions developed on *C. quinoa* 20 – 30 days after inoculation were transferred to papaya seedlings and observed for symptom expression under glass house. All the isolates selected were found severe except two designated as PRV HA 5-1 and PRV HA 6-1, which did not show any prominent symptoms. But, were strongly ELISA positive. They remained symptomless or showed diffuse mottling with no reduction in plant size, hence were considered as mild strains.

Mc Millan and Gonsalves (1987) tested PRV – HA 5-1 under field conditions in Florida, where mild strain was inoculated in large scale using spray gun equipped with 1.2 mm diameter nozzle at 8kg/cm^2 at a distance of 20cm and by hand with a cotton swab, both protected and unprotected blocks were maintained. The unprotected plants showed 6 per cent disease after 10 weeks, increasing up to 100 per cent after 30 weeks. The protected plants showed 0.8 per cent disease after 20 weeks increasing up to 9 per cent after 60 weeks. However, the protected plants produced 3.5 times more fruit than the unprotected plants.

Wang *et al.* (1987) studied same mutant mild strains PRV HA 5-1 and PRV HA 6-1 under glass house as well as field conditions in Taiwan. Both produced mild symptoms on Chenopodiaceae, *Cucurbitaceae* families and on papaya cultivar Tainung No.2. They provided high degree of protection against two severe strains of Taiwan under green house conditions. In field conditions under high disease pressure, unprotected plants showed severe symptoms at 2 –3 months after planting and protected plant showed at 4 – 7 months after planting. Under theses conditions cross protection did not provide economic benefit. However, in protected plants rouging once in every 10 days up to flowering stage reduced the incidence and showed 82 per cent higher yield compared to unprotected plants. Similarly Yeh *et al.* (1988) at Cornell University tested the effectiveness of the mild mutant PRV HA 5 – 1 to protect papaya plants against challenge inoculation with severe PRV HA strain under green house conditions. Papaya seedlings of 5 to 6 leaf stage were preinoculated with PRV HA 5 – 1 and challenge inoculated with severe strain at different intervals. A high proportion of plants remained symptomless even after 2 – 3 months of challenge inoculation, when the interval between protective and challenge inoculation was more than 26 days.

Gonsalves and Garnsey (1989) reviewed the results of initial field trails for control of PRV by cross protection that were conducted in three different locations in southern part of Taiwan. Results from Feng-shan and Kao-Shu locations showed that cross protection would not be economical in Taiwan under heavy pressure. About 80 per cent of the plants showed severe disease reactions within 3 – 5 months after transplantation. Although the occurrence of severe infection was delayed considerably in protected plants, the delay was not enough to give economic benefit. It was clear from these two trails that it would not be economical for a farmer to plant cross protected plants within or in close vicinity of an infected orchard and plants which show break down of cross protection before the flowering stage would generally have very few fruits of economical value.

Sheen *et al.* (1998) tested PRV HA 6-1 in glass house and field conditions. In glass house it provided high degree of cross protection by producing symptomless infection on test plants *viz., Chenopodiaceae, Cucurbitaceae* families and in papaya cultivars. Field trails showed a delay in symptom expression by severe strain up to 1 – 5 months compared to control. A recent approach of transgenic development was also employed by Bang *et al.* (2005) using engineered mild strains for cross protection against severe strain of PRSV in cucurbits.

Specificity of cross protection was evident in Taiwan where first large scale evaluations of the HA 5-1 strain were conducted (Wang *et al.*, 1987). In Hawaii, field trails of induced mild strain in several commercial papaya varieties demonstrated excellent protection (Mau *et al.*, 1990: Ferreira *et al.*, 1993). Trails in commercial farms indicated that a cross protected crop could be grown for 2 – 3 years. It will be economically viable despite high incidence in varying degrees. Since, protection is also dependent on variety being protected and the incidence of native strains in that area. The highest level of breakdown of cross protection by severe strains was noticed in 10 per cent of cross protected plants in 2.5 year.

The green house studies with severe strains from different regions showed that, protection with mild strain PRSV HA 5-1 is largely specific to the Hawaiian isolates of PRSV alone (Tennant *et al.*, 1994).

Cross protection by mild strain was incomplete but could give economic benefit only if applied to whole block of plants that were isolated from severely infected papaya trees. By 1991 more than 1700 hectares of cross protected orchards had been planted in Hawaii (Yeh and Gonsalves, 1984). However, the cross protection with the mild PRSV strain HA 5-1 is currently not recommended in Taiwan because growing papaya under screen enclosures is apparently a more economical practice (Yeh, personal communications). In Thailand, by contrast the mild strain HA 5-1 did not provide a useful level of disease control (Yeh and Gonsalves, 1984), this was probably due to strain differences. Since, comparative green house tests showed that HA 5-1 did not protect against the severe Thai strain, while it gave excellent protection to the severe Hawaiian strain. The green house and field results clearly showed that the specificity of the mild PRSV strain HA 5-1 is limited to Hawaii and Florida. Hence, the unavailability of mild strains under natural conditions, strains specificity in cross protection, reluctance of farmers to infect their trees with virus and adverse effect of mild strains on certain cultivars of papaya has limited their use over large scale area (Gonsalves, 1998). These lacunas in cross protection have triggered the invention of new methods of controlling the PRSV by employing the novel approaches of biotechnology for the development of transgenic plants in papaya which can resist PRSV.

Viral Coat Protein (*CP*) Mediated Resistance

The efforts to develop transgenic papaya to control PRSV begun shortly after the news spread that transgenic tobacco expressing the coat protein gene of tobacco mosaic virus was resistant or showed delay in symptom expression following inoculation with tobacco mosaic virus. In the similar fashion Fitch *et al.* (1992) transferred a cloned CP gene of PRSV into papaya and developed transgenic lines R0 that contained the CP gene showing varying degrees of resistance to PRSV and one line they found completely resistant and this was the first result demonstrated that CP mediated resistance can be executed to a tree species such as papaya. Tennant *et al.* (1994) developed transgenic papaya expressing the coat protein gene of mild papaya ringspot virus strain from Hawaii (PRV HA 5-1). The resultant transgenics showed high level of resistance against the severe strain PRV-HA. Inoculation with the high concentrations of the virus, multiple mechanical inoculations or graft inoculations failed to break the resistance of transgenic papaya.

The expressible coat protein (CP) gene of a Taiwan strain of Papaya ringspot virus (PRSV) was constructed in a Ti binary vector and transformed in to embryogenic callus of the papaya (Cheng *et al.*, 1996). Timothy and Robert (1999) first time in the world undertook a task of producing transgenic papaya seed in Hawaii resistant against PRSV using CP gene of PRSV in Sunset cultivar. They developed a first transgenic line "Sun up" and a second cultivar, F1 hybrid "UH Rainbow", by crossing "Sun up" with "Kapoho Solo". The seeds were developed in bulk and released to the Papaya Administrative Committee (PAC) for commercial cultivation.

The pathogen derived resistance in case of papaya was strain specific and it limits the application of a single transgenic line on large area. Huey *et al.* (2003) developed CP-transgenic papaya lines with broad spectrum resistance, which were potential for use in Taiwan and other geographic areas to control PRSV of different strains. Piblito *et al.* (2004) developed genetically engineered papaya using coat protein gene of Philippine PRSV and regenerated putative transgenic Ro plantlets, which were moderately susceptible but, R1 plantlets derived from R0 were completely resistant.

Integrated Disease Management

Papaya ringspot being a viral disease, it is impossible to cure the infected plants. The only option to reduce incidence or loss is by avoidance. Early infection of papaya will leads to very severe symptoms and maximum loss. Hence early infection should be avoided to prevent sterility and death of infected papaya plants. With these points in view Byadgi *et al.* (2004) suggested an Integrated Management schedule as detailed below.

1. Rouging must be followed right from the seedling stage. Infected plants should be uprooted and burnt to avoid further spread of virus in field.
2. Papaya should be cultivated in isolated area free from diseased papaya plants and collateral host.
3. Papaya nursery should be raised under insect proof net to avoid early infection on seedlings, which helps in improving the yield.
4. Yellow gummed labels and acetic acid traps should be placed in papaya orchards and along the borders. This helps in trapping aphid vectors, which get attracted to yellow color and fruity smell of acetic acid thus reducing their population and spread of the disease.
5. Rows of barrier crop like maize and sorghum should be raised along the borders. This prevents horizontal movement of aphids in orchard.
6. Boron @ 1.5 gm/l should be sprayed from seedling stage which helps in preventing spread of virus particles in host.
7. Spray insecticides around orchard to prevent entry of aphids in papaya plantation.
8. Spray Nimbecidine @ 4 ml/l and groundnut oil @ 20 ml/l. The layer of oil on host disturbs equilibrium between stylet of aphids and host cells and thus prevents aphid proboscis and consequently the disease spread.
9. PRSV infects cucurbitaceous plants. Thus their cultivation in and around papaya orchard could be prevented to avoid survival of PRSV and its spread.
10. Coat protein mediated transgenic and resistant varieties can be grown if available. Two transgenic Sun up and Rainbow hybrids are under commercial cultivation and performing well even at very high disease pressure.

References

Anonymous (2006). http: //www.agribusinessdwd.com/Papaya/Statistics.pdf

Anonymous (2009). National Horticultural Board, 2009, January, p. 42.

Badag, K.A. (2002)). Further Studies on cross protection ability of Papaya ringspot virus (PRSV) strains in papaya, *M.Sc. (Agri.) Thesis*, Marathawada Agricultural University, Parbhani, p. 60.

Bang, J. Y., Chu-Hui Chang, Li-Fang Chen, Wei-Chih Su and Shyi-Dong Yeh. (2005). Engineered mild strains of papaya ring spot virus for broader cross protection in cucurbits. *Phytopathology*, 95: 533–538.

Barbosa, F. R. and Paguio, O. R. (1982). Incidence and effect of Papaya ringspot virus on yield of papaya (*Carica papaya* L.). *Fitopathologia Brasiliera*, 7: 365 – 373.

Bateson,M, F., Henderson, J., Chaleeprom, W., Gibbs, A. J. and Dale, J. L. (1994).Papaya ringspot potyvirus: isolate variability and the origin of PRSV type P in Australia. *Journal of General Virology*, 75: 3547 – 3553.

Bhandari, M. M. (1952). Abnormal papaya fruits. *Biology and Medicine*, p. 43.

Byadgi, A. S., Kunklikar, S., Mokashi, A. N. and Kulkarni, V. R. (2004). *Papaya Ringspot Virus Disease and its Management.* Technical Bulletin– 1, Department of Plant Pathology, University of Agricultural Sciences, Dharwad 580 005, Karnataka, India, 19 pp.

Byadgi, A. S., Anahosur, K. H. and Kulkarni, M. S. (1995). Ring spot virus in papaya. *The Hindu*, 118: 28.

Capoor, S. P. and Verma, P. M. (1948). A mosaic disease on *Carica papaya* L. in the Bombay province. *Current Science*, 17: 265 – 266.

Capoor, S. P. and Verma, P. M. (1961). Immunity of papaya mosaic virus in genus *Carica*. *Indian Phytopathology*, 14: 96 – 97.

Cheema, S. S. And Reddy, R. S., 1885, Studies on the transmission of Papaya mosaic virus by *Rhopalosiphum maidis*. *Indian Journal of Virology*, 1: 49 – 53.

Cheng, H. Y., Yang, J.S. and Yeh, D. S. (1996). Efficient transformation of papaya by coat protein gene of papaya ringspot virus mediated by *Agrobacterium* following liquid –phase wounding of embryogenic tissue with carborundum. *Plant Cell Reports*, 16: 127–132.

Conover, R. A. (1962). Virus diseases of Florida. *Phytopathology*, 52: 6.

Conover, R. A. and Litz, R. E. (1981). Tolerance to Papaya ringspot virus in papaya (*Carica papaya* L.). *Phtyopathology*, 71: 868.

Conover, R. A., Litz, R. E. and Malo,S. E. (1988). *Cauliflora* – Papaya ringspot virus tolerant papaya from South Florida and Caribbean. *Horticultural Science*, 2: 1072.

De La Rosa and Lastra, R. (1983). Purification and characterization of Papaya ringspot virus. *Phytopathology Zurich*, 106: 329 – 336.

Ferreira, S. A., Mau, R. F. L., Manshardt, R., Pitz, K. Y. and Gonsalves, D. (1993). Field evaluation of papaya ringspot virus cross protection. *Proceedings of the 25th Annual Papaya Industry Association Conference*, pp. 14–19.

Fauquet, C. M., Mayo, M. A., Mainloff, J., Desserlberger, U. and Ball, L. A.. (2005). Virus Taxonomy: Eight Report of the International Committee on Taxonomy of Viruses. Elsevier, Academic Press, San Diego.

Fitch, M. M. M., Richard, M. M, Gonsalves, D., Jerry, L. Slightom and John, C. Sanford. (1992). Virus resistant papaya plants derived from tissue bombarded with the coat protein gene of papaya ringspot virus. *Biotechnology*, 10: 1466–1472.

Garga, R. P. (1963). Studies on virus diseases of plants in Madhya Pradesh: A serious virus disease of papaya. *Indian Phytopathology*, 16: 31 – 33.

Gonsalves, D. (1989).Cross protection techniques for control of plant viruses in the Tropics. *Plant Disease*, 73: 592 – 597.

Gonsalves, D. (1998). Control of Papaya ringspot virus: A case study. *Annual Review of Phytopathology*, 36: 415 – 437.

Gonsalves, D. And Garnsey, S. M. (1989). Cross protection techniques for control of plant virus diseases in tropics. *Plant Disease*, 73: 542 – 597.

Hegde, Prakash(1998). Further studies on Papaya ring spot virus., *M. Sc. (Agri) Thesis*, University of Agricultural Sciences, Dharwad, p. 87.

Holmes, F. O., Hendrix, J. W., Ikeda, W. Jensen, D. D., Lindner, R. C. And Storey, W. B. (1948).Ringspot of papaya (*Carica papaya*) in the Hawaiian islands. *Phytopathology*, 38: 310 –312.

Huey, J.., Ying–Huey Cheng, Tsong–Ann Yu, Jiu–Sherng Yang and Shyi–Dong Yeh (2003). Broad spectrum resistance to different geographic strains of papaya ringspot virus in coat protein gene transgenic papaya. *Phytoptaholog*y, 93: 112–120.

Hussain, S. and Varma, A. A. (1994). Occurrence of Papaya ringspot virus from Amritsar (Punjab), *India. Journal of Phytopathology*, 7: 77 – 78.

Jensen, D. D. (1949). Papaya virus disease with special reference to Papaya ring spot. *Phytopathology*, 39: 191 – 210.

Kalyankar, M. G. (2001). Studies on Papaya ringspot virus (PRSV) strains and their ability of cross protection in papaya, *M.Sc. (Agri) Thesis*, Marathawada Agricultural University, Parbhani, p. 47.

Kengnal, M. (2009). Transformation studies on papaya ring spot virus (PRSV) coat protein gene in Papaya. *Ph.D. Thesis*, University of Agricultural Sciences, Dharwad, p. 210.

Khupse, S., Hendre, R., Mascarenhas, A., Jagannathan, V., Thombre, M. and Joshi, A. (1980). Utilization of tissue culture of isolate interspecific hybrids in *Carica papaya* L. In: *National Seminar on manipulation of plant tissue culture, genetic manipulation and somatic hybridization of plant cell.* (Rao, P., Heblle, M. and Chade, M. eds.), Bhaba Atomic Research Center, Bombay, India, pp. 198 – 205.

Khurana, S. M. P. and Bhargava, K. S. (1970(a), Effect of plant extract on the activity of the three papaya viruses. *Journal of General Applied Microbiology*, 16: 225–230.

Khurana, S. M. P. And Bhargava, K. S. (1970(b), Induced apocarpy and "double papaya fruit" formation in papaya with distortion ringspot virus infection. *Plant Disease Reporter*, 54: 181–183.

Kunkalikar, S. (2003). Molecular characterization, cloning of coat protein gene, epidemiology and management of Papaya ringspot virus, *Ph.D. Thesis*, University of Agricultural Sciences, Dharwad, p. 109.Lin, C. C. (1980). Strain of Papaya ringspot virus and their cross protection. *Ph.D. Thesis*, National Taiwan University, Taipei. pp. 63–85.

Linder, R. C., JENSON, D. D. aND IKEDA, W. (1945). Ring spot new papaya plunderer. *Hawaii Farm and Home*, 8: 10 – 12.

Lokhande, N. M., Moghe, P. G., Matte, A. D. and Hiware, B. J. (1992). Occurrence of Papaya Ring Spot Virus (PRSV) in Vidharbha regions of Maharastra. *Journal of soils crops*, 2: 36 – 39.

Manshardt, R. M. And Wenslaff, T. F. (1989(a), Zygotic polyembryony in interspecific hybrids of *Carica papaya and C. cauliflora. Journal of American Society of Horticultural Science*, 114: 684 – 689.

Manshardt, R. M. And Wenslaff, T. F. (1989(b), Interspecific hybridization of papaya with other *Carica* species. *Journal of*

Mau, R. F. L., Gonsalves, D. and Bautista, R. (1990). Use of cross protection to control papaya ringspot virus at Waianae. *Proceedings of the 25th annual papaya industry association conference*, pp: 77–84.

Mc Kinney, H. H. (1929). Mosaic disease in the Canary islands, West Africa, and Gibralter. *Journal of Agricultural Research*, 39: 557 – 578.

Mc Millan, R. T. and Golsalves, D. (1987). Effectiveness of cross protection by a mild mutant of PRSV for control of ringspot diseases of papaya in Florida. *Proceedings of Florida State Horticultural Society*, 106: 146 – 147.

Mishra and Jha, A. (1955). Mosaic of papaya (*Carica papaya* L.) in Bihar. *Proceedings of Bihar Academy of Agricultural Science*, 4: 102 – 103.

Moore, G. And Litz, R. (1984). Biochemical markers for *Carica papaya* and *C. cauliflora* plants from somatic embryos of their hybrids. *Journal of American Society of Horticultural Science*, 109: 213–218.

Nishijima, W. T. (1999).Common names of plant diseases, Diseases of Papaya, www.aps.com

Parris, G. K. (1938). A new disease of papaya in Hawaii. *American Society of Horticultural sciences*, 36: 263 – 265.

Piblito, M. M., Loloita, D., Valencia, A., Ocampo, T. I. D., Reynaldo, T., T. and Violeta, N. V. (2004). Towards development of PRSV resistant papaya by genetic engineering: *Proceedings of the 4th International Crop Science Congress, Brisbane, Australia*, www. cropscience. org. au

Prucifull, D. E., Simone, G. W., Baker, C. A. And Hiebert, E. (1984).Papaya ringspot virus. *CMI/AAB Description of Plant viruses*, No. 292, p. 8.

Ramakrishna Rao, P. (1988).Studies on Papaya Ringspot Virus infecting *Carica papaya*. *Ph. D. Thesis*, Marathwada Agricultural University, Parbhani, p. 89.

Ranebennur, Hemavati(2005).Detection, Diagnosis and management of PRSV on cucurbits, *M. Sc. (Agri) Thesis*, University of Agricultural Sciences, Dharwad, p. 109.

Rezende, J. A. M., Costa, A. S. And Soares, N. B. (1981).Occurrence de UM isolado fraco de virus do mosail do mamoeiro Carica papaya L. *Fitopatologia Brasileria*, 6: 534–539.

Shaikh, B. (1996). Studies on Papaya ringspot virus, *M. Sc. (Agri) Thesis*, University of Agriultuarl Sciesnces, Dharwad, p. 71.

Sheen, T. F., Wang, H. L., Wang, D. N. and Iwahori, S. (1998).Control of papaya ringspot virus by cross protection and cultivation techniques. *Journal of the Japanese Society for Horticultural Science*, 67: 1232 – 1235.

Singh, R. K., Singh, D. and Singh, J. (2003).Incidence, distribution and detection of a virus infecting papaya (*Carica papaya* L.) in Eastern Uttar Pradesh. Indian *Phytopathology*, 21 (1 and 2): 51 – 56.

Sondur, N. S. (1994). Construction of genetic linkage map of papaya and mapping traits of horticultural importance. *Ph.D. Thesis*, University of Hawaii, Honolulu.

Surekha, S. K., Mathur, K. and Shukla, D. D. (1977). Virus disease of papaya (*Carica papaya*) in Udaipur. *Indian Journal of Mycology and Plant Pathology*, 7: 115 – 121.

Susan John, S. K. (1985). Studies on mosaic disease of papaya (*Carica papay*), *M. Sc. (Agri) Thesis*, Acharaya N.G Ranga Agricultural University, Hyderabad, p. 43.

Tennant, P. F., Gonsalves, C., Ling, K., Fitch, M., Manshardt, R. and Gonsalves, D. (1994). Differential protection against papaya ringspot virus isolates in coat protein gene transgenic papaya and classically cross–protected papaya. *Phytopathology*, 84: 1359–1366.

Thomas, J. E. and Dodman, R. L. (1993).The first record of Papaya rongspot virus type – P from Australia. Australian Plant Pathology, 22: 2 – 7.

Timothy, F. W. and Robert, V.O. (1999).Production of Transgenic Hybrid Papaya in Hawaii. *Tropical Fruit Report*, p. 1–4.

Wahab, N. (1991). Papaya ring spot virus disease is there in Malaysia. *MAPPS Newsletter*, 15: 36.

Wan, S. H. and Conover, R. A. (1983). Incidence and distribution of papaya virus in Southern Florida. *Plant Disease*, 67: 353 – 356.

Wang, D. N. (1982). Screening of papaya varieties for ringspot tolerance. *Journal of Agricultural Research*, 31: 162 – 168.

Wang, H. L. Yeh, S. D., Chiu, R. J. and Gonsalves, D. (1987). Effectiveness of cross protection by mild mutants of papaya ring spot virus for control of ringspot disease of papaya in Taiwan. *Plant Protection Bulletin*, 20: 160.

Wu, F. C., Peng, X. X. and Xu, S. H. (1983). Preliminary studies on identification, purification and properties of Papaya ringspot virus in South China. *Acta Phytopathologia sinica*, 13: 21 – 28.

Yeh, S. D. and Gonsalves, D. (1984). Evaluation of induced mutants of papaya ringspot virus for control by cross protection. *Phytopathology*, 74: 1086 – 1091.

Yeh, S. D. and Gonsalves, D. (1985). Translation of Papaya ringspot virus RNA. *In vitro* detection of a possible polyprotein that is processed for capsid protein, cylindrical inclusion protein and amorphous inclusion protein. *Virology*, 143: 260 – 271.

Yeh, S. D.,Gonsalves, D., Wang, H. L.,Namba, R. and Chiu, R. J. (1988).Control of papaya ringspot virus by cross protection. *Plant Disease*, 72: 373–380.

Yeh, S. D., Jan, F. J., Chiang, C. H., Doong, I. J., Chen, M. C., Chung, P. H. and Bau, H. J. (1992).Complete nucleotide sequence and genetic organization of Papaya ringspot virus RNA. *Journal of General virology*, 73: 2531 – 2541.

Yemewar, S. I. and Mali, V. R. (1980). On the identity of a sap transmissible virus of papaya in Marathawada. *Indian Journal of Mycology and Plant Pathology*, 10: 155 – 160.

Zee, F. (1985). Breeding of papaya virus tolerance in solo papayas, *Carica papaya* L. *Ph.D. Thesis*, University of Hawaii, Honolulu, p. 105.

Zhang De Yong, Wang,Zen Zong, Fan Huai Zong and Lin Kongxun (1995). Temporal dyanamics of papaya ring spot disease. *Journal of South Chaina Agricultural University*, 16: 69–73.

Plant Diseases Management in Horticultural Crops (2011) *Pages 385–397*
Editors: **Shahid Ahamad, Ali Anwar and P.K. Sharma**
Published by: **DAYA PUBLISHING HOUSE, NEW DELHI**

Chapter 30

Role of Nutrients in Plant Disease Management

***M.H. Chesti*[1]*, Anshuman Kohli*[1] *and Shahid Ahamad*[2]**

[1]*RARS, SKUAST-J, Rajouri*
[2]*Krishi Vigyan Kendra, Sher-e-Kashmir University of Agricultural Sciences and Technology, Jammu, Rajouri – 185 131, J&K*

Nutrients are essential for growth and development of all plants as well as microorganisms and they are important factors in plant disease control (Agrios, 2005). Plant nutrition, although frequently unrecognized, has always been a primary component of disease control. At the most basic level, plants suffering a nutrient stress will be less vigorous and more susceptible to a variety of diseases. In this respect, all nutrients affect plant diseases; however, some nutrient elements have a direct and a greater impact on plant diseases than others. There is no general rule, as a particular nutrient can decrease the severity of a disease but can also increase the severity of the incidence of other diseases or have a completely opposite effect in a different environment.

Disease resistance in plants is primarily a function of genetics. However, the ability of a plant to express its genetic potential for disease resistance can be affected by mineral nutrition. Plant species or varieties that have a high genetic resistance to a disease are likely to be less affected by changes in nutrition than plants only tolerant to the disease. Plants that are genetically more susceptible are likely to remain susceptible in nutritional regimes that greatly improve the disease resistance in less susceptible or tolerant plants. The severity of most plant diseases can be reduced by proper nutrition. The physiological functions of plant nutrients are generally well understood, but there are still many unanswered questions regarding the dynamic interaction between nutrients and the plant–pathogen system (Huber, 1996). It is important to manage nutrients availability through fertilizers or change the soil environment to influence nutrients availability, and in that way to control plant diseases in an integrated pest management system (Graham and Webb, 1991). The use of fertilizers produces a more direct means of using nutrients to reduce the severity of many diseases and together with cultural practices can affect the control of diseases. In addition, nutrients can affect the development of disease by affecting plant physiology or by affecting pathogens, or both of them. The level of nutrients can influence plant growth, which can affect microclimate, therefore affecting infection and sporulation of the pathogen (Marschner, 1995). Nutrients can affect the growth rate of the host which can enable seedlings to escape or avoid infection when they are at the most susceptible stages. In order to appreciate the effect that nutrients have on plant diseases, it is important to understand some aspects of how diseases infect and multiply within the plant host.

Fungal Diseases

The physical resistance presented by the strength and integrity of the cell wall and intercellular spaces is the plants first line of defense. Nutrients play a major role in the plants ability to develop the strong cell walls and other tissue. The germination of spores is stimulated by compounds exuded by the plants. The amount and composition of these exudates is affected by the nutrition of the plants. When plants have a low levels of nutrients, these exudates will contain higher amounts of compounds such as sugars and amino acids that promote the establishment of the fungus. As the plant gets infected by a fungus, its natural defenses are triggered. The infection causes increased production of fungus inhibiting phenolic compounds and flavonoids, both at the site of infection and in the other parts of the plant. The production and transport of these compounds is controlled in large parts by the nutrition of the plant. Therefore, shortages of key nutrients reduce the amount of the plants natural antifungal compounds at the site of infection.

Bacterial Diseases

Plants resistance to bacterial spread within the leaf is closely related to the structure and strength of the internal cells and intercellular space, as well as the plants ability to produce and transport antibacterial compounds. These disease fighting mechanisms are closely related to certain nutrients. Bacterial spread within the host plant is accomplished by the bacterial production of enzymes that cause the decomposition of pectin (a primary structural component of plant cells). The production and activities of these pectolytic enzymes is inhibited by some nutrients. Bacterial vascular diseases are spread by way of the xylem (the vessels that transport water and nutrients from roots to the leaves). Their presence leads to the formation of slime within the vessels, eventually blocking them and leading to wilting and death of leaves and stems. Certain plant nutrients play a role in blocking or reducing the bacterial ability to form the slime.

Viral Diseases

There is little data on effects of mineral nutrition on viral diseases. Viruses live and multiply within the plant cells and there nutrition is limited to amino acids and nucleotides found within the cells. The primary vectors carrying virus to a crop are sucking insects, such as aphids and fungi. It has been found that the nutrient status of a plant can affect the aphid population on plants. For instance, it was found that certain aphids tend to settle on yellow reflecting surfaces, such as chlorotic leaves caused by nutrient deficiency. Feeding intensity and reproduction by sucking insects tend to be higher on plants with higher amino acid content. This condition is typical of plants suffering certain nutrient stresses. This review aims at summarizing the most recent information regarding the effect of nutrients on disease resistance and tolerance.

Plant Nutrition and Disease Control

A common symptom of many soil borne pathogens is root infection, which reduces the ability of the root to provide the plant with water and nutrients (Huber and Graham, 1999). This effect is more serious when the levels of nutrients are marginal and also for immobile nutrients. One of the first observations of the effect of nutrients on disease development was that fertilization reduced disease severity when plants were under deficiency, as fertilization optimized plant growth. Since the interaction of the nutrients and disease pathogens is complex, the effect of each nutrient on certain diseases and also the possible mechanism for the tolerance or resistance to the particular pathogen is described as follows:

Nitrogen

One of the commonly assumed relationship of N to disease is that high N rates lead to more severe incidence of disease. There are various situations where excessive N uptake can promote conditions favorable to disease and insect damage.1) Excessive nitrogen may promote succulent growth and thinner cell walls, making plants more susceptible to infection. 2) Excessive N may promote increased plant density, thus more humid air around the plants that can favour diseases. 3) Excessive N may delay maturity, thus extend the time available for infection and disease development.4) Feeding intensity and reproduction by sucking insects tends to be higher on the plants with higher amino acid content.5) In addition, at higher levels of N, there is decrease in Si content which can affect the disease tolerance.

Despite the fact that nitrogen is one of the most important nutrient for plant growth, there are several reports of the effect of N on disease development that are inconsistent and contradict each other, and the real causes of this inconsistency are poorly understood. These differences may be due to the form of N nutrition of the host, rates of N levels, and type of pathogen: obligate vs. facultative parasites. For instance, high N supply increases the severity of infection of obligate parasites such as *Puccinia graminis* and *Erysiphe graminis* and decreases the severity of the infection by facultative parasites such as *Alternaria, Fusarium* and *Xanthomonas* sp. At high N rates there is a higher growth rate during the vegetative stage and the proportion of the young to mature tissue shifts in favour of the young tissues, which are more susceptible. Also, there is a significant increase in amino acids concentration in the apoplast and on the leaf surface, which promotes the germination and growth of conidia (Robinson and Hodges, 1981). Also, at high N rates some key enzymes of phenol metabolism have lower activity, the content of phenolics and lignin decreases- all these are part of the defense system of plants against infection. Therefore, main reason for the increased susceptibility of obligate parasites at high nitrogen levels is the various anatomical and biochemical changes together with the increase in the content of the low molecular weight organic nitrogen compounds which are used as substrates for parasites. It is believed that plants grown under the conditions of low N availability are better defended against pathogens because there is an increase in the synthesis of defense related compounds (Hoffland *et al.*, 2000). However, the response to the N level was different in the facultative parasites, as when the plants were grown under high levels of N, they were more resistant to the pathogens such as *Bacillus cineria*. In the case of obligate parasites such as *Pseudomonas syringae* pv.tomato, *Ustilago maydis* and *Oidium lycopersicum*, increased susceptibility was observed when plants were grown with high N supply (Hoffland *et al.*, 2000; Kostandi and Soliman, 1991). These reports indicate that disease susceptibility depends on N supply and that the effect of N supply on susceptibility is pathogen-specific. The form of N is also important in plant diseases and the presence of nitrification inhibitors is important too. There is evidence that the form of N, or the ratio of ammonium- N or nitrate- N can have an effect on certain diseases in some crops. There is evidence that the nitrate-N can decrease the disease in case of *Helminthosporium* in corn, *Aphanomyces* in soyabean, *Fusarium* in wheat and cotton, while ammonium-N aggravates these diseases. In contrast ammonium-N, however, disease is decreased in case of *Diplodia*, *Fusarium* and *Pythium* in corn, *Heterodera* in soyabean, Take-all in wheat, *Phymatotrichum* in cotton, while nitrate-N aggravates these diseases. In addition,work on use of nitrification inhibitor to maintain higher ratio of ammonium-N to nitrate-N has reduced the incidence of *Verticillium wilt*, but increased the incidence of *Rhizoctonia cancer* in potato.

Phosphorus

Phosphorus application seems to favour plant protection against diseases, either by correcting a deficiency in soil P, and thereby inducing better growth of the plant, or by speeding up the maturation process, disfavouring some pathogens like downy mildew that affects the young tissues. The role of phosphorus in disease resistance is variable and seemingly inconsistent (Kiraly, 1976). P has been shown to be most beneficial when it is applied to control seedlings and fungal diseases where vigorous root development permits plant to escape diseases (Huber and Graham, 1999). Phosphate fertilization disease of wheat can have a significant effect and almost eliminate economic losses from *Pythium* root rot (Huber, 1980). Similarly, in corn P application can reduce root rot especially when it is grown on soils deficient in P, and in other studies it can reduce the incidence of soil smut in corn (Huber and Graham, 1999; Potash and Phosphate Institute, 1988). A number of other studies have shown that P application can reduce bacterial leaf blight in rice, downy mildew, blue mould, leaf curl virus disease in tobacco, pod and stem blight in soybean, yellow dwarf virus disease in barley, brown stripe disease in sugarcane and blast disease in rice (Huber and Graham, 1999, Kirkegaard *et al.*, 1999; Reuveni *et al.*, 1998, 2000; Potash and Phosphate Institute, 1988). However,in other studies application of P may increase the severity of diseases caused by *Sclerotinia* in many garden plants, *Bremia* in lettuce and flag smut in wheat (Huber, 1980). Foliar application of P can induce local and systemic protection against powdery mildew in cucumber, wine roses, wine grapes, mango, and nectarines (Reuveni and Reuveni, 1998)

Potassium

Of all the nutrients essential for plant growth and function, potassium is most often associated with reducing disease severity. As a mobile regulator of enzyme activity, potassium is involved in essentially all cellular functions that influence disease severity. Potassium reduces bacterial and fungal diseases 70 per cent of the time, insects and mites 60 per cent of the time, and nematodes and viruses influences in a majority of cases. Potassium improves plant health in 65 per cent of the studies and was deleterious 23 per cent of the time. Some of the deleterious effects could have been due to the excess K inhibiting the uptake of Mg, Ca and NH_4^+ -N. Potassium deficiency symptoms such as thin cell walls, weak stalks and stems, sugar accumulation in the leaves, and accumulation of unused nitrogen encourage disease infection. Each of these reduces the ability of plant to resist entry and infection by fungal, bacterial and viral disease organisms. The high susceptibility of K deficient plant to parasitic infestations is due to metabolic functions of K in plant physiology. Under K deficiency, synthesis of high molecular weight compounds (proteins, starch, cellulose) is impaired and there is accumulation of low molecular weight simple N compounds such as amides which are used by invading plant pathogens.

Potassium also plays a central role in the development of thick cuticles, a physical barrier to infection or penetration by sucking Insects. The balance between K and other elements is also important. For instance, Stewart's wilt in corn is known to need or benefit from higher inorganic N in the tracheal sap of the plant. K deficiency reduces the ability of corn to metabolize N, thus increasing the amount of inorganic N in the tracheal sap. This increases the susceptibility to Stewart's wilt. In most plants, as inorganic N accumulates in the presence of low K levels, plant compounds that have fungicidal properties are rapidly broken down. Increased disease resistance by different varieties of the same plant species is sometimes related to the ability of the resistant variety to take up more K. This has been seen in species as widely divergent as flax and pine trees. It was found that a naturally wilt resistant variety of flax takes up and contains higher K levels than wilt susceptible varieties, while others were more susceptible, was associated with higher tissue K levels in the resistant varieties.

It has been frequently observed that K reduces the incidences of various diseases such as bacterial leaf blight, sheath blight, stem rot, sesamum leaf spot in rice, Helminthosporium leaf blight and black rust in wheat, sugary disease in sorghum, bacterial leaf blight in cotton, Cercospora leaf spot in cassava, tikka leaf spot disease in pea nut, red rust in tea, Cercospora leaf spot in mungbean and seedling rot caused by *Rhizoctonia solani* (Huber and Graham, 1999; Sharma and Duveiller, 2004; Sharma *et al.*, 2005). Many disease infections occur through open wounds and rapid healing of wounds tends to reduce infection. Work showed that grapes receiving higher K fertilization were less susceptible to *Botrytis cinerea*. This was attributed to more rapid healing of wounds and accumulation of compounds toxic to the fungus around the wounds. Adequate K nutrition can also reduce the severity of nematode infection. Application of K markedly reduced damage from root knot nematodes. Work with soybean cyst nematodes shows that the severity of infection and the yield loss caused by infection are both reduced by adequate K fertilization. However, work with potatoes and citrus trees showed that infection by common scab (*Streptomyces scabies*) in potatoes and Phytopthora root rot (*Phytopthora parasitica*) in citrus was increased by high levels of K fertilization. It was concluded in both cases that higher K uptake probably caused a shortage of calcium. The calcium shortage apparently resulted in improper formation or function of cell walls, leading to increased disease infection or spread within the plants. Gall diseases can also be affected by tissue K/Ca balance. Ensuring adequate K nutrition has reduced the incidence of several infection diseases of obligate and facultative parasites.

Fungal diseases where potassium has proven beneficial : Root rot (*Gibberella saubinetti*), Stalk rot (*Fusarium moniliforme, Gibberella zeae, Diplodia zeae*) and Stem rot (*Fusarium culmorum*) in Corn; Leaf spot (*Cercospora oryzae, Helminthosporium* spp.), Brown leaf spot (*Cochliobolus miyabeanus*), Sheath blight (*Corticium sasakii*), Stem rot (*Leptosphaeria salvinii*) and Blast (*Pyricularia oryzae*) in Rice; Leaf blotch (*Septoria tritici*), Glume blotch (*Septoria nodorum*), Take-all (*Gaeumannomyces graminis*), Stem rust (*Puccinia graminis*), Leaf rust (*Puccinia recondita*), Stripe rust (*Puccinia striiformis*) and Powdery mildew (*Erysiphe graminis*) in Wheat; Wilt (*Fusarium oxysporum, Verticillium alboatrum*), Seedling blight (*Rhizoctonia solani*) and Leaf blight (*Alternaria*) in Cotton; Powdery mildew (*Erysiphe graminis*) in Barley; Late blight (*Phytopthora infestans*) and Stem end rot (*Fusarium spp*) in Potato; Gray mold (*Botrytis cinerea*) in Cabbage; white mold (*Sclerotinia sclerotiorum*) in Pumpkin; Root rot (*Aphanomyces euteiches*) in Pea; Wilt (*Fusarium oxysporium*) and Leaf blight (*Alternaria solani*) in Tomato; Brown rot (*Sclerotinia fructocola*) in Apricot; Fusarium wilt (*Fusarium oxysporum*) in Banana; Fruit rot (*Botrytis cinerea*) in Grapes; Root rot (*Phytopthora cinnamomi*) in Pine apple.

Bacterial diseases where potassium has proven beneficial: Bacterial blight (*Pseudomonas syringae*) in Beans; Bacterial blight (*Xanthomonas oryzae*) in Rice; Stewarts wilt (*Erwinia stewartii*) in Corn; Angular leaf spot (*Xanthomonas malvacearum*) in Cotton; Soft rot (*Erwinia carotovora*) in Cabbage, Wilt (*Pseudomonas solanacearum*) and Gray wall (*Erwinia carotovora*) in Tomato; Angular leaf spot (*Pseudomonas lachrymans*) in Cucumber; Bacterial spot (*Xanthomonas pruni*) in Peach; Fire blight (*Erwinia amylovora*) in Pear.

Viral diseases where potassium has proven beneficial: Mosaic (*Tobacco mosaic virus*) in Beans and Potato; Leaf roll (*Potato leaf roll virus*) in Potato; Mosaic (*Tobacco mosaic virus*), Yellow mosaic (*Tobacco

Yellow mosaic virus) in Tobacco; Blotch ripening (*Tobacco mosaic virus*) in Tomato; Virus(*Tobacco ring spot virus*) in Squash.

Nematodal disease where potassium has proven beneficial : Root knot (*Meloidogyne incognita*) in Beans and Lima; Nematode (*Heterodera schachtii*) in Sugarbeet; Reniform (*Rotylenchulus reniformis*) in Cotton.

Calcium

Calcium is a structural component of cell walls and other plant membranes. As such, it plays a major role in the integrity and function of these structures. Calcium affects the susceptibility to diseases in two ways. Firstly, calcium is important for the stability and function of the plant membranes and when there is a calcium deficiency, there is a membrane leakage of low molecular weight compounds, eg. Sugars and amino acids, from the cytoplasm to the apoplast, which stimulates the infection by pathogens (Marschner, 1995). Secondly, calcium is an important component of the of cell wall structure as calcium polygalacturonates are required in middle lamella for cell wall stability. When calcium concentration drops, there is an increased susceptibility to fungi and bacteria which preferentially invade the xylem, release pectolytic enzymes that dissolve the cell walls of the conducting vessels, which leads to wilting symptoms. The activity of these enzymes is inhibited by the calcium ion. As the pathogen releases enzymes that dissolve plant tissue, K is lost from the tissue, with the likely loss of benefits that K provides.

Calcium plays a major role in improving the storage life of fleshy fruit. Calcium treatment of fruits before storage is an effective procedure for preventing losses both from physiological disorders and from fruit rotting. For example, apple varieties can be stored for longer periods when the fruit contains higher Ca levels. Fruit with lower Ca content tends to develop bitter-pit, which is the decomposition of the fruit tissue just under the skin. They are more likely to turn brown and begin to decompose much sooner than fruits with high Ca content. There is also a close correlation between the calcium content of the skin of potato tubers and bacterial soft rot of the tubers caused by the various species of the bacteria *Erwinia*. Adequate soil Ca is needed to protect peanut pods from infection by *Rhizoctonia* and *Pythium* and application of Ca to the soil eliminates the occurrence of disease (Huber, 1980). Calcium confers resistance against *Pythium, Sclerotinia, Botrytis* and *Fusarium*. (Graham, 1983). A putative mechanism by which Ca is believed to provide protection against *Sclerotinia sclerotiorum* is by binding of oxalic acid or by strengthening of the cell wall.

As with leaf spot disease, calcium plays a significant role with bacterial vascular disease. The severity of many of these diseases is often proportional to the degree of calcium deficiency in the plant. Excess calcium does not provide additional disease prevention benefits. Bacterial spread within the leaf is closely related to the calcium content of the leaf, where deficiencies greatly increase the spread and damage from the disease. In case of bacterial canker disease in tomato, both resistant and susceptible varieties benefit from adequate calcium supply,whereas with an increase in calcium supply and calcium content in plant tissue can greatly decrease the damage from the disease. There is not enough information available about role of sulphur and magnesium in plant diseases. Sulphur can reduce the severity of potato scab, whereas magnesium decreases the calcium content of peanut pods and may pre-dispose them to breakdown by *Rhizoctonia* and *Pythium* (Huber, 1980).

Micronutrients

The effect of micronutrients on reducing the severity of diseases can be attributed to the involvement in physiology and biochemistry of the plant, as many of the essential micronutrients are involved in processes that can effect the response of plants to pathogens (Marschner, 1995). Micronutrients also play an important role in plant metabolism by affecting the phenolics and lignin content and membrane stability (Graham and Webb, 1991). Micronutrients can influence disease resistance indirectly, as in deficient plants they become more suitable feeding substrates. Systemic acquired resistance may be involved in the suppression of plant diseases by micronutrients. Reduction in disease severity has been reported in many crops after a single foliar application of H_3BO_3, $CuSO_4$, $MnCl_2$, $KMnO_4$, which provided systemic protection against powdery mildew in cucumber plants (Reuveni *et al.*, 1997 a, b; Reuveni and Reuveni, 1998). Application of nutrients such as Mn, Cu and B can exchange and therefore release Ca^{2+} cations from cell walls, which interact with salicylic acid and activate systemic acquired resistance mechanism.

Manganese

Manganese is the most studied micronutrient about its effects on diseases, and is important for the development of resistance of plants to both root and foliar diseases (Graham and Webb, 1991; Huber and Graham, 1999; Heckman *et al.*, 2003). Manganese is known to contribute to the suppression of fungal and bacterial diseases, and has been one of the active ingredients of some fungicides. Manganese is required in much higher concentration by higher plants than by fungi and bacteria and there is an opportunity for the pathogen to exploit this difference in requirement (Marschner, 1995). Thus, it may be the case that by supplying high levels of Mn to some plants through fungicides or nutrition, we are causing a simple Mn toxicity for fungi and bacteria. Manganese plays a key role in the lignin biosynthesis, phenol biosynthesis, photosynthesis and several functions (Marschner, 1995; Graham and Webb, 1991). Both lignin and phenolic compounds are important biochemical barriers to pathogen invasion, since phenolic compounds are toxic to many disease pathogens and lignin is a physical barrier to penetration by disease organisms. Manganese also inhibits the induction of amino-peptidase, an enzyme which supplies essential amino acids for fungal growth and pectin methyl esterase, a fungal enzyme that degrades host cell walls.

Manganese fertilization can control a number of pathogenic diseases such as powdery mildew, downey mildew, take-all, tan spots and several others (Brennan, 1992, Huber and Graham, 1999; Heckman *et al.*, 2003; Simoglou and Dordas, 2006). It has been shown that soil applications of manganese reduce common scab of potato (Keinath and Loria, 1996). Manganese plays a role in the ability of wheat and barley to resist take-all (*Gaeumannomyces graminis*). This disease thrives in a neutral or alkaline soil pH, and is very sensitive to acidic soil. As the soil pH increases, the availability and uptake of Mn decreases significantly. This fungus also has a capability to oxidize soil Mn making it unavailable. The resulting drop in Mn uptake by the crop reduces its ability to synthesize lignin, which reduces its ability to resist infection. Despite the fact that Mn application can affect disease resistance, the use of Mn is limited, which is due to the ineffectiveness and poor residual effect of Mn fertilizers on most soils that need Mn supplements, and is because of complex soil biochemistry of Mn.

Manganese has been also beneficial in the control of the following diseases. Common scab (*Streptomyces scabies*) in Potato; Blast (*Pyricularia oryzae*) and Leaf spot (*Alternaria*) in Rice; Mildew (*Glumaria grammis* var. tritci) in Wheat; Wilt (*Verticillium alboatrum*) in Cotton; Root rot (*Pythium*) in Avocado.

Copper

The usefulness of copper for disease prevention and correction was well known by 1900. This was about thirty years before it was determined to be an essential nutrient. Actually, the work with the fungicidal properties of copper led to the discovery that it is an essential nutrient as well. Like Mn, copper is an essential nutrient for higher plants as well as fungi and bacteria. However, higher plant forms such as crops and ornamentals can tolerate much higher copper levels than lower forms, such as fungi and bacteria. This difference in tolerance enables growers to use copper as a disease treatment. Copper fertilization has decreased the severity of a wide range of fungal and bacterial diseases. While copper is known to control diseases when foliar applied, however many times foliar diseases were controlled with soil application of copper. Copper has been beneficial in the control of the following diseases: Mildew (*Glumaria grammis*) in Wheat; Leaf/stem spot (*Alternaria*) in Sunflower;Leaf rust (*Puccinia triticina*) in wheat;blast (*Pyricularia oryzae*) in Rice; wilt (*Verticillium albo-atrum*) in Cotton; Common scab (*Streptomyces scabies*) in Potato; Take-all (*Gaeumannomyces graminis* var. *tritci*), leaf/ glumes blotch (*Septoria*) in Wheat; Nematode (*Heterodera*) in Sugarbeet; wilt (*Verticillium dagliae*) in Cotton.

Zinc

Zinc has a number of different effects as in some cases it decreased, in others increased, and in others had no effect on plant susceptibility to disease (Graham and Webb, 1991; Grewal *et al.*, 1996). In most cases, the application of zinc reduced disease severity, which could be because of the toxic effects of zinc on the pathogen directly and not through the plants metabolism (Graham and Webb, 1991).The effect that zinc is an active ingredient in some fungicides is evidence that it is directly toxic to some pathogens.

Zinc plays an important role in protein and starch synthesis, and therefore a low zinc concentration induces accumulation of amino acids and sugars in plant tissue. As an activator of Zn-SOD, zinc is involved in membrane protection against oxidative damage through the detoxification of super oxide radicals (Cakmak, 2000). Impairment in membrane structure caused by free radicals led to increased membrane leakage of sugars on to the surface of leaf. This excess sugar, plus some leakage of sugars on to the plant surfaces, can enhance the successful invasion of the fungi and bacteria (Mengal and Krikby, 2001). Application of zinc to the soil reduced infection by *Fusarium* species and root rot diseases in wheat (Graham and Webb, 1991; Grewal *et al.*, 1996). Zinc is also beneficial in the control of following diseases; Mold (*Penicillium citrinum*) in Citrus; Wilt(*Verticillium*) and Root rot(*Phymatrotrichopsis omnivorum*) in Cotton, Nematode (*Rotylenchulus reniformis*) in Tomato; Root rot(*Fusarium*) in Chickpea; Root rot in (*Rhizoctonia bataticola*) in Peanut; Powdery mildew(*Oidium heveae*)in Rubber trees; Take –all (*Gaeumannomyces graminis* var.tritci) and head scab(*Fusarium graminearum*) in Wheat; Foot rot(*Sclerotium rolfsii*) in Soyabean; Powdery mildew (*Erysiphe polygony*) in Pea;Crown/root rot (*Phytophthora cinnamomi*) in Eucalyptus.

Iron

In plant nutrition and soil chemistry, the micronutrients Mn, Cu, Fe are antagonistic to each other. Additionally, many plants are somewhat sensitive to the Fe/Mn ratio in the tissue. Since we know that Mn, Cu and Zn have significant anti disease properties, it seems possible that surplus Fe may simply be depressing the activity of one of these other nutrients. Fe can control or reduce the disease severity of several diseases such as rust in wheat leaves, smut in wheat and *Colletotrichum musae* in banana (Graham and Webb, 1991; Graham, 1983). Foliar application of iron can increase resistance of apple and pear to *Sphaeropsis malorum* and cabbage to *Olpidium brassicae* (Graham, 1983). There is some evidence that Fe is active against some diseases. Several plant pathogens like *Fusarium*, have higher requirement for iron compared with higher plants. Therefore Fe differs from other micronutrients such as Mn, Cu and B, for which microbes have lower requirements. Addition of copper, manganese and boron to deficient soils generally benefits the host, whereas the effect of iron application is not as same as it can have a positive or negative effect on the host.

Iron can activate enzymes that are involved in the infection of the host by the pathogen or defense, which is why opposite effects were found (Graham and Webb, 1991). Some work has shown that increased availability or uptake can increase disease severity. For instance, in tomato, tolerance of *Fusarium* wilt was decreases by iron and also stimulated fungal spore germination. Supplemental iron increases the disease incidence and severity of *Take-all* in wheat and barley. It was observed that where *Take- all* fungus was present in a soil that normally suppresses this fungus and as iron was applied, the fungus became nearly as damaging as in the control sample of the conducive soil. Thus, the application of iron essentially negated the suppressive effects of the soil. Iron has been beneficial in the control of the following diseases. In most cases, the beneficial effects required foliar Fe applications, some at higher concentrations.Rust (*Puccinia recondita*) and Smut (*Tilletia* sp.) in Wheat; Anthracnose (*Coletotrichum musae*) in Banana; Black rot (*Sphaeropsis malorum*) in Apple and Pear; Virus vector (*Olpidium brassicae*) in Cabbage.

Chlorine

Chlorine is required in very small amounts for plant growth and chlorine deficiency has rarely been reported as a problem in agriculture. There are many examples fo chlorine fertilization being beneficial to crops, some crops can be damaged by chlorine levels that are beneficial to others. It has been known that for many years that tobacco, blue berries and some varieties of soybean are adversely affected by relatively low levels of chlorine. Other corps reported to have some sensitivity to high chlorine levels are alfalfa, potatoes, cotton, tomatoes and grain sorghum. Some of these same crops have also been shown to benefit from moderate chlorine application. Chlorine has shown to control a number of diseases such as stalk rot in corn, stripe rust in wheat, Take-all in wheat, northern corn leaf blight and downy mildew of millets, and Septoria in wheat (Graham and Webb, 1991; Mann *et al.*, 2004). It was suggested that chlorine can compete with NO_3 absorption and influences the rhizosphere pH, it can suppress nitrification and increase the availability of Mn. Further more, Chlorine ions can mediate reduction of Mn III and Mn IV oxides and increase Mn for the plant, increasing the tolerance to the pathogens. Chlorine has been beneficial in the control of various diseases: Stripe rust (*Puccinia striiformis*), Leaf rust (*Puccinia recondite*), Take-all(*Gaeumannomyces graminis*), Glume blotch(*Septoria nodorum*) and Root rot (*Cochliobolus*) in Wheat; Necrotic ringspots (*Leptosphaeria korrae*) in Turf; Leaf spot (*Pestalozzia palmarum*) in Palm tree; Fusarium yellows (*Fusarium*) in Celery; Stalk rot(*Gibberella zeae, Fujikuroi, Fusarium moniliforme*) in Corn; Sudden death syndrome(*Fusarium solani*) in Soybean; Downy mildew (*Pennisetum typhoides*) in Millet; Hallow heart and brown center (physiological disorder).

Boron

Boron has a direct function in cell wall structure/stability and has a beneficial effect on reducing disease severity. The function that B has in reducing disease susceptibility could be because of 1) Its role in cell wall structure 2) Its role in the cell membrane permeability or stability 3) Its role in the metabolism of phenolics,and a primary role in the synthesis of lignin. It has been demonstrated that when B is deficient, plant cell walls tend to swell and split, and to result in weakened intercellular space. This results in a weakened physical barrier to initial infection and expansion of the infection. Boron also plays a role in the production of disease protection compounds and structures within plants. Boron has been shown to reduce diseases caused by *Plasmodiosphora brassicae* in crucifers, *Fusarium solani* in bean, *Verticillium alboatrum* in Tomato and Cotton; Tobacco mosaic virus in Bean,

Tomato yellow leaf virus in Tomato (Graham and Webb, 1991). Boron has been also beneficial in the control of diseases caused by *Rhizoctonia solani* in Mungbean; *Rhizoctonia bataticola* in Peanut; *Synchytrium endobiotium* in Potato.

Silicon

Although silicon is not considered to be an essential nutrient for most plants, it is beneficial to many plants. Silicon has the potential to significantly decrease the susceptibility of certain plants to fungal pathogens and insects, and amelioration of abiotic stresses (Marschner, 1995; Epstein, 1994). Numerous studies have shown that disease resistance of rice increases in response to silicon fertilization. Neck blast and brown spot are major fungal diseases limiting rice production in Florida (Datnoff *et al.*, 1991). Adding silica as calcium silicate slag to Histosol, reduced the amount of next blast by 73–86 per cent and reduced the incidence of brown spot by 58 – 75 per cent. Remarkably this degree of disease control was not significantly different from that achieved by fungicides such as benomyl (Datnoff *et al.*, 1997). As with rice, fungal disease resistance of green house grown cucumber has been shown to increase substantially in response to Si fertilization (Belanger *et al.*, 1995)

Evidences indicate that one of the mechanism by which Si protects plants is by increasing the effectiveness of the mechanical barrier that plants present to infection. Soluble Si taken up by plants tend to accumulate in the apoplast, particularly in epidermal cell walls(Epstein, 1994; Marschner, 1995).This observation has led many investigators to hypothesize that Si inhibits fungal disease by physically inhibiting fungal germ tube penetration of the epidermis (Datnoff *et al.*, 1997; Belanger *et al.*, 1995).Some work has found that Si may induce accumulation of antifungal compounds such as flavonoid and diterpenoid phytoalexins which can degrade fungal and bacterial cell walls (Alvarez and Datnoff, 2001; Brescht *et al.*, 2004). Increased Si in plants has been shown to increase the difficulty of sucking insects, like aphids or leaf hoppers. Decreased feeding by these insects could also aid in the prevention of the spread of viral diseases that could be transmitted by insects. In addition to inhibiting fungal diseases, silicon has also been shown to ameliorate certain mineral imbalances and other diseases caused by abiotic stresses in plants. Several studies have found that Si can reduce or prevent Mn and Fe toxicity and may also have beneficial effects on Al toxicity (Marschner, 1995; Epstein, 1994).

Silicon can also reduce salinity stress and reduce transpiration in plants. Furthermore, in sugarcane, there is evidence that Si may play an important role in protecting leaves from ultraviolet radiation damage by filtering out the harmful ultraviolet rays (Tisdale *et al.*, 1993). Silicon has been beneficial in the control of the following diseases: Rice blast (*Piricularia oryzae*) and Brown spot (*Cochliobolus miyabeanus*) in Rice; Powdery mildew (*Erysiphe graminis*) in Barley; Powdery mildew (*Sphaerotheca fuliginea*) in Cucumber; Powdery mildew (*Uncinula necator*) in Grapes; Rust (*Uromyces phaseoli*) in Beans.

Molybdenum

Little is known about the effects that Mo may have on plant diseases. It has been reported that Mo applied to tomato roots reduced the symptoms of *Verticillium wilt* and can also reduce the production of a toxin *Myrothecium roridum*, a pathogen of muskmelon. A report indicated that Mo slightly decreased the reproduction of *Phytophthora cinnamomi* and *Phytopthora dreschleri* diseases of a variety of crops. A report indicated that soil application of Mo decreased the population of the nematode *Rotylenchulus reniformis*, which infects various crops.

Nickel

It is sometimes listed with essential elements, or as a potentially essential element. A 1985 study showed that the number of rust pustules on infected cowpeas was reduced more than 50 per cent by increasing the supply of Ni to the roots from 33 ppb to 3.3ppm. The resulting leaf concentration of only 1ppm was effective in reducing the level of rust infection. Ni has been found to be effective in protecting rice against rice blast fungus(*Pyricularia oryzae*) and brown spot (*Cochliobolus miyabeanus*).

Cadmium

It is not considered to be a essential element,and can be toxic to plants in low quantities. Cadmium at low concentration has been found to stimulate or enhance the formation of lignin in wheat, which is known to be a disease defense mechanism, and to suppress the effects of powdery mildews in a variety of crops. Cadmium was found to inhibit spore germination and development at a concentration of 3 ppm, which is not toxic but elicits a response to infection in the host. Cd and Hg can also promote synthesis of lignin in wheat (Graham and Webb, 1991).

Lithium

It is not considered to be an essential element, and can be toxic at low levels.The mechanism of Li is not known and it is quite possible that it catalyzes a metabolic pathway which can function in defense. Like Cd, Li has been shown to significantly suppress the effects of powdery mildews.

References

Agrios, N.G. (2005). *Plant Pathology*, 5th edn. Elsevier-Academic Press, p. 635.

Alvarez, J. and Datnoff, L.E. (2001). The economic potential of silicon for integrated management and sustainable rice production. *Crop Prot.*, 20: 43–48.

Belanger, R.R.,Brown P.A.,Ehret D.L. and Menzies J.G (1995). Soluble silicon: its role in crop and disease management of greenhouse crops. *Plant Dis.*, 79(4): 329–336.

Brennan, R.F. (1992). The role of manganese and nitrogen nutrition in the susceptibility of wheat plants to take-all in western Australia, *Fertilizer Res.*, 31: 35–41.

Brescht, M.O., Datnoff, L.E., Kucharek, T.A., Nagata, R.T. (2004). Influence of silicon and chlorothalonil on the suppression of grey leaf spot and increase plant growth in St. Augustinegrass. *Plant Dis.*, 88: 338–344.

Cakmak, I.M. (2000). Possible roles of zinc in protecting plant cells from damage by reactive oxygen species, *New Phytol.*, 146: 185–205.

Celar, F. (2003).Competition for ammonium and nitrate forms of nitrogen between some phytopathogenic and antagonistic soil fungi. *Biol.Control*, 28: 19–24.

Datnoff, L.E., Deren, C.W. and Snyder, G.H. (1997). Silicon fertilization for disease management of rice in Florida. *Crop Prot.*, 16(6): 525–531.

Datnoff, L.E., Raid, R.N., Snyder, G.H. and Jones, D.B. (1991). Effect of calcium silicate on blast and brown spot intensities and yields of rice. *Plant Dis.*, 75(7): 729–732.

Epstein, E. (1994). The anomaly of silicon in plant biology. *Proceedings of the National Academy of Sciences USA*, 91: 11–17.

Graham, D.R. (1983). Effects of nutrients stress on susceptibility of plants to disease with particular reference to the trace elements. *Adv. Bot. Res.*, 10: 221–276.

Graham, D.R. and Webb, M.J. (1991). Micronutrients and disease resistance and tolerance in plants. In: *Micronutrients in Agriculture*, 2nd edn, (Eds.) J.J. Mortvedt, F.R. Cox, L.M. Shuman and R.M. Welch. Soil Science Society of America, Inc. Madison, Wisconsin, USA, pp. 329–370.

Grewal, H.S., Graham, R.D. and Rengel, Z. (1996). Genotypic variation in zinc efficiency and resistance to crown rot disease (*Fusarium graminearum* Schw. Group 1). in wheat. *Plant Soil*, 186: 219–226.

Heckman, J.R., Clarke, B.B. and Murphy, J.A. (2003). Optimizing manganese fertilization for the suppression of take-all patch disease on creeping bentgrass. *Crop Sci.*, 43: 1395–1398.

Hoffland, E., Jegger, M.J. and van Beusichem, M.L. (2000). Effect of nitrogen supply rate on disease resistance in tomato depends on the pathogen. *Plant Soil*, 218: 239–247.

Huber, D.M. (1980). The role of mineral nutrition in defense. In: *Plant Disease: An Advanced Treatise, Vol 5: How Plants Defend Themselves*, (Eds.) J.G. Horsfall and E.B. Cowling. Academic Press, New York, pp. 381–406.

Huber, M.D. (1996). Introduction. In: *Management of Diseases with Macro- and Micro-elements*, (Ed.) W.A. Engelhard. APS Press, Minneapolis, USA, p. 217.

Huber, D.M. and Graham, R.D. (1999). The role of nutrition in crop resistance and tolerance to disease. In: *Mineral Nutrition of Crops: Fundamental Mechanisms and Implications*, (Ed.) Z. Rengel. Food Product Press, New York, pp. 205–226.

Keinath, P.A. and Loria, R. (1996). Management of common scab of potato with plant mutrients. In: *Mineral Nutrition of Crops: Fundamental Mechanisms and Implications*, (Ed.) Z. Rengel. Food Product Press, New York, pp. 152–166.

Kiraly, Z. (1976). Plant disease resistance as influenced by biochemical effects of nutrients and fertilizers. In: *Fertilizer Use and Plant Health, Proceedings of Colloquium 12*, Atlanta, GA: International Potash Institute, pp. 33–46.

Kirkegaard, J.A., Munns, R., James, R.A. and Neate, S.M. (1999). Does water and phosphorus uptake limit leaf growth of rhizoctonia-infected wheat seedlings? *Plant Soil*, 209: 157–166.

Kostandi, S.F. and Soliman, M.F. (1991). Effect of nitrogen rates at different growth-stages on corn yield and common smut disease [*Ustilage maydis* (DC). Corda]. *J. Agron. Crop Sci.*, 167: 53–60.

Mann, R.L., Kettlewell, P.S. and Jenkinson, P. (2004). Effect of foliar-applied potassium chloride on septoria leaf blotch of winter wheat. *Plant Pathol.*, 53: 653–659.

Marschner, H. (1995). *Mineral Nutrition of Higher Plants*, 2nd edn. Academic Press, London, p. 889.

Mengel, K. and Kirkby, E.A. (2001). *Principles of Plant Nutrition*, 5th edn. Kluwer Academic Publishers, Amsterdam, Netherlands, p. 847.

Potash and Phosphate Institute (PPI) (1988). Phosphorus nutrition improves plant disease resistance in PPI (Ed.), Better crops with plant food Fall 1988. Atlanta, Georgia, USA, pp. 22–23.

Reuveni, R. and Reuveni, M. (1998). Foliar-Fertilizer therapy: A concept in integrated pest management. *Crop Prot.*, 17: 111–118.

Reuveni, M., Agapov, V. and Reuveni, R. (1997a). A foliar spray of micronutrient solutions induces local and systemic protection against powdery mildew(*Sphaerotheca fuliginea*) in cucumber plants. *Eur. J. Plant Pathol.*, 103: 581–588.

Reuveni, M., Agapov, V. and Reuveni, R. (1997b). Controlling powdery mildew caused by *Sphaerotheca fuliginea* in cucumber by foliar sprays of phosphate and potassium salts, *Crop Prot.*, 15: 49–53.

Reuveni, M., Oppernheim, D. and Reuveni, R. (1998). Integrated control of powdery mildew on apple trees by foliar sprays of mono-potassium phosphate fertilizer and sterol inhibiting fungicides, *Crop Prot.*, 17: 563–568.

Reuveni, R., Dor, G., Raviv, M., Reuveni, M. and Tuzun, S. (2000). Systemic resistance against *Sphaerotheca fuliginea* in cucumber plants exposed to phosphate in hydroponics system, and its control by foliar spray of mono-potassium phosphate. *Crop Prot.*, 19: 355–361.

Robinson, P.W.and Hodges, C.F. (1981). Nitrogen-induced changes in the sugars and aminoacids of sequentially senescing leaves of *Poa pratensis* and pathogenesis by *Drechslera sorokiniana*. *Phytopathol. Z.*, 101: 348–361.

Sharma, R.C. and Duveiller, E. (2004). Effect of Helminthosporium leaf blight on performance of timely and late-seeded wheat under optimal and stressed levels of soil fertility and moisture. *Field Crop Res.*, 89: 205–218.

Sharma, S., Duveiller, E., Basnet, R., Karki, C.B. and Sharma, R.C. (2005). Effect of potash fertilization on Helminthosporium leaf blight severity in wheat, and associated increases in grain yield and kernel weight. *Field Crop Res.*, 93: 142–150.

Simoglou, K. and Dordas, C. (2006). Effect of foliar applied boron, manganese and zinc on tan spot in winter durum wheat. *Crop Prot.*, 25: 657–66.

Tisdale, S.L., Nelson, W.J. and Beaton, J.D. (1993). *Soil Fertility and Fertilizers*. McMillan Publishing Company, New York.

Index